ITALIAN STUDIES IN THE PHILOSOPHY OF SCIENCE

BOSTON STUDIES IN THE PHILOSOPHY OF SCIENCE

EDITED BY ROBERT S. COHEN AND MARX W. WARTOFSKY

VOLUME 47

ITALIAN STUDIES IN THE PHILOSOPHY OF SCIENCE

Edited by

MARIA LUISA DALLA CHIARA

University of Florence, Italy

Principal Translations by Carolyn R. Fawcett

D. REIDEL PUBLISHING COMPANY

DORDRECHT: HOLLAND/BOSTON: U.S.A.

LONDON: ENGLAND

Library of Congress Cataloging in Publication Data

Main entry under title:

Italian studies in the philosophy of science.

   (Boston studies in the philosophy of science; v.47)
   Some papers translated from Italian and French by C.R. Fawcett.
   Includes bibliographical references.
   1. Science—Philosophy—Addresses, essays, lectures. 2. Science—
History—Addresses, essays, lectures.
I. Dalla Chiara Scabia, Maria Luisa. II. Series.
Q174. B67 vol. 47 [Q175.3] 501s [501] 80–16665
ISBN 90–277–0735–9 (D. Reidel Pub. Co.)
ISBN 90–277–1073–2 (D. Reidel Pub. Co.: pbk.)

Published by D. Reidel Publishing Company,
P.O. Box 17, 3300 AA Dordrecht, Holland.

Sold and distributed in the U.S.A. and Canada
by Kluwer Boston Inc.,
190 Old Derby Street, Hingham, MA 02043, U.S.A.

In all other countries, sold and distributed
by Kluwer Academic Publishers Group,
P.O. Box 322, 3300 AH Dordrecht, Holland.

D. Reidel Publishing Company is a member of the Kluwer Group.

Papers 6–8, 10, 18, 20–22, 25–26 translated from the Italian, paper 19 from the French,
by C.R. Fawcett and revised by the authors; all other papers were written in English
and have been edited by C.R. Fawcett and R.S. Cohen.

Printed in The Netherlands

FOREWORD

The impressive record of Italian philosophical research since the end of Fascism thirty-two years ago is shown in many fields: esthetics, social and personal ethics, history and sociology of philosophy, and magnificently, perhaps above all, in logic, foundations of mathematics and the philosophy, methodology, and intellectual history of the empirical sciences. To our pleasure, Maria Luisa Dalla Chiara of the University of Florence gladly agreed to assemble a 'sampler' of recent Italian logical and analytical work on the philosophical foundations of mathematics and physics, along with a number of historical studies of epistemological and mathematical concepts. The twenty-five essays that form this volume will, we expect, encourage English-reading philosophers and scientists to seek further works by these authors and by their teachers, colleagues, and students; and, we hope, to look for those other Italian currents of thought in the philosophy of science for which points of departure are not wholly analytic, and which also deserve study and recognition in the world-wide philosophical community.

Of course, Italy has long been related to that world community in scientific matters. The list of Italian participants is well-known: Peano and his school, mainly devoted to rigorous logical and methodological foundations of mathematics, less to philosophy; the Peircean pragmatism of Vailati and Calderoni, linked to Mach's empiricist criticism of all scientific absolutes, and also anticipating our modern rigourous analysis of the language of science; Enriques' own link with Mach via Poincaré, with a Machian stress upon epistemological insight through the history of science; Rignano's broad and unified scientific scope of work and interest, which was for decades influential in his editorial direction of *Scientia*, the world journal of science and its inter-disciplinary and civilizational aspects (founded with Enriques in 1907); the severe neo-idealist Hegelian critique of science as inextricably 'positivist', as purely pragmatic and epistemologically vacuous, propounded by Croce and pressed further by Gentile during the inter-war period when Italian philosophy knew little of the Vienna Circle, Russell, Wittgenstein, nor of Husserl or Cassirer.

The primary and indeed essential role to be taken by the philosophy of science in modern civilization was first studied and profoundly perceived, in Italy, by Ludovico Geymonat, a student of Peano, to whom Professor Dalla

Chiara, and indeed all Italian philosophers, pays deserved respect. Not restricted to narrow or specialized studies in Geymonat's eyes, philosophy of sciences goes beyond the acknowledged values of such studies to encourage an understanding of the rationality of science *vis-à-vis* the recurring irrationalist temptations of personal and political life, and to stress the engagement of science in its social-historical situations.

Then, after Geymonat's first post-war writings and lectures, the new generation burst forth, from 1960, strongly influenced by a goodly number of studies abroad and by a surge of Italian translations of German and English language books in the philosophy and logic of the sciences. Some names of that generation are known throughout the scientific and scholarly world: Paolo Rossi on the origins of modern science; G. Toraldo di Francia on quantum optics and on the logic of physics; D. Costantini on the logical foundations of probability; Bruno di Finetti on the nature of probability; Ettore Casari as restorer of Italian logical studies of mathematics and the semantics of formal systems, and of non-classical logics; Casari's former student, Dalla Chiara, with her original work on modal logic; Evandro Agazzi on logic and the critique of any purely formal approach to the foundations of mathematics and physics; and these are only a few.

We believe this volume, selected from these writers and others, provides an intellectual feast, nutritious, savory, stimulating. For our readers and ourselves, we thank Professor Dalla Chiara and her splendid contributors.

Once again, it is our pleasure to express our gratitude to Carolyn R. Fawcett for her impeccable collaboration in the making of this book, for her fine bibliographical work, and for the preparation of the index.

<table>
<tr><td>Center for the Philosophy and History of Science</td><td>ROBERT S. COHEN</td></tr>
<tr><td>Boston University</td><td>MARX W. WARTOFSKY</td></tr>
<tr><td>October 1980</td><td></td></tr>
</table>

TABLE OF CONTENTS

PART III / HISTORY OF THE SCIENCES

PREFACE

Modern research in philosophy of science in Italy has a very young history. Throughout the world, many people working in foundational fields are familiar with the Italian tradition: the names of Giuseppe Peano (1858–1932), Cesare Burali-Forti (1861–1931), Alessandro Padoa (1868–1937), Eugenio Beltrami (1835–1900), Federico Enriques (1871–1946) occur in most textbooks concerning the foundations of logic, arithmetic, set theory and geometry. However, between this important tradition and the recent researches in logic and the philosophy of science, there is a very long gap, which was due to a number of different causes.

It is usually maintained that the main responsibility for this gap is to be attributed to the 'anti-scientific attitude' of the leading Italian philosophical school of the early twentieth century. Indeed, both versions of idealistic conceptions as proposed respectively by Benedetto Croce (1866–1952) and Giovanni Gentile (1875–1944), although antagonistic on many points,[1] nevertheless shared a common negative judgment about the philosophical interest of scientific theories. As is well known, Croce proposed a distinction which was rather successful for a long time, to the effect that whereas philosophy deals with concepts, science on the contrary can deal only with pseudo-concepts, which are in principle devoid of any theoretical relevance.[2]

One cannot simply identify the general ideology of the fascist movement with a form of idealistic philosophy (if only because the cultural level of fascism was well below any philosophical theory, which after all must satisfy some minimal standard requirements); nevertheless the anti-scientific attitude of Italian idealistic philosophers had, without any doubt, a considerable bearing on the fascist politics towards science. It is sufficient to recall the unbelievable blindness of the political class which caused the dispersion abroad of the group of physicists working with Enrico Fermi.

In spite of this, it would be too simple to attribute the decay of Italian philosophy of science after the First World War only to the faults of idealist philosophers and fascist politicians. Today it is often recognized

Maria Luisa Dalla Chiara (ed.), Italian Studies in the Philosophy of Science, ix–xi.
Copyright © 1980 by D. Reidel Publishing Company.

that some causes of this decay are also to be sought in some peculiar limitations of the Italian scientific tradition. For instance, in spite of their well-known merits, Peano and other scholars of his school (like Burali-Forti) did not fully understand some of the deepest aspects of the foundational analysis of mathematical theories; they maintained a somewhat narrow image of logical work, as a kind of *rigorization* and *symbolization* of the language of mathematicians. In other words, all the problems connected with inquiry into a *justification* of the whole building of mathematical theories (which represented, as is well-known, the *focus* of the Frege-Russell approach as well as of the Hilbert program) were completely overlooked by the Italian school. As a natural consequence of this attitude of the 'fathers' of mathematical logic in Italy, a kind of insensibility towards any foundational problem, accompanied by a misunderstanding of the actual role of modern logic, persisted for a long time and is, to a certain extent, still alive among many Italian mathematicians.

As to foundational researches in physics, one can recall the work of Enrico Persico (1900–1969), who diffused quantum theory in Italy[3] and discussed some logical problems concerning physical theories in general. Apart from this important exception, Italian physicists, between the two wars, were not much involved in the well-known international discussion about the foundations of relativity and quantum mechanics.

The revival of studies in foundational fields, which started after the second war, owes very much to the work of Ludovico Geymonat (born in 1908). In the thirties, Geymonat (philosopher and mathematician) frequented the Vienna Circle and introduced in Italy the main ideas of the neopositivistic philosophy of logical empiricism as well as the most important foundational questions concerning logical and mathematical theories.[4] His work had a large influence on the young scholars who were starting their research investigations in logic and foundations of mathematics.

Among other philosophers who, after the war, contributed to the diffusion of a 'new' attitude towards science, let us only recall the names of Nicola Abbagnano, Giulio Preti, Francesco Barone, Alberto Pasquinelli, Vittorio Somenzi.[5]

All this, in a sense, concerns the 'prehistory' of Italian philosophy of science. As to contemporary research, it is not my intention here to try to give a general survey of them. The essays collected in this volume are representative of recent work (in part) and of the fields of interest in Italy today.

The volume is not intended to represent a complete and exhaustive survey of recent philosophy of science in Italy. Most of the papers here collected share some common features, namely the attempt to analyze traditional as well as new epistemological problems concerning the foundations of formal and of empirical sciences by means of a systematic application of logical tools. Apart from this, there is no common underlying philosophy in the different essays, since the authors present a great variety of different philosophical approaches, and also belong to different research groups in logic, mathematics, physics, biology, and history of science.

University of Florence   MARIA LUISA DALLA CHIARA
March 1979

NOTES

[1] Besides their theoretical differences, Croce and Gentile also had different political attitudes: whereas Croce was an anti-fascist, Gentile represented, to his death, a kind of 'official philosopher' of the fascist regime.

[2] B. Croce, *Logica come scienza del concetto puro* (Bari, Laterza, 1908); translated by Douglas Ainslie as *Logic as Science of the Pure Concept* (London, Macmillan, 1917).

[3] In 1936 Persico published the volume *Fondamenti della meccanica atomica* (Bologna, Zanichelli), which, in a revised edition, was later translated into English as *Fundamentals of Quantum Mechanics*, translated and edited by Georges M. Temmer (New York, Prentice-Hall, 1950).

[4] Geymonat's first contributions were *La nuova filosofia della natura in Germania* (The New Natural Philosophy in Germany), (1934); *Nuovi indirizzi della filosofia austriaca* (New Trends of Austrian Philosophy), (1935); *Storia e filosofia dell'analisi infinitesimale* (History and Philosophy of Infinitesimal Calculus), (1947). His book, *Galileo*, has been translated into English by Stillman Drake (New York, McGraw-Hill, 1965).

[5] N. Abbagnano, *La fisica nuova*; *fondamenti di una teoria della scienza* (The New Physics. Foundations of a Theory of Science), (Naples, 1934); G. Preti, *Praxis ed empirismo* (Praxis and Empiricism), (Turin, 1957); F. Barone, *Il neopositivismo logico* (Logical Neopositivism), (Turin, 1953); A. Pasquinelli, *Introduzione alla logica simbolica* (Introduction to Symbolic Logic), (Turin, 1957); V. Somenzi, *Operazionismo in fisica* (Operationalism in Physics), (1958).

PART I

FOUNDATIONS OF LOGIC AND OF MATHEMATICS

ETTORE CASARI

POSITIVELY OMITTING TYPES*

0. As early as the Hannover Colloquium, 1966, P.H.G. Aczel explicitly contrasted the two possibilities before us when we try to make the concepts and the methods originally worked out within classical model theory fruitful for weaker logical frameworks.

Indeed, on the one hand, we have the possibility of a truly 'model-theoretical' approach to this task, i.e. the possibility of trying to adapt classical constructions to the concepts of model at our disposal for the new logical framework.

However, on the other hand, owing to the close interconnections which classically hold between models and theories, we have the possibility of a 'theory-theoretical' approach which consists in trying to make suitable syntactical reformulations of concepts and methods of classical model theory available for logically weaker formal systems.

Both possibilities seem worth being pursued, although the first one has apparently received the most attention in the last years. What follows lies, on the contrary, wholly in the second perspective and concerns some properties of formal systems suggested by the classical omitting types problem. The results should, after all, be useful in clarifying the extent to which the properties of classical theories are often hardly 'classical', i.e. truly dependent on classical assumptions.

1. We consider elementary languages whose terms and formulas are defined in the usual way by the logical symbols $\neg$, $\wedge$, $\vee$, $\rightarrow$, $\exists$, $\forall$, the identity $=$, the individual variables x_0, x_1, ... and *at most* denumerable sets of n-ary predicate constants P_0^n, P_1^n, ..., n-ary function constants f_0^n, f_1^n, ... and individual constants c_0, c_1, ... Closed terms are called *names* and closed formulas are called *sentences*. As metalinguistic variables for variables, terms, and formulas we use x, y, ...; t, s, ...; α, β, ...; respectively. We shall abbreviate the sequence x_0, x_1, ..., x_{n-1} by $\overset{n}{x}$ and designate by $\mathrm{Frl}(\overset{n}{x})$ the set of all formulas whose free variables are among x_0, ..., x_{n-1}. We use $\mathrm{Frl}(\overset{\circ}{x})$ for the set of all sentences. A set of formulas M for which there is an n such that $M \subseteq \mathrm{Frl}(\overset{n}{x})$ is said to be *limited* and the first n for which this happens is called the *diameter* of M. If M has no

3

Maria Luisa Dalla Chiara (ed.), Italian Studies in the Philosophy of Science, 3–11
Copyright © 1980 by D. Reidel Publishing Company.

diameter it is called *unlimited* and its *normal decomposition* shall be the chain of limited sets $M \cap \mathrm{Frl}(^{0}_{x})$, $M \cap \mathrm{Frl}(^{1}_{x})$, ... Every finite set is of course limited. For a finite set F, we designate by $\bigwedge F$ the conjunction of all elements of F. We shall abbreviate by $^{n}_{t}$ the finite sequence of terms $t_0, ..., t_{n-1}$ and by $^{\infty}_{t}$ the infinite sequence $t_0, t_1, ...$. Substitution in formulas of sequences of terms for sequences of variables must always be understood so that t_i is substituted for x_i. If α is a formula, then $\exists(\alpha)$ and $\forall(\alpha)$ shall be the existential and the universal closure of α respectively.

2. A *positive proof in a formal system SF on the language L* is a finite sequence of formulas of L each of which is either an element of a well-defined set of *specific axioms* of SF, or has one of the following forms:

A1.1	$\alpha \rightarrow (\beta \rightarrow \alpha)$,
A1.2	$(\alpha \rightarrow \beta) \rightarrow ((\alpha \rightarrow (\beta \rightarrow \gamma)) \rightarrow (\alpha \rightarrow \gamma))$;
A2.1	$(\alpha \wedge \beta) \rightarrow \alpha$,
A2.2	$(\alpha \wedge \beta) \rightarrow \beta$,
A2.3	$(\alpha \rightarrow \beta) \rightarrow ((\alpha \rightarrow \gamma) \rightarrow (\alpha \rightarrow (\beta \wedge \gamma)))$;
A3.1	$\alpha \rightarrow (\alpha \vee \beta)$,
A3.2	$\beta \rightarrow (\alpha \vee \beta)$,
A3.3	$(\alpha \rightarrow \gamma) \rightarrow ((\beta \rightarrow \gamma) \rightarrow ((\alpha \vee \beta) \rightarrow \gamma))$;
Q1	$\forall x\alpha \rightarrow \alpha(x/t)$,
Q2	$\alpha(x/t) \rightarrow \exists x\alpha$;
I1	$t = t$,
I2	$t = s \rightarrow (\alpha(x/t) \rightarrow \alpha(x/s))$,

or, finally, is obtained by preceding formulas of the sequence by means of one of the *rules of inference*:

$$\frac{\alpha \qquad \alpha \rightarrow \beta}{\beta},$$

$$\frac{\alpha \rightarrow \beta \qquad x \text{ not free in } \alpha}{\alpha \rightarrow \forall x\beta},$$

$$\frac{\alpha \rightarrow \beta \qquad x \text{ not free in } \beta}{\exists x\alpha \rightarrow \beta}.$$

As we are interested only in positive proofs in SF we will write $\vdash_{\mathrm{SF}} \alpha$ for 'α is positively provable in SF' and omit in all that follows the specification 'positive', also in many concepts related with provability.

Let M be a set of formulas in the language L of the formal system SF. By $SF + M$ we mean the formal system on L whose specific axioms are those of SF plus the formulas in M. If $M = \{\alpha_1, ..., \alpha_n\}$ we simply write $SF + \alpha_1 ... + \alpha_n$. We set by definition: $M \vdash_{SF} \alpha$ (to be read: α *is derivable from M in SF*) iff $\vdash_{SF+M} \alpha$. Clearly we have: $M \vdash_{SF} \alpha$ iff there is a finite subset F of M such that $\vdash_{SF} \forall(\bigwedge F) \to \alpha$. Always by definition we set: $M \Vdash_{SF} \alpha$ (to be read: α *is deducible from M in SF*) iff there is a finite subset F of M such that $\vdash_{SF} (\bigwedge F) \to \alpha$. Clearly, for $\alpha \in \mathrm{Frl}(^\circ_x)$, we have: $M \Vdash_{SF} \alpha$ iff there is a finite subset F of M such that $\vdash_{SF} \exists(\bigwedge F) \to \alpha$.

3. From now on M shall be a set of formulas in the language of the formal system under consideration and λ shall be a sentence in that language. We define:

(1) *SF is λ-consistent* iff $\vdash_{SF} \lambda$ does not hold;

(2) *SF is λ-complete* iff for every α in $\mathrm{Frl}(^\circ_x)$ it is either $\vdash_{SF} \alpha$ or $\vdash_{SF} \alpha \to \lambda$;

(3) *SF is rich* iff for every α such that $\exists x\alpha \in \mathrm{Frl}(^\circ_x)$ and $\vdash_{SF} \exists x\alpha$ there is a name t for which $\vdash_{SF} \alpha(x/t)$ holds;

(4) *M is deductively λ-consistent in SF* iff $M \Vdash_{SF} \lambda$ does not hold;

(5) *M is deductively λ-complete in SF* iff either $M \Vdash_{SF} \alpha$ or $M \Vdash_{SF} \alpha \to \lambda$ for all formulas α, in the case M is unlimited, and for all $\alpha \in \mathrm{Frl}(^n_x)$ in the case that M is limited and has diameter n;

(6) *M is deductively closed in SF* iff $M \Vdash_{SF} \alpha$ implies $\alpha \in M$ for all formulas α, in the case M is unlimited, and for all formulas α belonging to $\mathrm{Frl}(^n_x)$, in the case M has diameter n;

(7) *M is a λ-type in SF* iff M is deductively λ-consistent, deductively λ-complete and deductively closed in SF;

(8) *α is a λ-generator of M in SF* iff α is deductively λ-consistent in SF and $\vdash_{SF} \alpha \to ((\beta \to \lambda) \to \lambda)$ for every $\beta \in M$;

(9) *M is λ-principal in SF* iff either M is of diameter n and there is' an $\alpha \in \mathrm{Frl}(^n_x)$ which is a λ-generator of M in SF or M is unlimited and all members of its normal decomposition are λ-principal.

4. From the definitions it follows that a λ-generator α of a λ-type is in that type and that if M and N are λ-types of the same diameter such that $M \subseteq N$, then $M = N$. Moreover we have:

THEOREM 1. *If M is deductively λ-consistent in SF and has diameter n, then there is a λ-type T in SF of diameter n and such that $M \subseteq T$.*
Proof. By the familiar Lindenbaum's λ-completing procedure. In particular, having fixed an enumeration α_0, α_1, ... of all formulas in $\mathrm{Frl}\binom{n}{x}$, with n diameter of M, we define a sequence M_0, M_1, ... by

$$M_0 = M; \qquad M_{K+1} = \begin{cases} M_K + \alpha_K, & \text{if such a set is deductively} \\ & \lambda\text{-consistent in SF} \\ M_K, & \text{otherwise} \end{cases}$$

For $T = \sum M_K$ it is now easy to prove that T is a λ-type in SF. $\qquad \square$
The preceding theorem can obviously be proved for unlimited sets too.

5. We further define, for every set of formulas M:
(10) *M is locally exemplified in SF* iff for each finite subset F of M there is a sequence $\overset{\infty}{{}_t}$ of names such that $\underset{SF}{\vdash} \bigwedge F(\overset{\infty}{{}_t})$;

(11) *M is uniformly exemplified in SF* iff there is a sequence $\overset{\infty}{{}_t}$ of names such that $\underset{SF}{\vdash} \bigwedge F(\overset{\infty}{{}_t})$ for all finite subsets F of M;

(12) *M is λ-omitted in SF* iff for every sequence $\overset{\infty}{{}_t}$ of names there is a formula β in M such that $\underset{SF}{\vdash} \beta(\overset{\infty}{{}_t}) \to \lambda$.

The assumed infinity of the sequence of names is of course in many cases superfluous. In particular, uniform exemplification of a set of diameter n is secured by the existence of a sequence $\overset{n}{{}_t}$ of names such that $\underset{SF}{\vdash} \bigwedge F(\overset{n}{{}_t})$ for every finite subset F of M and omission of such a set by the existence, for every sequence $\overset{n}{{}_t}$ of names, of a β in M such that $\underset{SF}{\vdash} \beta(\overset{n}{{}_t}) \to \lambda$. Remark also that uniform exemplifiability of all members of its normal decomposition is in general a weaker condition than uniform exemplifiability of an unlimited set.

6. The 'omitting types theorem' is now the following assertion.

THEOREM 2. *Assume that SF is a λ-consistent formal system on a language L and that for every natural number i, M_i is a limited set of formulas in L which is not λ-principal in SF. Then there exists an extension SF* of*

SF (on an expansion by individual constants L^* of L) such that: (1) SF^* *is λ-consistent, λ-complete and rich;* (2) *for every i, SF^* omits M_i.*

Proof. Let C be a denumerable set of individual constants which are not in L. We define L^* as the expansion of L by means of C and SF^C as the expansion of SF on L^*. We then fix:

   (I) an enumeration $c_0, c_1, \ldots$ of C;

   (II) an enumeration $\alpha_0, \alpha_1, \ldots$ of all sentences in L^*;

   (III) for each i, an enumeration $\beta_0^i, \beta_1^i, \ldots$ of all elements of M_i;

   (IV) for each i, an enumeration $t_0^i, t_1^i, \ldots$ of all $n(i)$-tuples of names of L^*, where $n(i)$ is the diameter of M_i.

We then define a sequence $SF_0, SF_1, \ldots$ of formal systems on L^* by setting:

1. $SF_0 = SF^C$;

2. $SF_{3n+1} = SF_{3n}^n$,   where SF_{3n}^n is the last element of the sequence of formal systems SF_{3n}^j which is so defined for $j \leq n$

2.1 $SF_{3n}^0 = SF_{3n}$

2.2 $SF_{3n}^{k+1} = \begin{cases} SF_{3n}^k + \beta^k(t_{n-k}^k) \to \lambda, & \\ \\ SF_{3n}^k & \end{cases}$ where β^k is the first formula in the given enumeration of M_k for which $SF_{3n}^k + \beta^k(t_{n-k}^k) \to \lambda$ is λ-consistent, if such a formula exists; otherwise,

3. $SF_{3n+2} = \begin{cases} SF_{3n+1} + \alpha_n, & \\ SF_{3n+1}, & \end{cases}$ if such a system is λ-consistent, otherwise;

4. $SF_{3n+3} = \begin{cases} SF_{3n+2} + \gamma(x/c), & \\ \\ \\ \\ SF_{3n+2}, & \end{cases}$ if SF_{3n+2} is $SF_{3n+1} + \alpha_n$ and α_n is $\exists x\gamma$ and c is the first element in the given enumeration of C which does not occur in $\exists x\gamma$ and in the axioms of SF_{3n+2}, otherwise.

Finally we set $SF^* = \sum SF_n$.

It is now easy to show, by usual arguments that assertion (1) holds, i.e. that SF^* is λ-consistent, λ-complete and rich. As far as assertion (2) is concerned, let i be any natural number and $t_0, \ldots, t_{n(i)-1}$ be any $n(i)$-tuple of names of L^*. Such an $n(i)$-tuple shall occur at some place k in the given enumeration of all $n(i)$-tuples, i.e. it shall be t_k^i. Consider $SF_{3(i+k)+1}$. This system was defined using the auxiliary sequence of formal systems $SF_{3(i+k)}^0, \ldots, SF_{3(i+k)}^{i+k}$. In such a sequence the $n(i)$-tuple t_k^i was taken into account in the definition of $\mathrm{SF}^{i+1}_{3(i+k)}$ as we defined this system as $\mathrm{SF}^i_{3(i+k)}$ $+ \beta^i(t^i_{i+k-i}) \to \lambda$ (i.e. as $\mathrm{SF}^i_{3(i+k)} + \beta^i(t^i_k) \to \lambda$), where β^i was to be the first formula in the given enumeration of M_i for which such a system would be λ-consistent, under the hypothesis that at least one such formula would exist. To prove (2) it is then sufficient to prove that such a formula indeed exists. Suppose, therefore, that it does not exist, i.e. that for every β^i in M_i, we have $\underset{\mathrm{SF}^i_{3(i+k)}+\beta^i(t^i_k)\to\lambda}{\vdash} \lambda$ By construction, $\mathrm{SF}^i_{3(i+k)}$ is SF^C plus a finite number of new axioms. Let δ be the conjunction of all these new axioms. Then λ is provable in all formal systems $\mathrm{SF}^C + \delta + \beta^i(t^i_k) \to \lambda$ with $\beta^i \in M_i$ and so, for all $\beta^i \in M_i$, we have $\underset{\mathrm{SF}^C}{\vdash} \delta \to ((\beta(t^i_k) \to \lambda) \to \lambda)$. Let us write simply x^i for $\overset{n(i)}{x}$ (i.e. for $x_0, \ldots, x_{n(i)-1}$) and $x^i = t^i_k$ for $x_0 = t_0 \wedge \ldots \wedge x_{n(i)-1} = t_{n(i)-1}$. Then, obviously, $\underset{\mathrm{SF}^C}{\vdash} (x^i = t^i_k \wedge \delta) \to ((\beta^i(x^i) \to \lambda) \to \lambda)$. If $a_1, \ldots, a_s$ (briefly a) are the elements of C which occur either in t^i_k or in δ and if $y_1, \ldots, y_s$ (briefly y) are s 'new' variables then, by usual arguments, $\underset{\mathrm{SF}}{\vdash} (x^i = t^i_k(a/y) \wedge \delta(a/y)) \to ((\beta^i(x^i) \to \lambda) \to \lambda)$ and therefore also $\underset{\mathrm{SF}}{\vdash} \exists y(x^i = t^i_k(a/y) \wedge \delta(a/y))) \to ((\beta^i(x^i) \to \lambda) \to \lambda)$. But the antecedent of such a conditional belongs to $\mathrm{Frl}(x^i)$ and as it provably implies the 'double λ-negation' of every element of M_i it must, by hypothesis, be deductively λ-inconsistent in SF, i.e. we must have $\underset{\mathrm{SF}}{\vdash} \exists x^i(\exists y \ (x^i = t^i_k(a/y) \wedge \delta(a/y))) \to \lambda$. But then this formula can also be proved in $\mathrm{SF}^i_{3(i+k)}$ and since in this last system is obviously provable δ and δ logically implies $\exists x^i$ $(\exists y \ (x^i = t^i_k(a/y) \wedge \delta(a/y)))$ we get $\underset{\mathrm{SF}^i_{3(i+k)}}{\vdash} \lambda$ which has been excluded in proving assertion (1).

7. We have:

THEOREM 3. *If SF is λ-consistent and λ-complete and M is a λ-principal λ-type, then M is uniformly exemplified in every λ-consistent, λ-complete and rich extension of SF.*
Proof. Let SF and M be as in the hypothesis and let SF^* be any λ-consistent, λ-complete and rich extension of SF. We distinguish two cases.
Case 1. M is limited. Let n be the diameter of M. Then, by hypothesis, there is an $\alpha \in \mathrm{Frl}(\overset{n}{x})$ which is deductively λ-consistent in SF and such that $\underset{\mathrm{SF}}{\vdash} \alpha \to ((\beta \to \lambda) \to \lambda)$ for every $\beta \in M$. Then $\underset{\mathrm{SF}}{\vdash} \exists^n_x \alpha$ (were it not

so, then, by λ-completeness, it would be $\vdash_{SF} \exists^n_x \alpha \to \lambda$, contradicting the hypothesis that α be deductively λ-consistent in SF). Therefore $\vdash_{SF^*} \exists^n_x \alpha$ and also, by the richness of SF*, $\vdash_{SF^*} \alpha(^n_t)$ for some n-tuple of names ^n_t. But obviously $\vdash_{SF^*} \alpha \to ((\beta \to \lambda) \to \lambda)$ for each $\beta \in M$ and so, by substitution of ^n_t for ^n_x and by detachment $\vdash_{SF^*} (\beta(^n_t) \to \lambda) \to \lambda$. But then, by λ-consistency of SF*, not $\vdash_{SF^*} \beta(^n_t) \to \lambda$ and therefore, by λ-completeness, $\vdash_{SF^*} \beta(^n_t)$. So SF* uniformly exemplifies M.

Case 2. M is unlimited. It follows from definitions and hypothesis that for each n, $M \cap \mathrm{Frl}(^n_x)$ is λ-principal. We define by induction a sequence $^\infty_t$. As t_0 we take a name which uniformly exemplifies $M \cap \mathrm{Frl}(^1_x)$; such a name exists according to Case 1 under the assumption that $M \cap \mathrm{Frl}(^1_x)$ is λ-principal. Suppose now we have already defined $t_0, \ldots, t_{n-1}$ in such a way that it uniformly exemplifies $M \cap \mathrm{Frl}(^n_x)$. We show that there is a t such that $t_0, \ldots, t_{n-1}, t$ uniformly exemplifies $M \cap \mathrm{Frl}(^{n+1}_x)$. By hypothesis there is an $\alpha \in \mathrm{Frl}(^{n+1}_x)$ which is deductively λ-consistent in SF (and so, by the λ-completeness of SF, is such that $\vdash_{SF} \exists^{n+1}_x \alpha$) and for which $\vdash_{SF} \alpha \to ((\beta \to \lambda) \to \lambda)$ holds for every $\beta \in M \cap \mathrm{Frl}(^{n+1}_x)$. Clearly $\alpha \in M \cap \mathrm{Frl}(^{n+1}_x)$, because, as it is easy to see, the components of the normal decomposition of an unlimited λ-type are λ-types; but then, being $\vdash_{SF} \alpha \to \exists x_n \alpha$, it is also $\exists x_n \alpha \in M \cap \mathrm{Frl}(^{n+1}_x)$ and then, of course $\exists x_n \alpha \in M \cap \mathrm{Frl}(^n_x)$. By hypothesis, ^n_t uniformly exemplifies $M \cap \mathrm{Frl}(^n_x)$ in SF*. So $\vdash_{SF^*} \exists x_n \alpha(^n_t, x_n)$. As SF* is rich, we have $\vdash_{SF^*} \alpha(^n_t, t)$ for some name t. By substitution and detachment we then get $\vdash_{SF^*} (\beta(^n_t, t) \to \lambda) \to \lambda$ for every β in $M \cap \mathrm{Frl}(^{n+1}_x)$. But then, by λ-consistency and λ-completeness of SF*, $\vdash_{SF^*} \beta(^n_t, t)$ and so $t_0, \ldots, t_{n-1}, t$ uniformly exemplifies $M \cap \mathrm{Frl}(^{n+1}_x)$. As the $(n + 1)$th member of our intended sequence $^\infty_t$ we take precisely such a t. In this way we completely define a sequence $^\infty_t$ which obviously uniformly exemplifies M in SF*.

8. From the two preceding theorems we get

THEOREM 4. *A λ-consistent, λ-complete and rich formal system uniformly exemplifies all its λ-principal types and omits all other λ-types.*

Proof. Let SF be λ-consistent, λ-complete, and rich and let M be a λ-

principal λ-type in it. By Theorem 3, M is uniformly exemplified in SF. Let now M be a non-λ-principal λ-type. Suppose it is limited and of diameter n. By Theorem 2 we know that there is an extension of SF which omits M; then of course no n-tuple of names in SF can uniformly exemplify M in SF. Let then $\overset{n}{t}$ be any n-tuple of names in the language of SF. As it cannot uniformly exemplify M there must be a $\beta \in M$ such that $\beta(\overset{n}{t})$ is not a theorem of SF. But as SF is λ-complete, it must then be $\underset{SF}{\vdash} \beta(\overset{n}{t}) \to \lambda$. Next suppose that M is unlimited; then there must be an n such that $M \cap \mathrm{Frl}(\overset{n}{x})$ is not λ-principal. To such a λ-type we can apply the previous argument. Thus the desired conclusion follows immediately.

9. As expected we also have

THEOREM 5. *A λ-consistent formal system has only λ-principal λ-types iff for each natural number n there are only finitely many λ-types of diameter n.*

Proof. Assume that all λ-types of diameter n are λ-principal. Let M be the set of all formulas $\alpha \to \lambda$ where α is a λ-generator of a λ-type of diameter n. The diameter of M is of course n. It holds $M \underset{SF}{\Vdash} \lambda$. Were it not so, then, by Theorem 1, there would exist a λ-type of diameter n which includes M. As all such types are λ-principal it would have one of the α's as a λ-generator and so, containing both an α and its 'λ-negation' $\alpha \to \lambda$, it would be deductively λ-inconsistent, which is impossible. From $M \underset{SF}{\Vdash} \lambda$ it follows that for a finite number m of λ-generators $\alpha_1, \ldots, \alpha_m$ (which we suppose to generate the λ-types $T_1, \ldots, T_m$, respectively) it holds $\underset{SF}{\vdash} ((\alpha_1 \to \lambda) \wedge \ldots \wedge (\alpha_n \to \lambda)) \to \lambda$. Let now N be any λ-type in SF of diameter n. It is either $N \underset{SF}{\Vdash} \alpha_1$ or $N \underset{SF}{\Vdash} \alpha_1 \to \lambda$. If $N \underset{SF}{\Vdash} \alpha_1$, then, by easy arguments, $N = T_1$. If, on the contrary, $N \underset{SF}{\Vdash} \alpha_1 \to \lambda$, then, by detachment, $N \underset{SF}{\Vdash} ((\alpha_2 \to \lambda) \wedge \ldots \wedge (\alpha_m \to \lambda)) \to \lambda$. But now either $N \underset{SF}{\Vdash} \alpha_2$ (in which case $N = T_2$) or $N \underset{SF}{\Vdash} \alpha_2 \to \lambda$. In this latter case we apply another detachment and ... finally we get $N = T_1$ or $N = T_2$ or ... or $N = T_{m-1}$ or $N \underset{SF}{\vdash} (\alpha_m \to \lambda) \to \lambda$. As the last disjunct of course implies $N = T_m$ we finally have that every λ-type of diameter n must be one of the $T_1, \ldots, T_m$.

Now let T be a non-λ-principal λ-type of diameter n; we show that there exists infinitely many λ-types of diameter n. We take any formula $\alpha \in T$

and define a sequence of formulas by setting $\alpha_0 = \alpha$; where $\alpha_{n+1} = $ a particular formula of T such that $\vdash_{SF} (\alpha_0 \wedge \cdots \wedge \alpha_n) \to ((\alpha_{n+1} \to \lambda) \to \lambda)$ does not hold. Such a formula must always exist, because otherwise, $\alpha_0 \wedge \cdots \wedge \alpha_n$ would be a λ-generator of T. We then define, for each m, $M_m = \{\alpha_0, \ldots, \alpha_{m-1}, \alpha_m \to \lambda\}$. For each m, we have that M_m is deductively λ-consistent in SF, because, otherwise, it would be, contrary to the preceding construction, $\vdash_{SF} (\alpha_0 \wedge \cdots \wedge \alpha_{m-1}) \to ((\alpha_m \to \lambda) \to \lambda)$. By Theorem 1, each M_m can be extended to a λ-type T_m of diameter n. But for $r \neq s$, $T_r \neq T_s$; otherwise, for, say, $r < s$ it would be $T_s \Vdash_{SF} \alpha_r$ and $T_s \Vdash_{SF} \alpha_r \to \lambda$. Then there are infinitely many λ-types of diameter n.

NOTE

* The main content of this paper was presented at the 'Week on Non-Classical Logics' held at S. Margherita Ligure in June, 1975.

CARLO CELLUCCI

PROOF THEORY AND THEORY OF MEANING

1. A conjecture about intensional equality of purely logical proofs was formulated by Prawitz [1971, II.3.5.6]. It may be stated as follows, for derivations generated by purely logical rules satisfying the normalization property and uniqueness of normal form of derivations, such as those of classical or intuitionistic logic as formulated in Prawitz [1971, II.1.3].

Proofs and derivations are different sorts of objects. Derivations are formal objects: each derivation represents a proof. The problem then arises: when do two derivations represent *intensionally equal* proofs? The conjecture is:

(*) Two derivations represent intensionally equal proofs if and only if they reduce to the same normal form.

Of course the conjecture hinges on what we mean by two proofs being intensionally equal. An obvious explanation is: two proofs are said to be intensionally equal if and only if they are literally the same. By this explanation (*) is turned into the assertion:

(**) Two derivations represent the same proof if and only if they reduce to the same normal form.

Now, if the representation of proofs by derivations is to be really consistent and adequate, then (**) is clearly false. Only the following weaker (and rather trivial) assertion seems to be justified:

(***) Two derivations represent the same proof if and only if they are identical.

(By 'identical' we mean here 'identical up to changes of proper parameters'.) The 'if' part of (***) follows from the consistency, the 'only if' part of (***) follows from the adequacy, of the representation.

Of course the notion of intensional equality of proofs as explained above is quite natural. The intended meaning of intensional equality is that two proofs are intensionally equal if and only if they are given to us as the same proof. Here the stress is on 'given to us'. But if the only approach to proofs is through the way they are given to us, there is no

13

Maria Luisa Dalla Chiara (ed.), Italian Studies in the Philosophy of Science, 13–29.
Copyright © 1980 by D. Reidel Publishing Company.

practical difference between 'are given to us as the same proof' and 'are literally the same proof'.

At the other end of the scale we find the notion of *extensional equality* of proofs: two proofs are said to be extensionally equal if and only if their assumptions and conclusions are the same. This notion is also natural, but naturalness is no guarantee of mathematical usefulness.

In fact, the trouble with such notions as intensional or extensional equality is that they are too extreme. As Troelstra [1975, p. 308] points out:

The very fact that considering objects from a strictly intensional point of view means carrying along *all* information about these objects, that is, we are not permitted to abstract from any of their properties, clearly indicates that, in general, strict intensional equality is not mathematically useful.

Similarly, the fact that considering proofs from an extensional point of view means taking into account only their assumptions and conclusion – that is, we abstract from any of their properties but the trivial ones – clearly provides the same indication. The conclusion to be drawn is that, in order to make any significant advance, we must consider *non-extensional* notions of equality of proofs in the sense of Troelstra [1975], that is, notions intermediate between intensional and extensional equality.

2. A simple non-extensional notion is that of *equality up to reduction*, or *r-equality*, which is implicit in Feferman [1975] and Kreisel – Takeuti [1974, p. 38]: two proofs are said to be *r*-equal if and only if, when properly analyzed by eliminating what is not strictly required to derive the conclusion from the assumptions, they are turned into the same proof. (Of course we assume here that what is actually established by a proof explicitly appears in the conclusion.)

Then (∗) may be replaced by the following more plausible conjecture:

(+) Two derivations represent *r*-equal proofs if and only if they reduce to the same normal form.

The plausibility of the 'if' part of (+) depends on the consistency of the representation of proofs by derivations and on the fact that the reductions of derivations eliminate just what is not strictly required to derive the conclusion from the assumptions. The plausibility of the 'only if' part of (+) depends on the adequacy of the representation of proofs by derivations and on the fact that the reduction process of derivations may be considered as an adequate representation of an analysis of proofs which eliminates what is not strictly required to derive the conclusion from the assumptions.

3. The scope of $(+)$ may be widened by introducing the notion of *isomorphism* of proofs. As usual we assume that proofs are in tree form. Two proofs Π_1 and Π_2 are said to be isomorphic if and only if there is a one-to-one correspondence between their nodes and operations such that:

(i) The conclusions correspond.

(ii) If node μ in Π_1 corresponds to a node ν in Π_2, $\mu_1, \ldots, \mu_m$ (in that order from left to right) are the immediate ascendents of μ, $\mathcal{O}_1$ is the operation with arguments $\mu_1, \ldots, \mu_m$ and value μ, $\nu_1, \ldots, \nu_n$ are the immediate ascendents of ν, and $\mathcal{O}_2$ is the operation with arguments $\nu_1, \ldots, \nu_n$ and value ν; then $m = n$, for any i, $1 \leqslant i \leqslant n$, μ_i corresponds to ν_i, and $\mathcal{O}_1 = \mathcal{O}_2$.

(iii) The assumptions correspond, where corresponding assumptions are dischared by corresponding operations.

The notion of isomorphism of derivations is defined similarly to that of isomorphism of proofs. Of course in this case the operations $\mathcal{O}_1$, $\mathcal{O}_2$ are purely logical rules. For instance the derivations:

$$\mathscr{D} = (\to I)\dfrac{(\to I)\dfrac{\overset{\textstyle ①}{A}}{B \to A}}{A \to (B \to A)} - ①$$

and

$$\mathscr{D}' = (\to I)\dfrac{(\to I)\dfrac{\overset{\textstyle ①}{A \to A}}{(B \to B) \to (A \to A)}}{(A \to A) \to ((B \to B) \to (A \to A))} - ①$$

are isomorphic.

Two proofs are said to be *isomorphic up to reduction*, or *r-isomorphic*, if and only if, when properly analyzed by eliminating what is not strictly required to derive the conclusion from the assumptions, they are turned into isomorphic proofs. (Again we assume that what is actually established by a proof explicitly appears in the conclusion.) The notion of isomorphism up to reduction, or r-isomorphism, of derivations is defined similarly to that of r-isomorphism of proofs.

Then $(+)$ may be generalized as follows:

$(++)$ Two derivations represent r-isomorphic proofs if and only if they reduce to isomorphic normal forms.

Clearly $(++)$ is an extension of $(+)$ since every proof or derivation is isomorphic to itself. However $(++)$ is as plausible as $(+)$: the argument already mentioned for $(+)$ also applies to $(++)$.

4. As an application of $(++)$, let us consider the familiar negative translation of classical into intuitionistic logic. If the rules of classical and intuitionistic logic are formulated as in Prawitz [1971, II.1.3], then the negative translation associates to any formula A in $\{\lambda, \wedge, \rightarrow, \forall\}$ a formula A^- obtained by replacing every atomic part P of A different from λ by $\neg\neg P$.

Clearly any normal derivation $\mathscr{D}$ in classical logic of A from the empty set of assumptions may be transformed into a normal derivation $\mathscr{D}'$ in intuitionistic logic of A^- from the empty set of assumptions. For, since P and $\neg\neg P$ are formally equivalent in classical logic, so are also A and A^-. Hence if $\mathscr{D}$ is a normal derivation in classical logic of A from the empty set of assumptions, then by the normalization property, $\mathscr{D}$ can be transformed into a normal derivation $\mathscr{D}''$ in classical logic of A^- from the empty set of assumptions. On the other hand the reductions described in Prawitz–Malmnäs [1968, p. 223] transform $\mathscr{D}''$ into a normal derivation $\mathscr{D}'$ in intuitionistic logic of A^- from the empty set of assumptions. This establishes the result.

Now we may ask as in Kreisel [1971, p. 257] whether $\mathscr{D}$ and $\mathscr{D}'$ represent intensionally equal, or at least r-equal, proofs. It may be easily seen that the answer is: No. For instance the reductions which transform $\mathscr{D}''$ into $\mathscr{D}'$ consist in replacing subderivations of the form shown on the left below by derivations of the form shown on the right:

$$
(\rightarrow E)\ \frac{(\lambda_c)\dfrac{\overset{[\neg P]}{\overset{\mathscr{D}_1}{\lambda}}}{P} \quad \neg P}{\lambda} \qquad\qquad \overset{[\neg P]}{\overset{\mathscr{D}_1}{\lambda}}
$$

(Here the derivation ending with the major premiss of $(\rightarrow E)$ does not contain any (λ_c) which discharges an assumption.)

Clearly $\mathscr{D}''$ and $\mathscr{D}'$ are not isomorphic, hence they do not represent *r*-isomorphic proofs. Of course, this is relative to the specific reductions admitted, that is those by Prawitz [1971, II.3.3.1.1–3.3.1.5]. If we allowed also reductions of the form mentioned above, then $\mathscr{D}''$ and $\mathscr{D}'$ would represent *r*-isomorphic proofs.

5. A conjecture like $(++)$ may be viewed as providing a starting point for a theory of synonymy of proofs. We may also formulate a conjecture, expressed in terms of the same notions as $(++)$, which provides a starting point for a theory of synonymy of propositions.

Again we must distinguish between propositions and formulae. Formulae are formal objects: each formula represents a proposition. The problem then arises: when do two formulae represent *intensionally equal* propositions? A simple but rather trivial approach to the problem is embodied in the following conjecture:

(°) Two formulae represent intensionally equal propositions if and only if they are identical.

(By 'identical' we mean here 'identical up to changes of proper parameters'.) Of course the conjecture hinges on what we mean by two propositions being intensionally equal. An obvious explanation is: two propositions are said to be intensionally equal if and only if they are literally the same. By this explanation, (°) is turned into the assertion:

(°°) Two formulae represent the same proposition if and only if they are identical.

Clearly (°°) is plausible. In particular the 'if' part of (°°) follows from the consistency, the 'only if' part of (°°) follows from the adequacy, of the representation.

We may also introduce a notion of *extensional equality* of propositions: two propositions are said to be extensionally equal if and only if they are logically equivalent. Then we can formulate the conjecture:

(−) Two formulae represent extensionally equal propositions if and only if they are formally equivalent.

(Of course logical equivalence and formal equivalence refer to certain principles of reasoning, such as those of classical or intuitionistic logic. It is tacitly understood that (−) is relative to a given class of such principles.)

The plausibility of the 'if' part of (−) follows from the consistency of the rules which generate the derivations. The plausibility of the 'only if' part of (−) depends on whether the rules are complete with respect to logical validity. By an argument like that in Kreisel [1967, p. 154] such are the rules of classical logic. In the case of intuitionistic logic the situation is more complicated. For a discussion we refer to Troelstra [A].

Both intensional and extensional equality are of little use since they are too extreme. In particular extensional equality is too weak because all logically valid propositions are logically equivalent. Also, since all formulae derivable from the empty set of assumptions are formally equivalent, by (−) they represent extensionally equal propositions.

This applies for instance to the formulae $P \to P$ and $((P \to Q) \to P) \to P$, which are both derivable from the empty set of assumptions. Obviously there is no strict connection between these formulae since $P \to P$ represents an instance of a simple logical truth, the law of identity, whereas $((P \to Q) \to P) \to P$ represents an instance of a much more complex logical truth, Peirce's law.

6. Therefore, in order to make any significant advance in the subject of synonymy of propositions we must consider non-extensional notions of equality of propositions, that is notions intermediate between intensional and extensional equality. A simple non-extensional notion is that of *equality up to reduction*, or *r-equality*, of propositions: two propositions are said to be *r*-equal if and only if they are extensionally equal and for any proof of one of them from the empty set of assumptions there is an *r*-isomorphic proof of the other from the empty set of assumptions.

The notion of equality up to reduction, or *r*-equality, of formulae is defined similarly: two formulae are said to be *r*-equal if and only if they are formally equivalent and for any derivation of one of them from the empty set of assumptions there is an *r*-isomorphic derivation of the other from the empty set of assumptions.

Then we may formulate the following conjecture:

(×) Two formulae represent *r*-equal propositions if and only if they are *r*-equal.

(It is tacitly understood that (×) is relative to a given class of principles of reasoning.)

The plausibility of the 'if' part of (×) depends on the consistency, the plausibility of the 'only if' part of ($\times$) depends on the adequacy, of the representation of propositions by formulae.

Since we consider only derivations generated by rules satisfying the normalization property and uniqueness of normal form of derivations, ($\times$) may be replaced without loss in generality by the following more manageable assertion:

($\times \times$) Two formulae represent r-equal propositions if and only if they are formally equivalent and for any normal derivation of one of them from the empty set of assumptions there exists an isomorphic normal derivation of the other from the empty set of assumptions.

The fact that ($\times \times$) is more manageable than ($\times$) depends on the peculiar form of normal derivations from the empty set of assumptions as discussed by Prawitz [1971, II.3.2] .

Clearly ($\times \times$) does not present the difficulty implicit in the notion of extensional equality of propositions, that all logically valid propositions are extensionally equal. For example $P \to P$ and $((P \to Q) \to P) \to P$ are not r-equal because there is no normal derivation in classical logic of $P \to P$ from the empty set of assumptions isomorphic to:

$$
(\to I)\cfrac{(\to E)\cfrac{(\to I)\cfrac{(\land c)\cfrac{(\to E)\cfrac{\overset{\textcircled{1}}{P} \quad \overset{\textcircled{2}}{\neg P}}{\land}}{Q}}{P \to Q} - \textcircled{1} \quad (P \to Q) \to P \;\textcircled{3}}{(\land c)\cfrac{(\to E)\cfrac{P}{} \quad \overset{\textcircled{2}}{\neg P}}{\land}}{P} - \textcircled{2}}{((P \to Q) \to P) \to P} - \textcircled{3}.
$$

For, by the form of normal derivations mentioned above, there is one and only one normal derivation in classical logic of $P \to P$ from the empty set of assumptions, that is:

$$
(\to I)\cfrac{\overset{\textcircled{1}}{P}}{P \to P} - \textcircled{1}.
$$

On the other hand $P \to P$ and $A \to A$, for an arbitrary formula A, are r-equal. For, again by the form of normal derivations,

$$(\to I)\ \frac{\overset{\textcircled{1}}{A}}{A \to A}\ - \textcircled{1}$$

is the only normal derivation in classical logic of $A \to A$ from the empty
set of assumptions and is clearly isomorphic to the only normal derivation
in classical logic of $P \to P$ from the empty set of assumptions.

7. As an application of ($\times\times$) let us consider the following analogue to
the problem in Kreisel [1971, p. 257]: do a formula A in $\{\curlywedge, \wedge, \to, \forall\}$
and its negative translation A^- represent r-equal propositions? We can
easily see that the answer is: No. For instance let A be the formula $\neg\neg P$
$\to P$; then A^- is the formula $\neg\neg\neg\neg P \to \neg\neg P$. Ovbiously there is no
normal derivation in intuitionistic logic of $\neg\neg\neg\neg P \to \neg\neg P$ from the
empty set of assumptions isomorphic to the following normal derivation
in classical logic of $\neg\neg P \to P$ from the empty set of assumptions:

$$(E \to)\ \frac{\overset{\textcircled{1}}{\neg P}\ \ \overset{\textcircled{2}}{\neg\neg P}}{\quad}$$
$$(\curlywedge_C)\ \frac{\curlywedge}{P}\ - \textcircled{1}$$
$$(\to I)\ \frac{}{\neg\neg P \to P}\ - \textcircled{2}$$

For, here ($\curlywedge_C$) discharges an assumption, whereas no ($\curlywedge_C$) which discharges an assumption is permissible in intuitionistic logic.

8. We have considered so far only purely logical proofs and derivations
generated by purely logical rules. The only reason for that was that the
conjecture formulated by Prawitz [1971, II.3.5.6] was restricted to them.
Such a restriction seems to be quite reasonable as far as ($**$) is concerned,
even when one does not accept the conclusion that ($**$) is necessarily false
whenever applied to purely logical proofs and derivations generated by
purely logical rules.

For let us consider for instance derivations generated by the rules of
intuitionistic first order arithmetic as formulated in Troelstra [1973, 1.3.6].
Let $\mathscr{D}$ be a derivation which ends with $(\forall I)$ of $\forall x Mt(0, x, 0)$ from the
empty set of assumptions, followed by $(\forall E)$ to get an instance, say
$Mt(0, S^{5343}(0), 0)$, and let $\mathscr{D}$ reduce to $\mathscr{D}'$ in normal form. (Here Mt is a predicate constant for the graph of multiplication, that is $Mt(0, S^{5343}(0), 0)$
represents the proposition $0.5343 = 0$ in the intended interpretation.)

Clearly $\mathscr{D}$ provides more information than $\mathscr{D}'$, specifically information to the effect that $\forall x Mt(0, x, 0)$. For, by the form of normal derivations in intuitionistic first-order arithmetic of closed atomic formulae from the empty set of assumptions as discussed in Troelstra [1973, 4.2.19], $\mathscr{D}'$ simply gives a 'computation' which verifies $Mt(0, S^{5343}(0), 0)$. Now, although the computation contains among its steps instances of $\forall x Mt(0, x, 0)$, the instances do not tell of which general law they are instances.

This, as pointed out by Feferman [1975] and Kreisel (reported by Troelstra [1975, 5.3]), shows that $\mathscr{D}$ and $\mathscr{D}'$ do not represent the same proof. On the other hand by uniqueness of normal form of derivations as established by Troelstra [1973, 4.1.21], $\mathscr{D}$ and $\mathscr{D}'$ reduce to the same normal form. Hence $(**)$ is false.

Of course this argument does not apply to $(+)$ nor to $(++)$. In fact it provides the basic reason for replacing the notion of intensional equality of proofs by that of r-equality. Therefore the plausibility of $(+)$ and $(++)$ is not impaired by extension to non-logical proofs and derivations generated by non-logical rules, and the same applies to the plausibility of $(\times\times)$. This greatly widens the scope of $(\times)$ and $(\times\times)$.

9. In order to see that, let us discuss the problem of representation of self-referential propositions, for example propositions which assert the provability of their own negation. We consider derivations generated by rules $\tilde{f}$ which include the rules of intuitionistic first-order arithmetic as formulated in Troelstra [1973, 1.3.6], and possibly other non-logical rules. Such derivations are called *derivations in* $\tilde{f}$. The notation $\vdash A$ expresses that there is a derivation in $\tilde{f}$ of the formula A from the empty set of assumptions.

We assume that the rules of $\tilde{f}$ satisfy the normalization property and uniqueness of normal form of derivations, and that the form of normal derivations in $\tilde{f}$ from the empty set of assumptions is similar to that discussed by Troelstra [1973, 4.2.9]. We also assume that the rules of $\tilde{f}$ satisfy the usual conditions on Gödel's second incompleteness theorem as formulated in Kreisel–Takeuti [1974, A2], that is:

I. There are quantifier-free formulae $\mathrm{Sub}(a, b, c)$, $\mathrm{Neg}(a, b)$, $\mathrm{Der}(a, b)$ which satisfy the following conditions:

(a) If the formula which results from the formula with Gödel number m on substituting $S^n(0)$ for each occurrence of the proper parameter a_0 is the formula with Gödel number p, then $\vdash \mathrm{Sub}(S^m(0), S^n(0), S^p(0))$ and $\vdash \forall x(\mathrm{Sub}(S^m(0), S^n(0), x) \leftrightarrow x = S^p(0))$.

(b) If the negation of the formula with Gödel number m is the formula with Gödel number n, then $\vdash \mathrm{Neg}(S^m(0), S^n(0))$ and $\vdash \forall x(\mathrm{Neg}(S^m(0), x) \leftrightarrow x = S^n(0))$.

(c) If m is the Gödel number of a derivation in $\tilde{f}$ of the formula with Gödel number n from the empty set of assumptions, then $\vdash \mathrm{Der}(S^m(0), S^n(0))$ and $\vdash \forall x(\mathrm{Der}(S^m(0), x) \leftrightarrow x = S^n(0))$.

II. For any closed $A \in \sum_1^0$, $\vdash A \to \exists x \mathrm{Der}(x, \ulcorner A \urcorner)$.

III. For any two formulae A and B,

$$\vdash \exists x \mathrm{Der}(x, \ulcorner A \to B \urcorner) \to (\exists x \mathrm{Der}(x, \ulcorner A \urcorner) \to \exists x \mathrm{Der}(x, \ulcorner B \urcorner)).$$

Of course if a formula A_J represents a proposition which asserts the provability of its own negation, then A_J must be extensionally equal to the formula $\exists x \mathrm{Der}(x, \ulcorner \neg A_J \urcorner)$, where $\ulcorner \neg A_J \urcorner$ stands for $S^n(0)$ with n the Gödel number of $\neg A_J$. Therefore by $(-)$ we have:

$$(\square) \qquad \vdash A_J \leftrightarrow \exists x \mathrm{Der}(x, \ulcorner \neg A_J \urcorner).$$

However it seems rather doubtful whether ($\square$) is a condition not only necessary, but also sufficient, on a formula representing a proposition which asserts the provability of its own negation. For, the formula $\exists x \mathrm{Der}(x, \ulcorner \neg A_J \urcorner)$ does not represent a proposition which asserts the provability of its own negation since $\ulcorner \neg A_J \urcorner$ does not express the Gödel number of $\exists x \mathrm{Der}(x, \ulcorner \neg A_J \urcorner)$.

In order to avoid this difficulty let n be the Gödel number of $\exists x \exists y \exists z(\mathrm{Sub}(a_0, a_0, z) \wedge \mathrm{Neg}(z, y) \wedge \mathrm{Der}(x, y))$ and let B_J be short for $\exists x \exists y \exists z(\mathrm{Sub}(S^n(0), S^n(0), z) \wedge \mathrm{Neg}(z, y) \wedge \mathrm{Der}(x, y))$. Clearly B_J is the formula which results from the formula with Gödel number n on substituting $S^n(0)$ for each occurrence of the proper parameter a_0. Hence by I (a) we have:

$$(1) \qquad \vdash \forall x(\mathrm{Sub}(S^n(0), S^n(0), x) \leftrightarrow x = \ulcorner B_J \urcorner).$$

On the other hand $\neg B_J$ is the negation of B_J, hence by I (b) we obtain:

$$(2) \qquad \vdash \forall x(\mathrm{Neg}(\ulcorner B_J \urcorner, x) \leftrightarrow x = \ulcorner \neg B_J \urcorner).$$

Thus B_J represents a proposition P which asserts the provability of the proposition represented by $\neg B_J$. Since such a proposition is the negation of P, B_J represents a proposition which asserts the provability of its own negation.

10. A simple solution to the problem of representation of propositions which assert the provability of their own negation is provided by the notion of *intensional representation*: we say that a formula A_J intensionally represents a proposition which asserts the provability of its own negation if and only if A_J and B_J represent intensionally equal propositions. However, by $(°)$, such a solution seems to be too extreme because it requires A_J to be identical to B_J.

An alternative solution may be obtained by introducing the notion of *extensional representation*: we say that a formula A_J extensionally represents a proposition which asserts the provability of its own negation if and only if A_J and B_J represent extensionally equal propositions. One may easily see that if A_J extensionally represents a proposition which asserts the provability of its own negation, then A_J satisfies $(\square)$.

First we show that B_J satisfies $(\square)$. Assume B_J and $\mathrm{Sub}(S^n(0)$,

$S^n(0)$, $c) \wedge \mathrm{Neg}(c, b) \wedge \mathrm{Der}(a, b)$. Then $\mathrm{Sub}(S^n(0)$, $S^n(0)$, $c)$, $\mathrm{Neg}(c, b)$ and $\mathrm{Der}(a, b)$. From $\mathrm{Sub}(S^n(0)$, $S^n(0)$, $c)$ by I (a) we obtain $c = \ulcorner B_J \urcorner$. From $\mathrm{Neg}(c, b)$ and $c = \ulcorner B_J \urcorner$ we obtain $\mathrm{Neg}(\ulcorner B_J \urcorner, b)$, hence by I (b) $b = \ulcorner \neg B_J \urcorner$. From $\mathrm{Der}(a, b)$ and $b = \ulcorner \neg B_J \urcorner$ we obtain $\mathrm{Der}(a, \ulcorner \neg B_J \urcorner)$, hence $\exists x \mathrm{Der}(x, \ulcorner \neg B_J \urcorner)$. Therefore $\vdash B_J \rightarrow \exists x \mathrm{Der}(x, \ulcorner \neg B_J \urcorner)$.

Conversely assume $\exists x \mathrm{Der}(x, \ulcorner \neg B_J \urcorner)$ and $\mathrm{Der}(a, \ulcorner \neg B_J \urcorner)$. Since B_J is the formula which results from the formula with Gödel number n on substituting $S^n(0)$ for each occurrence of the proper parameter a_0, by I (a) $\vdash \mathrm{Sub}(S^n(0)$, $S^n(0)$, $\ulcorner B_J \urcorner)$. Since $\neg B_J$ is the negation of B_J, by I (b) $\vdash \mathrm{Neg}(\ulcorner B_J \urcorner, \ulcorner \neg B_J \urcorner)$. Thus $\mathrm{Sub}(S^n(0)$, $S^n(0)$, $\ulcorner B_J \urcorner) \wedge \mathrm{Neg}(\ulcorner B_J \urcorner, \ulcorner \neg B_J \urcorner) \wedge \mathrm{Der}(a, \ulcorner \neg B_J \urcorner)$. So $\exists x \exists y \exists z(\mathrm{Sub}(S^n(0)$, $S^n(0)$, $z) \wedge \mathrm{Neg}(z, y) \wedge \mathrm{Der}(x, y))$, that is B_J. Therefore $\vdash \exists x \mathrm{Der}(x, \ulcorner \neg B_J \urcorner) \rightarrow B_J$.

Next we prove that A_J satisfies $(\square)$. Since A_J and B_J represent extensionally equal propositions, by $(-)$ we have $\vdash A_J \leftrightarrow B_J$. Hence $\vdash \neg A_J \rightarrow \neg B_J$. Let k be the Gödel number of a derivation in $\tilde{f}$ of $\neg A_J \rightarrow \neg B_J$ from the empty set of assumptions. By I (c) $\vdash \mathrm{Der}(S^k(0), \ulcorner \neg A_J \rightarrow \neg B_J \urcorner)$, hence $\vdash \exists x \mathrm{Der}(x, \ulcorner \neg A_J \rightarrow \neg B_J \urcorner)$. So by III $\vdash \exists x \mathrm{Der}(x, \ulcorner \neg A_J \urcorner) \rightarrow \exists x \mathrm{Der}(x, \ulcorner \neg B_J \urcorner)$. On the other hand, since B_J satisfies $(\square)$, $\vdash \exists x \mathrm{Der}(x, \ulcorner \neg B_J \urcorner) \rightarrow B_J$. Hence $\vdash \exists x \mathrm{Der}(x, \ulcorner \neg A_J \urcorner) \rightarrow B_J$. Now from $\vdash A_J \leftrightarrow B_J$ we get $\vdash B_J \to A_J$. Therefore $\vdash \exists x \text{Der}\,(x,\ \ulcorner \neg A_J \urcorner)$ $\to A_J$. Similarly we may prove that $\vdash A_J \to \exists x \text{Der}\,(x,\ \ulcorner \neg A_J \urcorner)$.

11. In terms of the notion of extensional representation Gödel's second incompleteness theorem may be restated as follows. Let Con be short for $\neg\, \exists x \text{Der}\,(x,\ \ulcorner 0 = S(0) \urcorner)$. Then:

($\square\square$) (i) For any formula A_J which extensionally represents a proposition asserting the provability of its own negation, $\vdash \neg A_J \leftrightarrow \text{Con}$.

(ii) If the rules of $\tilde{f}$ are consistent, then not $\vdash \text{Con}$.

In order to prove ($\square\square$) (i), let A_J be any formula which extensionally represents a proposition asserting the provability of its own negation. Since A_J satisfies ($\square$), $\vdash \neg A_J \to \neg\, \exists x \text{Der}\,(x,\ \ulcorner \neg A_J \urcorner)$. Let k be the Gödel number of a derivation in $\tilde{f}$ of $\neg A_J \to \neg\, \exists x \text{Der}\,(x,\ \ulcorner \neg A_J \urcorner)$ from the empty set of assumptions. By I (c) $\vdash \text{Der}\,(S^k(0),\ \ulcorner \neg A_J \to \neg\, \exists x \text{Der}(x,$ $\ulcorner \neg A_J \urcorner) \urcorner)$, hence $\vdash \exists x \text{Der}\,(x,\ \ulcorner \neg A_J \to \neg\, \exists x \text{Der}(x,\ \ulcorner \neg A_J \urcorner) \urcorner)$. Then by III:

(3) $\qquad \vdash \exists x \text{Der}\,(x,\ \ulcorner \neg A_J \urcorner) \to \exists x \text{Der}\,(x,\ \ulcorner \neg \exists x \text{Der}\,(x,\ \ulcorner \neg A_J \urcorner) \urcorner)$.

On the other hand by II:

(4) $\qquad \vdash \exists x \text{Der}\,(x,\ \ulcorner \neg A_J \urcorner) \to \exists x \text{Der}\,(x,\ \ulcorner \exists x \text{Der}\,(x,\ \ulcorner \neg A_J \urcorner) \urcorner)$.

Assume $\exists x \text{Der}\,(x,\ \ulcorner \neg A_J \urcorner)$. Then by (4) $\exists x \text{Der}\,(x,\ \ulcorner \exists x \text{Der}\,(x,\ \ulcorner \neg A_J \urcorner) \urcorner)$ and by (3) $\exists x \text{Der}\,(x,\ \ulcorner \neg \exists x \text{Der}\,(x,\ \ulcorner \neg A_J \urcorner) \urcorner)$, that is $\exists x \text{Der}\,(x,$ $\ulcorner \exists x \text{Der}\,(x,\ \ulcorner \neg A_J \urcorner) \to 0 = S(0) \urcorner)$. So by III $\exists x \text{Der}\,(x,\ \ulcorner 0 = S(0) \urcorner)$. Thus $\vdash \exists x \text{Der}\,(x,\ \ulcorner \neg A_J \urcorner) \to \exists x \text{Der}\,(x,\ \ulcorner 0 = S(0) \urcorner)$, hence $\vdash \text{Con} \to \neg \exists x \text{Der}\,(x,\ \ulcorner \neg A_J \urcorner)$. On the other hand, since A_J satisfies ($\square$), $\vdash \neg \exists x \text{Der}\,(x,\ \ulcorner \neg A_J \urcorner) \to \neg A_J$. Therefore $\vdash \text{Con} \to \neg A_J$.

Conversely $\vdash 0 = S(0) \to \neg A_J$. Let k be the Gödel number of a derivation in $\tilde{f}$ of $0 = S(0) \to \neg A_J$ from the empty set of assumptions. Then by I (c) $\vdash \text{Der}\,(S^k(0),\ \ulcorner 0 = S(0) \to \neg A_J \urcorner)$, hence $\vdash \exists x \text{Der}\,(x,$ $\ulcorner 0 = S(0) \to \neg A_J \urcorner)$. So by III:

(5) $\qquad \vdash \exists x \text{Der}\,(x,\ \ulcorner 0 = S(0) \urcorner) \to \exists x \text{Der}\,(x,\ \ulcorner \neg A_J \urcorner)$.

Assume $\neg A_J$ and $\exists x \mathrm{Der}(x, \ulcorner 0 = S(0) \urcorner)$. Since A_J satisfies $(\square)$, $\vdash \neg A_J \rightarrow \neg \exists x \mathrm{Der}(x, \ulcorner \neg A_J \urcorner)$, hence $\neg \exists x \mathrm{Der}(x, \ulcorner \neg A_J \urcorner)$. On the other hand by (5) $\exists x \mathrm{Der}(x, \ulcorner \neg A_J \urcorner)$. So we obtain a contradiction. Therefore $\vdash \neg A_J \rightarrow \neg \exists x \mathrm{Der}(x, \ulcorner 0 = S(0) \urcorner)$, that is $\vdash \neg A \rightarrow \mathrm{Con}$.

In other to prove $(\square\square)$ (ii), suppose that $\vdash \mathrm{Con}$. Then by (i) $\vdash \neg A_J$. Let k be the Gödel number of a derivation in $\tilde{f}$ of $\neg A_J$ from the empty set of assumptions. They by I (c) $\vdash \mathrm{Der}(S^k(0), \ulcorner \neg A_J \urcorner)$, hence $\vdash \exists x \mathrm{Der}(x, \ulcorner \neg A_J \urcorner)$. On the other hand, since A_J satisfies $(\square)$, $\vdash \exists x \mathrm{Der}(x, \ulcorner \neg A_J \urcorner) \rightarrow A_J$. So $\vdash A_J$. From this and $\vdash \neg A_J$, that is $\vdash A_J \rightarrow 0 = S(0)$, we obtain $\vdash 0 = S(0)$. Therefore if the rules of $\tilde{f}$ are consistent, then not $\vdash \mathrm{Con}$.

12. In spite of its interest, the notion of extensional representation does not provide an adequate approximation to the idea of actually representing a proposition which asserts the provability of its own negation. For, as we have seen, B_J satisfies $(\square)$. So $\exists x \mathrm{Der}(x, \ulcorner \neg B_J \urcorner)$ extensionally represents a proposition which asserts the provability of its own negation. On the other hand $\ulcorner \neg B_J \urcorner$ does not express the Gödel number of $\exists x \mathrm{Der}(x, \ulcorner \neg B_J \urcorner)$. Therefore the notion of extensional representation fails to discriminate between $\exists x \mathrm{Der}(x, \ulcorner \neg B_J \urcorner)$ and B_J.

In order to avoid this difficulty we must consider non-extensional notions of representation. An obvious candidate is the notion of *representation up to reduction*, or *r-representation*: we say that a formula A *r*-represents a proposition which asserts the provability of its own negation if and only if A is *r*-equal to B_J.

Clearly this notion is sufficiently sharp to discriminate between $\exists x \mathrm{Der}(x, \ulcorner \neg B_J \urcorner)$ and B_J. For, let $\mathscr{D}$ be a normal derivation in $\tilde{f}$ of $\exists x \mathrm{Der}(x, \ulcorner \neg B_J \urcorner)$ from the empty set of assumptions. Then, by the form of normal derivations in $\tilde{f}$, we have:

$$\mathscr{D} = (\exists\, I)\ \frac{\overset{\textstyle \mathscr{D}_1}{\mathrm{Der}(S^k(0), \ulcorner \neg B_J \urcorner)}}{\exists x \mathrm{Der}(x, \ulcorner \neg B_J \urcorner)}\ .$$

On the other hand there is no normal derivation in $\tilde{f}$ of B_J from the empty set of assumptions isomorphic to $\mathscr{D}$. For, again by the form of normal derivations in $\tilde{f}$, any such derivation $\mathscr{D}'$ has the form:

$$\mathscr{D}_1' \qquad\qquad \mathscr{D}_2' \qquad\qquad \mathscr{D}_3'$$

$$(\wedge I)\ \frac{\mathrm{Sub}(S^n(0),S^n(0),\ulcorner B_J\urcorner)\quad \mathrm{Neg}(\ulcorner B_J\urcorner,\ \ulcorner\neg B_J\urcorner)}{\mathrm{Sub}(S^n(0),S^n(0),\ulcorner B_J\urcorner)\wedge \mathrm{Neg}(\ulcorner B_J\urcorner,\ \ulcorner\neg B_J\urcorner)\qquad \mathrm{Der}(S^h(0),\ \ulcorner\neg B_J\urcorner)}$$

$$(\wedge I)\ \frac{}{\mathrm{Sub}(S^n(0),\,S^n(0),\,\ulcorner B_J\urcorner)\wedge \mathrm{Neg}(\ulcorner B_J\urcorner,\ \ulcorner\neg B_J\urcorner)\wedge \mathrm{Der}(S^h(0),\ \ulcorner\neg B_J\urcorner)}$$

$$(\exists I)\ \frac{}{\exists z(\mathrm{Sub}(S^n(0),\,S^n(0),\,z)\wedge \mathrm{Neg}(z,\ \ulcorner\neg B_J\urcorner)\wedge \mathrm{Der}(S^h(0),\ \ulcorner\neg B_J\urcorner))}$$

$$(\exists I)\ \frac{}{\exists y\exists z(\mathrm{Sub}(S^n(0),\,S^n(0),\,z)\wedge \mathrm{Neg}(z,\,y)\wedge \mathrm{Der}(S^h(0),\,y))}$$

$$(\exists I)\ \frac{}{\exists x\exists y\exists z(\mathrm{Sub}(S^n(0),\,S^n(0),\,z)\wedge \mathrm{Neg}(z,\,y)\wedge \mathrm{Der}(x,\,y))}$$

Hence by $(\times\times)$ $\exists x \mathrm{Der}(x,\ \ulcorner\neg B_J\urcorner)$ and B_J do not represent r-equal propositions.

13. Instead of the formula B_J one might very well consider a different formula B_J'. For example let m be the Gödel number of
$\exists x\exists y\exists z\exists w(\mathrm{Sub}(a_0,\ a_0,\ w)\wedge \mathrm{Ad}(w,\ 0,\ z)\wedge \mathrm{Neg}(z,\ y)\wedge \mathrm{Der}(x,\ y))$.
Then we may take $\exists x\exists y\exists z(\mathrm{Sub}(S^m(0),\ S^m(0),\ z)\wedge \mathrm{Neg}(z,\,y)\wedge \mathrm{Der}(x,\,y))$ as B_J'. For, B_J' is the formula which results from the formula with Gödel number m on substituting $S^m(0)$ for each occurrence of the proper parameter a_0. Thus by an argument similar to that already used for B_J we see that B_J' represents a proposition which asserts the provability of its own negation.

Clearly B_J and B_J' do not represent r-equal propositions. In fact any normal derivation $\mathscr{D}''$ in $\tilde{f}$ of B_J' from the empty set of assumptions has a form similar to that of the derivation $\mathscr{D}'$ mentioned above. But $\mathscr{D}''$ contains a subderivation:

$$\mathscr{D}_3''$$
$$\mathrm{Der}(S^j(0),\ \ulcorner\neg B_J'\urcorner)$$

with $j\neq h$ and $\ulcorner\neg B_J'\urcorner\neq \ulcorner\neg B_J\urcorner$, which is in general not isomorphic to $\mathscr{D}_3'$. Therefore $\mathscr{D}''$ is not isomorphic to $\mathscr{D}'$.

This shows that the notion of r-representation is relative to a specific formula B_J. The same applies to the notion of intensional or extensional representation. It seems to be open whether the relation between B_J and B_J' might be analyzed in terms of some significant non-extensional notion of equality of propositions.

14. The non-extensional notions considered so far are not necessarily the most useful. Indeed for some applications they may turn out to be too strong. On the other hand it is not difficult to envisage possible alternatives. For example an alternative to the notions of r-isomorphism of

derivations and r-equality of formulae may be introduced in terms of the notion of faithful translation.

Let $\tilde{f}_1$ and $\tilde{f}_2$ be (not necessarily distinct) rules, and let τ be a function which maps formulae of $\tilde{f}_1$ into formulae of $\tilde{f}_2$. For any rule:

$$\mathcal{R} = \frac{\begin{array}{cc}[A_1] & [A_n]\\ B_1 \ \ldots \ B_n\end{array}}{C}$$

of $\tilde{f}_1$ we put:

$$\tau(\mathcal{R}) = \frac{\begin{array}{cc}[\tau(A_1)] & [\,(\tau A_n)]\\ \tau(B_1) \ \ldots \ \tau(B_n)\end{array}}{\tau(C)}.$$

We say that τ is a *faithful translation* of $\tilde{f}_1$ into $\tilde{f}_2$ if and only if, for any rule $\mathcal{R}$ of $\tilde{f}_1$, $\tau(\mathcal{R})$ is a derived rule of $\tilde{f}_2$.

For example let $\tilde{f}_1$ be the rules of classical logic, $\tilde{f}_2$ the rules of intuitionistic logic and τ the negative translation of $\tilde{f}_1$ into $\tilde{f}_2$, i.e. for any formula A in $\{\curlywedge, \wedge, \rightarrow, \forall\}$, $\tau(A) = A^-$.

One may easily show that τ is a faithful translation of $\tilde{f}_1$ into $\tilde{f}_2$. The only non trivial case is $(\curlywedge_c)$. Clearly:

$$\tau(\curlywedge_c) = \frac{\begin{array}{c}[\neg\,\neg\,\neg P]\\ \curlywedge\end{array}}{\neg\,\neg P}.$$

In order to see that $\tau(\curlywedge_c)$ is a derived rule of $\tilde{f}_2$ assume that:

$$\begin{array}{c}[\neg\,\neg\,\neg P]\\ \mathcal{D}_1\\ \curlywedge\end{array}$$

is a derivation in $\tilde{f}_2$. Then:

$$(\rightarrow I)\cfrac{(\rightarrow I)\cfrac{(\rightarrow E)\cfrac{\overset{①}{\neg P}\quad\overset{②}{\neg\,\neg P}}{\curlywedge}-②}{\begin{array}{c}[\neg\,\neg\,\neg P]\\ \mathcal{D}_1\\ \curlywedge\end{array}}}{\neg\,\neg P}-①$$

is a derivation in $\tilde{f}_2$, as desired.

Let τ be a faithful translation of $\tilde{f}_1$ into $\tilde{f}_2$. The notion of *isomorphism with respect to* τ, or *isomorphism*$_\tau$, of two derivations in $\tilde{f}_1$ and $\tilde{f}_2$, respectively, is defined similarly to that of isomorphism of two derivations, except that in clause (ii) the condition $\mathcal{R}_1 = \mathcal{R}_2$ is to be replaced by $\tau(\mathcal{R}_1) = \mathcal{R}_2$.

Two derivations in $\tilde{f}_1$ and $\tilde{f}_2$, respectively, are said to be *r-isomorphic with respect to* τ, or *r-isomorphic*$_\tau$, if and only if, when properly analyzed by eliminating what is not strictly required to derive the conclusion from the assumptions, they are turned into isomorphic$_\tau$ derivations.

Two formulae A and B of $\tilde{f}_1$ and $\tilde{f}_2$, respectively, are said to be *r-equal*$_\tau$ if and only if $\tau(A) = B$ and for any derivation in $\tilde{f}_1$ of A from the empty set of assumptions there is an *r*-isomorphic$_\tau$ derivation in $\tilde{f}_2$ of B from the empty set of assumptions, and viceversa.

As an application of *r-equality*$_\tau$ we show that, for any formula A in $\{\wedge, \wedge, \rightarrow, \forall\}$, A and its negative translation $\tau(A) = A^-$ are *r-equal*$_\tau$. Let $\mathcal{D}$ be any derivation in classical logic of A from assumptions Γ. We prove by induction on the length k of $\mathcal{D}$ that there is a derivation $\mathcal{D}'$ in intuitionistic logic of $\tau(A)$ from $\tau(\Gamma) = \{\tau(B): B \in \Gamma\}$ which is *r*-isomorphic$_\tau$ to $\mathcal{D}$. If $k = 0$, the result is trivial. If $k > 0$, we must consider cases according to the form of the last rule of $\mathcal{D}$. For example, if the last rule if ($\wedge_c$), then:

$$\mathcal{D} = (\wedge_c) \; \frac{\begin{matrix} [\neg P] \\ \mathcal{D}_1 \\ \wedge \end{matrix}}{P}.$$

By induction hypothesis there is a derivation $\mathcal{D}'_1$ in intuitionistic logic of $\wedge$ from $[\neg \neg \neg P]$ which is *r*-isomorphic$_\tau$ to $\mathcal{D}_1$. We put:

$$\mathcal{D}' = \tau(\wedge_c) \; \frac{\begin{matrix} [\neg \neg \neg P] \\ \mathcal{D}'_1 \\ \wedge \end{matrix}}{\neg \neg P}.$$

Clearly $\mathcal{D}'$ has the desired property. Conversely let $\mathcal{D}$ be any derivation in intuitionistic logic of $\tau(A)$ from assumptions $\tau(\Gamma)$. By a procedure similar to that used above we may easily prove that there is a derivation $\mathcal{D}'$ in classical logic of A from Γ which is *r*-isomorphic$_\tau$ to $\mathcal{D}$. For $\Gamma = \emptyset$ this yields the result.

REFERENCES

Feferman, S. 1975, 'Review of Prawitz [1971]', *J. Symbolic Logic* **40**, 232–234.

Kreisel, G. 1967, 'Informal rigour and completeness proofs', in I. Lakatos (Ed.), *Problems in the Philosophy of Mathematics*. Amsterdam, North-Holland, pp. 138–171.

Kreisel, G. 1971, 'Review of Szabo [1969]', *J. Philosophy* **68**, 238–265.

Kreisel, G. and Takeuti, G. 1974, 'Formally self-referential propositions for cut free classical analysis and related systems', *Dissertationes Mathematicae* **118**, 50.

Prawitz, D. 1971, 'Ideas and results in proof theory' in J.E. Fenstad (Ed.), *Proceedings of the Second Scandinavian Logic Symposium*. Amsterdam, North-Holland, pp. 235–307.

Prawitz, D. and Malmnäs, P.-E. 1968, 'A survey of some connections between classical, intuitionistic and minimal logic' in K. Schütte (Ed.), *Contributions to Mathematical Logic*. Amsterdam, North-Holland, pp. 215–229.

Szabo, M.E. (Ed.) 1969, *The Collected Papers of Gerhard Gentzen*. Amsterdam, North-Holland.

Troelstra, A.S. 1973, (Ed.) *Metamathematical Investigation of Intuitionistic Arithmetic and Analysis*. Berlin, Springer-Verlag.

Troelstra, A.S. 1975, 'Non-extensional equality', *Fundamenta Mathematicae* **82**, 307–322.

Troelstra, A.S. 'Completeness and validity for intuitionistic predicate logic', Department of Mathematics, University of Amsterdam, Report 76–05.

E. BENCIVENGA[1]

FREE SEMANTICS

The valuation of sentences containing non-denoting singular terms is obviously the fundamental question to be solved in the construction of a semantics for free logic.[2] Besides being fundamental, this question is very difficult, for the notion of *truth* commonly applied in contemporary logic, i.e., the notion of truth as correspondence with *reality*, does not fit sentences cortaining references to 'non-existing objects',[3] which, as such, cannot be constituents of reality. For this reason, a free semantics should provide, first of all, for a new conception of truth, or at least for a suitable generalization of the correspondence theory; but the efforts made until now have not achieved results which we can regard as completely satisfactory from a philosophical point of view. As a matter of fact, these efforts resulted substantially into accepting one of the following three theses:

(a) An atomic sentence is true if it corresponds to a fact, and false otherwise. But an atomic sentence containing some non-denoting singular terms cannot correspond to any fact; hence it is false.[4]

(b) Non-denoting singular terms are not truly non-denoting; they really refer to some special sort of objects, i.e., purely possible objects, null entities, and so on.[5] (In an interesting variant of this position, non-denoting singular terms 'denote' themselves.)[6]

(c) A sentence *A* containing some non-denoting singular terms cannot *prima facie* receive any truth-value. We can, however, evaluate *A* if and only if–regardless of the truth-values we imagine assigned to its atomic constituents containing non-denoting singular terms–we always obtain the same truth-value for the total sentence.[7]

To show the intuitive implausibility of these three theses, let us consider a sentence containing a non-denoting singular term but whose truth can scarcely be questioned, i.e.,

(1)   Pegasus = Pegasus,

and ask ourselves how we can deal with it by each of (a)–(c).

By (a) we must conclude that (1) is false, contrary to our intuitions. By (b) we can state its truth, but the reason is not convincing, for it consists

31

Maria Luisa Dalla Chiara (ed.), Italian Studies in the Philosophy of Science, 31–48.
Copyright © *1980 by D. Reidel Publishing Company.*

in the *fact* that the 'fictitious object', Pegasus, is identical with itself. We cannot deny that the contrast with the correspondence theory is thus settled by straining the notion of 'fact'. Lastly, by (c) we cannot decide whether (1) is true or false, but we can decide on the truth of

(2)   Either Pegasus = Pegasus or Pegasus ≠ Pegasus

and on the falsity of

(3)   Pegasus = Pegasus and Pegasus ≠ Pegasus.[8]

In my opinion, the thesis (c), however, is the most promising one for the construction of that generalization of the correspondence theory we are looking for. By choosing it, we can say that a sentence is true not only when it corresponds with reality (i.e., when it is *factually* true), but even when every 'mental experiment' of a certain sort makes it true (what we can call its being *formally* true). The fact is that such mental experiments are defined by (c) at the level of unanalyzed sentences, and thus do not allow us to validate schemata as the one instantiated by (1), whose validity would depend essentially on the analysis of sentences into terms. My aim in the present paper is precisely to develop the suggestions contained in (c) at the level of *analyzed* sentences, and to solve some serious difficulties which are strictly connected with the fulfillment of this task.

1. LANGUAGE AND ITS INTERPRETATION

The primitive symbols of our language L are:
(a) a denumerably infinite set of *individual variables*;
(b) for every n, a denumerably infinite set of *n-ary predicates*;
(c) the *identity symbol* $=$;[9]
(d) the *existence symbol* E!;
(e) the two *connectives* ¬ and &;
(f) the *universal quantifier* ∀;
(g) the two *parentheses* (and).[10]

All these symbols are autonomous. We will use x, y, z (possibly with subscripts) as metavariables on the set (a), and P^n (which will become simply P when no confusion is possible) as metavariable on the sets (b). Furthermore, we will suppose some alphabetical order being defined on the individual variables.

Every string of primitive symbols (of L) is a formula (of L).[11] In the set of the formulas we isolate as usual the set of the *wffs* (well-formed formulas) by the following recursive definition:

(i) every formula of one of the three forms $P^n x_1 \ldots x_n$, $x = y$, $E!x$ is an (atomic) wff;

(ii) if A and B are wffs, then $(\neg A)$, $(A \& B)$ and $(\forall x A)$ are wffs (we will usually omit as many parentheses as possible);

(iii) nothing is a wff if not by virtue of (i)-(ii).

We will use A, B, C, possibly with superscripts or subscripts, as metavariables for wffs. We will presuppose common definitions of the connectives $\vee$, $\supset$ and $\equiv$ and of the existential quantifier $\exists$, of the notions of a *free* or *bound* occurrence of an individual variable in a wff, and of a *wf* (well-formed) *part* of a wff. If x does not occur free in any wf part of A of the form $\forall y B$, by A^y/x and $A^y//x$ we will mean the result obtained by replacing every free occurrence of x in A with an occurrence of y, and any result obtained by replacing zero or more free occurrences of x in A with an occurrence of y, respectively. If x occurs free in some wf part of A of the form $\forall y B$ (let all these wf parts be $\forall y B_1, \ldots, \forall y B_n$), by $A^y/x(A^y//x)$ we will mean the (any) result obtained by making the above replacement in the wff A' obtained from A by replacing every occurrence of y in every $\forall y B_i$ (where i is such that $1 \leqslant i \leqslant n$) with an occurrence of the first individual variable (in the alphabetical order) not occurring (either free or bound) in A.

And now for the semantics. As is well-known, the two fundamental notions of standard semantics are those of a model-structure and of the truth of a wff in such a structure. The first one can be easily adapted to the case in which some singular terms of the language (in our particular case, some individual variables) are non-denoting: for this purpose, it is sufficient to establish that a *model-structure for L*[12] is an ordered pair $M = \langle D, f \rangle$, where D is a set, *possibly empty*, to be called the *domain*, and f is a unary function (to be called the *function of interpretation*), total on the set of the predicates of L and *partial* on the set of the individual variables of L, which assigns to every predicate of L a set of ordered n-tuples of members of D and to every individual variable of L for which it is defined a member of D. (Obviously, the function of interpretation will be defined for no individual variable when the domain is empty.)

Much more complex is the adaptation of the second fundamental notion of standard semantics: the notion of the truth of a wff in a model-structure M. What is clear at present is only this: according to the approach we have chosen, we must distinguish between factual truth in M and formal truth in M, the last being defined in a sense (and we will have to make precise in what sense) on the ground of some 'mental experiments' made starting from M.

The notion of factual truth is clearly less problematical, but its definition is not without difficulties, for it requires that we give an answer at least to the following questions:

(a) Is the factual truth of a wff A a sufficient condition for the *factual* truth of all the wffs which are tautological consequences of A? For instance, is the factual truth of a wff of the form Px a sufficient condition for the *factual* truth of all the wffs of the form $Px \lor Py$, *even when y is non-denoting*?

(b) Is every wff of the form $E!x$ *factually false* (rather than, for example, without any factual truth-value) when x is non-denoting?

(c) Is every wff of the form $x = y$ *factually false* (rather than, again, without any factual truth-value) when exactly one of x and y is denoting?

Personally, I am inclined in all these cases to give a positive answer, while being conscious, thus, of differing from other authors.[13] From an intuitive point of view, the grounds of this attitude consist (i) in my assimilating the two notions of 'factual truth in a model-structure M' and 'truth which can be inferred just from the information given by M', and (ii) in my regarding as (intuitively)[14] valid the inference-schemata which are instantiated by the following examples:

$$(4) \quad \frac{\text{Snow is white}}{\text{Snow is white or Pegasus is white}}$$

$$(5) \quad \frac{\text{'Pegasus' is a non-denoting singular term}}{\text{Pegasus does not exist}}$$

$$(6) \quad \frac{\begin{array}{l}\text{Pegasus does not exist}\\\text{Jimmy Carter does exist}\end{array}}{\text{Pegasus is not Jimmy Carter}}$$

Obviously, this is not an argument, but at most the formulation of a heuristic procedure. However, I can strengthen my position by pointing out that, while a positive answer to (b) and (c) is to a certain extent essential to the approach I will propose here, a positive answer to (a) is not really essential. If someone does not agree with it, and believes that a wff of the form $Px \lor Py$ is always factually truth-valueless when y is non-denoting (I do not see any other reasonable alternative position), I ask him only to make the necessary changes in what follows, and if he accepts the remainder of my construction, he will be able to obtain exactly the same results.

We are now ready to define, relative to any model-structure M, a valuation V_M^* which makes precise the notion of factual truth in M. The definition is as follows:

(7) The *primary auxiliary valuation* V_M^* relative to a model-structure $M = \langle D, f \rangle$ is the partial unary function W from the set of all the wffs of L to the set $\{T, F\}$ such that

(a) if A is of the form $Px_1 \ldots x_n$ and $f(x_i)$ is defined for every i such that $1 \leqslant i \leqslant n$, then $W(A) = T$ if $\langle f(x_1), \ldots, f(x_n) \rangle \in f(P)$, and otherwise $W(A) = F$;

(b)(1) if A is of the form $x = y$ and both $f(x)$ and $f(y)$ are defined, then $W(A) = T$ if $f(x) = f(y)$, and otherwise $W(A) = F$;

(b)(2) if A is of the form $x = y$ and exactly one of $f(x)$ and $f(y)$ is defined, then $W(A) = F$;

(c) if A is of the form $E!x$, then $W(A) = T$ if $f(x)$ is defined, and otherwise $W(A) = F$;

(d) if A is of the form $\neg B$ and $W(B)$ is defined, then $W(A) = T$ if $W(B) = F$, and otherwise $W(A) = F$;

(e)(1) if A is of the form $B \,\&\, C$ and $W(B) = W(C) = T$, then $W(A) = T$;

(e)(2) if A is of the form $B \,\&\, C$ and either $W(B) = F$ or $W(C) = F$, then $W(A) = F$;

(f)(1) if A is of the form $\forall x B$ and $W(B^y/x) = T$ for every individual variable y such that $W(E!y) = T$, then $W(A) = T$;

(f)(2) if A is of the form $\forall x B$ and $W(B^y/x) = F$ for at least one individual variable y such that $W(E!y) = T$, then $W(A) = F$;

(g) $W(A)$ is not defined if not in virtue of (a)–(f).

Turning now to formal truth in M, we have first of all to make precise the notion of a mental experiment. As we want to work at the level of analyzed sentences, such an experiment must be defined as the assignment of denotations to non-denoting singular terms (rather than as the assignment of truth-values to truth-valueless atomic sentences); but to what non-denoting singular terms? Here we have two options, i.e., we can either

(i) assign denotations only to the non-denoting singular terms occurring in the single wff we have to evaluate, or

(ii) assign denotations to all the non-denoting singular terms. On the other hand, the Coincidence Theorem of standard semantics very strongly suggests that the two options are likely to give the same results;[15] so to simplify things we can choose the more radical one, i.e., (i). According to this choice, the notion of a mental experiment can be made precise by the following definition:

(8) A model-structure $M' = \langle D', f' \rangle$ is a *completion* of a model-structure $M = \langle D, f \rangle$ if and only if (i) D' is a non-empty (possibly improper) superset of D; (ii) for every predicate P of L, $f'(P)$ is a (possibly improper) superset of $f(P)$; and (iii) $f'(x)$ is defined for every individual variable x of L, and is identical with $f(x)$ whenever $f(x)$ is defined.

At this point, we could think that formal truth in M is to be defined as truth in all completions of M, just as in the semantics of supervaluations, formal truth in a valuation V is defined as truth in all the 'classical valuations' constructed over V. But this simple course would give rise to the serious difficulties we mentioned at the beginning of the present paper. For let us consider the following schema:

(9) $Py \supset \exists x Px.$

As is well-known, such a schema is provable in standard logic, but is not provable in free logic; so in a free semantics we should be able to invalidate it. But by the above course, invalidating (9) would amount to finding a model-structure M such that either an instance of (9) is factually false in M or an instance of (9) is factually truth-valueless in M but factually false in a completion of M, and it is easy to see that both cases are impossible.[16] So by the course in question we are very near a collapse into a standard semantics![17]

We cannot develop our approach without finding a device capable of avoiding the above collapse. For this purpose, it will be useful to make first of all the following remark. When a particular instance of (9) – let it be called A – is factually truth-valueless in a model-structure M, A is not *totally* truth-valueless in M from a factual point of view. There is always a wf part of A – i.e., its consequent – which *has* a factual truth-value, and such a truth-value in some cases – i.e., when it is T – is regarded as sufficient to assign a truth-value to the same A.[18] However, when the truth-value of the consequent is F, the above course suggests forgetting this information completely and turning our attention only to the completions of M. There is clearly an asymmetry of behavior here, and perhaps by eliminating this asymmetry we can solve our difficulties. In other words, what we need perhaps is to retain *in all cases* the information coming from M, thus emphasizing the fact that we are solving a problem – to be precise, the problem of the evaluation of a wff – *relative to M*, hence that the consideration of *other* model-structures (in particular, of the completions of M) is purely functional to the solution of this problem relative to M.

How can we obtain such a result in practice? Let M be a model-structure in which the consequent of A is (factually) false and the antecedent of A is (factually) truth-valueless, and let M' be a completion of M in which the antecedent of A (as well as its consequent, of course) is (factually) true. The situation may be sketched as follows:

	Py	$\exists x Px$
M	–	F
M'	T	T

This sketch points out a contrast between the information coming from M and the information coming from M',[19] and makes clear that our difficulty can be viewed as the difficulty of settling such a contrast. The course we are criticizing favors M', but all of the above discussion has revealed that the prevailing source should instead be M. To establish such a prevalence we can define, on the grounds of the primary auxiliary valuations of M and M', a third valuation, giving the truth-value of A in M' *relative to M*, in which all the wf parts of A having a (factual) truth-value in M retain this truth-value, and then make this valuation – not simply the primary auxiliary valuation of M' – relevant for the decision on the formal truth (or falsity, or truth-valuelessness) of A in M. This explains the following definitions.

(10) The *secondary auxiliary valuation* $V^{**}_{M'(M)}$ relative to a model-structure M and to a completion M' of M is the (total) unary function W from the set of all the wffs of L to the set $\{T, F\}$ such that

(a)(1) if A is an atomic wff and $V^{*}_{M}(A)$ is defined, then $W(A) = V^{*}_{M}(A)$;

(a)(2) if A is an atomic wff and $V^{*}_{M}(A)$ is not defined, then $W(A) = V^{*}_{M'}(A)$;

(b) if A is of the form $\neg B$, then $W(A) = T$ if $W(B) = F$, and otherwise $W(A) = F$;

38   E. BENCIVENGA

(c) if A is of the form $B \,\&\, C$, then $W(A) = \mathrm{T}$ if $W(B) = W(C) = \mathrm{T}$, and otherwise $W(A) = \mathrm{F}$;

(d) if A is of the form $\forall x B$, then $W(A) = \mathrm{T}$ if $W(B^y/x) = \mathrm{T}$ for every individual variable y such that $W(\mathrm{E}!y) = \mathrm{T}$, and otherwise $W(A) = \mathrm{F}$.

(11) The *valuation* V_M relative to a model-structure M is the supervaluation constructed on all the secondary auxiliary valuations $V^{**}_{M'(M)}$, where M' is a completion of M; that is to say, it is the partial unary function W from the set of all the wffs of L to the set $\{\mathrm{T}, \mathrm{F}\}$ such that

(a) if $W'(A) = \mathrm{T}$ for every secondary auxiliary valuation W' relative to M and to a completion M' of M, then $W(A) = \mathrm{T}$;

(b) if $W'(A) = \mathrm{F}$ for every secondary auxiliary valuation W' relative to M and to a completion M' of M, then $W(A) = \mathrm{F}$;

(c) $W(A)$ is not defined if not by virtue of (a)–(b).[20]

We are not in a position to define the notion of a wff being *valid*. This definition will follow in three steps.

(12) A wff A is *verifiable* (or *falsifiable*, or *not completely determinable*) if and only if $V_M(A) = \mathrm{T}$ (or $V_M(A) = \mathrm{F}$, or $V_M(A)$ is not defined) for at least one model-structure M.

(13) A wff A is *invalid* if and only if A is either falsifiable or not completely determinable.

(14) A wff A is *valid* if and only if A is not invalid.

(As a result, a wff A is valid if and only if $V_M(A) = \mathrm{T}$ for every model-structure M.)

We have thus substantially completed the definition of a semantics.[21] Now we must define a formal system of free quantification and identity theory, and show that it is adequate for the semantics in question.[22] This will be the subject of the next section.

2. THE FORMAL SYSTEMS

There are many equivalent formal systems of free logic in the literature, hence our choice of one of them (or our definition of a new one) cannot but be largely arbitrary. Such an arbitrariness, however, will be restricted, at least to a certain extent, by my wish to make quantification theory independent of identity theory. For this reason I introduced E! as a primitive symbol, while many authors define it by making use of = ; and for the same reason I will now give two independent sets of axiom-schemata for the two symbols.[23] As a result, my system will certainly not be the most economical one, but in the present context I am less interested in economy than in the deepness and perspicuity of philosophical analysis.

Now let us come to the point. *FLI* is the formal system containing the following axiom-schemata and rules of inference:[24]

(A0) A, where A is a tautology;
(A1) $(\forall xA \ \& \ E!y) \supset A^y/x$;
(A2) $\forall xE!x$;
(A3) $\forall x(A \supset B) \supset (\forall xA \supset \forall xB)$;
(A4) $A \supset \forall xA$, where x does not occur free in A;
(A5) $x = x$;
(A6) $x = y \supset (A \supset A^y//x)$;
(A7) $\forall xA$, where A is an axiom;[25]
(R1) B can be inferred from A and $A \supset B$.

The notions of a *proof*, of a *derivation* from a set of premises, of a *theorem* and of a *deductive consequence* of a set of premises can be defined for *FLI* in the usual way.

Our aim is to prove that *FLI* is adequate for the semantics of Section 1, i.e., that its theorems are all and only the valid wffs of the above semantics. For this purpose, our strategy will be as follows. First of all, we will define a system of semantic tableaux, which will be called *STI*. Then we will show that, if *Val* is the set of valid wffs, T_{STI} is the set of theorems of *STI* and T_{FLI} is the set of theorems of *FLI*,

   (a) $T_{STI} \subseteq T_{FLI}$;
   (b) $T_{FLI} \subseteq Val$;
   (c) $Val \subseteq T_{STI}$.

As a simple corollary of (a)–(c), the three sets *Val*, T_{STI} and T_{FLI} will coincide, and thus it will be proved that *FLI* (as well as *STI*, of course) is adequate for our semantics.[26]

Let us begin with the definition of *STI*. We presuppose common definitions of a (*finitely generated*) *tree*, of a *point* and of a *branch* in a tree, and of a point in a tree being the *origin* or the *successor* of another point. We establish that a point in a tree is a *last point* if and only if it has no successors, that a branch *closes* if and only if it contains a last point, and that a tree *terminates* if and only if all its branches close. A *semantic tableau* relative to the language L and to the system *STI* is then a tree whose points are (occurrences of) wffs of L^{27} and in which the property of being a last point and the relation of being a successor are determined by the following rules (S1)–(S7). (To understand these rules, note that an expression of the form A_1 & ... & A_n (where $n \geqslant 0$) stands for any wff of L obtained by adding parentheses to the same expression, that such an expression is to be called *conjunction* of A_1, ..., A_n and that A_1, ..., A_n are to be called *conjuncts* in it, that lastly F and G stand for any conjunctions (possibly empty) such that no conjunct in F is of one of the five forms $x = y$ (where x and y are different individual variables), $\neg\neg A$, $\neg(A$ & $B)$, $\forall xA$, $\neg\forall xA$.)

(S1) A point is a last point if and only if (a) for certain A_1, ..., A_n (where $n > 0$), it is the conjunction of A_1, ..., A_n, and (b) either, for a certain i such that $1 \leqslant i \leqslant n$ and for a certain individual variable x, $A_i = \neg(x = x)$ or, for certain i, j such that $1 \leqslant i \leqslant n$ and $1 \leqslant j \leqslant n$, $A_i = \neg A_j$.

(S2) Every non-last point of the form F & $\neg\neg A$ & G has as its only successor F & A & G.

(S3) Every non-last point of the form F & $\neg(A$ & $B)$ & G has exactly the two successors F & $\neg A$ & G and F & $\neg B$ & G.

(S4) Every non-last point of the form F & $\forall xA$ & G has as its only successor F & A^{y1}/x & ... & A^{yn}/x & G & $\forall xA$, where y_1, ..., y_n are all and only the individual variables z such that $E!z$ is a conjunct in F or in G.[28]

(S5) Every non-last point of the form F & $\neg\,\forall xA$ & G has as its only successor F & $E!y$ & $\neg A^y/x$ & G, where y is the first individual variable (in the alphabetical order) not occurring either in F or in $\neg\forall xA$ or in G.

(S6) Every non-last point of the form F & $x = y$ & G (where x and y are different individual variables) has as its only successor $(F$ & $x = y$ & $G)^y/x$ & F & G & $y = x$ & $x = y$.

(S7) Every non-last point of the form F has as its only successor F.[29]

To conclude our characterization of *STI*, we establish that the semantic tableau *for* a wff A is the semantic tableau whose origin is $\neg A$, and that the *theorems* of *STI* are all and only the wffs A such that the semantic tableau for A terminates.

Now let us prove the above-mentioned results.

THEOREM 1. $T_{STI} \subseteq T_{FLI}$.

Proof. Let us call *degree* of a point C in a semantic tableau the number of points which (a) have at least one branch in common with C and (b) do not precede C in any branch.[30] We will show, by induction on n, that every point of any degree n of a terminating semantic tableau has as a deductive consequence in *FLI*, a contradiction. This will allow us to assert that the origin of any terminating semantic tableau is a wff refutable in *FLI*, hence that the wff for which the tableau is constructed (which is tautologically equivalent to the negation of the origin) is a theorem of *FLI*.

Then let us suppose that the desired result holds whenever $n < k$, and prove it for $n = k$. As the tableau is terminating, there cannot be any point in it which results from an application of (S7);[31] hence we may limit ourselves to considering the following four cases.

Case 1: C is a last point. Trivial.

Case 2: C has one successor (let it be called C') constructed by one of the rules (S2), (S4) and (S6). Then C' is an easy deductive consequence (in *FLI*) of C. But by the induction hypothesis C' has, as a deductive consequence (in *FLI*), a contradiction; then by the transitivity of the relation of deductive consequence (in *FLI*)[32] the same contradiction is a deductive consequence of C (in *FLI*).

Case 3: C has two successors (let them be called C_1 and C_2) constructed by the rule (S3). Then by the induction hypothesis both C_1 and C_2 have, as a deductive consequence (in *FLI*), a contradiction. But both $(C_2 \supset (A \, \& \, \neg A)) \supset (C_2 \supset B)$ and $((C_1 \supset B) \, \& \, (C_2 \supset B)) \supset (C \supset B)$ are theorems of *FLI*, and by making use of them we can easily prove that a contradiction is a deductive consequence of C (in *FLI*).

Case 4: C has one successor (let it be called C') constructed by the rule (S5). Then C is of the form $F \, \& \, \neg \forall x A \, \& \, G$; C' is of the form $F \, \& \, E!y \, \& \, A^y/x \, \& \, G$ (where y is as specified in (S5)), and by the induction hypothesis C' has, as a deductive consequence (in *FLI*), a contradiction. Hence by the Deduction Theorem and other easy steps, $F \, \& \, G$ has, as a deductive consequence (in *FLI*), $(E!y \, \& \, \neg A^y/x) \supset (B \, \& \, \neg B)$, where, without loss of generality, we can suppose that y does not occur free in B, hence $F \, \& \, G$ has, as a deductive consequence (in *FLI*), $\exists y(E!y \, \& \, \neg A^y/x) \supset (B \, \& \, \neg B)$, hence, $\neg \forall x A \supset \exists y(E!y \, \& \, \neg A^y/x)$ being a theorem of *FLI*, C has, as a deductive consequence (in *FLI*), $B \, \& \, \neg B$.   Q.E.D.

THEOREM 2. $T_{FLI} \subseteq Val$.
Proof. A wff C of L is a theorem of *FLI* if and only if either it is an axiom of *FLI* or it is inferred from other theorems of *FLI* by (R1); so we will have proved Theorem 2 if we can show that (a) if C is an axiom of *FLI*, then C is valid, and (b) if C is obtained by applying (R1) to two valid wffs, then C is

valid. Hence we have to distinguish nine cases, but in the present context we will prove only two of them, leaving the others (which are simply routine) to the reader.

Case 2: C is of the form (A1). Then, for a certain wff A and certain individual variables x, y, $C = (\forall x A \,\&\, E!y) \supset A^y/x$. Let us suppose that C is invalid, which is the same as supposing (as the reader can easily see) that, for a certain model-structure M and a certain completion M' of M, $V^{**}_{M'(M)}(C)$ = F. Then, by the clauses (b) and (c) of (10), $V^{**}_{M'(M)}(\forall x A) = V^{**}_{M'(M)}(E!y)$ = T and $V^{**}_{M'(M)}(A^y/x)$ = F, hence, by the clause (d) of the same (10), $V^{**}_{M'(M)}(A^z/x)$ = T for every individual variable z such that $V^{**}_{M'(M)}(E!z)$ = T, hence in particular (as $V^{**}_{M'(M)}(E!y)$ = T) $V^{**}_{M'(M)}(A^y/x)$ = T. Thus the initial assumption that C is invalid entails a contradiction.

Case 9: C is inferred from two valid wffs of the form C' and $C' \supset C$, respectively. If, for a certain model-structure M and a certain completion M' of M, $V^{**}_{M'(M)}(C')$ = T and $V^{**}_{M'(M)}(C)$ = F, then, by the clauses (b) and (c) of (10), $V^{**}_{M'(M)}(C' \supset C)$ = F. But this is to say that, for every model-structure M and every completion M' of M, if $V^{**}_{M'(M)}(C') = V^{**}_{M'(M)}(C' \supset C)$ = T then $V^{**}_{M'(M)}(C)$ = T. Hence in particular, if $V^{**}_{M'(M)}(C') = V^{**}_{M'(M)}(C' \supset C)$ = T for every model-structure M and every completion M' of M (i.e., if C' and $C' \supset C$ are valid), then $V^{**}_{M'(M)}(C)$ = T for every model-structure M and every completion M' of M (i.e., C is valid).   Q.E.D.

THEOREM 3. $Val \subseteq T_{STI}$.

Proof. Let us consider a wff C of L. We will prove that C is invalid if it is not a theorem of *STI*. More precisely, we will prove that if the semantic tableau for C contains at least one branch which does not close, then there are two model-structures M and M' such that (a) M' is a completion of M and (b) $V^{**}_{M'(M)}(A)$ = T for every wff A cocurring as a conjunct in the above branch. As the origin $\neg C$ belongs to all branches, this will entail that $V^{**}_{M'(M)}(C)$ = F, hence that C is invalid.

Then let us suppose that the semantic tableau for C contains at least one branch X which does not close, and define in the set of all the individual variables of L the following binary relation I: xIy if and only if either x and y are the same individual variable or $x = y$ occurs as a conjunct in some point of X. It is easy to see that I is a relation of equivalence; let us agree to call the *identity-class* of any individual variable x the equivalence class with respect to I to which x belongs. Now let M be the model-structure such that (i) its domain D is the set of all and only the individual variables z such that (a) z is the first member of its identity-class which occurs in some point of X, and (b) $\text{E!}z$ occurs as a conjunct in some point of X; (ii) its function of interpretation f assigns to every individual variable z the member of the identity-class of z which belongs to D, if this member exists, and is not defined for all the individual variables such that no member of their identity-classes belongs to D; and (iii) f assigns to every n-ary predicate P the set of all the ordered n-tuples $\langle x_1, ..., x_n \rangle$ of members of D such that $Px_1 ... x_n$ occurs as a conjunct in some point of X. Moreover, let M' be the model-structure such that (i) its domain D' is the set of all and only the individual variables z such that either z is the first member of its identity-class which occurs in some point of X or z is the only member of its identity-class; (ii) its function of interpretation f' assigns to every individual variable z the member of the identity-class of z which belongs to D'; and (iii) f' assigns to every n-ary predicate P the set of all the ordered n-tuples $\langle x_1, ..., x_n \rangle$ of members of D' such that $Px_1 ... x_n$ occurs as a conjunct in some point of X.

It is easy to see that M' is a completion of M, hence it remains to show that $V^{**}_{M'(M)}(A) = \text{T}$ for every wff A occurring as a conjunct in some point of X. This part of the proof will be carried out by induction on the number n of connectives and quantifiers occurring in A.

Then let us suppose that the desired result holds whenever $n < k$, and prove it for $n = k$. We will distinguish two cases.

Case 1: A is atomic. We will distinguish three subcases.

Subcase 1a: A is of the form $Px_1 ... x_n$. Then, by the definition of M and M', either $V^*_M(A) = \text{T}$ or $V^*_M(A)$ is not defined and $V^*_M(A) = \text{T}$. In both cases $V^{**}_{M'(M)}(A) = \text{T}$.

Subcase 1b: A is of the form $x = y$. Then, by the definition of M and M', either $f(x) = f(y)$ or neither $f(x)$ nor $f(y)$ is defined and $f'(x) = f'(y)$. In both cases $V^{**}_{M'(M)}(A) = \text{T}$.

Subcase 1c: A is of the form $\text{E!}x$. Then, by the definition of M, $V^*_M(A) = \text{T}$, which entails $V^{**}_{M'(M)}(A) = \text{T}$.

Case 2: A is not atomic. We will distinguish three subcases.

Subcase 2a: A is the negation of an atomic wff B. Then we can easily prove that B does not occur as a conjunct in any point of X (otherwise, X would close), and obtain the desired result from the definitions of M and M'.

Subcase 2b: A is a conjunction B_1 & B_2. Then B_1 and B_2 are also conjuncts in X, and our result follows from the induction hypothesis.

Subcase 2c: A is neither a conjunction nor the negation of an atomic wff. Then, as A occurs in some point of X, in some (other) point(s) of X the result(s) of applying to A one of the rules (S2)–(S5) has (have) to occur. Simple applications of the induction hypothesis will then suffice to obtain in every (sub-sub) case the desired result. Q.E.D.

As we anticipated earlier, a simple consequence of Theorem 1–3 is

THEOREM 4. $T_{STI} = T_{FLI} = Val$,
which establishes the adequacy of the systems FLI and STI to our semantics.

NOTES

[1] I want to thank my friends Bas van Fraassen and Hans Herzberger for their useful comments on an earlier draft of this paper.

[2] From now on, simply a 'free semantics'.

It is important to note that the present paper can be read in two different ways: from left to right and from right to left, as it were. In the first sense, it is to be regarded as an attempt to construct a satisfactory semantics for a free language, and then to axiomatize the resulting set of valid wffs: in this sense, of course, it depends heavily on my intuitions about what counts as 'satisfactory' in the present context. In the second sense, it can be conceived as a semantical analysis of some *existing* axiomatic systems of free logic, that is to say, as an attempt to interpret such systems without the awkward results of the alternative semantics which have been proposed for *the same* systems, and in this second sense what is important is the comparative strength of the above intuitions. In other words, the approach I will propose here must be contrasted with both the semantics for different free logics (such as the one arising from the thesis (a) below) and different semantics for (essentially) the same free logic (such as those grounded on the theses (b) and (c) below). Thus if someone does not agree with my intuitions (for example, if he does not believe that the sentence (1) below is to be counted as true), I ask him to read the paper in the second sense (hence to consider that the sentence in question is an instance of a schema which is *provable* in most free logics).

[3] As the reader will see, one of my fundamental purposes in the construction of a new free semantics has been the elimination of the notion of a non-existing object, which I regard as very obscure from a philosophical point of view, and the attempt to replace it everywhere with the notion of an *object existing in another world*. However, this will not prevent me from sometimes using the *phrase* 'non-existing object' (between quotation-marks, to emphasize its impropriety), for this use is only colloquial and does not express any commitment of any kind whatsoever.

[4] For this thesis, see Schock (1964, 1968).

[5] For this thesis, (which gives rise to what we may call 'semantics of outer domains'), see Leblanc and Thomason (1968).

[6] See Meyer and Lambert (1968).

[7] For this thesis (which gives rise to what we may call 'semantics of supervaluations'), see van Fraassen (1966a, 1966b).

[8] As a matter of fact, the semantics of supervaluations *does* regard (1) as true, but only by virtue of an *ad hoc* device (a good instance of the 'awkward results' referred to in Note 2).

[9] We will use $=$ as a metatheoretical symbol, too, but the context will always prevent any confusion.

[10] Notice that L, in contrast with many other free languages but in agreement e.g. with Leonard (1956), Lambert (1963, 1967), Meyer and Lambert (1968), does not contain individual constants. The reason is that I (as at least two of the above authors; see the discussion contained in Meyer and Lambert, 1968) regard any individual constant on which we make no assumptions as behaving simply as an individual variable, hence I deem useless the presence in this context (i.e., in 'pure' logic) of two categories of singular terms. The most important consequence of the choice of such a course will be the attribution of a double role to the individual variables, according to their being free or bound, and then the necessity of rejecting the Closure Theorem provable for standard logic (as for many free logics, too).

[11] In what follows, we will often omit this qualification.

[12] From now on, we will simply say a 'model-structure'.

[13] Meyer and Lambert (1968), for example, regard as 'factually' truth-valueless a wff of the form $E!x$ whenever x is non-denoting.

[14] This qualification is essential, for otherwise I would be involved in a circle, one of my aims being the definition of a notion of validity. This is precisely the reason why (as I say below) this is not an argument, but only another (I hope more perspicuous) way of stating my position.

[15] We have to refer here to the Coincidence Theorem of standard semantics because we have not yet defined a free semantics. When this definition is complete, we will be in a position to prove a variant of such a theorem, but we will leave this simple task to the reader.

[16] For if the antecedent of an instance of (9) is factually true (in M or in a completion of M), its consequent is factually true as well.

[17] This collapse, however, is not complete, for some theorems of standard logic (such as the ones of the forms $\forall xPx \supset \exists xPx$, $\exists x(x = x)$, $\exists x(x = y)$) could be invalidated in the above semantics on purely 'factual' grounds. These standard theorems, on the other hand, would be formally true in every model-structure (given the above definition of 'formally true'), hence even in order to invalidate them we should establish a criterion of priority among sorts of truth (which is to say, among sources of information), similar to (but less complicated than) the one we will propose later.

[18] This part of the argument will not be accepted by anyone who gave a negative answer to the question (a) of p. 34. I said above that he could obtain my results, but this was not to imply that he could be convinced to the same extent (or at least for the same reasons) of their plausibility.

[19] The contrast is only relative to the consequent; as to the antecedent, there is no contrast at all, but only an increase of information.

[20] To sum up, our semantics is substantially grounded on the acceptance of the following principle (which I would like to call *Principle of the Prevalence of Reality*):

(*) When we are deciding on the truth-value of a sentence A in a world M, the information coming from M prevails over all other information (in particular, over all information coming from a 'mental experiment' carried out starting from M).

The reader will see that the principle (*) (embodied in the definition of a secondary auxiliary valuation) allows us to completely avoid 'non-existing objects', while on the contrary such objects are present not only in the outer domains of Leblanc and Thomason (1968) but also in the *logical points* of Meyer and Lambert (1968) (where 'there are' x such that $E!x$ is true and x such that $E!x$ is false). For accepting this principle we take the liberty of considering several 'possible worlds' (which do belong to the ontology of most semantics for standard and free logics) related in some way to the world assumed as real (i.e., to the world in which we want to evaluate a certain wff), but do not admit more than one sense in which an object can belong to such a possible world. Degrees of existence are thus altogether banished.

It is also interesting to note that by accepting (*) (hence the particular definition of a secondary auxiliary valuation) we could introduce into our semantical construction the simplification promised at p. 35. For even though $Px \lor Py$ is regarded as factually truth-valueless in the case in question, it will be formally true anyway, so that nothing will change in practice.

Finally, the particular form of (11) deserves some comment, too. From the previous discussion, we could have expected that the definition of a valuation V_M had the following form; if $V_M^*(A)$ is defined, then $V_M(A) = V_M^*(A)$, and otherwise ... . In this way indeed we would have emphasized the difference between factual and formal truth, but (11) is simpler and furthermore it allows us to *prove* easily that if $V_M^*(A)$ is defined, then $V_M(A) = V_M^*(A)$. On the other hand, if we did not want a complete definition of factual truth in a model-structure (which as a matter of fact we *do* want, for philosophical reasons), the particular form of (11) would allow us to obtain, instead of the above-mentioned proof, a great simplification of (7). For the reader can easily see that the only clauses of (7) that are really used in the determination of the final truth-value of a wff A are the ones relative to atomic wffs.

[21] The qualification 'substantially' can be made more explicit in the following way: we have defined the two fundamental notions of a formal semantics, and some of the derived ones. Other derived notions (first of all, the notion of semantic entailment) will not be dealt with in the present context.

[22] We will prove this adequacy only in the weak sense, i.e., we will prove that the theorems of the system are all and only the valid wffs of our semantics. To prove the adequacy in the strong sense we should introduce the notion of semantic entailment, which we will not do here (as we said in the previous note). Anyway, under any reasonable definition of such a notion, the system we will propose is not strongly adequate, and for substantial reasons. For this purpose, see the discussion contained in van Fraassen (1966b).

[23] This would allow us to derive easily from our general proposal a satisfactory semantics for a free language without identity (if the paper is read, and accepted, in the first of the senses referred to in Note 2) or a semantical analysis of some existing systems of free logic without identity (if the paper is read in the second sense).

[24] Among the systems of free logic proposed in the literature, the most similar to *FIL* has been hinted at by Meyer and Lambert (1968), in a footnote. It can be obtained from

FLI by simply replacing (A1) with its tautological equivalent $\forall xA \supset (E!x \supset A^y/x)$ and dropping (A5)–(A6) (but elsewhere in the paper Meyer and Lambert consider the possibility of adding to their system common axioms for identity). Systems equivalent to *FLI* (apart from some notational differences) are also given by Lambert (1963, 1967), van Fraassen (1966a), Leblanc and Thomason (1968), Meyer and Lambert (1968, main text) and other authors.

[25] The practice of replacing the rule of Universal Generalization with something similar to (A7) originates with Fitch (1948).

[26] As the reader will see, this is nothing but an adaptation of a common procedure used to prove the weak completeness of standard logic. Such a procedure was first adapted to free logic (but for a different semantics, and a different language) by van Fraassen (1966a).

[27] We have to require that the points of a semantic tableau be *occurrences of* wffs, and not simply wffs, because the same wff(-type) may constitute different points of the tableau. In what follows, however, we will disregard this qualification.

[28] Obviously, if there is no such z, the successor of $F \& \forall xA \& G$ is just $F \& G \& \forall xA$.

[29] The rule (S7) is necessary to make sure that only those branches close in which we actually reach a refutable conjunct. This result is obtained in other treatments (such as, for example, the one contained in van Fraassen, 1966a) by a slightly different definition of a branch closing, which involves the introduction of metatheoretical devices like crosses or underlinings.

[30] Obviously, a point C of a semantic tableau *precedes* a point C' (of the same tableau) if and only if either (a) C' is a successor of C, or (b) there is a point C'' in the tableau such that C precedes C'' and C' is a successor of C'.

[31] For every application of (S7) will be obviously followed by an infinite number of applications of the same rule.

[32] This auxiliary result (as the Deduction Theorem referred to below) can be proved in a usual way. We leave this simple task to the reader.

REFERENCES

Fitch, Frederick B. 1948, 'Intuitionistic Modal Logic with Quantifiers', *Portugaliae Mathematica* **7**, 177–85.

Lambert, Karel 1963, 'Existential Import Revisited', *Notre Dame Journal of Formal Logic* **4**, 288–92.

Lambert, Karel 1967, 'Free Logic and the Concept of Existence', *Notre Dame Journal of Formal Logic* **8**, 133–44.

Leblanc, Hugues, and Thomason, Richmond H. 1968, 'Completeness Theorems for Some Presupposition-Free Logics', *Fundamenta Mathematicae* **62**, 125–64.

Leonard, Henry S. 1956, 'The Logic of Existence', *Philosophical Studies* **7**, 49–64.

Meyer, Robert K., and Lambert, Karel 1968, 'Universally Free Logic and Standard Quantification Theory', *Journal of Symbolic Logic* **33**, 8–26.

Schock, Rolf 1964, 'Contributions to Syntax, Semantics, and the Philosophy of Science', *Notre Dame Journal of Formal Logic* **5**, 241–89.

Schock, Rolf 1968, *Logics Without Existence Assumptions*. Stockholm, Almqvist and Wiskell.

van Fraassen, Bas C. 1966a, 'The Completeness of Free Logic', *Zeitschrift für mathematische Logik und Grundlagen der Mathematik* **12**, 219–34.

van Fraassen, Bas C. 1966b, 'Singular Terms, Truth-Value Gaps, and Free Logic', *Journal of Philosophy* **67**, 481–95.

SERGIO BERNINI

A TEMPORALIZATION OF MODAL SEMANTICS

The Barcan Formula (BF) is a critical principle for modal semantics. By BF we mean the following schema

$$\forall x \Box A(x) \rightarrow \Box \forall x A(x).$$

(If all are necessarily A, then necessarily all are A).

As is well known, Kripke found a very interesting semantics for sentential modal logic in terms of possible worlds connected by an accessibility relation [1]. This semantics admits of a straightforward extension to the predicative case, where BF turns out to be always valid. However, at the same time BF cannot be proved as a theorem in every modal system. In order to overcome this difficulty, which excludes the possibility of proving a completeness theorem, Kripke considered a natural extension of this semantics, where each possible world has its own domain of individuals. Because of that, he was forced to take an axiomatization of modal logic based on a non-standard codification of first-order logic[2]. On the other hand, if one wants to preserve standard first-order logic, one has to change the original semantical definition of [2], by postulating a semantical principle such as 'every actual individual is necessary', which appears to be intuitively somewhat unnatural.

Both solutions describe as critical the transition from *sentences* to open *formulas*, where individual parameters occur. However, it seems to me that it is perhaps more natural to see the source of the difficulties in the transition from *finite* logical situations (connectives) to *transfinite* ones (quantifiers). This view can be justified by the following consideration. Take a system S of infinitary logic, that is, a logic which admits of infinitely long conjunctions $A_0 \wedge A_1 \wedge A_2 \wedge \ldots$. Add modal axioms and rules which characterize a system where BF is not provable. We see that the infinitary schema BF^+ corresponding to BF is not provable in S

$$(BF)^+ \quad \Box A_0 \wedge \Box A_1 \wedge \Box A_2 \wedge \ldots \rightarrow \Box (A_0 \wedge A_1 \wedge A_2 \wedge \ldots).$$

On the contrary the finite corresponding schema is provable in S for every n:

$$\Box A_0 \wedge \Box A_1 \wedge \ldots \wedge \Box A_n \rightarrow \Box (A_0 \wedge A_1 \wedge \ldots \wedge A_n).$$

49

Maria Luisa Dalla Chiara (ed.), Italian Studies in the Philosophy of Science, 49–58.
Copyright © 1980 *by D. Reidel Publishing Company.*

If one takes the usual Kripke truth definition one gets the universal validity of BF$^+$. The difficulties we find in this particular independence problem and in many other similar questions seem to suggest that Kripke semantics cannot always be considered a natural one beyond the finitary propositional case of the classic modal systems. In fact, in spite of its analogy with BF, BF$^+$ has clearly nothing to do with individual parameters. Hence we may prefer more abstract semantics where no hypotheses are put on individuals. Following [3] we could consider Montague semantics. On the other hand, Kripke semantics seems to have a more intuitive interpretation. We give here a modelling which may be regarded as a kind of reformulation of Montague semantics in the way of a temporalization of it. In (I) we informally introduce the semantics. (II) will be more formal. Intuitionistic logic will be also considered. (III) contains sketches of completeness proofs.

I. The structures we are concerned with can be visualized as in Figure 1. The points represent 'situations' and each arrow represents a change from a given situation, say s, to another one which is 'possible' or 'accessible' with respect to s. The question about the meaning of 'possible situation' or 'accessibility' might be left open. However, we find useful to consider our concept of possibility as non-univocal. At the same time we require that this liberal conception always include a particular one. A situation may be reached as a possibility from another situation in many ways: logically, temporally, spatially and so on; we postulate that temporal

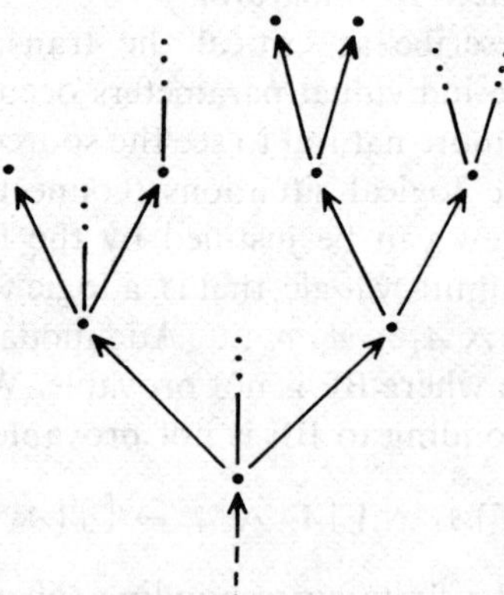

Fig. 1

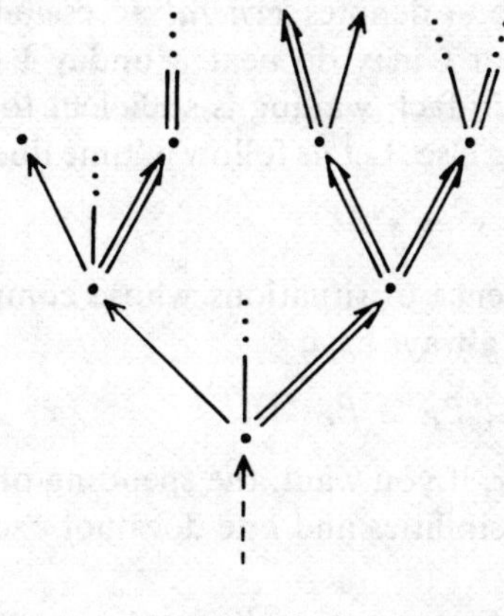

Fig. 2

accessibility is always present. Each situation always has a temporal evolution that is among its own relative possibilities.

Let P_s be the set of situations accessible from s, namely the points of Figure 1 that are connected upward to a given point s by an arrow. Let T_s be the set of situations which are temporally accessible from s. We have $T_s \subseteq P_s$. Furthermore we put: for every s, if $P_s \neq \emptyset$ then $T_s \neq \emptyset$. If there exist possibilities, there exist temporal possibilities. Instead of $\rightarrow$, write $\Rightarrow$ to denote temporal accessibility. In this way Figure 1 becomes, for instance, as shown in Figure 2.

We now list four conditions on $\Rightarrow$ to characterize its 'temporality'. As the first two conditions take

(1)   reflexivity: for every s, $s \Rightarrow s$

(2)   transitivity: for every s, s', s'', if $s \Rightarrow s'$ and $s' \Rightarrow s''$ then $s \Rightarrow s''$.

It may be they are not neutral with respect to the philosophy of time but they are rather natural. To complete the list take the following two conditions that are stronger:

(3)   antisymmetry: for every s, s', if $s \Rightarrow s'$ and $s' \Rightarrow s$ then $s = s'$

(4)   mixed transitivity: for every s, s', s'', if $s \Rightarrow s'$ and $s' \rightarrow s''$ then $s \rightarrow s''$.

By (3) we see that time is not cyclic, namely, to come back is to stay. A little more about (4). It says that accessibility is closed under future accessibility. Suppose $\rightarrow$ denotes *general* accessibility. Then (4) asserts, for instance, that what I may do next Monday I find among my actual general possibilities; in fact, waiting is sufficient to reach that. Moreover (4) suggests something else. Let us follow a 'time line'

$$\ldots \Rightarrow s \Rightarrow s' \Rightarrow s'' \Rightarrow \ldots$$

that is a certain sequence of situations whose components are related by $\Rightarrow$. Because of (4) we always have

$$\ldots \supseteq P_s \supseteq P_{s'} \supseteq P_{s''} \supseteq \ldots$$

So a temporal flow or, if you want, the spending of time has the property of not increasing possibilities and one does not exclude having an actual decrease of them.

Because of our assumptions each point s now determines a certain temporal structure $\alpha = (Z_s, \Rightarrow)$ where Z_s is the set of all the points s' which are $\Rightarrow$ – or $\Leftarrow$ – related to s. This α can be considered the world, s is an image of it at a given instant.

Take a language L whose set of individual terms is a countable set D of constants and attribute to each s a certain set $V(s)$ of atomic propositions of L to be interpreted as true in that situation. Then truth for a modal language could be defined on each s in a Kripke style:

$$s \models A \text{ iff } A \in V(s), \text{ for atomic } A,$$
$$s \models \neg A \text{ iff not } (s \models A),$$
$$s \models A \wedge B \text{ iff } s \models A \text{ and } s \models B,$$
$$s \models \forall x A \text{ iff for every } c \in D, s \models A(x/c),$$
$$s \models \Box A \text{ iff for every } s' \text{ such that } s \rightarrow s', s' \models A.$$

Already in this definition we see that truth is 'temporalized'. In fact every formula which is true in a situation is temporally true because every situation s can be considered as a point of a given temporal evolution $(Z_s, \Rightarrow)$. On the other hand we should want to make explicit the role of time and to do it in the following way: consider each true formula as *temporalized*, disregarding the *exact* instant of that truth. That practically means to refer truth to a whole temporal complex α instead of referring it to a singular point s; truth is to be truth *modulus* temporal transformations. We could get that by defining $\models$ in the following way:

$$s \models A \text{ iff there is an } s' \in Z_s \text{ such that } A \in V(s'), \text{ for atomic } A$$
$$s \models \Box A \text{ iff there is an } s' \in Z_s \text{ such that, for every } s'' \leftarrow s',$$
$$s'' \models A \text{ for } \neg, \wedge, \forall, \text{ take } \models \text{ as in the former definition.}$$

By this definition we have: for every s, $s \models A$ iff $\exists s' \in Z_s(s' \models A)$; that is, truth disregards time.

Modelling modal logic can now be obtained by putting a series of further conditions. As a *basic* condition we take

(5)   time is causal-relativistic: for every s, s' there is always an s'' such that $s \Rightarrow s''$ and $s' \Rightarrow s''$.

This condition plays the same role of 'closure under finite intersections' in [3]. By means of it we get the validity of the modal law $\square(A \to B) \to (\square A \to \square B)$.

One could also put a stronger condition

(5′)   time is linear: for every s, s', $s \Rightarrow s'$ or $s' \Rightarrow s$.

It is clear that if time is linear, then it is causal-relativistic. We shall assume (5) but we shall actually work with linear time. Furthermore we put the usual conditions on $\to$ to get models which are adequate for the usual modal systems T, S4 (both without BF), B, S5: reflexivity for T, reflexivity and transitivity for S4, reflexivity and symmetry for B, reflexivity, transitivity, symmetry for S5. To get models for the corresponding systems *with* BF, put another condition on $\Rightarrow$: time has a last instant, i.e. for every s there is an $s'' \in Z_s$ such th$s$, for every $s' \in Z_s$, $s' \Rightarrow s''$. For this purpose put, for instance, $Z_s = \{s\}$, for every s, that is, nullify all our discussion about time.

In what follows we shall introduce a more manageable reformulation of this semantics. We shall start from objects α, β, ... to be thought as, say, complexes Z_s. Points s are not explicitly mentioned. Instead of them *instants* are introduced by attributing to each α a structure $(I_\alpha, \leqslant_\alpha)$ where I_α is the set of the instants relative to the world α and $\leqslant_\alpha$ is the temporal order on I_α. This attribution can be intuitively taken in such a way that if α is thought as a structure $(Z_s, \Rightarrow)$, then $(I_\alpha, \leqslant_\alpha)$ and $(Z_s, \Rightarrow)$ are isomorphic.

II. Let L be a language with the following logical signs: $\neg$ (negation), $\wedge$ (conjunction), $\forall$ (universal quantifier), $\square$ (necessity operator). Let the set D of closed terms consist only of individual constants. We shall use A, B, C, ... for *sentences* (closed formulas) of L. By a *modal structure* (for L) we mean a structure $(K, I, \leqslant, V, P, D, \models)$ where: (a) $-$ K is a not-void set, I, $\leqslant$, V are functions on K such that, for every $\alpha \in K$, I_α is a not-void set, $\leqslant_\alpha$ is a relation on I_α, V_α is a set of atomic sentences of L; (b)$-$$P$ is a function on $K \times U_{\alpha \in K} I_\alpha$ such that, for every $\alpha \in K$ and $i \in I_\alpha$, P_α^i is a subset of K; (c) — D is the set of closed terms of L and $\models$ is a relation between elements of K and sentences of L.

A modal structure is said to be a *model* (for L) if the following two groups of five conditions hold:

(1.1)   if $i \in I_\alpha$ then $i \leqslant_\alpha i$,

(1.2)   if $i, j, k \in I_\alpha$ and $i \leqslant_\alpha j, j \leqslant_\alpha k$ then $i \leqslant_\alpha k$,

(1.3)   if $i, j \in I_\alpha$ and $i \leqslant_\alpha j, j \leqslant_\alpha i$ then $i = j$,

(1.4)   if $i, j \in I_\alpha$ then there exists a $z \in I_\alpha$ such that $i \leqslant_\alpha z$ and $j \leqslant_\alpha z$,

(1.5)   if $i, j \in I_\alpha$ and $i \leqslant_\alpha i$ then $P_\alpha^j \subseteq P_\alpha^i$.

(2.1) $\alpha \models A$ iff $A \in V_\alpha$, for atomic A,

(2.2) $\alpha \models A \wedge B$ iff $\alpha \models A$ and $\alpha \models B$,

(2.3) $\alpha \models \neg A$ iff not $(\alpha \models A)$,

(2.4) $\alpha \models \forall x A$ iff, for every $c \in D$, $\alpha \models A(x/c)$,

(2.5) $\alpha \models \Box A$ iff there is an $i \in I_\alpha$ such that, for every $\beta \in P_\alpha^i$, $\beta \models A$.

K corresponds to the Kripke universe of possible worlds, I_α is the universe of the instants of α, $\leqslant_\alpha$ is the ordering on I_α, V_α are the atomic truths in α, P_α^i is the set of the worlds which are possible with respect to α at the instant i, D are the names of all individuals. Because of this absoluteness of the domain with respect to $\alpha, \beta, \ldots$ we could consider them as 'possible states' of the individuals D instead of 'worlds'.

If L admits of infinitary conjunctions $\bigwedge \mathscr{A}$, where $\mathscr{A}$ is a not-void set of sentences whose cardinality does not exceed a given k, we substitute (2.2) with

(2.2′) $\alpha \models \bigwedge \mathscr{A}$ iff, for every $A \in \mathscr{A}$, $\alpha \models A$.

A sentence A is *true* in a given model $(K, I, \leqslant, V, P, D, \models)$ iff, for every $\alpha \in K, \alpha \models A$.

Let us consider the following three conditions on P:

(a) if $i \in I_\alpha$ then $\alpha \in P_\alpha^i$,

(b) if $i \in I_\alpha, j \in I_\beta, \beta \in P_\alpha^i$ then $P_\beta^j \subseteq P_\alpha^i$,

(c) if $i \in I_\alpha, j \in I_\beta, \beta \in P_\alpha^i$ then $\alpha \in P_\beta^j$.

They correspond to the conditions of reflexivity, transitivity, symmetry for the relation of accessibility, respectively. If (a) holds in a model M then M is said to be a *T-model*. If (b) holds in a T-model M then M is said to be an S4-*model*. If (c) holds in an S4-model M then M is said to be an S5-*model*. Finally if (b) holds in a T-model M then M is said to be a *B-model*.

Note that (c) is equivalent to the destruction of any effective reference to the instants. In fact from (c) we immediately get that, for every $i, j \in I_\alpha$, $P_\alpha^i = P_\alpha^j$.

Let us take the following axiomatization of the classical modal systems we are concerned with, namely T, S4, B, S5:

T is the classical predicate or infinitary logic plus the rule 'from A infer $\Box A$' and the axiom schemata $\Box A \rightarrow A$, $\Box(A \rightarrow B) \rightarrow (\Box A \rightarrow \Box B)$.
S4 is T plus the axiom schema $\Box A \rightarrow \Box\Box A$.
B is T plus $\neg A \rightarrow \Box\neg\Box A$.
S5 is S4 plus $\neg A \rightarrow \Box\neg\Box A$.

The reader can show the validity of these systems with respect to the corresponding models. We know that BF is not provable in T and S4. Let us construct a counterexample to BF in a T-model. Take a model $M = (K, I, \preccurlyeq, V, P, D, \models)$ where $K = \{\beta_0, \beta_1, \beta_2, ...\}$, I_{β_n} is the set ω of natural numbers, for every n, and $\preccurlyeq_{\beta_n}$ is the usual less-or-equal relation on ω, $D = \{c_0, c_1, c_2, ...\}$, V is such that $V_{\beta_n} = \{A(c_m) : m \neq n + 1\}$, for every n. Take the following two conditions on P

(1) for every n, m, $\beta_m \in P_{\beta_0}^n$ iff $m \geqslant n$
(2) for every r, n and m, if $m \neq 0$, $\beta_r \in P_{\beta_m}^n$ iff $r = m$.

It is easy to show that M is a T-model. We want also to show that $\beta_0 \models \forall x \Box A(x)$ but $\beta_0 \not\models \Box \forall x A(x)$, that is $\beta_0 \not\models$ BF. In fact from the definition of V we have that for every c_n there is an m such that $A(c_n) \in V_\gamma$, for every $\gamma \in P_{\beta_0}^m$: take $m = n$. By the definition of $\models$ we get so that $\beta_0 \models \forall x \Box A(x)$. On the other hand, because of the definition of V, there can not be any m such that, for every c_n and every $\gamma \in P_{\beta_0}^m$, we have $A(c_n) \in V_\gamma$. So $\beta_0 \not\models \Box \forall x A(x)$.

Essentially the same counterexample holds for BF$^+$. Instead of $\forall x A(x)$ take the infinite conjunction of $\{A(c_0), A(c_1), ...\}$.

Application to Intuitionistic Logic

One of the most important classical laws which are not intuitionistically provable is $\forall x(A \vee B(x)) \rightarrow A \vee \forall x B(x))$, where x is not free in A. Let us call that $(+)$.

Intuitionistic logic admits of faithful translation in S4. See [5] or [6], p. 43. It turns out that the *modal* reasons for the independence of $(+)$ and BF are the same. That is to say that the translation of $(+)$ is not provable in S4 because of the lack of BF.

Kripke found a very interesting modelling of intuitionistic logic which is strictly related to his models for S4. In particular we have there the same hypothesis we had in the modal case: each actual individual is necessary. This condition is essential for the non-validity of $(+)$. On the other hand, the same problem of independence may be raised for other forms of $(+)$. For instance, infinitary form: $(A \vee B_0) \wedge (A \vee B_1) \wedge ... \to A \vee (B_0 \wedge B_1 \wedge ...)$. The idea of actual-necessary individuals does not work here. The reader is invited not to seek arguments against the use of this particular example, by assuming its senselessness for the intuitionists. In fact an infinitary language is explicitely used by Brouwer in [7], p. 5. So, models without any trick about individuals are generally preferable. Take, for instance, Beth model. See [8] §145. By means of a slight reformulation of our S4-models a topological model for intuitionistic logic, say, as in [9], is obtained directly.

Let us introduce a new class of models. By S4'-*models* we mean T-models where we have

(1) $I_\alpha = \omega$ and $\leqslant_\alpha = \leqslant$, for every $\alpha \in K$

(2) if $m \leqslant n$, $\beta \in P_\alpha^n$ then $P_\beta^m \subseteq P_\alpha^n$.

The main difference with respect to S4-models is (2): inclusion between P_β^m and P_α^n holds under the condition that m is sooner than n.

If you take models where K is a subset of M^N with $M \neq \emptyset$ and (1) holds, then, by putting $P_\alpha^n = \bar{\alpha}(n)$, where $\bar{\alpha}(n)$ is the set of the sequences in K whose first n components are equal to $\alpha(0)$, $\alpha(1)$, ..., $\alpha(n-1)$, respectively, you clearly get S4'-models.

By *intuitionistic models* we mean models of the latter type where $\models$ is substituted by the following relation $\Vdash$:

(a) $\alpha \Vdash A$ iff $\exists n \forall \beta \in \bar{\alpha}(n)(A \in V(\beta))$, for atomic A,

(b) $\alpha \Vdash A \vee B$ iff $\alpha \Vdash A$ or $\alpha \Vdash B$,

(c) $\alpha \Vdash A \wedge B$ iff $\alpha \Vdash A$ and $\alpha \Vdash B$,

(d) $\alpha \Vdash \neg A$ iff $\exists n \forall \beta \in \bar{\alpha}(n)$ not $(\beta \Vdash A)$,

(e) $\alpha \Vdash A \to B$ iff $\exists n \forall \beta \in \bar{\alpha}(n)$ (if $\beta \Vdash A$ then $\beta \Vdash A$),

(f) $\alpha \Vdash \exists x A(x)$ iff $\exists c \in D(\alpha \Vdash A(c))$,

(g) $\alpha \Vdash \forall x A(x)$ iff $\exists n \forall \beta \in \bar{\alpha}(n) \forall c \in D(\beta \Vdash A(c))$.

We are forced to consider only models where if $\alpha \Vdash \forall x (A \to B(x))$ then $\alpha \Vdash A \to \forall x B(x)$ and if $\alpha \Vdash \bigwedge_i (A \to B_i)$ then $\alpha \Vdash A \to \bigwedge_i B_i$.

If the language admits of infinitary conjunctions $\bigwedge \mathscr{A}$ and disjunctions $\bigvee \mathscr{A}$, substitute (b) and (c), respectively, with (b') $\alpha \Vdash \bigvee \mathscr{A}$ iff $\exists A \in \mathscr{A}(\alpha \Vdash A)$ and

(c') $\alpha \Vdash \bigwedge \mathscr{A}$ iff $\exists n \forall \beta \in \bar{\alpha}(n) \forall A \in \mathscr{A}(\beta \Vdash A)$.

Truth in a model is defined as in modal case.

III. *Completeness*. Let L be a language which admits of countable conjunctions $\bigwedge$ and disjunctions W. We say that a finite set S of sentences of L is $\mathscr{L}$-*consistent*, where $\mathscr{L}$ is a logic (T or $S4$), if $\neg \bigwedge S$ is not a thesis of $\mathscr{L}$. Suppose now L is based on a countable set D of individual constants and let A be a sentence of L in which only a finite subset of D occurs. Let $S(A)$ be any countable set of sentences of L with the following properties: (1) $A \in S(A)$; (2) every sentence in $S(A)$ contains only a finite subset of D; (3) $S(A)$ is closed under subsentences (i.e. subformulas closed by substitution of free variables with elements of D); (4) if $W\mathscr{B}$ or $\bigwedge \mathscr{B} \in S(A)$ then $W\{\neg B | B \in \mathscr{B}\}$ and $\bigwedge \{\neg B | B \in \mathscr{B}\}$ belong to $S(A)$; (5) if $\forall x B(x)$ or $\exists x B(x) \in S(A)$ then $\forall x \neg B(x)$ and $\exists x \neg B(x)$ belong to $S(A)$. Suppose now $\{S_i\}$ is an enumeration of $S(A)$ and let α, β, ... be increasing sequences of finite $\mathscr{L}$-consistent subsets of $S(A)$ such that, for every n, (1) $S_n \in \alpha_n$ or $\neg S_n \in \alpha_n$, (2) if $W\mathscr{B} \in \alpha_n$ then $B \in \alpha_{n+1}$, for some $B \in \mathscr{B}$, (3) if $\exists x B(x) \in \alpha_n$ then $B(c) \in \alpha_{n+1}$, for some $c \in D$. Such α's will be said *Barwise $\mathscr{L}$-complete* sequences.

Consider now the following ternary relation $(\mathscr{L})$:

if $\mathscr{L} = T$ then $(\alpha, n, \beta) \in (\mathscr{L})$ iff $\alpha = \beta$ or $\{A \mid \Box A \in \alpha_n\} \subseteq \beta_0$

if $\mathscr{L} = S4$ then $(\alpha, n, \beta) \in (\mathscr{L})$ iff $\alpha = \beta$ or $\{\Box A \mid \Box A \in \alpha_n\} \subseteq \beta_0$.

Take the following structure $M(\mathscr{L}) = (K(\mathscr{L}), \omega, \leqslant, V, P(\mathscr{L}), D, \models)$ where $K(\mathscr{L})$ is the set of the Barwise $\mathscr{L}$-complete sequences, V_α is the set of the atomic sentences of α and $\beta \in P(\mathscr{L})^n_\alpha$ iff $(\alpha, n, \beta) \in (\mathscr{L})$. It is easy to realize that $M(\mathscr{L})$ is an $\mathscr{L}$-model and to show, by induction on the complexity of B, that, for every α, if $B \in \alpha$ then $\alpha \models B$. Now, if $B \in S(A)$ is *not* a thesis of $\mathscr{L}$, the set $\{\neg B\}$ is $\mathscr{L}$-consistent. We can construct an α with $\alpha_0 = \{\neg B\}$. So $\alpha \not\models B$.

REFERENCES

[1] S. Kripke, 'Semantical analysis of modal logic 1', *Zeitschrift für mathematische Logik und Grundlagen der Mathematik*, **9** 67–96, (1963).

[2] S. Kripke, 'Semantical considerations on modal logic', *Acta Philosophica Fennica* **16** 82–94 (1963).

[3] Dov M. Gabbay, 'Montague type semantics for modal logics with propositional quantifiers', *Zeitschrift für mathematische Logik und Grundlagen der Mathematik*, **17** 245–249, (1971).

[4] G. E. Hughes and M. J. Cresswell, *An Introduction to Modal Logic*. London, Methuen, 1968.

[5] K. Gödel, 'Eine Interpretation des intuitionistischen Aussagenkalkuels', *Ergebnisse eines mathematischen Kolloquiums*, Vol. 4, pp. 39–40. Vienna, Deuticke, 1933.

[6] M. C. Fitting, *Intuitionistic Logic, Model Theory and Forcing*. Amsterdam, North-Holland, 1969.

[7] L. E. J. Brouwer, 'Points and spaces', *Canadian Journal of Mathematics* **6** 1–17, (1954).

[8] Evert W. Beth, *The Foundations of Mathematics*. 2nd ed. Amsterdam, North-Holland, 1968.

[9] Dana Scott, *Completeness Proofs for the Intuitionistic Sentential Calculus*. Summaries of the talks presented at the Summer Institute for Symbolic Logic, Cornell University 1957. Princeton, Princeton University Press, 1960, pp. 231–241.

GISÈLE FISCHER SERVI

SEMANTICS FOR A CLASS OF INTUITIONISTIC
MODAL CALCULI

Most of the papers dealing with modal extensions of the intuitionistic propositional calculus (IC) have been written by R. A. Bull (see [1–3]). Since these papers appeared in print, there seems to have been little discussion of the subject. It might be argued that research on intuitionistic modal logic has received little impetus because there are difficulties in defining some plausible intuitive modal concepts for intuitionistic logic. Some progress might be made if we had, as in the classical case, a class of modal axiomatic systems which could be discussed and compared in terms of their semantical theories. In [4] we attempted to solve part of this problem by proposing a general criterion which enables us to define a rather large number of intuitionistic modal calculi. The idea that lies behind this criterion is that of presenting a uniform rule by means of which we can find the 'intuitionistic analogues' for some of the most usual classical modal systems.

The aim of the present paper is to give successful Kripke-type semantics for a class of intuitionistic modal calculi, obtained by means of the criterion mentioned above. Hence we shall concentrate on the methods employed in defining modal intuitionistic Kripke models with respect to which completeness results are forthcoming. Since the emphasis of the paper is on semantical considerations we will omit the completeness proofs, restricting ourselves to giving a general idea of the techniques used, when thought to be necessary.

Thus, in Section 1 we are concerned with the concepts involved in the definition of certain intuitionistic modal calculi. These systems are seen to be determined by a special class of theorems derived in the so-called 'bimodal calculi' (classical modal systems having two modal operators), and the type of definition used is shown to be applicable to a rather large number of cases. In Section 2, we analyze the interconnections between algebraic and Kripke-type model theory for bimodal calculi and obtain model theoretic characterizations of both kinds. Since the deductive structure of each intuitionistic modal calculus is defined through the auxiliary of a bimodal system, the Kripke-type semantics for the latter suggests how to find an adequate class of models for the former. Thus in Section 3 we are able to give a semantic counterpart to the original

59

Maria Luisa Dalla Chiara (ed.), Italian Studies in the Philosophy of Science, 59–72.
Copyright © 1980 *by D. Reidel Publishing Company.*

definitions of intuitionistic modal calculi and using bimodal semantics we prove a Kripke-type completeness theorem for a whole class of modal intuitionistic logics.

1. INTUITIONISTIC MODAL CALCULI

Let $(*)$-C be a modal system containing the full propositional calculus (C) as its base. We shall define its 'intuitionistic analogue' $(*)$-IC by means of a transform which assigns, to formulas from the language of $(*)$-IC, formulas of the bimodal calculus $(S^4, *)$-C. The particular concept of 'intuitionistic modal analogue' which emerges is rather natural if one bears in mind the well-known relationship between the intuitionistic propositional calculus and Lewis' S^4. In view of this result, it is reasonable to assume that a $(*)$-IC calculus could be correlated with a classical sentential calculus with two modal connectives, one being a sort of S^4 operator and the other a sort of $(*)$-C operator. Thus, let $(S^4, *)$-C be the *bimodal calculus* with $\neg$, $\rightarrow$, L_1, L_2 as primitive connectives (M_1 being defined by $\neg L_1 \neg$ and M_2 by $\neg L_2 \neg$), having the following axiom schemas and rules:

(b_0) *classical propositional base*, including Modus Ponens and the usual definition for $\wedge$ and $\vee$;

(b_1) S^4 *axiom schemas and rules on* L_1:

$$L_1\alpha \rightarrow \alpha,$$
$$L_1(\alpha \rightarrow \beta) \rightarrow (L_1\alpha \rightarrow L_1\beta),$$
$$L_1\alpha \rightarrow L_1L_1\alpha,$$
$$\frac{\vdash \alpha}{\vdash L_1\alpha}.$$

(b_2) $(*)$-C *axiom schemas and rules of* L_2;

(b_3) *connecting axioms*:

$$M_2L_1\alpha \rightarrow L_1M_2\alpha,$$
$$M_2M_1\alpha \rightarrow M_1M_2\alpha.$$

The two axioms in (b_3) have an intuitive meaning which we can exemplify once we abandon the rather abstract level of $(S^4, *)$-C calculi. Let us then consider an (S^4, S^5)-C system. As usual, we can imagine the collection of all possible worlds to be divided into equivalence classes by the S^5 accessibility relation. Thus in each equivalence class all worlds are

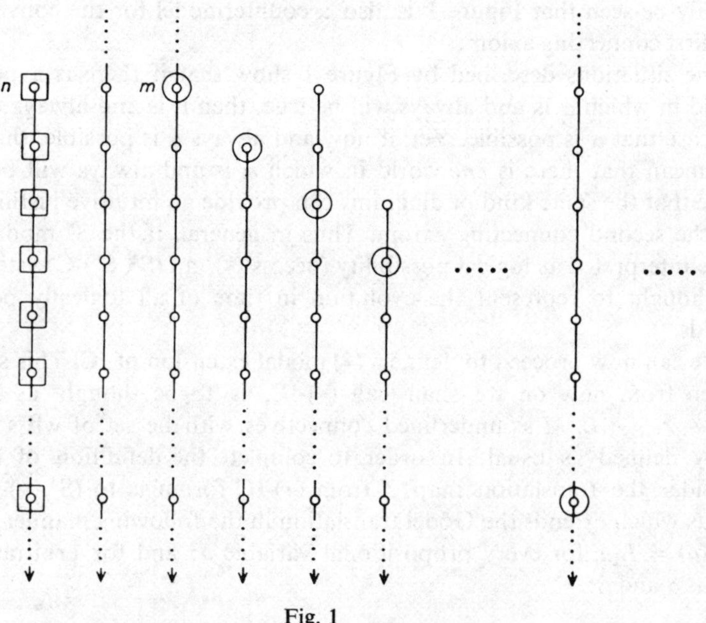

Fig. 1

accessible one to the other. Now let each world of each equivalence class determine a set of worlds ordered by an S^4 time relation, which for simplicity's sake we shall assume to be a linear order[1]. Assume, furthermore, that the future of two distinct worlds belonging to the same equivalence class cannot ever coincide. This situation can be described by Figure 1.

In Figure 1, we understand the arrows to indicate the direction of time and postulate that all worlds on the same horizontal belong to one equivalence class. Now let $L_1\alpha$ mean 'it is and always will be the case that α' and interpret $M_1\alpha$ as 'it is or could later become the case that α'. In order to understand the plausibility of the first axiom in (b$_3$), suppose that $M_2L_1\alpha$ is true in the world m. This means that there exists a time line determined by a world n such that n belongs to the same equivalence class as m and α is true in the present and all futures of n. Let us translate this hypothesis on Figure 1, by squaring all points in time in which α holds. Then it is immediate that $L_1M_2\alpha$ is also true in m. Vice versa, if the circled dots indicate the moments of time in which β is true, it can readily be seen that Figure 1 is also a countermodel for the converse of our first connecting axiom.

The situations described by Figure 1 show that if there is a possible world in which α is and always will be true, then it is and always will be the case that α is possible. Yet, if now and always α is possible, this does not mean that there is *one* world in which α is and always will be true. Note that the same kind of diagrams can provide an intuitive justification for the second connecting axiom. Thus in general, if the S^5 modality is to be interpreted as logical possibility (necessity), an (S^4, S^5)-C system can be thought to represent the evolution in time of all logically possible worlds.

We can now proceed to define a $(*)$ modal extension of IC. This system, which from now on we shall call $(*)$-IC, is to be thought as having $\neg, \vee, \wedge, \rightarrow, L, M$ as underfined connectives, with the set of wff's inductively defined as usual. In order to complete the definition of $(*)$-IC, consider the translation map T from $(*)$-IC formulas to $(S^4, *)$-C formulas which extends the Gödel translation in the following manner:

$T(a) = L_1 a$, for every propositional variable a; and for arbitrary formulas α and β:

$$T(\alpha \wedge \beta) = T\alpha \wedge T\beta,$$
$$T(\alpha \vee \beta) = T\alpha \vee T\beta,$$
$$T(\neg\alpha) = L_1 \neg T\alpha,$$
$$T(\alpha \rightarrow \beta) = L_1(T\alpha \rightarrow T\beta),$$
$$T(M\alpha) = M_2 T\alpha,$$
$$T(L\alpha) = L_1 L_2 T\alpha.^2$$

Then define the theorems of $(*)$-IC to be those formulas whose T-translates are theorems of $(S^4, *)$-C. In other words:

(A) $|_{\overline{(*)-IC}}\, \alpha$ iff $|_{\overline{(S^4,*)-C}}\, T\alpha.$

Let us introduce the following definitions: if α is a wff of the language of $(*)$-IC, α shall be called a *modal*, wff, while if α belongs to a bimodal calculus, we shall say that α is a *bimodal* wff.

Now, if (A) is to express a plausible interpretation of the concept of 'intuitionistic analogue' it is necessary that the $(*)$-IC calculi obtained through (A) satisfy the following conditions:

(a) $(*)$-IC is a subsystem of $(*)$-C,

(b) adding the excluded middle to $(*)$-IC yields a logic equivalent to $(*)$-C.

It is possible to show that there is a substantial class of classical modal calculi, whose intuitionistic analogues satisfy the properties required in (a) and (b). For instance let the $*$ in $(*)$-C range over the following class $\mathscr{C}$ modal calculi (considered in [5] and [6]): C2, D2, E2, T(C), T(D), E3, ET, E4, T, S^4, E2(S), EB, E5, B, S^5, PC, E, L. Proving that $(*)$-IC systems satisfy (a) requires a general type of argument. First let $C\alpha$ be the modal formula obtained from a bimodal wff α by collapsing the L_1, M_1 operators and replacing L_2, M_2 by L and M, respectively. Then it is easy to verify that for every axiom schema α of (S^4, $*$)-C, $C\alpha$ is a theorem of $(*)$-C and for each rule of inference α/α', if $C\alpha$ is a theorem of $(*)$-C, then $C\alpha'$ is one too. Hence, using criterion (A), we get $\vdash_{(*)-C} CT\beta$ for every $(*)$-IC theorem β. But induction on the complexity of formulas can be used to prove $CT\beta = \beta$ for every modal wff β. Consequently, every theorem of $(*)$-IC is also a theorem of $(*)$-C.

Condition (b), on the other hand, requires a separate proof for each $(*)$-C. What needs to be shown is that for each axiom schema β of $(*)$-C, β is a theorem of the system obtained by adding the excluded middle and the rules of $(*)$-C to $(*)$-IC. Criterion (A) suggests that we might prove a stronger result, viz.,

$$\text{if } \beta \text{ is a modal axiom of } (*)\text{-C, then } \vdash_{(S^4,*)-C} T\beta$$

and thus ultimately have β a theorem of $(*)$-IC alone. Actually in some cases, as for the characteristic B axiom, the stronger result does not hold and the addition of the excluded middle proves to be crucial for the proof of (b).

Note that all modal systems belonging to $\mathscr{C}$ contain the rule for the substitution of equivalents. Our conjecture is that this rule is needed to render criterion (A) plausible in the sense of (a) and (b). But so far we have not obtained any result which decides the question in one way or the other.

Finally it is interesting to note that the system MIPC, which R. A. Bull shows to be a sort of S^5 extension of the intuitionistic propositional calculus, not only satisfies (a) and (b), but actually coincides with the S^5-IC system obtained by means of criterion (A) (see [1] and [4]).

2. BIMODAL CALCULI

In the remaining part of this paper, $(*)$-C calculi will be understood to range over those modal calculi which have the rule of necessitation as a primitive rule and belong to $\mathscr{C}$, i.e. T(C), T(D), T, S^4, B, and S^5. The aim of this section is to describe the semantics of $(S^4, *)$-C calculi. Since the methods for handling these systems are identical we may consider a generic $(S^4, *)$-C calculus, define the concept of an $(S^4, *)$-C Kripke-type model and proceed to prove semantic completeness in such a general setting. This final result will be obtained by employing and extending Lemmon's techniques (see [5] and [6]) for deducing Kripke completeness theorems from algebraic results. Thus we first establish a correlation between $(S^4, *)$-C calculi and $(*)$-*bimodal algebras*. Second we define the concept of *bimodal model on a $(*)$-double model structure* $((*)$-d.m.s.$)$. Third we introduce two functions, one of which tells us how to construct a $(*)$-bimodal algebra from a $(*)$-d.m.s., and another which enables us to do the converse. Finally we state a representation theorem for $(*)$-bimodal algebras which yields the necessary link between algebraic and relational model theory. An immediate consequence of this central theorem is semantic completeness for $(S^4, *)$-C calculi.

Thus, let a *bimodal algebra* be a triple $\mathscr{B} = (B, K_1, K_2)$ such that B is a Boolean algebra, K_1 a topological closure operator on B, K_2 an operator on B satisfying the following conditions:

$$K_2 K_1 x \leqslant K_1 K_2 x, \tag{1}$$

$$(x \in B)$$

$$K_2 I_1 x \leqslant I_1 K_2 x, \tag{2}$$

where $I_1 x =_{\mathrm{df}} -K_1 - x$.

If (B, K_2) is of the variety required for algebraic completeness of a $(*)$-C calculus, $\mathscr{B} = (B, K_1, K_2)$ is said to be a $(*)$-*bimodal algebra*. Using the Lindenbaum matrix, it is easy to verify that the $(S^4, *)$-C theorems are just the bimodal wff's true[3] in all $(*)$-bimodal algebras.

On the other hand, a *double model structure* (d.m.s.) is a triple $\mathscr{M} = (S; R_1, R_2)$, where S is a non-empty set, R_1 and R_2 are relations on S such that R_1 is reflexive and transitive and

$$R_1 \circ R_2 \subseteq R_2 \circ R_1, \tag{3}$$

$$(\forall m, n, p \in S)(m R_1 n \mathbin{\&} m R_2 p \Rightarrow (\exists q \in S)(n R_2 q \mathbin{\&} p R_1 q)). \tag{4}$$

In addition we shall say that a d.m.s. $\mathscr{M}$ is a $(*)$-*double model structure* $((*)$-d.m.s.$)$ if R_2 is an accessibility relation known to be appropriate for $(*)$-C.

A *bimodal model v on a d.m.s. $\mathscr{M}$* is a function defined on all ordered pairs (a, m), where a ranges over arbitrary propositional variables, m over elements of S, whose values lie in the set $\{0, 1\}$.

Given a bimodal model v on $\mathcal{M}$, we can inductively define a value $v'(\alpha, m) \in \{0, 1\}$ for an arbitrary bimodal formula α. Thus we stipulate:

$$v'(a, m) = v(a, m), \qquad \text{if } \alpha \text{ is a propositional variable } a, \qquad \text{(i)}$$

supposing that $v'(\alpha, m')$ and $v'(\beta, m')$ have already been defined, for all $m' \in S$, then:

$$v'(\neg\alpha, m) = 1 \text{ iff } v'(\alpha, m) = 0, \qquad\qquad \text{(ii)}$$
$$v'(\alpha \to \beta, m) = 1 \text{ iff } v'(\alpha, m) = 0 \text{ or } v'(\beta, m) = 1, \qquad \text{(iii)}$$
$$v'(L_1\alpha, m) = 1 \text{ iff } v'(\alpha, m') = 1 \text{ for all } m' \in S \text{ such that} \qquad \text{(iv)}$$
$$m\,R_1\,m',$$
$$v'(L_2\alpha, m) = 1 \text{ iff } v'(\alpha, m') = 1 \text{ for all } m' \in S \text{ such that} \qquad \text{(v)}$$
$$m\,R_2\,m'.$$

We say that *a bimodal model v on a d.m.s. $\mathcal{M}$ verifies a bimodal wff α in m* iff $v'(\alpha, m) = 1$. Then, α is said to be *bimodally valid in $\mathcal{M}$* iff for every $m \in S$, α is verified in m by all bimodal models v on $\mathcal{M}$. Furthermore we shall say that α is *$(S^4, *)$-C valid* iff α is bimodally valid in all $(*)$-d.m.s.'s.

In order to grasp the meaning of (3) and (4), we may, as usual, interpret S as a set of possible worlds and think of mR_1n and mR_2n (with $m, n \in S$) as meaning that 'n is (1)-accessible to m' and 'n is (2)-accessible to m' respectively. If we represent by single arrows the (1)-accessibility relation and by double arrows the (2)-accessibility relation, then (3) and (4) simply say that Figures 2 and 3 can be completed to yield Figure 4.

It is interesting to note that if the (2)-accessibility relation is symmetric (as in the case of (S^4, S^5)-C and (S^4, B)-C), then (3) and (4) are equivalent. Consider Figure 5. It is obvious that in view of the symmetry of R_2, Fig. 5 can alternatively be interpreted as describing the completion, via broken lines, either of Fig. 2 (with m as a departure point) or of Fig. 3 (with n as a departure point). Consequently, (3) holds if and only if (4) does.

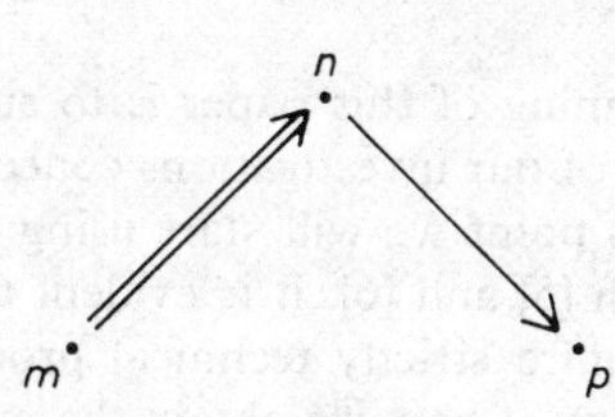

Fig. 2

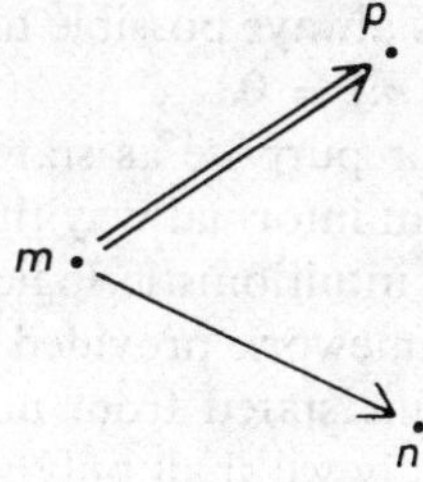

Fig. 3

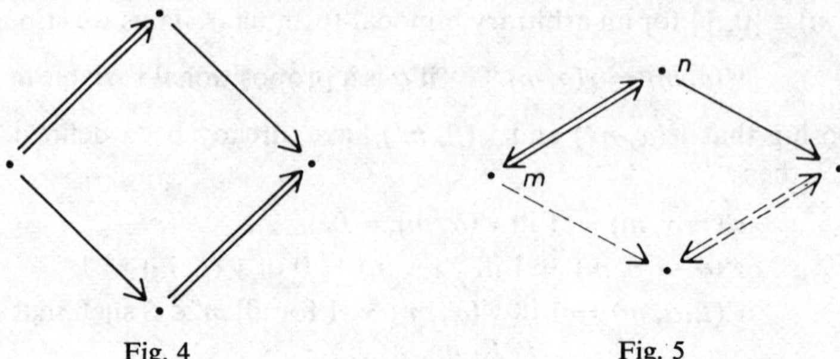

Fig. 4   Fig. 5

It is clear that our problem is to postulate a connection between relations R_1 and R_2 on S, which can adequately be described by the axioms in (b₃). Now, using the above diagrams, it can be shown that conditions (3) and (4) are just the relational properties that will serve our purpose. For, suppose that (3) and (4) hold in a d.m.s. $\mathcal{M} = (S, R_1, R_2)$ and that a bimodal model v on $\mathcal{M}$ verifies $M_2M_1\alpha$ in a world m. This implies that there are two worlds n and p belonging to S, which are related in the way indicated by Fig. 2 and such that $v'(\alpha, p) = 1$. But Fig. 4 tells us that there always is a world q such that mR_1q and qR_2p. Thus $v'(M_2\alpha, q) = 1$ and ultimately $v'(M_1M_2\alpha, m) = 1$. Conversely, suppose that Fig. 2 cannot be completed in the way indicated by Fig. 4. This means that p does not belong to the set of worlds r, (2)-accessible to worlds q, which are (1)-accessible to m. Thus it is possible to define a model v_0 which verifies the propositional variable a in p, but such that $v_0(a, r) = 0$ for every world r belonging to the set above. Hence the instance $M_2M_1a \rightarrow M_1M_2a$ of (b₃) is not bimodally valid in $\mathcal{M}$. Analogous methods can be used to show that if Fig. 3 is always completed to Fig. 4, then $M_2L_1\alpha \rightarrow L_1M_2\alpha$ is bimodally valid in $\mathcal{M}$ and vice versa, if Fig. 3 does not complete to Fig. 4, then it is always possible to chose v_0 and α in such a way that $v_0'(M_2L_1\alpha \rightarrow L_1M_2\alpha, m) = 0$.

Now, our purpose as stated in the beginning of this paper is to summarize in an informal way the main results of our investigations concerning modal intuitionistic logic. Since at this point we will start using the general framework provided by Lemmon in [5] and [6], it is evident that the theorems stated from now on will require strictly technical proofs. Consequently we shall only give a rough idea of how we obtain the most important results, postponing to another paper a more detailed discussion of the subject.

Before we state our main theorem, we introduce a function Φ which assigns to each bimodal algebra $\mathscr{B} = (B; K_1, K_2)$ a relational structure $\Phi(\mathscr{B}) = (S; R_1, R_2)$ constructed in the following way:

$$S \text{ is the set of all maximal filters of } B; \tag{5}$$

$$mR_1n \text{ holds iff } (\forall x \in B)(x \in n) \Rightarrow K_1(x) \in m), \tag{6}$$
$$(m, n \in S);$$

$$mR_2n \text{ holds iff } (\forall x \in B)(x \in n \Rightarrow K_2(x) \in m), \tag{7}$$
$$(m, n \in S).$$

Conversely, given any d.m.s. $\mathscr{M} = (S; R_1, R_2)$, we can form an algebraic structure $\Psi(\mathscr{M}) = (B; K_1, K_2)$ by defining:

$$B \text{ tobe the power set of } S \text{ with the usual set theoretic} \tag{8}$$
operations $\cup, \cap, -$;

$$K_1(X) = \{m \in S: (\exists n \in X)(mR_1n)\}, \tag{9_1}$$
$$(X \subseteq S).$$

$$K_2(X) = \{m \in S: (\exists n \in X)(mR_2n)\}. \tag{9_2}$$

Using these constructions we have the following:

THEOREM 1. *Every (∗)-bimodal algebra $\mathscr{B} = (B, K_1, K_2)$ is isomorphic to a subalgebra of $\Psi(\Phi(\mathscr{B}))$.*

The isomorphism in question is in fact the Stone isomorphism, i.e. the function $\varphi: B \mapsto \mathscr{P}(S)$ which assigns to each element x of B the set of maximal filters of B to which x belongs. The proof of Theorem 1, however, turns on the following:

$$\text{if } \mathscr{B} \text{ is a } (\ast)\text{-bimodal algebra, then } \Phi(\mathscr{B}) \text{ is a } (\ast)\text{-d.m.s.;} \tag{10}$$

$$\text{if } \mathscr{M} \text{ is a } (\ast)\text{-d.m.s., then } \Psi(\mathscr{M}) \text{ is a } (\ast)\text{-bimodal algebra.} \tag{11}$$

As in the case studied by Lemmon, the proof of (10) makes use of the properties of maximal filters. But it is worth mentioning that in order to prove properties (3) and (4) for $\Phi(\mathscr{B})$, we must in addition take into consideration the topological features of the Stone space S associated with the Boolean algebra B.

Now, it is possible to show that function Ψ preserves validity: *if $\mathscr{M}$ is a (∗)-d.m.s., then α is bimodally valid in $\mathscr{M}$ if and only if α is true in $\Psi(\mathscr{M})$.*

This fact together with Theorem 1 and algebraic completeness for $(S^4, *)$-C yields, through considerations similar to the ones made in [6], pp. 61, 62, the desired result:

THEOREM 2. *A bimodal formula is a thesis of* $(S^4, *)$-C *if and only if it is* $(S^4, *)$-C *valid.*

3. KRIPKE SEMANTICS FOR $(*)$-IC CALCULI

In this section, the bimodal semantic concepts introduced in Section 2 are used to define Kripke-type models for modal intuitionistic calculi. This is not surprising since each $(*)$-IC is determined, via translation T, by $(S^4, *)$-C. Thus it can be expected that there is, in correspondence to translation T, a semantic relation which yields the necessary link between $(S^4, *)$-C and $(*)$-IC Kripke models. So first we define the concept of an *intuitionistic model* on a double model structure. Second we describe the precise connection between bimodal and intuitionistic validity in a $(*)$-d.m.s. Finally, using completeness results for $(S^4, *)$-C, we establish that the set of theses derived in $(*)$-IC is coextensive with the set of $(*)$-IC valid wff's.

Thus, let $\mathcal{M} = (S; R_1, R_2)$ be a d.m.s. An *intuitionistic model* w on $\mathcal{M}$ is a function defined on all pairs (a, m), where a is a propositional variable, m an element of S, whose range is $\{0, 1\}$ and which satisfies the following condition:

$$\text{if } w(a, m) = 1, \text{ then } w(a, m') = 1 \text{ for every } m' \in S \quad (12)$$
$$\text{such that } mR_1m'.$$

For each intuitionistic model w, let $\bar{w}$ be the unique function which extends w, with values in $\{0, 1\}$ and such that, if α, β are modal wff's and $m \in S$, then:

$$\bar{w}(\neg\alpha, m) = 1 \text{ iff } \bar{w}(\alpha, m') = 0 \text{ for every } m' \in S \text{ such} \quad \text{(i)}$$
$$\text{that } mR_1m';$$

$$\bar{w}(\alpha \vee \beta, m) = 1 \text{ iff } \bar{w}(\alpha, m) = 1 \text{ or } \bar{w}(\beta, m) = 1; \quad \text{(ii)}$$

$$\bar{w}(\alpha \wedge \beta, m) = 1 \text{ iff } \bar{w}(\alpha, m) = 1 \text{ and } \bar{w}(\beta, m) = 1; \quad \text{(iii)}$$

$$\bar{w}(\alpha \rightarrow \beta, m) = 1 \text{ iff } \bar{w}(\alpha, m') = 0 \text{ of } \bar{w}(\beta, m') = 1 \text{ for} \quad \text{(iv)}$$
$$\text{every } m' \in S \text{ such that } mR_1m';$$

$$\bar{w}(M\alpha, m) = 1 \text{ iff there is an } m' \in S \text{ such that } mR_2m' \quad \text{(v)}$$
$$\text{and } \bar{w}(\alpha, m') = 1;$$

$$\overline{w}(L, \alpha, m) = 1 \text{ iff } \overline{w}(\alpha, m'') = 1 \text{ for every } m'' \in S \text{ for} \qquad \text{(vi)}$$
$$\text{which there is an } m' \in S \text{ with } mR_1m'$$
$$\text{and } m'R_2m''.$$

We say that *an intuitionistic model w on a d.m s. $\mathcal{M}$ verifies α in m* iff $\overline{w}(\alpha, m) = 1$. Then, α is *intuitionistically valid in a d.m.s. $\mathcal{M}$* iff for every $m \in S$, α is verified in m by all intuitionistic models w on $\mathcal{M}$. Finally. α will be said to be *(∗)-IC valid* iff α is intuitionistically valid in all (∗)-d.m.s.'s.

At this point it might be useful to describe a class of (∗)-d.m.s.'s. Although this class does not cover all possible cases of (∗)-d.m.s.'s, it is helpful to have a substantial set of examples that are readily available. Thus, consider two structures (X, R') and (I, R''), where X and I are sets, R' is a reflexive and transitive relation on X and R'' is a (∗) relation on I. Define S to be a subset of the Cartesian product of X and I such that[4]

$$\text{if } (x, i) \in S \text{ and } xR'y, \text{ then } (y, i) \in S. \qquad \text{(13)}$$

Thus S is made up of all final segments (modulo R') of X. Now, let R_1, R_2 be the two relations on S defined by:

$$(x, i)\, R_1\, (y, k) \quad \text{iff } xR'y \text{ and } i = k; \qquad \text{(14)}$$
$$(x, i)\, R_2\, (y, k) \quad \text{iff } x = y \text{ and } iR''k. \qquad \text{(15)}$$

One can easily verify that $(S; R_1, R_2)$ is a (∗)-d.m.s. Consider, for instance, the characteristic condition (3). By hypothesis we have Figure 6. From (14) it is clear that $xR'y$ and thus by (13) $(y, i) \in S$. Moreover, (y, i) is just the element which allows us to complete Fig. 7 in the way described by Fig. 5. From now on we shall call this type of (∗)-d.m.s. a *Cartesian* (∗)-d.m.s.

That not all d.m.s.'s are Cartesian is easily seen once we consider Figure 7. Nonetheless, if we want to analyze the modal intuitionistic concepts that emerge from conditions (i)–(vi), it is helpful to bear in mind these particular structures. Such a context suggests, in fact, that the above

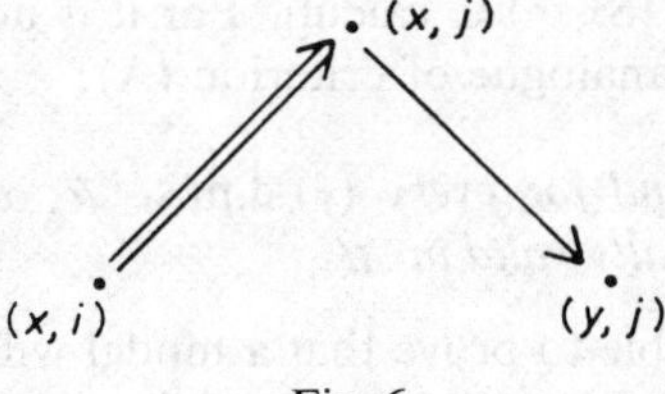

Fig. 6

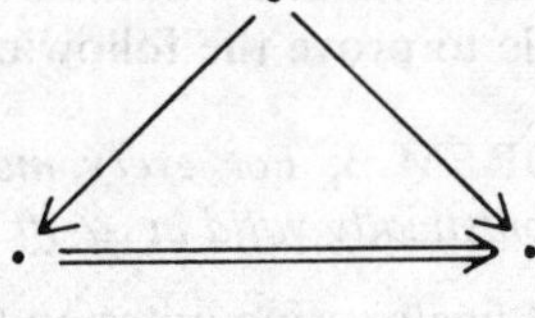

Fig. 7

semantics could be interpreted as involving an objective possibility for individuals to share and compare mental experiences. To see this, interpret the R' relation as ordering all conceivable states of consciousness of an idealized individual X and think of R'' as a relation on a set I of individuals with mental structures that are analogous to those of X. We could be more precise and say that each individual $i \in I$ participates more or less completely in a common nature which is given by the mental development of our prototype X. Now, participation contains a factor of identity as well as a factor of non-identity. Cartesian d.m.s.'s imply in fact that in each individual i, each mental state (x, i) 'corresponds' to a well-determined conceivable state of consciousness x of our prototype X, while the type of information acknowledged in (x, i) may be quite different from the one accepted in (x, j), $(i, j \in I)$. In particular if $M \alpha$ is to be acknowledged by an individual in a state of consciousness m, then there must exist another individual (maybe only himself) who acknowledges in the 'corresponding' mental state m' that α is true. On the other hand, if an individual i accepts $L \alpha$ in a state of consciousness m, then α must be thought to be true by all other individuals in all states of consciousness which 'correspond' to the mental states that i will develop from stage m on.

Let us proceed to describe in a precise fashion the connection which can be seen to exist between bimodal and intuitionistic validity in (∗)-d.m.s.'s. It turns out that given an intuitionistic model w on a (∗)-d.m.s. $\mathcal{M}$, we can define a bimodal model v on $\mathcal{M}$ such that

$$\bar{w}(\alpha, m) = v'(T\alpha, m) \qquad (\alpha \text{ a modal wff, } m \in S). \qquad (16)$$

In fact the bimodal model v defined by $v(a, m) = w(a, m)$ can be proven, by induction on the height of α, to satisfy (16). Conversely, if $\mathcal{M}$ is as above and v is a bimodal model on $\mathcal{M}$, then by putting $w(a, m) = v'(Ta, m)$, we can prove that w is an intuitionistic model on $\mathcal{M}$ satisfying condition (16).

Using this piece of information, we can finally reveal the model-theoretic connection between (∗)-IC and $(S^4, ∗)$-C calculi. For it is now possible to prove the following semantic analogue of criterion (A):

THEOREM 3. *For every modal wff α and for every (∗)-d.m.s. $\mathcal{M}$, α is intuitionistically valid in $\mathcal{M}$ iff $T\alpha$ is bimodally valid in $\mathcal{M}$.*

And finally using criterion (A) we are able to prove that a modal wff is (∗)-IC valid if and only if it is a theorem of (∗)-IC. Let α be a theorem of

(∗)-IC. Then criterion (A) tells us that $T\alpha$ is a theorem of (S⁴, ∗)-C, hence $T\alpha/$is bimodally valid in all (∗)-d.m.s.'s. By theorem 3, this entails in turn that α is intuitionistically valid in all (∗)-d.m.s.'s, i.e. α is (∗)-IC valid. The other way round is established in a similar manner and thus:

THEOREM 4. *The class of* (∗)-IC *valid formulas coincides with the class of theorems of* (∗)-IC.

It is interesting to note that the models with respect to which Bull obtains completeness for MIPC (our S⁵-IC) belong to the class of models on Cartesian (S⁵)-d.m.s.'s. It follows that Bull's models form a proper subclass of our models on (S⁵)-d.m.s.'s.

The fact that there are two classes of structures with respect to which an S⁵ version of Theorem 4 holds is slightly surprising. It seems that the situation is formally similar to the one revealed by Kripke for intuitionistic logic, viz. that models on tree model structures are sufficient to obtain completeness for IC. In our case we do not have as yet a method for transforming models on (S⁵)-d.m.s.'s in the models described by Bull, but Theorem 4 and Bull's completeness result imply that the class of models on (S⁵)-d.m.s.'s can be represented by one of its proper subclasses.

These results, however, are parenthetical to the main theme of this paper, which was to find a suitable Kripke-type modelling for a whole class of intuitionistic modal calculi.

NOTES

[1] Note that we are not interested here in giving a precise description of (S⁴, S⁵)-C models. We merely want to suggest an intuitive interpretation of the connecting axioms. Thus we take the liberty of making an additional assumption (linear order) even though it is known that in this particular case, the class of valid formulas coincides with the set of theorems of a system stronger than S⁴.

[2] Since modal operators bear some structural similarity to quantifiers, the lack of duality between L and M corresponds to the same characteristic feature of the intuitionistic quantifiers.

[3] We assume the usual definition of truth in an algebraic structure.

[4] If the (∗) relation is one known to be adequate for the system T(D), it is necessary to add the following: if $(x, i) \in S$ and $iR''k$ then $(x, k) \in S$.

REFERENCES

[1] R. A. Bull, 'A modal extension of intuitionistic logic', *Notre Dame Journal of Formal Logic* **6**, 2, 142–146 (1965).

[2] R. A. Bull, 'MIPC as the formalization of an intuitionistic concept of modality',
J. Symb. Logic **31**, 4, 609–616 (1966).

[3] R. A. Bull, 'Some modal calculi based on IC', in *Formal Systems and Recursive Functions*, edited by J. N. Crossley and M. A. E. Dummett. Amsterdam, North Holland,
1966.

[4] G. Fischer Servi, 'On modal logic with an intuitionistic base', *Studia Logica* **36**, 2
(1977).

[5] E. J. Lemmon, 'Algebraic semantics for modal logics, I', *J. Symb. Logic* **31**, 1, 46–
65 (1966).

[6] E. J. Lemmon, 'Algebraic semantics for modal logics, II', *J. Symb. Logic* **31**, 2, 191–
218 (1966).

'SINCE', 'EVEN IF', 'AS IF'

1. A persistent source of difficulty in analyzing conditionals lies in the fact that they are asserted against a variable background of knowledge which, even if it is not part of what is being asserted, is partially reflected in the grammatical form of the conditional itself. This paper will examine the difficulties arising from formalizing propositions (usually classified as conditionals in the literature on the subject) introduced in ordinary language by the locutions 'since', 'even if', 'as if'. As will become clear, the problems posed in the formal analysis of these sentences depend to a great extent on the fact that it is difficult to distinguish what is being asserted with sentences in this form from what is presupposed by their use. The analysis which will be put forward for these conditionals in the following pages will in no way contradict the theory, well known before the sixties, that a conditional is true if and only if there is a certain connection between protasis [antecedent] and apodosis (consequent) such that the apodosis can be inferred from the protasis accompanied by a *ceteris paribus* clause. However, this theory will not be discussed here and will be taken for granted in the course of the paper.

2. It is important to make clear at once that what is meant by 'presupposition' in the following pages does not correspond to a generic notion of 'pre-existing knowledge' and above all has nothing in common with the various forms of logical presupposition characterized by contemporary logic. A satisfactory definition of nonlogical presupposition was recently put forward by R. Stalnaker in (1973), but for our purposes it will suffice to adopt, with some modification, A. Pap's definition of *causal* presupposition of (1958):

A proposition q is causally presupposed by an assertion of proposition p if p would not have been asserted unless q had been believed (p. 203).

This definition is adequate for our purposes only if it is corrected in this way:

A proposition q is causally presupposed by a given sentence S if S would not have been used unless q had been believed.

73

Maria Luisa Dalla Chiara (ed.), Italian Studies in the Philosophy of Science, 73–87.

The difference between Pap's formulation and the one put forward here can be clearly seen at once in the logical analysis of conditionals. An indicative conditional and a subjunctive constitute invariably different sentences, but if the only difference between them is the verbal mood, they express the same proposition and therefore are to be formalized in the same way. The difference between verbal moods, in fact, regards the sentence and not the proposition since it reflects at the most the existence of different presuppositions regarding the truth of clauses. However, as to exactly what presuppositions are indicated by the grammatical form of the conditional, logicians are very uncertain. From here on in we shall adopt the convention of calling the conditionals of which speaker and listener know that the protasis is false, *counterfactuals*; the conditionals of which they know that the protasis is true, *profactuals*; and the conditionals of which they do not know the truth value of the protasis, *afactuals*. According to Stalnaker, past subjunctive conditionals "presuppose the falsity of the antecedent and perhaps also of the consequent" (1973, pp. 447–48): hence they would all be counterfactuals. But this observation contradicts the fact that there are past subjunctive conditionals which are not counterfactuals, such as those sometimes used by detectives. In fact, in a 'detective' sentence such as 'If Jones had swallowed strychnine, he would show the exact symptoms he is showing now,'[1] the consequent of the conditional is asserted as being true and it can be presumed that the premises for inferring that the antecedent is also true are actually set. In conditionals, then, the use of verbal moods is not such that it reflects univocally what is known about the truth value of clauses. In Table I

TABLE I

True	Indeterminate	False
Indicative	Indicative or Subjunctive	Subjunctive

the verbal mood used in the presence of belief in question is placed beneath the indication of what is known about the truth of the protasis. Generally it cannot be determined from the fact that the speaker uses the subjective in the protasis of the conditional that he knows that the protasis is false, even if it can be excluded that he holds it to be true. An analogous holds true, *mutatis mutandis*, for the use of the indicative. Verbal moods, then, are not in and by themselves sufficient to determine whether the use of a sentence presupposes one thing rather than something else; to characterize presuppositions we must therefore consider, in addition to the verbal mood, the meaning of the propositions expressed and eventually other grammatical characteristics such as verb tense, adverbial particles, and so on.

The preceding observations allow us to question a remark in Goodman (1947) where he maintains that the conditional

(°)   If that piece of butter had been heated to 150°F., it would have melted

is the same as

(°°)   Since that butter did not melt, it wasn't heated to 150°F.

The *prima facie* difference between (°) and (°°) lies in the fact that while (°) is compatible with sentences such as 'it is not known as to whether that piece of butter which was in the refrigerator has been heated', this cannot be said of (°°), which clearly suggests that the butter has not melted and has not been heated. But it seems that the difference does not rest solely in the pre-existing knowledge known to the one who utters one or the other of the two sentences. If Goodman maintains logical equivalence between (°) and (°°), the reason lies probably in his conviction that (°°) is the contrapositive of (°), and therefore is equivalent to the standard contrapositive of (°), that is, 'if it were true that that piece of butter has not melted, it would be true that it has not been heated to 150°F.' But this last sentence, if we accept the equivalence posited by Goodman between 'if it were true that p, it would be true that q' and 'since non-q, non-p', expresses the same proposition of 'since that piece of butter has been heated it has melted': it is difficult, however, to maintain that this sentence expresses a proposition equivalent to that expressed by (°°) and above all by (°).

In correctly analyzing since-conditionals, which Goodman calls *factual* conditionals, the fundamental problem to solve is substantially this: through the use of the particle 'since', is the presupposition indicated that the clauses are true, or is it asserted that the clauses are true in conjunction with the conditionals of which they are part?[2]

In the remainder of this paper I shall symbolize with $\Box Pq$ any conditional with the protasis p and apodosis q, and I shall indicate with $(r, s)\Box Pq$ a conversational situation in which $\Box Pq$ is asserted using a sentence which presupposes r and s. Now if we must adhere to what Ramsey wrote in (1931), 'since p, q' is simply a variant of 'if p, q' when it is known that p is true; and inasmuch as 'since' is used only when the protasis is known to be true, we conclude from this that 'since' indicates that the protasis is known to be true, and thus, following Ramsey, 'since p, q' would correspond to $(p)\Box Pq$. But this formula is suitable also for yielding those profactual conditionals in which the use of the indicative mood and the meaning of what is being asserted are such that they indicate that the protasis is known to be true: for example, 'If I got to the top of this mountain in this weather, anyone can get here.' Ramsey's proposal thus tends toward the identification of since-conditionals and profactuals of the kind mentioned above. Nevertheless it is necessary to observe a subtle but perceptible difference between these two kinds of conditionals. In fact, a person who said, 'Anyone can get to the top of this mountain since I got to it in this weather', could be contradicted by an interlocutor who objected: 'but you have *not* yet reached the top of this mountain.' This objection is nevertheless ineffective against the above-mentioned profactual. Whoever uttered the latter can simply defend himself by correcting not the proposition but simply the grammar of the sentence and saying, for example, 'if I *had got* to the top of the mountain in this weather, any one could get there.' This sentence, however, does not express a different proposition from the preceding sentence, but unlike the first, simply suggests that the speaker is presupposing what corresponds to the truth (and that is that he has not reached the top of the mountain).

The foregoing leads us to think that in the since-conditional, unlike the profactual of the kind seen above, there is expressed, in addition to the conditional true and proper, the fact that the protasis and apodosis are both true: in symbols, $p \wedge q \wedge \Box Pq$. A proposition equivalent to 'since p, q' might thus be expressed also by 'if it is true, as it is true, that p, it is true that q, and it is true that q.'

A counterproof of the correctness of this analysis can be had by trying to characterize the behavior of a logical operator which symbolizes 'since' (which we can render with the symbol ' $\rightarrowtail$ '). We realize right away that $p \rightarrowtail p$ cannot be a logical truth in that it turns out to be intuitively false when it is false that p (for example, think of '2 + 2 = 5 since 2 + 2 = 5'), and it can be verified that many simple logical properties cannot be attributed to this operator whose behavior therefore does not correspond to that of the conditional operator.

The problem of establishing what is meant by denying a since-conditional confirms the complex character of propositions of this type which share the characteristics of conditional as well as non-conditional propositions. What is meant, for example, by a sentence such as 'is it not true that it is raining since there has been a long drought?' Is one denying that it is raining while being willing to admit that an eventual rainfall would be attributed to a long drought? Or is one denying that the fact that it is raining can be inferred from the drought or is one denying that there has been a long drought? This disparity of interpretations is not surprising if we accept the hypothesis advanced above concerning the meaning of 'since'. In fact in such a case the negation of '$p \dashrightarrow q$' is the same as the disjunction $\sim \Box Pq \lor \sim p \lor \sim q$, which holds true for 2^3-1 assignments of value to its disjuncts.

An interesting problem regarding the since-conditional is not only its relation to simple conditionals but its relation to propositions expressed by sentences such as 'p, and therefore q' and 'p, and in fact q' (which we are assuming for our purposes as being equivalent without further discussion). In sentences in this form, as in since-conditionals, one also asserts that p and q are true, but the problem here is whether by means of 'and in fact' one asserts or simply presupposes also the conditional with antecedent p and consequent q. I know of no conclusive evidence in favor of the symbolization $(\Box Pq)p \land q$ for 'p, and in fact q' instead of $\Box Pq \land p \land q$. An argument in favor of the first is perhaps that there are ironic nuances in 'in fact' (for example, 'Rossi was treated by Dr. Fiorelli and in fact he is much worse now.') which are lost when these sentences are transformed into since-conditionals: and the irony arises from the fact that 'in fact' suggests the presence of a connection between stated facts, while this connection is made very explicit by the use of 'since.'

3. In (1973), J.L. Mackie proposed classifying some sentences beginning with the words 'even if' among what he calls *factual* conditionals. This proposal may seem reasonable if one thinks of an example such as 'even if America was discovered by Columbus, the name comes from Vespucci.' Through sentences such as these one does not presuppose, but one asserts, that the clauses are true and therefore it is justifiable to stress their analogy with since-conditionals. However, it is difficult to render the meaning of 'even if' formally in this accepted meaning, since 'even if' has the same value as 'notwithstanding', 'in spite of', or simply 'but'. More precisely, 'even if p, q' in this accepted meaning is not so much the same as 'q, but p' as 'p, but q'; and 'p, but q' indicates that it is true that p and it is true that q when one knows that, knowing p, it would be reasonable to expect that $\sim q$ is true. In order to render the conversational situation corresponding to 'p, but q' one would think that $(\sim \Box Pq)p \land q$ is sufficient, which constitutes the formula obtained from that proposed

above for 'q, in fact p', denying the presupposition $\Box Pq$. But this proposal is clearly inadequate. The fact that we are now in 1977 cannot be inferred from the fact that the pyramids are in Egypt but this does not authorize us to make assertions such as 'even if the pyramids are in Egypt we are in 1977.' The use of 'but' or 'notwithstanding' in the preceding case in fact indicates the presupposition that from the fact that the pyramids are in Egypt it is reasonable to expect that we are not in 1977. It is very difficult to render the notion of 'it is reasonable to expect p' unless it is reduced to some proposition containing operators (moreover not yet treated formally in logical literature) such as 'almost always', 'very probably', and so on.

It is interesting to observe that 'even if' has the same meaning as 'notwithstanding that' for reasons having nothing whatsoever to do with the presence of the indicative mood in the clauses. In fact, the indicative can be used with 'even if' in contexts in which this expression does not have the same meaning as 'notwithstanding that' or 'but'. This depends on the fact that the indicative, as is apparent in Table I, is used correctly even in the presence of uncertainty surrounding the truth value of the protasis. For example it could be said of a match that one believes that it is in another room and one does not know whether it has been struck or not: 'Even if that match has been struck, it did not ignite.' The conditional of the above cannot imply the truth of 'the match did not ignite' which would happen if 'even if' meant 'notwithstanding'.

Since the even-if sentence formulated above could be expressed just as well using the subjunctive in the protasis and the conditional in the apodosis, we are clearly in the presence of the propositions Goodman in (1947) called *semifactuals*. Since Goodman calls propositions which we have rendered with $p \wedge \Box Pq \wedge q$ *factuals*, his terminology makes us think that the *semi*factual, obviously excluding $\Box Pq \wedge p$, is expressed by $\Box Pq \wedge q$. This formulation is exactly the one proposed for 'even if' conditionals by J.L. Pollock in (1975). But however much the 'even if' conditional may be asserted in the vast majority of cases in the presence of the fact that the consequent is known to be true, the example given above, and others of the same kind, show that it does not indicate this presupposition.

What is the correct formalization of 'even if' conditionals? According to N. Goodman and R. Chisholm in their first works on counterfactuals in 1946–47 an 'even if' conditional does not assert a connection between protasis and apodosis but denies a connection between the protasis and the negation of the apodosis. According to R. Stalnaker this would be a "baldly *ad hoc* maneuvre" to salvage the theory according to which a conditional expresses an asymmetrical connection between the two clauses (see (1975), p. 174). Semifactual conditionals, to be sure, present no problem for Stalnaker's theory and for possible worlds semantics in general, which Stalnaker and later David Lewis associated with conditional logic. By virtue of this semantics, in fact, a conditional is true at a possible world if and only if the consequent is true at all the worlds (for Stalnaker only one) which are the value of a selection function whose arguments are the antecedent and the given world at which the conditional is asserted to be true. By virtue of this theory the truth conditions for a conditional do not depend on the fact that the protasis is relevant for the apodosis, and still less on the fact that it is accompanied by the word 'even'. Like Stalnaker I am convinced that the word 'even' has nothing to do with what is being asserted but, if anything, with what is being presupposed; all the same, I do not maintain that it is necessary for this reason to abandon the theory that a conditional expresses a certain connection between the two clauses.

First it is well to stress again what has emerged from the foregoing analysis of 'even if' indicatives: that is, that the use of the word 'if' does not in itself guarantee that the molecular sentence in which it appears expresses a conditional. Another example is provided by sentences, very common in Italian, of the kind: 'if it is true that it rained today, it is also true that the weather cleared up'; these sentences express pure adversative conjunctions and not conditionals true and proper. Other idiomatic forms of the 'if ... then' kind cannot be easily classified as conditionals without periphrases or complements: 'if I have to be sincere, I was bored at the concert'; 'if you want cigarettes, they are in their place'; 'if the Devils win, I am a Dutchman'. On the other hand, it is true that in everyday language authentic conditionals often have the exact 'if-then' form: for example, 'if I were he, then I would telephone'.

With regard to 'if-then' sentences the problem which seems most difficult and most important is that of establishing whether they express a conditional true and proper or whether they are used to make the weaker assertion that the negation of the apodosis does not follow from the protasis. I shall call these propositions, which can be represented formally as $\sim \Box p \sim q$ or $\Diamond pq$, *semiconditionals*. This difficulty stands out clearly when we try to formalize an 'if-then' sentence in which a negating adverb appears after 'then'. To clarify the source of the difficulty it is appropriate to recall briefly the peculiarity of the negation connective. By now there is increasing common acceptance of the distinction between an exclusive or 'external' negation by means of which it is excluded that the proposition which is being denied is true, and an 'internal' negation, which R.H. Thomason in (1972) proposed to call *choice negation*, in which only the predicate is denied. While the distinction in a propositional logic can be outlined only by introducing two kinds of negation and therefore adopting a trivalent semantics, in a first-order language, the distinction can be made more simply by discriminating, using adequate formation rules, between $\sim(Px)$ (external negation and $(\sim P)x$ (internal negation). In sum, the negation behaves as a modality which can be *sensu composito* or *sensu diviso*. This distinction is useful in analyzing conditionals. The most recent investigations into conditionals, arising principally from the work of R. Stalnaker and D.K. Lewis, lead one to conclude that conditionals are propositionally indexed modalities and that in them it is asserted that the consequent is necessarily true *relative* to the antecedent, which can thus be interpreted as an index (not necessarily the only one) with respect to which the consequent is asserted as necessary. Now let us introduce a diadic operator R of the kind proposed by A.N. Prior (see, for example, his (1968)) such that, if m is a possible world, Rmp is read as 'p is true in m'. By means of mp let us symbolize the notion of 'world selected with respect to m and p, in which p is true'. According to Stalnaker's theory it is then plausible to regard a conditional $\square Pq$ as something equivalent to $\forall mP\, RmPq$. Since $\forall mP RmPq \exists mP RmPq$ but not the converse must hold true for first-order logic, one concludes, in the light of the obvious equivalence in bivalent R-systems between $Rm\sim p$ and $\sim Rmp$, that $\square Pq \sqsupset \sim \square P \sim q$ must be a logical truth for conditionals but not the converse. In (1976) I presented a system where the first formula, which I call "Boethius' conditional thesis", is a thesis but not the second. In Stalnaker's system, on the other hand, the first formula holds true only on the condition that the protasis is consistent while the converse, which D.K. Lewis has called "conditional excluded middle", is a thesis and describes the highly improbable semantic presupposition that there is only one selected world – or as Lewis says, a 'more similar' one with respect to the given world. If we accept the law of the conditional excluded middle together with Boethius' conditional thesis, the $\square Pq$ conditional collapses on the semiconditional $\sim\square P\sim q$, and thus $\sim\square Pq$ collapses on $\square P\sim q$. Thus there is no difference between denying a conditional and denying its consequent. If we want to consider, somewhat contrivedly, the conditional's consequent as one of the antecedent's predicates, this is the same as dropping every distinction between *choice negation* and *exclusion negation*, while this difference can be perceived in all systems where the law of the conditional excluded middle does not hold.

The preceding considerations allow us to analyze 'even if' conditionals with precision, beginning with the example proposed by Pollock in (1975): 'even if the witchdoctor were to do a rain-dance, it would not rain'. Here the difficulties of formalization do not depend on the presence of the word 'even', which affects not what is being asserted but what is being presupposed, but on the fact that the adverb of negation occurring in the sentence is ambiguous. The question is whether it must be understood as expressing a negation *sensu composito* ('if the witchdoctor were to do a rain-dance, *it is false* that it would rain', and thus '*it is false* that if the witchdoctor were to do a rain-dance it would rain') or *sensu diviso* ('if the witchdoctor were to do a rain-dance *it would be false* that it rains'): the difference is between $\sim \Box P q$ and $\Box P \sim q$, which, in the light of what was said above, can be plausibly considered logically independent. Now the choice between these two alternative ways of formalizing should be made by looking not at the superficial form of the sentence but at the speaker's intentions which very often emerge only from an analysis of the context. If I know from the context that the rain-dance has no influence on climatic conditions I will interpret the negation as negation *sensu composito* and therefore I will use the $\sim \Box P q$ formula. But let us move to a more far-fetched context in which the witchdoctor's rain-dance is in reality a signal for a meteorological warfare station to bring about a long period of drought in the area. Then the negation is understood *sensu diviso* and the sentence is formalized as $\Box P \sim q$.

It may seem that the distinction between conditionals and semiconditionals is possible when an adverb of negation appears in the sentence to be formalized. But observe the difference in meaning between two superficially similar sentences as 'if the bus had had more gas, it would have continued its run' and 'if the seas were more salty, the stars would continue in their course'. In most usual contexts the first sentence expresses a conditional, while the second expresses a semiconditional. That there is no adverb of negation in the sentence to be formalized in both cases is not in itself important. A sentence such as 'Leo is unhappy' presents no adverb of negation and yet it maximizes the discriminating capacity of formal language by rendering it as $\sim p$ (where p stands for 'Leo is happy') rather than with a simple atomic variable; in the same way, 'X continues to do

A' is better formalized as $\sim p$ (where p stands for 'X stops doing A') than by a simple atomic variable. The usefulness of this procedure is evident in conditionals where many ambiguities are eliminated by placing the negation outside of or inside the complex formula.

Contrary to what one might think, the use of the adverb 'even' before a conditional does not resolve the ambiguities of 'if then' sentences but, if possible, introduces others. For example, let us take a sentence such as 'Even if the other pilot had been flying the plane, it would have crashed'. What exactly is being asserted with a sentence like this? Is something being asserted which could be rendered equally well by 'not even if the other pilot had been flying the plane would the plane crash have been avoided', or is it something which could be expressed otherwise by 'It is even true that if the other pilot had flown the plane, it would have crashed'? The first paraphrase is convincing on a background of contexts in which 'normally' something of this kind is being asserted: the plane suffered damage caused by neither of the two pilots and it was impossible to repair it during the flight. But there could be another context: the plane crashed because of the pilot's inability to control a critical situation and the other pilot on board was more incapable than the first. Obviously in the first case I am asserting a semiconditional and in the second a conditional true and proper.[3]

This analysis of 'even if' leads us therefore to the conclusion that it is inexact to assert that the supporters of the connection theory are forced to use an *ad hoc* expedient to formalize 'even if' conditionals. The truth is that in general sentences of the form 'if p, then q' have no univocal meaning in ordinary language, above all when they are preceded by the word 'even'. Naturally we can ask ourselves why ordinary language prefers to use the form 'even if p, q' when it would be less ambiguous to use 'not even if p, not q'. The study of these idiosyncrasies, however, concerns more linguistics than logic; in this regard it will simply be enough to remember that the problems of formalization analogous to those presented by conditionals appear in other branches of modal logic. If 'F' stands for the temporal monadic operator for the future such that Fp is read as 'it will be true that p', it is very often only the context which helps us understand whether expressions of the type 'it will be true that p' correspond to the form Fp or to the form $\sim F \sim p$ (for example, think of 'the earth will continue to turn'). The more serious objection which can be raised against this interpretation of 'even if' perhaps is another, namely that there is no reason to follow David Lewis suggestion to read what I call semiconditionals (and which Lewis calls *might-counterfactuals*) as 'if p were true, it *might* be the case that q is true'. Even if it cannot be denied that this reading often effectively yields formulas of the $\sim \Box \, P \sim q$ form, it seems that in many other cases the formalization provided by $\Box P \Diamond q$ or by $\Diamond \Box Pq$ is more suitable for these propositions. In my opinion L. Åqvist is much closer to the truth in this matter when, defining $\Box \, Pq$ as $\boxdot \Box (*p \sqsubset q)$ (where $*$ is an operator for the notion *ceteris paribus*, $\boxdot$ is an operator for 'actual' necessity, $\Box$ an operator for physical necessity), he adds:

Its dual, defined by $\sim \boxdot \Box (*p \sqsupset \sim q)$, i.e. $\diamondsuit\!\!\!\! \diamond \Diamond (*p \wedge q)$, could perhaps be read as 'it *might* be the case that q when p' where the 'when' conceals some idea of conjunction rather than of implication (1971, p. 12).

If we take up the preceding example it cannot be denied that as far as the first interpretation is concerned, it could just as well have been read as 'the plane could have been flown by the other pilot and could have crashed all the same'. Note that this interpretation is in accordance with the preceding interpretation of the factual 'even if': while the latter had been formalized by means of a conjunction, the hypothetical 'even if' can be reduced to the assertion that a given conjunction is possible.

To conclude this discussion of the hypothetical 'even if', we must describe the role played by words such as 'even' before the protasis, or 'notwithstanding' before the apodosis in suggesting what the speaker believes true. The most common answer given to this question is that with these words one indicates that the apodosis is believed true, but this is in contrast to the fact that the truth value of the apodosis is sometimes unknown notwithstanding the use of 'even if'. The more general response which can be given to this question is perhaps that in using 'even if' one presupposes that the apodosis can be inferred from some proposition known to be true. In a language extended with propositional quantifiers, the conversational situation would be thus describable as $[\exists r(r \wedge \Box^r q] \sim \Box^P \sim q$ or, in the less frequent conditional case, as $(\exists r(r \wedge \Box^r q)) \Box Pq$. However, it could be maintained that the formulation for the second case is not general enough. Since the conditional 'even if' is often used as an appendix to other conditionals previously asserted with the same apodosis, it is plausible to maintain that in the case of the conditional 'even if' the most general rendering of the conversational situation is given simply by $(\exists r \Box^r p) \Box Pq$.

4. One way to introduce the problem of the correct formalization of the propositions expressed by sentences containing the words 'as if' is to start from a question which was left open in the preceding paragraphs and that is: given that the formula $\square\underline{}^{p}q \wedge q$ does not stand for an 'even if' conditional and not even for a factual conditional, to propositions of what kind, if there are any, does it correspond? The answer can be easily obtained by observing sentences in ordinary language which faithfully reflect propositions in the form $q \wedge \square^{Pq}$ for example, 'The tracks in the snow are still visible, and this would occur if the bear had just passed.' A sentence like this in ordinary language, however, is indistinguishable from a sentence such as 'the tracks in the snow are still visible, as if the bear had just passed': in this context, in fact, 'as if' is the same as 'which thing would happen if' or 'as would happen if'. I must hasten to add that this is a fairly unusual use of 'as if'. It is not always the same thing to say 'p, as if q' and 'p as if q'. Even if the distinction cannot always be perceived in spoken language, it must be understood that the proposition which precedes 'as' when there is no comma is in some way incomplete and 'as' has the same meaning as 'in the same way in which'. For example, if I say that 'Physical system P behaves as if it were under the influence of a magnetic force' I mean that 'if physical system P were under the action of a magnatic force it would behave in the same way in which it is actually behaving'. It is rather obvious that the resources of propositional conditional logic are inadequate for the formalization of propositions containing 'as if' in this second sense. It is only within a first-order language that notions such as 'x owns property P in way m' and 'the way in which x owns P is equal to the way in which y owns Q' can be expressed. One could, however, hypothesize that by appropriately extending propositional language it is possible to define a propositional operator, symbolized, for instance, by '$\circ$' such that $p \circ q$ can be read as 'p in the same way as q', either when p and q are such that they admit an object of mode after themselves, or in the opposite case (in the latter case the meaning of $p \circ q$ would be simply 'p is true in the same way that q is true'). In such an extended language, also further enriched with Åqvist's operator $\boxdot$ for 'it is actually true that', assertions in the form 'p as if q' in the light of the proposed analysis would be formalized as $\square^{q}(p \circ \boxdot p)$.

The preceding considerations by no means exhaust the problems arising in the analysis of 'as if' conditionals. As is known, the 'as if' constitutes an important module in Kantian philosophy in which it appears in terms which are perhaps still more stimulating than those in which it appears in Hans Vaihinger's *Philosophy of 'As If'* (1911). Vaihinger's book is at

any rate a valuable source of reflections on 'as if' and it will be interesting
to see whether the analysis which we have tentatively proposed above
enables us to have a better understanding of, or to reorganize, Vaihinger's
formulations. Let us take for example the assertions, which Vaihinger
considers essential for scientific research, of the sort '*a* is to be considered
as if it were *b*', and which he calls *fictional judgments*. For the moment let
us leave aside the prescriptive character of fictional judgments and analyze
just their descriptive variants of the form '*a* is considered by x as if it were
b'. For the proposed analysis '*a* is considered by x as if it were *b*' would
equal 'if *a* were *b*, *a* would be considered by x in the same way in which he
actually considers it'. But this analysis does not reflect exactly what is
meant by the fictional judgment. Suppose, for example, that a scientist
considers entities of type *a* as if they were entities of type *b*, by mistake
or more simply because of lack of information about entities of type *a*.
The consequences of such suppositions are maybe more perspicuous in
relation to propositions in an analogous form such as, for example 'x sees
a as if it were *b*'. For example, let us suppose that Joe is color blind and
confuses red with green and vice versa. Then 'Joe sees red as if it were
green' cannot mean 'if red were green, Joe would see it as in fact he does
see it' (in such a case he would in fact see red): what is meant is, if anything, something like 'if Joe were to perceive colors correctly and red were
green, he would see red as he now sees it'. This example leads us to paraphrase 'x considers *a* as if it were *b*' by 'if x had correct knowledge of
phenomena and *a* were *b*, x would consider *a* as he now considers it', or
more simply, by 'if x knew that *a* is *b* he would consider *a* as he now considers it'.

Fictional judgments of the form '*a* is to be considered by x as if it were
b', which for Vaihinger express the fundamental part of theoretical knowledge, are prescriptive in character. Their correct analysis would then be 'if
x knew that *a* is *b* he would consider *a* as he actually is to consider it'. In
fictional judgments it is prescribed that one must behave as if definite
supplementary information were possessed which in reality is not. In this
way Vaihinger expresses the practical and regulatory character of theoretical knowledge (but note that it has been authoritatively maintained that
in general any subjective conditional is a rule of inference and thus a pre-
scription: for example, see Sellars (1953)).

'As if' conditionals could be appropriately called *fictional conditionals*.
The sentences which express them in ordinary language are invariably in
the subjunctive mood and thus indicate the presupposition that the an-

tecedent is false or undetermined: invariably then, 'as if' introduces a hypothesis which is the antecedent of an afactual or counterfactual conditional.

It would be interesting to discover the logical relations occurring between the various kinds of conditionals and pseudoconditionals characterized in the preceding pages, with the view of drawing out more hidden aspects of conditional reasoning. However, the study of these interrelations goes beyond the aims of this paper.

NOTES

[1] The example is found in Anderson (1951).

[2] In what follows it will always be taken for granted that 'since' is not to be understood in a specifically causal sense which would be better expressed by the English 'because' (Italian *perché*). In the sense meant here 'since' is synonomous with 'given that', 'as', 'inasmuch as'.

[3] The distinction between these two kinds of 'even if' is already clearly formulated in Reichenbach (1954) where the 'even if's of the first kind are called 'noninterference counterfactuals' and are subjected to an independent probabilistic treatment.

REFERENCES

Åqvist, L. (1971), 'Modal logic with subjunctive conditionals and dispositional predicates', *Filosofiska Studier*. Reprinted in *Journal of Philosophical Logic*, **2**, 1–76 (1973).

Anderson, Alan Ross (1951), 'A note on subjunctive and counterfactual conditionals', *Analysis* **12**, 35–38.

Goodman, Nelson (1947), 'The problem of counterfactual conditionals', *Journal of Philosophy* **44**, 113–128. Reprinted in *Fact, Fiction, and Forecast*. 3rd edition. Indianapolis, Bobbs-Merrill, 1973, pp. 3–27.

Lewis, D. K. (1973), *Counterfactuals*. Oxford, Basil Blackwell.

Mackie, J. L. (1973), *Truth, Probability and Paradox*. Oxford, Oxford University Press.

Pap, A. (1958), 'Disposition concepts and extensional logic' in *Minnesota Studies in the Philosophy of Science* **2**, 196–224.

Pizzi, C. (1977), 'Boethius' thesis and conditional logic', *Journal of Philosophical Logic*, **6**, 283–302.

Pollock, J. (1975), 'Four kinds of conditionals', *American Philosophical Quarterly* **12**, 51–59.

Prior, A. N. (1968), 'Modal logic and the logic of applicability', *Theoria* **34**, 183–202.

Ramsey, F. P. (1931), 'General propositions and causality' in his *The Foundations of Mathematics and Other Logical Essays*. London, K. Paul, Trench and Trubner.

Reichenbach, H. (1954), *Nomological Statements and Admissible Operations*. Amsterdam, North-Holland. Reprinted as *Laws, Modalities, and Counterfactuals*. Berkeley, University of California Press, 1976.

Sellars, W. (1953), 'Inference and meaning', *Mind* **62**, 313–338.

Stalnaker, R. (1968), 'A theory of conditionals' in *Studies in Logical Theory*, edited by N. Rescher. *American Philosophical Quarterly Monograph* **2**, Oxford, Blackwell; and in *Causation and Conditionals*, edited by E. Sosa. London, Oxford University Press, 1975, pp. 165–179.

Stalnaker R. (1973), 'Presuppositions', *Journal of Philosophical Logic* **2**, 447–457.

Thomason, R. H. (1972), 'A semantic theory of sortal incorrectness', *Journal of Philosophical Logic* **1**, 205–258.

Vaihinger, H. (1911), *Die Philosophie des Als Ob*. Berlin, Reuther and Reichard. (The philosophy of *'As If'*. Translated by C. K. Ogden. London, K. Paul Trench and Trubner, 1924.)

CORRADO MANGIONE

WHAT IS CONTEMPORARY LOGIC
TALKING ABOUT?[1]

If we take a look, even a superficial one, at the literature of logic, pure and/or applied, extending, for example, from model theory to linguistics, from proof theory to deontic logic or intensional logics in general; if, in short, we reflect on the logical works developing in close contact with either mathematics or completely different disciplines, arriving finally at so-called 'philosophical logic', we can well ask ourselves to what extent all this can qualify as logic. *One* way to answer this question (considered by some as *the* way) is to indicate the entire historical course leading up to the formation of contemporary logical practice: the 'necessity' of recent developments should result from the interrelation of themes and techniques which have led to them historically. Certainly an argument of this kind is plausible; in fact, the understanding, let's say, of the categorical approach to logic would be inconceivable without taking into account the concrete development of mathematical practice. History, including the history of science, certainly has its own 'necessity' but God is not always with the strongest battalions, so to speak. Outside of metaphors, it seems to us that the *formation* of a discipline does not univocally impose a way of utilizing and contextualizing the same, hence it does not impose the acceptance of determined tendencies as inevitable.

Behind this problem lies a more general one which we have no intention of tackling here, but which cannot be avoided: this is the problem of scientific tradition. Certainly all scientific activity, like all human activity in general, is part of a network of institutions, interdisciplinary ties, connections established between problems, hierarchies of techniques, socially determined relations which somehow 'impose' certain choices rather than others. In our opinion, however, an historical understanding of these determinations makes sense and is useful insofar as it allows a *choice* to operate at present within the constituted tradition, which does not supinely accept this same tradition as the natural place of present and future research, but which, on the contrary, recomposes this past on the basis of norms of rationality, to clear up tensions arising from the present. The comparison with the past is one way of *testing* present choices, not of *imposing them.*

In following this kind of argument there is the risk of yielding to the

Maria Luisa Dalla Chiara (ed.), Italian Studies in the Philosophy of Science, 89–111.
Copyright © *1980 by D. Reidel Publishing Company.*

temptation, ever alive today, of postulating an external, metahistorical norm for scientific activity, hence for logic in particular, on the basis of which historical development is to be judged. In our case this temptation is translated into an appeal to an *a priori* notion of logic which has led many to see in contemporary research a kind of betrayal of fundamental problems, a technical regimentation, substantially sterile, to which is contrasted a philosophical logic which would instead remain faithful to the basic problems, which would be the 'true' heir to the tradition.

As far as we know, it was Nicholas Rescher who, in 1968, theorized that this distinction derived from the need to emphasize a whole series of other research problems which, in his opinion, had been obscured, particularly in the last twenty years, by the steady stream of results of 'mathematical logic', which would have pushed into the background "the phenomenal growth spurt of logic in directions related to philosophical considerations". Hence this subdivision of Rescher seems to refer not so much to mathematical logic – which as an instrument would remain constant in both cases – as to the subject which is taken into consideration from time to time.

However useful and fruitful this distinction may turn out to be for practical purposes[2], it seems somewhat artificial: it proposes once again longstanding dichotomies of an almost always conservative and reactionary kind, so well outlined by Frege in the two opposing attitudes toward *metaphysica sunt non leguntur* or *mathematica sunt non leguntur* which he first undertook to combat with the philosophical approach of a typical traditional 'mathematical' program. Furthermore to us it calls to mind the fruitless practical/theoretical (Crocean) dichotomies and the proposal to distinguish between 'exact sciences' and 'philosophy', which – in our opinion – really belongs to the most deleterious and obstructive way of looking at philosophy. In reality what we believe must result from the examination of the development of contemporary logic is that it is precisely mathematical logic which has contributed heavily to partially filling in this arbitrary and sterile breach which was (and is) present between philosophy and mathematics; it has allowed us to throw up a bridge between two 'activities', as Wittgenstein says, which can eventually have differing methods and problems but certainly not cultural 'dignity' or 'depth' of argument. If, in most philosophy journals today, a conceptualization of a logico-mathematical kind is increasingly used, this depends especially on the fact that in its process of evolving, this discipline, besides being given an imposing 'technical' orchestration, raised problems whose concern with 'philosophy' is a purely traditional fact: in other words, perhaps a little simplified, we hold that logic today is 'mathematical' for a whole series of complex historical reasons which we have tried to illustrate, but not 'mathematical' as opposed to 'philosophical'.

Hintikka asks:

Is there such a thing as 'philosophical logic'? Basically, my answer is 'no'. There does not seem to be much intrinsic difference in philosophical interest between the different conventional compartments of logic. Much of the recent work in the more esoteric parts of mathematical logic possesses, it seems to me, a great deal of relevance to philosophical inquiry. It is true that most of this work has not caught the eye of philosophers, or has done so only in those relatively rare cases in which logicians have themselves called philosophers' attention to their problems. However, much of the work that has been done in such areas as recursive function theory, model theory, and metamathematics is concerned with the explication and development of concepts and conceptual problems which are of the greatest interest and relevance to a philosopher's pursuits.[3]

On the other hand we must stress at this point that we are not at all of the opinion that the intervention of the logico-mathematical apparatus ends by conferring 'rigor' on the formulation of positions and problems which are, in the first instance, intuitive. We are expressly opposed to a thesis, sometimes credited among philosophers – but not only among them – according to which the task of logic (and of science in general) is to make an argument *rigorous*, while it is characteristic of philosophy to make an argument *truthful*. In our opinion the task of mathematical logic is not at all limited to that of conferring rigor; rather, it offers conceptualizations and methods which are indeed technical and rigorous, but which become operative and fruitful only when they give rise to new formulations and intuitive interpretations to which they must furnish articulations but never purely an external covering. Put in slightly different terms: we are convinced that logical analysis does *not* generally exhaust the extensive understanding of the content and potentialities of a theory; but we are equally convinced that logical analysis is necessary in that it often leads to new generalizations which are indeed quite precise but which are not less fruitful precisely because logical analysis shows up fundamental content-related points.

Here I should like to recall what C.C. Chang said in his intervention at the 1971 Tarski symposium. He asks: "Are our models sufficient for the study of all 'deductive sciences'?" Even if, historically, things could be so at the time when Tarski introduced this theory and can still be so for mathematics, "this is clearly not true of the term 'models' as understood and used by a large number of people doing research in such diverse fields as ... economics, biology, psychology ... These researchers are all crying out for models. Everyone wants to find a model, but no one seems to know what a model is. Everyone also agrees that models play or will play an important role in his science." Now continues Chang,

> we model theorists all know exactly what a model is and we can tell these people again and again. But, our models do not satisfy them. This may be because their science is still not deductive, or their language is not quite formal enough, or (perhaps more importantly) because there are inherently more complicated things that they call models . . . Whatever their notions of models may be, they are not the models we have been studying so intensively.[4]

We seemed to be able to recognize a first result of this direction in the proposal of an 'empirical' theory of models put forward by M. Luisa Dalla Chiara and G. Toraldo di Francia in their 'The logical analysis of physical theories', *Rivista del nuovo cimento* 3, (1973) 1.

Naturally we do not mean by this that modern mathematical logic has become a kind of panacea with which to tackle and resolve 'everything': *nothing is further from our intention*. Instead we mean exactly the opposite, namely that it is perfectly useless to dream of the formation of strange 'philosophical logics' with which one would think one could solve every problem; and that every historical period has its own *global* instruments of investigation which are taken up precisely in order to be able to surpass, improve and refine them continually, and towards which it is, at least by now, absurd to have a culturalist repulsion toward 'the technical'. In our opinion this entire set of problems is a further proof of the philosophical relevance that mathematical logic (or if one prefers, the mathematical treatment of logic), understood as the *contemporary way of doing logic* (whatever kind this may then be), has as such, without there being any need to bother with a 'philosophical logical' pretense except as a purely convenient label.

We believe that the only way out of this abstract dichotomy between the necessity of historical development and the metahistorical norm is to broaden the range of analysis and to take into explicit account the ties which logic historically has maintained and still maintains with other scientific practices, and to see to what extent it is in a position to respond to the demands these practices impose. In other words, it is a question of recognizing that logical research 'must' deal with the foundations of theoretical practices and this 'must' arises from the opportunity, which today is increasingly noticed but which was also very much alive in the past, to give a general framework to theoretical work, noting at the same time that, historically speaking, this has not always happened. There have been exclusions, privileged theoretical practices, ideological superimpositions, the study of which, in our opinion, constitutes one of the tasks of the historian of logic who does not want to limit himself to recording the emergence and development of the concepts at the heart of present day research; it is, however, precisely this acceptance of so broad, so 'philosophical' a concept of logic which places in the foreground the concrete relation established historically between the various theoretical practices and logical reflection. This last is realized only to the extent to which it can enter into an effective exchange with these practices, and its philosophical value lies in the concrete, material and non-ideal realization of its project: if, in fact, logic exists as a project, as a particular way of tackling theoretical work, it has no contents which are not simultaneously contents of other sciences or in any case, human activities: *logicus purus asinus est* [the pure logician is an ass], the Renaissance saying cautions us.

Undeniably the most profound relationship logic has achieved with other theoretical practices in the last 100 years is that which gave rise to the investigation into the foundations of mathematics. The obvious philosophical interest of these studies seems to us to confirm what was said above. The period between the early years of this century and 1930 marked the attempt – articulated in different directions – to re-establish, through logic, the search for a unity of mathematics which the developments of the preceding century had seemed to render increasingly precarious. This search for unity was not 'doctrinaire', and even if it often assumed the guise of philosophical positions borrowed from traditional settings, this was in large measure determined more by the necessity imposed by the *lack* of new and adequate conceptualizations than by free choice. It would be a serious mistake – in our opinion – to believe that the motives which pushed Russell, Brouwer, Hilbert, etc., to attempt a systematization of mathematics were the same which a few years earlier had pushed the exponents of the various philosophical currents to make room in the respective systems for the developments mathematics was recording.

The breaking point, in fact, lay in presenting mathematics for the first time in a certain sense as *pure mathematics*, able to find its own motivations within itself; perhaps the clearest – and also the most pathetic – expression of this phenomenon is Cantor's appeal to the intrinsic *liberty* of mathematical thought, while its most conscious assertion, on the other hand, is obviously that of Hilbert. Faced with this outgrowth, which undermined the old connective modules within mathematics itself which were projecetd toward external reality, the problem posed was precisely that of how to explain the possibility of new relations, how to justify them, even in the impossibility, just at the onset of research, of showing concrete examples of them. It is in this sense that research on foundations, besides the largely fortuitous motivation offered by paradoxes, constitutes a general theoretical attempt to become attached once more to external reality, however this attempt may come about, through logic or intuition, forms or contents.

Through the most diverse paths, the most disparate characterizations, the most heterogeneous approaches, this process, in the present state of research, seems to have almost reached its climax, its breaking point. The developments of mathematics, which foundations research had posed as the goal of justification, had in large measure realized this relation; but at the same time, dialectically, they had shown that the conceptualization, on the basis of which it had to be justified, was in fact surpassed. It does not seem risky to state that while the traditional 'technique' of logical research continues to produce important results in every field, today it appears too 'coarse', so to speak, at its most significant and advanced points. It seems that a rigid true-false dualism must be replaced with a more flexible consideration of the possible states of things; extensionality seems to reshape itself more and more before intensionality: it is as if that 'rethinking' which the physical sciences had to carry out in our century for an ever greater accomodation of reality, had to be imposed on the more formal, abstract sciences as well.

But we shall continue this discussion of the foundations of mathematics below, in the light of the most recent developments. The problem posed is rather that of seeing whether the richness of this connection is exclusive, so to speak, that is, limited to the relation between logic and mathematics or whether it can instead be extended to other theoretical activities. If in fact this turns out to be the case, we would then pose the more general question of justifying this privileged position of mathematics in comparison with the other sciences. (It will be remembered that it is a question already posed at the emergence of mathematical logic with Boole and Frege.) It seems to us that in order to avoid preconceived solutions, it is necessary to analyze more closely what is meant by logical problematics. Some 'applications' of logical techniques to experiences which cannot be reduced to mathematical ones are well known, be they linguistic or physical or pertaining to the philosophy of law, etc. In all honesty we do not feel inclined to maintain that in each of these cases an optimal or particularly fertile relation was attained: undeniably some applications of logic to specific problems can turn out to be superficial; but what is significant is that the logicians themselves, who are concerned with the various problems, have often realized this. A discussion about deontic logic which took place ten years ago between Alan Ross Anderson and Georg von Wright[5] can furnish a valuable example of this.

According to Anderson the purpose of deontic logic is to regulate and introduce criteria of *correctness* into moral and juridical argument; according to von Wright on the other hand, the goal is to allow an articulation of the argument about the *ontology of norms*, that is, about the fundamental concepts which lie at the bottom of normative argument. In our opinion this last position is the one most conscious of the intrinsic limits of a purely logical approach which tends to isolate one 'specific logic' within each practice, independently of the general configuration of the discipline under examination. It is this point which seems to us to deserve broader reflection.

When we speak of the logical problematic within a specific discipline, we tend too often to concentrate the argument on pure linguistic data and on criteria of correctness: it is a heritage of undoubtedly very long standing which sees in language the primary given from which the logical dimension arises. Now, behind Anderson's position lies precisely this assumption which is, however, extremely reductive, from both the general point of view (the origin of logical speculation) and the specific point of view. In other words we do not feel great need for a logic of juridical discourse as a norm of correctness, while inversely we see that the specific concepts with which juridical norms are analyzed, present problems not of correctness, but of explicitation. To be more precise: what appears important to us is making explicit the presuppositions in a language which allows comparison with other practices rather than the pure individuation of formal criteria of correctness. To return to the problem of rigor discussed above: the purpose of logical research is *not* rigor as an end in itself but the individuation of fundamental concepts which can happen only by broadening the horizon, not by concentrating exclusively on the linguistic aspect.

This argument holds true in general. Undoubtedly the analysis of the linguistic apparatus in which a theoretical practice is concretized is important but it is not the exclusive object of logical analysis. Once again the example of the foundations of mathematics is illuminating. The problem of correctness was the last reason to push Boole or Frege to undertake such an analysis: the real purpose was the individuation of fundamental concepts, and the other problem was posed only as subordinate to it, as the minimum natural presupposition. This obviously involves a much deeper and more direct contact with the specific practical problematic of every discipline: examples of this broadening of prespective, in which the logical problematic is substantially indissoluble as content from the specific one, seem to be research on quantum logic and the requirements put forward in Chang's general discussion of model theory, etc. In another direction, an analogous example is furnished precisely by von Wright's works on the foundations of law and partially at least by the recent development of the semantics of natural languages.

We spoke above of the complex relationship between logic and natural language and of how too often the logical relevance of a set of problems is reduced to the analysis of the linguistic aspect. It seems to us that this *a priori* conception of the role of logic is at the origin of another misunderstanding, very serious for us, concerning the 'plurality' of logics: this is an argument to which we shall briefly return below. Here we want to emphasize one aspect: much of the research concerning the logic of practices other than mathematics makes use of non-classical notions of truth which are often presented as more nuanced, more adequate for rendering the flexibility of the field under examination. This is particularly the case concerning theories dealing with the 'human sciences' or common discourse. Some then have reached the conclusion that while classical logic *is the* logic of mathematics or generally of the natural sciences concerned with rigidly determined concepts, when we move on to talk of Man with a capital M and cultural phenomena, this logic turns out to be inadequate precisely because it is too rigid, and more subtle logics are needed, if it is ever possible to pinpoint logics of this kind. Now in our opinion the only drawback to a position like this is that unfortunately even mathematics is 'too human' and its logic is not necessarily the classical one. Intuitionism, topoi, etc., well illustrate the fact that it is mathematical practice itself which imposes the necessity of 'nuances' and hence it makes no sense to interpret – or really establish – the distinction between various logics on the basis of a hypothetical breach between the human world, irremediably irregular and subtle, and the scientific natural world, equally irremediably determined. Once again it seems appropriate to us to stress that logical research does not mean the imposition of an external logic to regulate discourse, but the enunciation of fundamental concepts and articulations which depend on the field under examination and on the depth of analysis, not on irreducible ontological differences between nature and history at the bottom of which there is a substantially *a priori* metaphysical assumption. It seems that one of the philosophically important outcomes of recent research is precisely this.

The argument can be extended by noting a tendency present in recent years, at least in Italy, which has led to a recrudescence of the antithesis between logic, seen precisely as the imposition of rigid schemes, and human language, which is rich in unanalyzable articulations or in any case articulations opposed to logical structures. One aspect of these debates concerns the contrast between rhetoric and logic. One very clear exposition of this contrast is to be found in *The New Rhetoric: A Treatise on Argumentation* by Chaim Perelman and Lucie Olbrechts-Tyteca.[6] Perelman's basic thesis is that in Western culture, at least beginning with the seventeenth century (read: Descartes) a breach was created between a rationalism, which sees in the rigid application of logical principles directed at pinpointing the True and the False the only rational norm, and a more flexible view which examines the concrete articulation of human lines of argument, accepting rhetoric as a kind of logic of probable or persuasive discourse. According to Perelman, the acceptance of the first position would have involved and would still involve the abandonment of decisions (taken in a community) to pure will: this is because, if the search for the True and the False is accepted as the only criterion for rationality, even in contexts where this is not possible but must be limited to opinions, one ends by rejecting a distinction between degrees of plausibility which allow one to substitute a reasonable line of argument for pure will. Rhetoric would be precisely the logic, the classification of these arguments.

The argument clearly concerns the very concept of rationality and it is interesting precisely in that it presents a clear conception of the role of logic in the formation and analysis of reasonable behavior. What appears unacceptable to us in this position is not the exhortation to study rhetoric, which we hold to be extremely interesting and useful, but rather the very antithesis between rhetoric and logic. Even in this case the mechanism underlying the traditional idea of logic as regulation of language bursts forth. Logic, as we have already said several times above, is not this, and what assigns this role to logic is only an *ideology* of logic which – let us say parenthetically – cannot be necessarily imputed to Cartesian thought as Perelman maintains. In reality we maintain that the problem dealt with by Perelman is enormously more complex and regards not so much the criteria for evaluating arguments and hence their *form*, as the *contents* of decisions and arguments. In other words we do not believe that in a deliberating collective gathering or assembly it must be rhetorical analysis which furnishes a guide to choices but rather the analysis of contents: the real problems posed relate to information and conflicts of interest and in the analysis of these realities the 'old' logic seems to go very well to the extent that it is substantiated with an analysis of specific contents, with a social and political analysis of the assembly, etc. The opposition is not between types of arguments but between types of arguers, between their interests and their awareness of them, their knowledge, etc. The risk of Perelman's position, it seems to us, is that of seeing rhetoric a bit as the logic of the poor.

We think that rationality is not measured by the *form* of arguments but by the richest possible exploitation of information and investigative methods, so that the problem of a collective rationality, which operates socially, does not seem to concentrate on the analysis of the use of language, which takes place socially, but on the contents of decisions taken and on the forces, obviously, which such decisions translate into practice. To conclude, we should like to observe that the adherence to the idea of logic as having been founded on the analysis of natural language today constitutes for some philosophers the last beachhead against the necessity of passing finally from abstract discussions to the analysis of practices and specific problems. There is a kind of continuity between the positions which oppose logical language to natural language – seeing in the first an improverishment of the second – and Perelman's general positions on the concept of rationality: it seems to us that both positions are in fact conservative in that they interrupt the exchange relations between scientific research and everyday life.

Having rejected, as we have tried to show, the thesis on the prevalence of the linguistic aspect in logical investigation, it is then easier to retackle the problem from which we started, dealing with the relations between logic and mathematics. We believe that the fecundity and 'priority' of the reciprocal relations between logic and mathematics is not casual for the very simple reason that—in a Kantian sense if you will – mathematics is an essential component in *every* discussion concerning the scientific world; to the extent that logic concerns not only language but the fundamental concepts of scientific analysis, it is clear that there is a basic solidarity between these two activities. Let us repeat that it does not matter whether mathematics is the only exact language but rather that it is the

source of an extraordinarily large number of concepts which are fundamental to the scientific understanding of the world. Here too the link
between logic and language seems to have led to an erroneous interpretation of the role of mathematics which finds its most explicit form in the
neopositivist positions that mathematics is the *language* we use to articulate our scientific discussion of the world. At the base of this position lies
the acceptance of a basic solidarity between logic and mathematics which,
as was said above, completely convinces us. What is not convincing is, as
usual, the transition from logic to language based on the linguistic view
of logic criticized above. The role of this position within a general, empirical set of problems is clear: for the neopositivists, language on the one
hand is conventional, hence the choices between one language and another
must not be justified on extralinguistic presuppositions; on the other hand,
it is intersubjective and this guarantees that 'neutrality' of mathematics
which is necessary to give a plausible explanation of the role it plays in
scientific theorization.

An implicit consequence of this attitude, broadly developed in practice,
is the ambiguous role assigned to mathematics: on the one hand it turns
out to be indispensable in every scientific discussion; on the other hand,
however, it is denied the possibility of a direct contact with reality which
is not mediated by other theoretical practices. One of the most widespread
forms of this conception is that according to which, as mathematics must
serve physics – which would be the basic study of external reality – the
proof of the plausibility of given mathematical concepts would result
from their applicability within physics. This 'physical chauvinism' lies at
the bottom of more than a few discussions often heard on the abstractness
and vacuity of at least some aspects of modern mathematics and conversely it allows the 'pure' mathematician freedom in his gilded cage which,
in the final analysis, turns out to be suffocating in that his theorizing
would come under patronage with respect to the empirical sciences.

It is precisely from the establishment of these practical consequences in
the neopositivist position that we can move on to a theoretical critique of
it. As Einstein said, "for many physicists, experience is the theory they
learned until they were eighteen; the rest is pure theoretical speculation".
Naturalla the argument concerns not only physicists but all those who see
in mathematics a language with or without a logic, as Feynman maintains,
the choice of which is conventional and substantially arbitrary, bound as it
is only to criteria of practical opportunity (besides, of course, to 'internal'
criteria of consistency). Undoubtedly a mathematics which, in the final

analysis, does not assume a concrete form in new applications would be sterile; but the point is that the relations between mathematics and the empirical sciences go in two directions: a mathematical concept can be sterile not in the sense that it is inadequate but rather that it is inadequately utilized. There are two alternatives: either we hypothesize, for students of the empirical sciences, a superhuman capacity to use every 'good' mathematical concept fruitfully and hence we delegate to them the task of deciding on 'good' or 'bad' mathematics; or we admit that the criteria for choice must be different and in the final analysis mathematics can be measured directly with reality. We believe the untenability of the first alternative lies in the fact that every empirical theorizing which exploits mathematics is in part determined precisely by the mathematical concepts it uses, and consequently not even it has a pure contact with external reality but rather a mediated contact. We further believe that the fact that theory is empirically justified and 'functions' is one proof that also the mathematics on which the theory is based 'functions' and hence it neatly incorporates external reality. It follows from this that mathematical theorizing to the extent that it is not sterile, deals directly with the external world and it must be measured directly with it. In this way every discussion about the conventionality of mathematical axioms fails and hence also the role of the logical analysis of the foundations of mathematics is formed differently.

In reflecting on the motives which led to the development of the investigation of foundations at the end of the last century, it seems correct to stress the role played by the presentation of mathematics as 'pure' science. One could say that at the bottom of these investigations lies precisely the attempt to justify mathematical research which, although it moves more freely than the traditional, does not lose contact with the rest of scientific knowledge, but on the contrary establishes new contacts. In a certain sense, the development of twentieth century mathematics has shown the inadequacy of the general concepts which at the end of the nineteenth century had justified its rise. Perhaps the most important event from this point of view has been the new relation which geometry has established with the rest of mathematical practice. The emergence of set theory and the development of the axiomatic method were the point of departure for the foundations problem and this coincided with a reversal of course on the part of research which gradually led to the separation of geometric research from the rest of the new mathematics which was in the process of being constructed. An echo of this reversal of trend, as Shafarevich observes, can be found in what Felix Klein wrote in 1926:

> When I was a student, Abelian functions [a topic which today is an important aspect of geometry] – as an after-effect of Jacobi's tradition – were regarded as the undisputed summit of mathematics, and each of us, as a matter of course, had the ambition to forge ahead in this field. And now? The young generation hardly know what Abelian functions are.[7]

New fields of investigation were constituted by function theory, abstract algebra based on a broader conception of mathematical practice, which found their most natural foundation in set theory seen as a description of the universe of mathematical objects.

Behind this conception lay the idea, expressed more or less explicitly in the mathematical works in the field, and very clearly in the works of logicians, that mathematics was essentially the creation of abstract conceptual schemes whose practical justification was to be sought in richness, that is, in the capacity to systematize the problems under examination (think, for example, of Hilbert's famous speech given in 1900)[8] but whose logical justification (foundation) was found in the *reduction* to basic concepts which could be those of general logic, finitist mathematics, intuitive constructions. In other words a breach was created between the justification of the *correctness* of a mathematical theory and the justification of its *richness*. This can explain, at least in part, the uneasiness of some mathematicians when faced with logical research and also, in some cases, the decided attitude of repulsion. From this perspective (the intuitionist position aside), which constitutes the most decisive attempt to avoid creating a breach between nineteenth century mathematics and the new mathematics, set theory is seen as the theory of mathematical concepts, the basic 'reality' from which single specific notions are isolated from time to time. The acceptability of the axioms of set theory can be justified on the basis of general logical notions (Frege, Russell, etc.) or on the basis of finitist consistency proofs (Hilbert) but the central point does not change: the entire mathematical edifice is seen as a structure of concepts, schemes, which finds its unity in this conceptual nature of the notion of set. In this sense the unsuccessful relation between geometry and the new developments of abstract mathematics bears witness to the impossibility of integrally and adequately reconstructing the notions given us by external physical experience into pure conceptual schemes. It seems to us that the opposition between Hilbert and Frege regarding the foundations of geometry, at least in part, reflects this problem.

Naturally on the level of concrete mathematics, the question is different and concerns more specifically the lack of conformity on the part of traditional geometric research to the new standards of rigor (and in the final analysis, to the new conception of mathematics) posited by set theory and axiomatic method. This breach was healed in the twenties and thirties, with the developments of abstract algebra, with the work of the school of Emmy Nöther and Emil Artin made possible by a fertile algebraicization of classical geometry. The works of Bartel L. van der Waerden, Oscar Zariski and André Weil lay the foundations for an algebraic geometry which perfectly and richly rose to the new standards of rigor and, using an algebraic base, allowed the treatment of central concepts of geometric investigation. In other words it is not a question of a connection of principle but a real connection which involves the very practice of research. It is from the successive developments in geometric research that a counter-tendency arose which once again put geometry in a central position. The works of Leray, J.P. Serre and Grothendieck, together with the notion of sheaf and scheme, have widened the horizon enormously, thus allowing a deep unification among geometry, arithmetic and algebra. Geometric language increasingly pervades these fields of research so that geometry is still set up as the natural context within which a large part of mathematics develops.

In this new perspective, category theory has a central role in that it constitutes the basis of a renewed connection, furnishing geometric concepts with a sufficiently general scope and adhering to the basic problems. Discussion is clearly simplified, but from the point of view of the problem of foundations – which is what interests us here – it has a precise meaning in that from it derives the possibility of a different choice of fundamental concepts. Lawvere's research has brought to light a profound relation between logic and geometry which replaces the earlier tie between logic and set theory. The theory of elementary topoi as the universes of discussion, functorial semantics, analyses of logical rules as adjointness between functors, etc., are not simply an application of the conceptual store of category theory to traditional logical problems, but bring to light the possibility and opportunity of substituting the concept of set with that of topos in the foundational context.

The way in which the fundamental logical structure of mathematical objects is characterized is shown to be not only more general but in many cases much more 'natural' and direct than that of set bases. Topos theory as the 'theory of continuously variable objects' amply testifies to this fact. The discussion can be generalized by drawing a 'philosophical' moral from these developments. The substitution of topoi for the universe of sets also means the abandonment of a *conceptualistic* theory of mathematical objects, for at least two reasons: in the first place the notion of topos arises from considerations which are at once geometric and logical and hence finds its justification not in a theory of mathematical *concepts* but in a theory of real *objects*, geometric objects; in the second place the acceptance of a plurality of basic universes is implicit. Different topoi exist and this diversity does not signify the impossibility on our part of pressing closer the only universe to which we intend to refer, by means of a linguistic description: the plurality of elementary topoi is a basic given which reflects the possibility of different contexts, different totalities of objects.

The situation is completely different from that of set theory where the unity of the universe is sought insofar as it is conceived as the universe of mathematical concepts. From this point of view the existence of undecidable set-theoretical propositions has a fundamental significance which we believe provokes a crisis for the idea of the universe of sets as the only universe of all concepts. On the other hand, to accept the plurality of set theories, as some have done, seems to involve necessarily the abandonment of the very notion of set as a fundamental notion. It is not apparent why, once sets are no longer seen as concepts, one must necessarily *reconstruct* every mathematical object as set-theoretical object. A reduction of this kind makes sense if, by this, one means that every mathematical object is a conceptual construct, but if the universe of sets is not the universe of concepts, one cannot understand what this can mean. There remains the idea of conceiving the various models of set theory as the totality of *collections*, but at this point it is concrete mathematical practice which shows us how the essential properties of an object cannot necessarily be clearly studied by considering its elements. The most conspicuous example is provided by algebra: in studying groups or rings it is not absolutely necessary to see them as collections of elements on which relations and/or operations are defined: their properties result from the study of the morphisms which link them to other objects of the category in which we situate them, and from transformations between this category and others. Not only is a study of this kind possible but, as modern algebraic practice shows, it allows one to pinpoint basic common structures which are difficult to grasp when mathematical objects are seen as collections. To base algebra on set theory means *reconstructing it*, but not through this pin-pointing of its fundamental components. The concept of topos, or, in general, that of category, allows instead a *direct* study of mathematical objects which takes account of the fact that their properties are the result of the consideration of morphisms and transformations. In this sense the set-theoretical 'foundation' acquires a more reduced role which is that of the *reconstruction* of objects and not of their *direct study*.

The result is the *idealistic* nature of the set foundation, where the word 'idealistic' has a precise meaning and refers to the fact that in this kind of foundation the only way to justify the reconstruction of mathematical objects in terms of sets is to conceive them as concepts and to see the study of mathematics as the study of concepts. Hence Lawvere's conception of foundation as a *materialistic* mode of studying mathematics seems justified and not just by opposition[9]. We believe that there are at least two aspects in Lawvere's discussion which justify this attribution. The first concerns the non-reduction of the objects of study to concepts: from the categorical point of view an object is not reconstructed but its relations with other objects are simply examined without making assumptions of a reductive nature. The second concerns the general conception of elementary topoi as the theory of continuously variable objects. Classically the analysis of movement takes place in terms of instants, that is, objects in movement are considered by analyzing their position at single instants. On the other hand, this analysis is based on a reduction in that it assumes the possibility of being able to speak of instants which are not given by experience but which are constructed conceptually. In elementary topos theory this does not happen, and the movement of objects does not result from their reconstruction in terms of instants, but globally, and the different types of variations which we are considering result from the property of the topos' internal logic.

These two aspects are closely connected. The idea that the basic given of mathematics is a reality in movement means that every reconstruction of it, whatever the basic conceptions assumed may be, is necessarily artificial in that it obviously does not admit the possibility of a deepening of the knowledge of objects under examination based on elements which do not appear in the original reconstruction.

In a certain sense it can be said that this way of conceiving foundations constitutes a systematic and consistent application of the axiomatic method, no longer seen as constitutive but as descriptive: axioms represent the fundamental facts which we succeed in characterizing in the reality under examination. The possibility of this rests precisely on the adoption of the concept of topos. Classically, one should think of Hilbert and the axiomatics of geometry: the problem posed by a consistently axiomatic view of mathematics was given by the fact that one did not know how to justify the choice of logic, or in general, of axioms, without having recourse to a previous conceptual reconstruction of the objects to be described: in other words an axiomatic theory was seen as the description of a structure or a class of structures within set theory (hence it was constitutive). The only alternative to this conception was a rigidly formalist view of axiomatic systems which could not justify the choice of axioms in every case. Hilbert's difficulties in justifying classical logic in the face of Brouwer's objections seem to stem from this, and the recourse to the Cantorian paradise was precisely the role of making the choice of classical logic plausible under the form of an 'as if', so that the problem, in the final analysis, shifted on that of the consistency of set theory.

The concept of elementary topos with its internal logic allows us to avoid this difficulty in that every theory can be interpreted in a topos and the determination of a topos – think of the form of axioms – does not presuppose any logic. In this way it becomes possible to avoid recourse to external elements not directly suggested by the structure of the domain one wants to study and hence this consistently axiomatic conception of mathematical theories is justified.

According to Lawvere – at least as we interpret him[10] this acceptance of basic reality as reality in movement (in the double sense of reality which varies and of the knowledge of reality which is modified) constitutes a confirmation of the *dialectical* nature of mathematical thought. In a certain sense it seems to us that this dialectical nature is implicit in the very foundational use of the concept of category in that a foundation in these terms sees as the central nucleus of mathematical knowledge not the *substance*, that is, the internal composition of objects, but their *mutual relations*, their transformations. In this way a conception of the development of mathematics is presented which it will be instructive to compare with the one outlined by Hilbert at the Paris Congress of 1900[11], which constitutes the most consistent and subtle formulation of the (then) new mathematical practice. For Hilbert the growth of mathematics was caused by the emergence of new problems and by the elaboration of theories to solve them; these theories constituted the starting points, the hypotheses for their solution, and Hilbert ended by postulating a kind of principle of the resolvability of every mathematical problem. The difficulty was to select suitable hypotheses and to justify their correctness on the basis of consistency proofs. From our perspective the situation changes in the sense that it is the contradictions between the concepts in play, more than single problems, which constitute the driving force behind mathematical growth. These contradictions arise both from the conditions of mathematical work itself and the variations in the knowledge of the objects of study, and the sense of mathematical progress lies in pinpointing the level of discussion and in devising the problems in which the contradictory terms are both accepted without reducing one to the other and where mutual relations are analyzed.

For Lawvere one basic contradiction implicit in mathematical work is that between formal and conceptual,[12] between computation procedures, linguistic presentations, and structures. We believe that twentieth century logic contains one of the most conspicuous, if not always conscious, examples of this conception, namely model theory, which has precisely the theme of the interrelation between the linguistic given and the structures which realize linguistic conditions. The content of a good part of the central results of model theory regards neither the theories alone nor the structures alone but their interrelation.

An even more significant prototype of this contradiction and of its positive exploitation is found in geometry, where the contradiction between linguistic conditions, polynomial equations and varieties, classes of objects which satisfy the conditions posed by them, is systematically exploited. Lawvere's functorial semantics and more generally functorial model theory constitute, from the foundational point of view, the most explicit formulation of this conception in which the oppostion between formal and conceptual is resolved and analyzed in terms of adjointness between functors. In the background, as a setting, there is an even more general contradiction: that between geometry and logic, which constitutes the most natural context for seeing topos theory, which is presented precisely as the unitary analysis of the interrelations between geometric and logical concepts and their basic unity.

In our opinion the great interest of this foundational perspective constitutes the most important recent event in research. Naturally this does not mean that the more traditional problems have suddenly lost all meaning or that the categorical approach constitutes the definitive answer to foundational problems; on the other hand, we do not want to assume a neutral attitude, in that the greater philosophical profundity in this second conception seems undeniable, and equally undeniable, conversely, is the progressive drying up of some of the research done along more traditional lines. In other words we believe the most vital line of research today is the categorical one, and that it is necessary to take into consideration the developments of logical research and not only logical research, from this perspective, by isolating its important elements and carrying them ahead. We believe that in this new perspective facts and problems, which, in the traditional perspective turn out to be unimportant or philosophically non-germane, can acquire deeper significance. One can find an example of what we mean by considering the different philosophical weight acquired by model theory research along Robinson's lines, after several years when, implicitly or no, it was asserted that in the background of a traditional conception of foundations, it was mathematically interesting but philosophically scarcely significant. The recent works of A. Joyal, Makkai and Reyes[13] – which go along the lines of Lawvere's program – seem to show amply how Robinson's conceptions acquire an importance in this new context which was always denied them by the traditional scholars of foundations.

The undeniably new element seems to be the decisive abandonment of the idea of foundation as the justification for correctness, in favor of the much more fertile idea of foundation as investigation and individuation of fundamental ideas. Taken in itself this assertion is perhaps generic, but in the light of what was said above, it has a precise meaning. For too long, following a conception which is, in the final analysis, Aristotelian, every idea was considered fundamental if, on its basis, the other specific ideas of a research field could be defined. Turning to a dichotomy suggested by Kreisel, it does not matter whether a distinction is finally made between organization and foundation, seeing this last as being centered on validity rather than efficiency as is the case of organization. At most, the distinction only accentuates the separation between concrete practice and foundational analysis: in the end this only reduces the analysis of foundations to a Linus-like security blanket – if you will allow us the expression – for mathematicians.

Once again this perspective has a sharp idealistic flavor. It is not that one idea, in terms of which we can define other ideas, necessarily expresses a more profound aspect of reality: such an idea certainly has a priority in exposition, or at most in knowledge but, unless one maintains that the only important thing is the *way* we know rather than *what* we know, it cannot be assumed *ipso facto* as more fundamental than others. To give an example: an algebraic structure can be seen as a particular type of set,

but this does not mean that the concept of structure is less fundamental
than that of set since in a categorical context we can study the second
without necessarily presupposing the first. On the other hand, sets could
really be conceived as degenerate structures, by inverting the order of
priority. We can say, if even generically, that the fundamental ideas are
those which reflect invariant aspects, recurring structures of the largest
possible number of fields: in this sense one no longer creates that breach
between foundation and organization as a matter of principle, from which
we started above, and the characterization of fundamental ideas is not
indifferent to mathematical practice and its organization but instead has
constant relations with it.

This has another aspect which it is important to stress and to which
we referred implicitly above: the admission of a plurality of fundamental
ideas. When all is said and done, we did not make the world, and if
different ideas, different logics, are needed to understand it, that is not
our fault. Only if we conceive ideas and logic as arbitrary fruits of the
human intellect must we have reductionistic uniformity. Naturally this
does not mean, as was said above, that scientific development is to a
great extent caused by the desire to recompose a unity by resolving the
contradictions posed by different concepts; but this does not have to
mean at all an artificial and aprioristic reduction to a single base. We
have spoken of plurality of logics. From what was said above it seems
clear to us that one of the most significant results of recent research lies
precisely in the non-aprioristic justification of this plurality of logics,
adjusted from time to time to the specific 'environmental conditions' in
which they are to be applied. The establishment of intensional logics
with their intrinsic flexibility in adapting themselves to different 'states
of knowledge'; the development of 'empirical' logics as their specification
which substantially exploits the mediation of the neat True/False oppostion; the same non-arbitrariness of logic to be applied in the study of
mathematical theories brought to light by categorical analysis: all these
facts make it highly possible to maintain that it is completely gratuitous
or in any case unjustified to want to impose *a priori* a single rigid logical
scheme, valid for all cases.

Note that it is not a question of having to substitute one logic for
another, by imposing, for example, a choice between classical logic and
intuitionist logic which would somehow resolve the dispute (which, as we
saw, arose at the beginning of this century) in favor of intuitionist logic.
The results of the interpretations of topoi could certainly make one

incline toward this solution by assigning the victory palm in the age-old controversy to intuitionist logic and at the same time the ingrained halo of *absoluteness* to classical logic. But I believe that, among others, the results concerning quantum logic, with the individuation of 'real' logical systems *which cannot be compared* with intuitionist logic, are sufficient in order to have it understood that the question is much more complex than it seems, and in no way allows one to pass from absolute to absolute, so to speak. What is doubtful is precisely the whole basic argument concerning logic which considers logic completely *a priori*, separated from experience and fixed once and for all. The results we have cited would seem in fact to involve the consequence that 'reality' somehow chains and conditions logic, which comes to assume an 'empirical' dimension hence forth completely unthinkable and scornfully refused by an entire philosophical tradition. Naturally this poses a philosophical problem of particular importance – which we certainly cannot analyze here – of the relations between specific logics and natural language. One point seems to us undeniable: the decisive rejection of every attempt to subordinate logic to natural language and still less to use adherence to the latter as a criterion for justifying logic. The 'choice' of logic depends on the kind of practice in which one is engaged and hence, substantially, on the kind of objects studied.

We have returned to the discussion of practice to which we have appealed more than once in the course of these pages. It seems right to conclude with some considerations about this. As was said at the beginning, a practice is closely tied to the conditions, even political and social, in which it develops. Hence it may seem that the desire to increasingly restrict foundational research – and more generally logical investigation – to the various theoretical practices constitutes a kind of acceptance of the status quo, a renunciation at any attempt at critical analysis. Sociologically, at least, this may be partially true. In fact too many appeals to practice mask a substantial conservatism. In the case of logic, traditionally situated on the border between philosophy and mathematics, this appeal has often had the significance of refusal: research on foundations was rejected as irrelevant for practice, as speculative, as much by philosophers as by mathematicians. But practice is not only conservation: within it – think of what was said above about mathematics – arise contradictions, tensions, which push one to new analyses; the task of the logician is precisely that of taking this fact into account. In other words there exists a tension within specific scientific practice between adherence to old schemes and framing new perspectives. This provokes a juggling of time, so that often the 'good' practice of the future is the pure ideology of the present. It seems to us that this ideological role should be accepted in that it alone allows a possible transcendence, at least if the world is not conceived as rigidly determined. The only problem is to make certain that this ideological attitude does not become purely defensive and finally mystifying; the important thing, it seems to us, is to maintain the tension between real, present practice, its criticism, and new perspectives. There is no recipe for finding a proper equilibrium and only continual, free, and intellectually honest debate can guarantee it. In order for these conditions to be fulfilled materially, a radical change in the organization of scientific work is needed, particularly in the schools, but more generally in the social and political community, so that, in the final analysis, the criticism of science, without losing itself in (and confusing itself with) it, is certainly and positively grafted onto the criticism of society.

NOTES

[1] This article is taken from a long chapter devoted to a critical/historical presentation of twentieth-century logic appearing in Ludovico Geymonat's *Storia del pensiero filosofico e scientifico*. 9 Vols. Milan, Garzanti, 1976–. The section translated here – with slight modifications – can be found in volume 9, Chapter 3, pp. 253–273, and is printed here by the kind permission of Ludovico Geymonat and Garzanti publishers.

[2] It can have a pure and simple conventional classificatory value with the aim, for example, of 'ordering' a bibliography or bringing to light what may be recent applications of logic in this or that cultural field; or, at the most, of giving a name to a journal, or other. But – and here I fully agree with Hintikka (see below) – it has neither theoretical value nor theoretical justification.

[3] We can easily add to Hintikka's examples all the problems related to the theory of categories in its reflection on logic and on the very conception of mathematics as well as the problems related to the logical analysis of the empirical sciences, etc. It is worth reporting the conclusion of Hintikka's argument:

> This lack of interest of the philosophical community in some of the most truly philosophical parts of logic is of course partly inevitable. It is due to the difficulty of mastering a rapidly growing and in many cases highly technical field. What is really disconcerting is not so much the undiluted ignorance of some philosophers' comments on such old subjects as Gödel's results as their failure (or reluctance) to follow Socrates and recognize the extent of their ignorance.

Jaakko Hintikka, *Logic, Language-Games and Information; Kantian Themes in the Philosophy of Logic* (Oxford, Clarendon Press, 1973), pp. 1–2.

[4] 'Model theory 1945–1971' in *Proceedings of the Tarski Symposium; an international symposium held to honor Alfred Tarski on the occasion of his seventieth birthday*. Edited by Leon Henkin, *et al.* Proceedings of Symposia in Pure Mathematics, Volume 25. (Providence, American Mathematical Society, 1974), pp. 173–186.

[5] cf. Davis, W., *et al.* (eds.), *Philosophical Logic* (Dordrecht, Reidel, 1969).

[6] (Notre Dame, University of Notre Dame Press, 1969). This is a translation by John Wilkinson and Purcell Weaver, of *La nouvelle rhétorique: traité de l'argumentation*. (Paris, Presses universtaires de France, 1958).

[7] This quotation is taken from Igor S. Shafarevich, *Basic algebraic geometry* (Berlin, Springer, 1973), p. v. (This volume is a translation, by K. A. Hirsch, of *Osnovy algebraicheskoi geometrii*. Moscow, Nauka, 1972.) The original German version of the Klein quotation can be found in his *Vorlesungen über die Entwicklung der Mathematik im XIX Jahrhundert* (Berlin, Springer, 1926), p. 312.

[8] 'Mathematische Probleme'. Speech given at the Internationale Mathematiker-Kongress, Paris 1900, and published in the *Nachrichten von der Königlichen Gesellschaft der Wissenschaften zu Göttingen*, 253–297; and, with some additions, in *Archiv der Mathematik und Physik*, 3rd series 1 (1901), 44–63, 213–237.

[9] For example, the following works of Lawvere can be consulted: 'Functorial semantics of algebraic theories', *Proc. Nat. Acad. Sci.* (U.S.A.) **50** (1964), 869–872; 'An elementary theory of the category of sets', *ibid.* **51** (1964), 1506–1510; 'Adjointness in foundations', *Dialectica* **23** (1969), 281–296; 'Quantifiers and sheaves', *Actes* du Congrés International des Mathématiques, 1970, Volume 1, p. 239; 'Continuously variable sets; algebraic logic geometric logic', in *Logic Colloquium '73* (Amsterdam, North Holland, 1975), pp. 135–156; 'Variable quantities and variable structures in topoi', in *Algebra Topology and Category Theory. A Collection of Papers in Honor of Samuel Eilenberg*, edited by A. Heller and M. Tierney (New York, Academic Press, 1976), pp. 101–131.

[10] To our knowledge there is no organic presentation which illustrates all the motives and articulations of Lawvere's perspective. Hence every reference to his positions can turn out to be deviant or reductive. This is particularly the case in our present discussion which has as its sole objective that of bringing to light from the point of view of present-day logical research, some of the reasons why Lawvere's work is, in our opinion, significant. It is important to confirm this point because, as can be ascertained in the succinct bibliography cited above in Note 9, Lawvere's general point of view does not assume logic as foundational instrument, as the pole of his own research.

[11] See Note 8.

[12] On this point, cf. for eample, Lawvere's article in *Dialectica*, cited above in Note 9.

[13] M. Makkai and G. Reyes, *First-Order Categorical Logic*, Lecture Notes in Mathematics (Berlin, Springer, 1977), p. 611.

GIULIO GIORELLO

INTUITION AND RIGOR:
SOME PROBLEMS OF A 'LOGIC OF DISCOVERY' IN MATHEMATICS[1]

1. INTRODUCTION

In a famous intervention at the Second International Congress of Mathematicians (Paris, 1900), Poincaré divided research mathematicians, from Euclid onwards, into two groups: those who are *rigorous* (or 'analysts') and leave nothing to chance, and those who are intuitive (or 'geometricians') and make rapid, but often precarious conquests. While recognizing that "both analysis and synthesis have their legitimate role" and that consequently mathematical research, in order to make progress, cannot do without either group, he added, however, that "*in becoming rigorous, mathematics assumes an artificial character* [. . .]; *it forgets its historical origins.*"[2] Regressive results, in research not less than in didactics, could be avoided only by carefully distinguishing between the roles of the two groups: with 'analysis' it is mathematical proof; with 'synthesis', on the other hand, it is mathematical invention.

Poincaré's distinction between rigorization of mathematics and historical understanding of the origin of problems and theories follows undoubtedly from his conviction that 'absolute rigor'[3] had been reached and the history of rigor was concluded. Incidentally, it is worth remembering here, that it was the very investigation following the 'foundations crisis' which finally brought to light the presence of "different logical structures, proof procedures, and types of rationality"[4] in mathematics, thus testing the idea of a single logic and an absolute rigor. As we know, an analysis of 'justificationism', implicit in the original programs aiming at giving mathematics a foundation as secure as an 'eternal rock' (to use Frege's words), has been developed by Imre Lakatos from 1962 onwards. He was in a good position to remark that Russell's logicism, as well as other paths (such as the one embarked on by Hilbert's formalism), called for *strategic retreats*. But if we recognize that the 'ultimate test' of the admissibility of a method must *obviously* consider the question of 'whether this method is *intuitively convincing*', it is very difficult to give a satisfactory answer to the question: 'Why foundations, if they are admittedly subjective?' Only a methodological decision eludes infinite regress and allows

113

Maria Luisa Dalla Chiara (ed.), Italian Studies in the Philosophy of Science, 113–135.
Copyright © 1980 *by D. Reidel Publishing Company.*

us to stop at a given stage. But in this way the *fallible* character of the very same mathematics is recognized by attributing to it, as well, the tentative nature which Pierre Duhem noticed only in 'theoretical physics'; the tentative nature which even the early twentieth-century realists were forced to attribute to theoretical physics (and more generally to every empirical theory) pervades mathematics as well.

On these premises, it is easy to understand why we intend to give some consideration to Lakatos' views: (1) his criticism of justificationism, as far as it concerns foundations, brings out the risks of reducing all philosophical reflection on mathematics (i.e., the philosophy of mathematics) to the foundations of mathematics. This tendency, in fact, *reduces* the scope of philosophical reflection on mathematics and thereby encourages, albeit unwittingly, all mathematicians whose practice is outside such a scope, and who are inclined to envisage a sort of intuition as the key-element of discovery, eluding *a priori* any kind of rational investigation. (2) Lakatos' 'logic of mathematical discovery' allows us a flexible view of the intuition/rigor dichotomy. On this basis, in Section 2 we shall discuss the possibility of applying Lakatos' methodology of scientific research programs to mathematics; in Section 3 we shall deal with a historical case study (Riemann's research program in complex functions); finally in Section 4 we shall discuss some problems in Lakatos' logic of discovery.

2. THE LOGIC OF MATHEMATICAL DISCOVERY AND THE METHODOLOGY OF SCIENTIFIC RESEARCH PROGRAMS

According to Lakatos (1962), mathematics also is "fallible knowledge, knowledge without foundations". The affinity of this view with Popperian fallibilism is only too obvious. But Popper never developed a logic of *mathematical* discovery comparable in breadth and systematicity to the *Logic of Scientific Discovery*, even if many of his works contain stimulating references to mathematical 'knowledge'. The quotation marks are not haphazard: is such a term really appropriate for mathematics? For a Popperian, mathematics is 'non-empirical' in Popper's own sense (1959). This poses a new problem.

Let us go back to Poincaré for a moment: according to him logic offers only tautologies.[5] If – as for instance for some logical positivists – mathematics is then reduced to logic, it officially becomes *non*-knowledge. So that we can admit, as Hans Hahn does, that "at first glance it is difficult to believe that the whole of mathematics, with its theorems that it cost such labor to establish, with its results that so often surprise us, should admit of being resolved into tautologies"; but such a difficulty can be avoided by Hahn in the following way: we are not omniscient [*allwissend*], and while "an omniscient being, indeed, would at once know everything that is implicitly contained in the assertion of a few propositions", we on the other hand, in order to take any mathematical fact into account, must go through a series of formal writings.

For Hahn however, even the most important theorem is indeed a conquest, but only to the extent to which, since the chain of proofs is long, "it may prove quite surprising to us that in asserting a few propositions we have implicitly also asserted a proposition which seemingly is entirely different from them".[6] This argument in certain respects recalls the famous passage in Laplace which illustrates universal determinism and introduces the calculus of probabilities in the sciences of nature. In a *weak* sense, it is a word of caution: when accepting a given principle, look at its consequences; the methods of logic put you in a position to take in the most unexpected consequences and eventually 'exorcise' the pathologies with appropriate stratagems. In a *strong* sense, it asserts that mathematical propositions "say nothing about their objects" and allows us to conclude that, "while we do not know whether any alleged natural law has really the value of law" insofar as we are dealing with hypotheses, it is permissible to state that "in expressing hypotheses, we are expressing much more than we are saying" (making this explicit is the task of logic and mathematics). It is precisely with regard to the problem from which we started in Section 1 – how to characterize the role of intuition in an increasingly rigorized mathematics – that Hahn's argument, *in the strong sense*, does not at all forbid reevaluations of intuition as a 'short cut' which the *long* road of logic does not exclude in principle. Hardly a satisfying outcome, even from a positivist point of view.

In a bitter polemic with the classical 'dogmas of empiricism', Lakatos the 'dialectician' maintains, *on the other hand*, that, in abandoning the dichotomies of 'static rationality', it is possible to assert the 'quasi-empirical' character of mathematics. Certainly no one will claim that mathematics is empirical in the sense that its potential falsifiers are singular spatio-temporal statements, as it is the case with the empirical theories considered by Popper in his *Logic*; however, it will be possible to determine statements which for mathematical theories play a role analogous to that of their empirical counterparts.

We will not go into details about their determination.[7] However, it should be observed that in a 'non-empirical' (or better 'quasi-empirical') discipline, heuristic falsifiers *à la* Lakatos are limited only to suggesting 'falsifications' and, as we know, suggestions can be ignored. In reality, they are only rival hypotheses. But this does not separate mathematics from physics in as clear cut a fashion as would appear *prima facie*: Popper's basic statements are also rival hypotheses.

Furthermore Lakatos has not only underlined the importance of grasping the function of (mathematical) *proofs* in building science, but, following Polya, has also insisted on the necessity of noting that the *concept of proof* in mathematics *has evolved.* It is the very modalities of such changes that Lakatos proposed to examine in particular in *Proofs and Refutations*, in which the initial stimulus is provided by the case study of the various ramifications of mathematical research on Euler's conjecture about polyhedra.

Without becoming entangled in this particular case study, it is worth remembering that Lakatos' reason for such an approach was a critical need first articulated in the great works of nineteenth century mathematicians. Lakatos remarks that often a proof's assumptions are not suitably explained; when a counterexample is presented, it turns out to be difficult to explain its scope. The critical teaching of mathematicians like Weierstrass consisted in actually realizing the need for rigor, "not hiding a lemma [...] but assuming it publicly". This is confirmed by Lakatos: "the revolution of rigor" in nineteenth-century analysis constituted a formidable stimulus to the development of a critical method which succeeded in overcoming the reactionary opposition of those who mistook prejudices for intuition, and complained of the 'artificiality' of rigorous investigation. We shall return to this aspect of the question when we come to the case study in Section 3. These first sketches are starting points for reexamining the themes outlined in Section 1, especially in relation to Poincaré's judgments and their fortune with many mathematicians.

Let us add, following Lakatos (1963–64; 1976) that Weierstrass and the other creators of this revolution not only found 'proofs' of important statements, but also delineated 'a proof analysis'. The historian of mathematics then cannot neglect those 'conflicts' (which can involve formal or informal mathematical theories, metaphysical speculations, different conceptions of the physical world, etc.) which for Lakatos represent the moments, in mathematics as well as in the empirical sciences, in which knowledge grows. Once this pluralistic viewpoint has been adopted both in the problems of foundations (Lakatos (1962; 1967; 1976b) and in proof analysis (Lakatos (1963–64; 1976a), we are faced in these contexts too with the basic problem of falsificationism: in the conflicts (between formal and informal theorems, between the increase of content and the demand for rigor, etc.) which have flourished in the history of mathematics since the time of the Pythagoreans, what *part* of the 'system' is the true target of 'refutation'? Here too we can certainly emphasize *the necessity of methodological decisions* in order to be able to promote a counterexample (which may be avoided with suitable adjustments) to the rank of a far-ranging falsifier of a mathematical theory, by adopting a 'resolute' Popperian strategy.

But is not such a strategy, here as well as in the empirical sciences, *too* bold? The history of mathematics in the second half of the nineteenth century offers many examples of researchers who in some cases heroically adhered to a code of scientific honesty, rejecting some directions in research in the name of standards of rigor commonly accepted as 'background knowledge', but who then repented when confronted with more unprejudiced, successful research, and sometimes recognized the 'narrowness' of their original criteria. We can then side with that mathematician who stated that if the most prestigious mathematical theories resulted in a conflict with the most familiar logical laws, for example the *tertium non datur*, *the latter* should be dropped. Within the context of mathematical discovery as well it is necessary to take into account a certain *continuity* in research.

This is the reason why the methodology of scientific research programs seems promising: even in this context it should be possible to specify the 'hard cores' and 'protective belts' of the various programs and apply those criteria which will allow us to evaluate their progressive or regressive character, as Lakatos (1970) did for the empirical disciplines. A typical development might be the following: a program which successfully solves the problems it was originally intended to solve will introduce new languages and new theories into which it must *translate* already existing and autonomously formed disciplines, in order to profitably assimilate them (that is, in order to use the translations of the traditional theorems of such theories): our interest will be focused on *non-trivial* translations wherein the development of the program opens *new* problems; the extent to which the heuristic of the program provides a solution to such problems will allow a first appraisal of the program itself, etc. As the reader expects by now, those who adopt Lakatos' methodology in mathematics will look for 'case studies', which in this context play a role analogous to that played by 'Prout's program', and 'Bohr's program' in Lakatos (1970). It is a question of showing that the standards of the methodology of scientific research programs are at the same time closer to the actual research practice than the strictly Popperian ones, and are more satisfactory, in terms of an 'objectivist' perspective (in Popper's sense), than the considerations at once "descriptive and normative"[8] advanced by Thomas Kuhn.

Ever since the first edition of *The Structure of Scientific Revolutions* moreover, there have been various attempts to consider significant turning points in mathematical thought in a Kuhnian perspective as well. We shall not discuss here the accusations of irrationalism against Kuhn's conceptions of the *function of dogma, normal science, progress through revolutions*, etc. Our criticism will be limited to the fact that in the growth of knowledge, *concatenations* between scientific achievements are observed which no typology of scientific change should ignore, but strangely enough, the author of *The Structure of Scientific Revolutions* seems to pay little attention to this.[9] In our opinion, (in the context of mathematical discovery) *Lakatos' perspective provides more adequate tools than Kuhn's*; the aim of this paper is to consider a relevant case study along the lines of the methodology of scientific research programs, and thus to explain some problems in the light of the intuition/rigor dichotomy from which we began.

3. A CASE STUDY: RIEMANN'S RESEARCH PROGRAM

This historical case study is suggested to us by Poincaré's intervention with which we opened this paper (Section 1). In fact, for Poincaré, Riemann is the prototype of the *intuitive* mathematician;[10] and besides, among Poincaré's examples (1900) of how intuition can deceive, there is that '*summary proof*' [*démonstration sommaire*] of the so-called Dirichlet principle which Weierstrass was the first to point out as a weak spot in Riemann's approach to the theory of functions of a complex variable. It seems interesting to test Lakatos' approach (1963–64), revisited in the light of his methodology of scientific research programs, by selecting an example[11] of *discovery* and *criticism* in mathematics which does not concern the problem of foundations in the strict sense, but affects various areas of mathematics. Of course, we are not proposing here to give a complete picture of Riemann's ideas or results; we will limit ourselves to some particular aspects[12] which provide suitable *historical* references for the concluding discussion of Section 4.

3.1 *Riemann and the Dirichlet Principle*

Let us consider Riemann's approach to functions of a complex variable. His definition of a single-valued analysis function

$$w = f(z) = u + i v \quad \text{where} \quad z = x + i y$$

states that the function is analytic in a point and its neighborhood if it is continuous and differentiable and satisfies what we now call the Cauchy–Riemann equations. From this it follows, assuming that the derivatives appearing in such equations are differentiable with respect to x and y, that u and v satisfy the two-dimensional potential equation; the complex function w can then be determined once and for all throughout its domain of existence: in particular it is a matter of determining the harmonic function u, since, in view of the Cauchy–Riemann equations, v is also determined up to an additive constant.[13] A well-known 'fact' in Riemann's time (Cauchy) provides the motive for a considerable section of the heuristics of Riemann's program, that is, a 'reductionist' approach relating the foundations of his theory of complex analytic functions to the study of harmonic functions. This explains, among other things, why Riemann applies potential theory in the bi-dimensional case and not, like Dirichlet and other predecessors, in the three-dimensional case. Furthermore, in this way Riemann finds the means for determining u in what he christens the *Dirichlet principle*, having learned it from Dirichlet.[14]

It is then in the constellation of mathematical ideas of the time (the tradition of the Cauchy–Riemann equations, Dirichlet's teaching, etc.) that Riemann's approach has its origins: *the theories which make up a great research program do not come to light unexpectedly, but find rational motives in 'constellations' of this kind*. A research program "displays a potentiality for generating a succession of new theories in accordance with its heuristic, *which may have been articulated from the start*":[15] the links of such a process to previous or contemporary research results constitute, so to speak, the boundary conditions of the process; it is necessary to take them into account in order to understand the course of the process even if they are insufficient to explain its entire development. But let us return to the Dirichlet principle. In short, it states that the function u, which minimizes the so-called Dirichlet integral, $D(u)$, satisfies the potential equation; the latter is in fact the necessary condition for annulling the first variation of the Dirichlet integral. Now, like Gauss and Dirichlet, Riemann observed that $D(u)$ is never negative; and concluded from this that there must be an 'admissible' function – as we say today – by which $D(u)$ reaches its greatest lower bound, thus providing a solution to the problem. This last passage makes Riemann's reasoning a 'summary proof' [*démonstration sommaire*].

It is worth analyzing closely the arguments adopted by Weierstrass on the non-admissibility [*Unzulässigkeit*] of the Dirichlet principle. In the report 'Uber das sogenannte Dirichlet'sche Prinzip'[16] (read in 1870), Weierstrass' criticism is based on an example which shows that there are variational problems in which a greatest lower bound need not be a minimum value.[17] Even if this criticism immediately generated a certain mistrust of Dirichlet's principle (some mathematicians refused to consider the results obtained with Riemann's method), it actually excludes only one step in Riemann's reasoning, and does not in itself constitute a 'refutation' of the Dirichlet principle. In Lakatos' terminology (1963–64; 1975), *Weierstrass' counterexample would be 'local', not 'global'. At most, it can be concluded that the validity of such a principle is doubtful.*[18] Nor is it less interesting that after 1870 some mathematicians believed that it was possible to prove the validity of the Dirichlet principle in an *indirect* way for a *large* class of domains: once the Dirichlet principle is challenged, it is a matter of determining the existence of the desired harmonic function by *other* means; every success in this direction results in an *indirect* confirmation of the principle itself. Only in 1906 did an example of Hadamard show that this fact was not generally verified; however, we are not so much concerned here with pursuing the criticism of the Dirichlet principle,[19] as rather with observing that the attempts which we have mentioned represent a continuation of Riemann's program, though with appreciable adjustments.

The idea we are testing in our case study is central to Lakatos' entire approach, according to which even highly progressive research programs succeed in reabsorbing counterevidence and counterexamples – sometimes at the cost of considerable adjustments – only gradually; notwithstanding this (or maybe *for this very reason*), *research lines are carried out without paying 'excessive' attention to refutations*. Unfortunately we do not have Riemann's reactions (he died in 1866) to Weierstrass' criticism of 1870! But in his lecture on Riemann's work given in Vienna in 1894, Felix Klein observed that if indeed the Riemann edifice *prima facie* "threatens to collapse into ruins" because of Weierstrass' criticism, *such pessimism is in reality unjustified*, insofar as "the fertile results obtained by Riemann are completely exact as Karl Neumann and Schwarz later proved with perfect rigor".[20] In this *comedy*[21] the acts follow one another according to a scheme dear to Klein (first "physical intuition", then the results established "with perfect rigor"). But the 'rational reconstruction' which Klein proposes is nonetheless extremely interesting: it was only after Riemann had "drawn his theorems from physical intuition, which once more confirmed the method of discovery [*heuristisch*]", that he *then* recurred to the Dirichlet principle "with the aim of making a set of mathematical procedures homogeneous". In the same passage *in a footnote* Klein specifies that "contrary to a very widespread mode of expression", the term *principle* should be understood as the 'concatenation' of mathematical reasoning and not as the result which is obtained. We could compare the role of Klein's 'principles' to the idea of certain art historians, that the ribbed vault would have served to support the Gothic structure more in its various phases of construction than when it was completed and the play of thrusts and counterthrusts were able to hold up the building. And, going outside the metaphor, according to the methodology of scientific research programs *this is really the function of heuristics*: we aim at overcoming the traditional attempt of mathematicians "to present the crystallized product of their thought as amounting to general deductive theory while relegating the individual mathematical phenomenon to the rank of examples".[22]

That inductivist residue still present in Klein's scheme is not very important: the interesting point lies not so much in the fact that, psychologically, Riemann's use of the Dirichlet principle derives from 'physical facts' as in the fact that such 'facts' *are adopted to defend* the use of the principle itself. In this rational reconstruction, *even fragmentary information about the behavior of researchers results in something more than simple anecdotes*. Klein himself tells us that Riemann was perfectly aware of the difficulty of his approach but did not worry, "seeing that his closest circle and Gauss himself had admitted similar reasonings in analogous cases".[23] Must we then deplore "the end of critical rationality" together with the 'naive' (read: Popperian) falsificationist?[24] It does not seem to us that Riemann's and Klein's defenses authorize such a drastic conclusion. *In the long run, the behavior of a 'Lakatosian' mathematician seems much more rational than that of a 'Popperian' one.*

3.2 *The Definition of Mathematical Entities in Terms of their Behavior in the Infinitely Small*

In his Vienna talk, Klein also recalled that Riemann, who grew up "in the great tradition symbolized by the names of both Gauss and Wilhelm Weber" but was also "influenced by Herbart's philosophy", always worked *to find a mathematical form in which he could express, in a unified way, the laws which govern physical phenomena*. Even if the traces of such research are to be sought above all in the posthumous fragments, much of Riemann's work, according to Klein, is still guided by an idea close to that "adopted today by the new school of physicists who follow Maxwell with regard to his electromagnetic theory of light": *the hypothesis according to which space is filled with a continuously spread fluid which is the medium for the manifestations of light, electricity, and gravity*. Even if "today" (1894), Klein remarks, considerations of this kind "are only of historical interest", it must still be emphasized that *"it is in these ideas that one must look for the source of developments in pure mathematics due to Riemann"*: the role played in physics by the negation of action at a distance and the explanation of phenomena in terms of "forces within an ether filling space", is played instead in mathematics by the definition of functions *in terms of their behavior in the infinitely small* and, consequently, *in particular in terms of the differential equations which they satisfy*.

The guiding principle of Riemann's research which emerges from Klein's talk is, in our opinion, an essential element of the *hard core* of his research program. For instance, it will be useful to recall that such a principle is also obviously at work in Riemann's research in differential geometry, in particular in 'Über die Hypothesen, welche der Geometrie zu Grunde liegen' ('On the hypotheses which lie at the foundations of geometry') where the definition of the linear element ds is given as the square root of a differential quadratic form of n variables. With this remark, we do not wish to discuss Riemann's contributions to differential geometry, but only to pinpoint one aspect of that deep *unity* which historians of mathematics now acknowledge in the various contributions of the German mathematician (we shall take up this principle in Section 4).

3.3 *Riemann's Therapy for the Prejudices 'Inadvertently Transmitted through Language'.*

Not only did Riemann not fear extremely general investigations, but he was also aware of the conceptual relevance inherent in research of this kind. This very research – as he states in 'Über Hypothesen' – operates so that "*der Fortschritt im Erkennen des Zusammenhangs der Dinge*" ['progress in perceiving the connection of things'][25] is not hindered by the prejudices of tradition. And in the posthumous writing, 'Erkenntnisstheoretisches',[26] this point of view is taken up again and scrutinized: it is precisely in order to fight against such prejudices *inadvertently transmitted with language* that the researcher must recur to systems of concepts endowed with maximum generality and hence free from the constraints of ordinary language.

We must remember that, like the observations quoted from 'Über die Hypothesen', the epistemological analysis sketched in the fragment is, for Riemann, also inherent both in mathematics and in the sciences of nature; this depends on the contextualization of geometry in the framework of 'experience', in a very broad sense of the word; and in Riemann all this is functional to the accepted considerations of differential geometry as a start toward the awareness of the abstractness of all mathematics. But the particular philosophical *therapy* proposed by Riemann is closely linked to two theses which correspond to two classical 'metaphysical' questions-and-answers: "When is our conception of the world true? When the connection of our representations corresponds to the connection of things. [. . .] On what should the connection of things be founded? On the connection of appearances". For Riemann then, "the connections must correspond to one another" and further, "the representation of objects of speculation in determinate spatio-temporal relations is that which is met with in the course of speculation on nature or which is given by itself."[27]

It seems that 'metaphysical' (in the broadest sense of the term) theses also find a place in the hard core of Riemann's program. They seem *prima facie* to characterize Riemann's own *image of science*; it is remarkable that such theses (consider particularly the phrase *Zusammenhang der Dinge* ['connection of things'] refer to the conclusions of such a great scientific contribution as "Über Hypothesen'.

3.4 *Conclusions about Riemann's Program*

Let us draw some conclusions from what we have seen in sections 3.1, 3.2, and 3.3.

(1) Section 3.2 explains in terms of mathematical discovery, why Riemann's conception of analytic functions differs from that of both Cauchy and Weierstrass.[28] This is obviously one of the elements which can help to clarify the reasons for Weierstrass' harsh criticisms of Riemann (an ex-

ample in point is the criticism of 1870 mentioned in Section 3.1). Two different mathematical *styles* are being compared: Poincaré observed that Weierstrass' approach to analytic functions was part of a reduction of the entire analysis to "*a kind of extension of arithmetic*", and recalled that his works do not contain a single figure while "*Riemann immediately calls geometry to his aid*".[29] But, *in nuce*, the *antithesis* is already in Weierstrass himself;[30] it was not by chance, for example, that in 1864, he stated, with regard to results on elliptical integrals that "he understood Riemann *because he already had the results of the research*" and also that "Riemann's pupils were wrong in attributing everything to their master while many [discoveries] had already been made by and were due to Cauchy, . . . and Riemann *only dressed them up in his own way as he pleased*". More generally, "in Berlin, Riemann's work caused difficulties."[31] What Poincaré in 1900 presented as the antithesis between 'rigor' and 'intuition' is rather *a contrast between extremely different views of mathematical work, and also between different types of rigor* (as was seen in Section 3.2, Riemann too has his own conception of the nature and function of mathematical rigor).

(2) Why then has such an antithesis been *reinterpreted* by Poincaré in the way we have seen? The quotations we gave in Section 3.3 illustrating '*der Zusammenhang der Dinge*' ['the connection of things'] do not seem too far from assertions which are repeated, for instance, in *La science et l'hypothèse* (which state that what the sciences can reach are not the things-in-themselves but exclusively the relations between things) even if such assertions are sometimes forced by Poincaré in a conventionalist sense. But if we recall Poincaré's perplexities in using that "*langue hypergéométrique*" which constitutes one of the most interesting features of Riemann's 'bold' program,[32] as well as his reluctance to appeal to intuition *on this ground*, we can understand that what appeared to Riemann as *rigor* was for Poincaré barely tested (and in principle, barely testable) *intuition*.

(3) That said, we have pinpointed *at least three* important 'principles' in Riemann's program: (a) the Dirichlet principle (see Section 3.1); (b) the definition of mathematical entities by means of their behavior in the infinitely small (see Section 3.2); (c) the 'metaphysical' theses mentioned in Section 3.3. What can we still say about (a) – (c) using the methodology of scientific research programs? To what extent does it allow us to grasp the *specificity* of the role of each principle? In Section 3.1 we already anticipated something about the heuristic function of (a). Can we now

conclude that (b) and (c) are rightfully part of the program's *metaphysical hard core*?

4. THE DIFFERENT TYPES OF RIGOR AND THE ROLE OF INTUITION

In our opinion, we cannot be satisfied with a simple affirmative answer; it is instead a matter of emphasizing the 'fine structure' of the research program's hard core. But in order to do this we shall refer (within the limits of mathematical discovery) to a notion drawn from mathematical practice as well as daily experience, that of a *translation scheme* (we warn you immediately that the term is used here in a slightly Pickwickian sense).

4.1 *The Function of 'Dictionaries' (or 'Translation Schemes').*

We refer, for example, to the consistency of proofs concerning non-Euclidean geometries, etc.[33] But we can go beyond this restricted accepted meaning. As Lakatos himself observed, "in establishing a dictionary" (i.e. a translation scheme in our terminology) *prima facie* incomparable theories can be *made* comparable; furthermore, with suitable *reinterpretations* two incompatible theories can be made compatible in the usual interpretation of their descriptive terms. In considerations of this kind the reader will easily recognize a generalization of what happened precisely for non-Euclidean geometries. We can push a bit further still by reading in this perspective the observations of 'Über Hypothesen' concerning the fact that the ordinary Euclidean metric of space is only *one* of the infinite number of metrics which the expression of the linear element ds (cf. Section 3.2) allows for $n = 3$. We can now understand how the definition of mathematical entities in terms of their behavior in the infinitely small (principle (b)) *also* allows us to rediscover the traditional conception as a particular case, and at the same time to "free ourselves completely from the chains of Euclid"[34] opening up the possibility of an *infinite* number of geometries. A translation scheme then, is that which shows an old and familiar theory to be a particular instance of a more abstract and more general conception; the whole of these linguistic indications (for instance, those which we find in 'Über Hypothesen' in the passages which account for the geometric tradition 'from Euclid to Legendre'), which allows us to link the principles of heuristics (and/or the 'results' one reaches from these principles) with the 'background' of commonly accepted assertions, enables us to overthrow "any arbitrarily chosen pillar of established knowledge".[35] In this sense, *dictionaries allow us to evaluate rationally theoretical innovation, to specify – in Kuhn's language – the significance of a 'scientific revolution'*; but in and of themselves they are neither a new 'paradigm' nor the whole 'hard core' of a new research program: they are only the tools for the rational comparison of different theoretical proposals.[36]

4.2 *The Problematic Character of Translation Schemes*

In this way we discover the dialectic of intuition and rigor. In our sense, the translation scheme allows us to actually show the limits of an old theory (and with them "the prejudices which are transmitted with language") and in so doing provides us with *standards of rigor*. But the most 'abstract' theory is a 'work in progress'; the possibilities of tradition can be *prima facie* more than one, the auxiliary theories to which one has recourse highly problematic, etc. The methodology of scientific research programs only tells us that, of the different paths which a program can take, compared with older, traditional theories, some will be better than others to the extent that they will give 'novel facts' which have passed those tests set by the various standards of rigor, etc. But all this can be ascertained only with hindsight: how *rational* are the 'rational' evaluations which translation schemes seem to allow researchers? The introduction of a translation scheme among the categories of the logic of mathematical discovery is *considerably problematic*. In fact it involves such a logic in all the well-known difficulties surrounding the notion of translation. It is an old story: LeRoy against Poincaré at one time, Kuhn and Feyerabend against Lakatos today. But we do not want to discuss here the broad group of themes regarding this question in general. We shall instead limit ourselvesto our case study, trying to determine the role of sentences of type (c).

The whole 'Erkenntnisstheoretisches' fragment no less than the analogous observations of 'Über Hypothesen', provide further clarifications. Precisely where he specifies the conditions under which we can assert that our 'image of the world' [*Bild von der Welt*] is true, Riemann also notes that "the truth of the images is independent of the degree of the fineness of the image" [*Die Wahrheit des Bildes ist unabhängig von dem Grade der Feinheit des Bildes*]. It is precisely for this reason that when one element of this image is replaced by one or several 'more precise' elements, we can still speak of 'progress' in our comprehension of the connection of things, "but without having to declare the earlier conception false" [*aber ohne dass die frühere Auffassung für falsch erklärt werden müsste*].

A few lines above, Riemann had outlined the course of scientific knowledge with the following words:

The systems of concepts which lie at their basis originated through gradual transformation of older systems of concepts, and the foundations, which pushed toward new modes
of explanation, are always brought back to the contradictions or improbabilities which
characterize the older, modes of explanation.[37]

Up to now, we have arrived in a different way to a conclusion which we
had established through a re-examination of the role of the Dirichlet
principle in Riemann's heuristics: the promoters of the great research
programs (in mathematics) unfailingly avoid the difficulties of naive
falsificationism. But now the same possibility of a growth of mathematical
knowledge is involved: consequently we are looking for some 'principle'
to characterize this growth. One could have recourse *prima facie* to some
sort of Popperian "principle of correspondence";[38] but Riemann did
not speak of *Begriffsysteme* ['systems of concepts'] haphazardly: if we
assume that it is the basic *concepts* (also in mathematics or in highly mathematized theories) which change from one theoretical system to another,
such a principle cannot correlate the concepts of different theoretical
systems. We can push a bit beyond our interpretation to bridge this gap.
Riemann not only speaks of the completion and integration of old conceptual schemes by new ones (*Ergänzung* and *Verbesserung* are the German
mathematician's terms) but also insists on the gradualness of this process.
Implicit then is also a principle of charity,[39] so to speak, *which we can
formulate explicitly by our translation schemes*: the more the selected
translation scheme allows one to save assertions of the old context, the
more the new system 'integrates' the old, whereas the corrections appear
all the more drastic (and 'pitiless') the more assertions are dropped. Now,
whatever the claims of the naive falsificationist, in our opinion we should
not label the attitude of those who liberally use the principle of charity
'conservative' or 'dogmatic', and that of those who instead pitilessly
destroy, 'revolutionary' (or 'critical'). *In cases of this kind the traditional
Popperian dichotomies* (dogmatic attitudes vs. critical attitude) or *Kuhnian
dichotomies* (normal science vs. revolutionary science) *do not say much.
We need a model of the growth of mathematical knowledge, which takes into
account the fine structure of the process to which Riemann refers.*

   We can understand this by examining an apparent difficulty in the use
of principles such as those above. Naturally the truth-value assignments

are not invariant under the translation process. So, in 'Über die Hypothesen', Riemann, in opening up the possibility of a finite and unlimited
universe, observes that one must distinguish unlimitedness [*Unbegrenzheit*]
from infinity [*Unendlichkeit*].[40] It is this splitting up of fundamental
concepts which makes our translation schemes highly problematic and
which poses, once again, the question of the relationships between different
languages, including the relationships between the language of a considerably mathematized theory and ordinary language; in this instance,
the relations between a language (*before* Riemann) in which 'infinite'
and 'unlimited' are almost synonymous and one (*with* and *after* Riemann)
in which they are *rigorously* held distinct. Precisely because the languages
in which we formulate various theories present 'obscure zones', translation
schemes are of necessity never difinitive, always shifting and not less
hypothetical than the theories they bridge together.[41] *But this does not
constitute a difficulty*: since we are not dealing only with beliefs (or as
Riemann would say, with 'prejudices') but with scientific theories (in this
instance, mathematical ones), *just as falsification in the strong sense was
an abstract 'code of intellectual honesty', the ideal of conservation in a
strong sense is likewise an abstract code of dishonesty*: and as we have
already shown, it is at this level of the mathematician's practice that
rigor, which is a guarantee against such regressive results, intervenes. So,
by another path we find the 'moral' which we had drawn from the examination of Riemann's ideas and Weierstrass' criticism: *the significant
conflicts in the philosophy of mathematics* – in the broad sense in which this
term is employed in the present work – *are not so much between 'rigor'
and 'intuition', considered as antithetical poles, as between different levels of
rigor and different levels of intuition. This does not mean that the contrast
between 'intuitive' and 'rigorous' mathematicians makes no sense; but it is
a shifting one*; it is, furthermore, *a secondary product of the struggles in
which the 'old guard' is liquidated by the 'new'*. To forget these deeper
contrasts can lead to clichés which are functional to a certain 'image of
science' (as in Poincaré), whereas if we take them into account we will
perhaps be able to appreciate, at least to a certain extent, what Kaufmann
considered one of the most fascinating characteristics of the study of
history: "the perception of discontinuity in the context of continuity".[42]
Our translation schemes – precisely when similar 'splitting up' phenomena
are found – *make us 'feel' how researchers like Riemann can be profoundly
influenced by others and yet be strikingly original and even revolutionary.*

4.3 *The Role of Intuition*

A translation scheme – in our sense – indicates a mediation: but obviously there are identifiable 'zones' of the scheme which change with time, mediations which vary in intensity. *A historical element cannot then be eliminated from the translation scheme*; the 'interactions' between its different components are functions of time; a modification which takes place in a given zone can induce slight variations in another zone or radical restructurings in apparently far-off levels. Hence, not only must the historian analyze translation schemes, but also the most attentive researcher (like Riemann, like Klein, etc.) must look at the history of science in order to reconstruct "retrospectively [. . .] the course taken by our knowledge". This said, *all these considerations remain to some extent paradoxical*: we are faced with the problem of a 'good' translation just when we must specify what its criteria are; but before applying them, how will we know which propositions of other theories will be preserved in a particular theory of our program? We think that they are maintained or abandoned *in another way*, not by virtue of a direct comparison but by means of reference to mathematical 'facts' which in some way provide test-assertions.

We can consider the question in the following way, finding *a role for intuition* which in part rehabilitates the statements of mathematicians like Poincaré, Enriques, Severi, etc., concerning the 'particular' nature of mathematical intuition, without, however, having to conclude that rigor marks the end of creativity or having to concede something to the characterization (cf. Section 2) of mathematical intuition as a "short cut to reach a certain result". *Intuition now denotes that set of heuristic principles which leads to preferring one mode of 'translating' rather than another, to constructing one translation scheme rather than another.*

But have we not taken up again, with some cue from Poincaré, all the voluntarism implicit in much of French conventionalism? In inserting intuition into 'the third world of Plato and Popper' have we not re-evaluated too much the 'ingenuity' of the individual researcher? Have we not reconciled intuition and rigor at too high a price?

These questions do not worry us much. In the effort to comprehend mathematical discovery in its various aspects, *the problem is not to reduce one of these aspects to the other, but to try and understand how they are articulated and combined to construct a unique human phenomenon*, which appears in history in different ways. In this specific case, intuition is no longer something mysterious (or only "*the affair of the psychologist and the metaphysician*" as Poincaré would have said) but can at least be characterized by means of the various translation schemes produced continuously by history.

Furthermore: we can now distinguish between the roles of (b) and (c) (cf. Section 3.4): *principles like b appear to be the object of the mediation of the translation scheme; assertions like c have the function of justifying such a mediation.*[43]

Finally, in the perspective adopted here, if the problem of comparing results expressed in different languages turns out to be essential for the mathematical researcher (not less than for the historian of mathematics), *the assertions which can be labelled as 'influential metaphysics' or 'metaphysical framework' really belong to what we could call, in the best sense of the term, 'rhetoric of scientific communication'.*

But is their inclusion in the hard core of a research program in Lakatos' sense then justified? In a sense, sentences of type (c) are still elements of a 'metaphysical hard core' if these are characterized, as in Koertge (1973), as the sentences which are preserved 'as long as possible;' however, it is trivial to say that Riemann referred to 'philosophical' considerations (of which those in Section 3.3 are an example) as to the last sentences which he would have abandoned; in other aspects they do not seem to be so much sentences which are in some way 'testable' as elements of a general 'image of science' or *Weltanschauung* and therefore *traditionally* metaphysical: but both characterizations should not make us forget that, in Riemann's program, they have the function of justifying the transition to higher levels of abstraction – this feature is common to all Riemann's research and, since his inaugural dissertation of 1851, had been expressed by the thesis according to which "in order to make the field of research easier" one must examine points "which belong to more general domains". From a certain point of view, therefore, considerations of this kind should be 'shifted' from the 'hard core' of the program to the farthest 'periphery' of its 'protective belt', precisely in view of their *rhetorical function.*

Do we hold then that one of the more distinctive characteristics of the methodology of scientific research programs – the distinction between the metaphysical hard core and the protective belt, in characterizing the heuristics – cannot be applied to case studies such as the one under consideration, or that it should be abandoned? As an absolute scheme, this distinction, in our opinion, leads to considerable difficulties. However, if it is assumed as a working hypothesis, in a flexible, non-schematic way, it will allow us to characterize assertions such as those of type (c), in our example as at once 'metaphysical' and 'auxiliary' according to our interests in the reconstruction of the research program.

It goes without saying that our remarks in no way reduce the importance – both scientific and cultural – of 'philosophical' theses such as those of Riemann discussed here: on the contrary, they are so important that it can be rightly maintained that, precisely because scientific discovery cannot be reduced to the private domain but is a public fact, Riemann's articulated heuristics is not at all intelligible without them. Consequently, a logic of discovery which excludes them programmatically, relegating them to purely 'external' motivations, would give an extremely 'poor' rational reconstruction. That the re-evaluation of passages such as those quoted from 'Erkenntnisstheoretisches' derives from a perspective which partly recognizes the gradualistic nature of discovery and can therefore be interpreted *also* as a rehabilitation of a *cumulative growth* in mathematics, can surprise only an orthodox falsificationist. The growth outlined here is 'cumulative' in a rather Pickwickian sense: as Riemann's words indicate, the path which leads away from the *'Oberfläche der Erscheinungen'*, from the 'surface of appearances' is not marked by the static contrast between intuitive and rigorous researchers, but by the tortuous dialectic between various kinds of rigor and intuition.

NOTES

1 While writing this essay, I have been influenced by many stimulating discussions with several colleagues, who advised me during the draft and after the completion of the manuscript. I wish to thank in particular Umberto Bottazzini, Ludovico Geymonat, Stefano Magistretti, Roberto Maiocchi, Marco Mondadori, Fulvio Papi.
2 "en devenant rigueureuse, la Science mathématique prend un caractère artificiel [. . .]; elle oublie ses origines historiques." Poincaré (1900), pp. 118 and 123, respectively.
3 "la rigueur absolue". Poincaré (1900), p. 122.
4 Geymonat (1953), p. 64.
5 Poincaré (1900), p. 121 in particular.
6 Hahn (1933).
7 We are referring in particular to Lakatos (1967; 1976b) and Howson (1975).
8 Feyerabend (1970).
9 Cf. the pertinent criticism directed at Kuhn by Russo (1974), in particular p. 362, Note 1.
10 As pinpointed in Mooij (1966), pp. 59–60 and 115–116, Poincaré, especially in his (1900), distinguishes between three forms of intuition: *sensorial intuition* based on the

images of the senses; *generalizing intuition* "shaped, so to speak, on the processes of the experimental sciences", and the *intuition of pure number*, the only one which does not deceive us. On the ambiguities of such a distinction, cf. Mooij (1966), especially p. 60.

[11] Here I refer again to some starting points from my (1975a).

[12] For a general picture of Riemann's mathematical work, see Kline (1972), in particular pp. 655–66; 691–93; 722–24; 889–99. For an extremely stimulating view of Riemann's contributions to algebraic geometry which brings out clearly the profound significance of Riemann's abstraction and generalization, cf. Dieudonné (1974), in particular pp. 42–57.

[13] For a more detailed treatment see Monna (1975), in particular pp. 68 ff.

[14] Cf. Kline (1972), p. 659.

[15] Urbach (1975), [Italics mine.]

[16] Weierstrass. Karl: *Mathematische Werke*, 7 Vols, Berlin, Mayer and Müller, 1895–1924. See Vol. 2, pp. 49–54.

[17] For a detailed analysis of Weierstrass' objection see Monna (1975), p. 36.

[18] Cf. Monna (1975), pp. 36–37.

[19] For further details, see Monna (1975), pp. 37 ff.

[20] ". . . sind die weitreichenden Resultate, welche Riemann auf das genannte Prinzip stützt, alle richtig, wie dies Carl Neumann und Schwarz durch strenge Methoden später ausführlich gezeigt haben." Klein (1894), p. 492.

[21] The term is from Monna (1975): "a mathematical comedy of errors".

[22] Courant (1950), p. vii.

[23] ". . . aber in Hinblicke darauf, dass er die Schlussweise in analogen Fällen von seiner Umgebung, selbst von Gauss, anstandslos angenommen sah . . . " Klein (1894), p. 492.

[24] Lakatos (1976a), p. 54.

[25] *Gesammelte mathematische Werke* (New York, Dover, 1953), p. 286. All my references to Riemann's works are to this volume, a reprint of the second edition (Leipzig, Teubner, 1892) and the 1902 supplement (Nachträge) edited by Heinrich Weber. (An English version – by Henry White – of 'Über die Hypothesen' can be found in *A Source Book in Mathematics* compiled by David Eugene Smith (New York, McGraw-Hill, 1929), pp. 411–425; see p. 425 in particular.)

[26] *Ibid.*, pp. 521–25; see also Geymonat (1957).

[27] "Wann ist unsere Auffassung der Welt wahr? Wenn der Zusammenhang unserer Vorstellungen dem Zusammenhange der Dinge entspricht. . . . Woraus soll der Zusammenhang der Dinge gefunden werden? Aus dem Zusammenhange der Erscheinungen. . . . Die Verbindungen müssen einander entsprechen . . . Die Vorstellung von Sinnendingen in bestimmten räumlichen und zeitlichen Verhältnissen ist dasjenige, was beim absichtlichen Nachdenken über die Natur vorgefunden wird oder für dasselbe gegeben ist." *Ibid.*, p. 523.

[28] Manning (1975) insists on this difference which is reflected, from a historical point of view, by the fact that the various approaches originate from different sources; see pp. 358–59. Manning's work is devoted in particular to the emergence of Weierstrass' approach.

[29] Poincaré (1900), p. 117.

[30] See, for example, Dugac (1973), in particular pp. 88–90.

[31] These quotations are taken from the – as yet unpublished – notes of Felice Casorati concerning his meeting with Weierstrass on 17 October, 1864. I am particularly grateful to Umberto Bottazzini for referring me to this important source. At the time when I was writing this paper Bottazzini was working on these notes within the framework of detailed research into the connections between the Italian and German mathematical milieux in the second half of the nineteenth century.

[32] Cf. Bourbaki (1969), pp. 198–99.

[33] Cf., for example, Poincaré (1902), Chapter III.

[34] Fano (1929), p. 472.

[35] Lakatos (1970), p. 188.

[36] For this problem, see Giorello-Mondadori (1978).

[37] "Die Begriffssysteme, welche ihnen jetzt zu Grunde liege, sind durch allmähliche Umwandlung älterer Begriffssysteme enstanden, und die Gründe, welche zu neuen Erklärungsweisen trieben, lassen sich stets auf Widersprüche oder Unwahrscheinlichkeiten, die sich in den älteren Erklärungsweisen herausstellten, zurückführen." Riemann, *Gesammelte mathematische Werke*, p. 521.

[38] "The demand that a new theory should contain the old one approximately, for appropriate values of the parameters of the new theory, may be called (following Bohr) the *'principle of correspondence'*." Popper (1972), p. 202.

[39] We are closely following Hesse (1976).

[40] Riemann, *Gesammelte mathematische Werke*, p. 284.

[41] Cf. Giorello – Mondadori (1978).

[42] Kaufmann, quoted by Koertge (1973), p. 167.

[43] In my (1975b) I intended to investigate the 'fine structure' of the hard core of a program of mathematical research in the light of the intuition/rigor dichotomy without explicitly using the terminology of translation schemes. I think that the results of that work can be easily presented again using this terminology.

BIBLIOGRAPHY

Bourbaki, N. (1969), *Eléments d'histoire des mathématiques*. 2nd edition, revised and enlarged. Paris, Hermann.

Courant, R. (1950), *Dirichlet's Principle, Conformal Mapping, and Minimal Surfaces*. New York, Interscience Publishers.

Dieudonné, J. (1947), *Cours de géométrie algébrique*. (Volume I: *Aperçu historique*), Paris, Presses Universitaires de France.

Dugac, P. (1973), 'Eléments d'analyse de Karl Weierstrass', *Archive for the History of Exact Sciences* **10**, 41–176.

Fano, G. (1929), 'Geometrie non euclidee e non archimedee', in *Enciclopedia delle matematiche elementari e complementi* edited by L. Berzolari, G. Vivanti, D. Gigli. Volume II. Milan, Hoepli, 1929 and later years: the citations are from the reprint, Milan, 1958, pp. 435–511.

Feyerabend, P.K. (1970), 'Consolations for the specialist', in *Criticism and the Growth of Knowledge*, ed. by Imre Lakatos and Alan Musgrave. Cambridge, at the University Press, pp. 197–230.

Geymonat, L. (1953), *Saggi di filosofia neorazionalistica*. Turin, Einaudi.

Geymonat, L. (1957), 'Matematica ed esperienza', *Il pensiero* 2, 332–339, reprinted in Geymonat (1960), pp. 181–188.

Geymonat, L. (1960), *Filosofia e filosofia della scienza*. Milan, Feltrinelli.

Giorello, G. (1975a), 'Il problema di Dirichlet e il metodo della proiezione ortogonale', from *Seminario di storia della fisica*, 1–2 June, 1974, edited by E. Bellone. Genoa, 1975. pp. 133–165.

Giorello, G. (1975b), 'Archimedes and the methodology of research programmes', *Scientia* 110, 125–135.

Giorello, G. and Mondadori, M. (1978), 'Dinamica della conoscenza scientifica', in *La razionalità scientifica*, ed. by U. Curi. Padua, Francisci.

Hahn, H. (1933), *Logik, Mathematik und Naturerkennen*. Vienna, Gerold. (The first four sections of this pamphlet have been translated, by Arthur Pap, as 'Logic, Mathematics and Knowledge of Nature' in *Logical Positivism*, ed. by A.J. Ayer. Glencoe, The Free Press, 1959. pp. 147–161. The passages quoted in this paper are on p. 159.)

Hesse, M. (1976), 'Truth and the growth of scientific knowledge', in Proceedings of the 1976 Biennial Meeting of the Philosophy of Science Association, ed. by F. Suppe and P.D. Asquith. East Lansing, Philosophy of Science Association, Vol. 2.

Howson, C. (1975), 'Methodology in non-empirical disciplines' in *Criteria of scientific progress: a critical rationalist view* by members of the Department of Philosophy, London School of Economics.

Klein, F. (1894), 'Riemann und seine Bedeutung für die Entwicklung der modernen Mathematik. Amtlicher Bericht der Naturforscherversammlung zu Wien', *Gesammelte mathematische Abhandlungen*. 3 Vols. Berlin, Springer, 1921–1923. See Volume 3, pp. 482–497.

Kline, M. (1972), *Mathematical Thought from Ancient to Modern Times*. New York, Oxford University Press.

Koertge, N. (1973), 'Theory change in science' in *Conceptual Change*, edited by G. Pearce and P. Maynard. Dordrecht, Reidel.

Lakatos, I. (1962), 'Infinite regress and the foundations of mathematics' *Aristotelian Society Supplementary Volume* 36, 155–84; reprinted in Lakatos' *Philosophical Papers*, 2 Vols. Ed. by John Worrall and Gregory Currie. Cambridge, Cambridge University Press, 1978. Vol. 2: *Mathematics, Science and Epistemology*, pp. 3–23.

Lakatos, I. (1963–64), 'Proofs and refutations', *The British Journal for the Philosophy of Science* 14, pp. 1–25, 120–39, 221–43, 296–342. See also Lakatos (1976a).

Lakatos, I. (1967), 'A renaissance of empiricism in the recent philosophy of mathematics?' (abstract) in *Problems in the Philosophy of Mathematics*. Proceedings of the International Colloquium in the Philosophy of Science, Bedford College, 1965. Volume I. Edited by Imre Lakatos. Amsterdam, North-Holland, 1967. Republished in a much expanded form as Lakatos (1976b).

Lakatos, I. (1970), 'Falsification and the methodology of scientific research programmes' in *Criticism and the Growth of Knowledge*. Proceedings of the International Colloquium in the Philosophy of Science, London, 1965, Volume 4, ed. by Imre Lakatos and Alan Musgrave. Cambridge, Cambridge University Press, pp. 91–196.

Lakatos, I. (1976a), *Proofs and Refutations; the Logic of Mathematical Discovery*, ed. by John Worrall and Elie Zahar. Cambridge, Cambridge University Press. Lakatos (1963–64) is reprinted as Chapter I of this posthumous revised and enlarged edition.

Lakatos, I. (1976b), 'A renaissance of empiricism in the recent philosophy of mathematics?', *The British Journal for the Philosophy of Science* **27**, 201–23; reprinted in Lakatos' *Philosophical Papers*. 2 Vols, ed. by John Worrall and Gregory Currie, Cambridge, Cambridge University Press. Vol. 2: *Mathematics, Science and Epistemology*, pp. 24–42.

Manning, K.R. (1975), 'The emergence of the Weierstrassian approach to complex analysis', *Archive for the History of Exact Sciences* **14**, 297–383.

Monna, A.F. (1975), *Dirichlet's Principle*. Utrecht, Oosthoek, Scheltema and Holkema.

Mooij, J.J.A. (1966), *La philosophie des mathématiques de Henri Poincaré*, Translated into French by H. Helding. Paris, Gauthier-Villars.

Poincaré, H. (1900), 'Du rôle de l'intuition et de la logique' in *Compte rendu du Deuxième congrès international des mathématiciens*. Paris, Gauthier-Villars, 1902, pp. 115–130.

Poincaré, H. (1902), *La science et l'hypothèse*. Paris, Flammarion; edition ed. by J. Vuillemin. Paris, Flammarion, 1968. (*Science and Hypothesis*, Trans. by W.J. Greenstreet. London, Walter Scott, 1905.)

Popper, K.R. (1959), *The Logic of Scientific Discovery*. London, Hutchinson.

Popper, K.R. (1972), *Objective Knowledge. An Evolutionary Approach*. Oxford, Clarendon Press.

Russo, F. (1974), 'Typologie du progrès des connaissances scientifiques', *Revue des questions scientifiques*, **145**, 345–363 and 479–502.

Urbach, P. (1975), 'The objective promise of a research programme' in *Criteria of Scientific Progress: A Critical Rationalist View* by members of the Department of Philosophy, London School of Economics.

PIERO TOSI

INTUITIVE PROOFS AND FIRST-ORDER DERIVATIONS: SOME NOTES ON THE METAMATHEMATICS OF FIRST-ORDER NUMBER THEORY

O. INTRODUCTION

0.1. The method of syntactical transformations (cut elimination and normalization) in the metamathematics of first-order number theory has known, basically, two different treatments: the first one ((i) for short, in Gentzen's style) establishes a (partial) *Hauptsatz* for a formal system; the second one ((ii) for short) establishes the full *Hauptsatz* for a semiformal (or: infinitary) system. Examples of (i) are in Gentzen (1935) and (1938); Scarpellini (1969), Jervell (1971), Prawitz (1971), Troelstra (1973). Examples of (ii) are in Schütte (1951), (also in Mendelson (1964)) and (1960), Prawitz (1971).

In this paper, we will consider how the two different approaches work, by examining in their main lines a representative of (i) and a representative of (ii), in Sections 1 and 2. Such an examination, though faithful, is made in such a way that it will suggest a third possible treatment, contained in Section 3. In Section 4 we will make some remarks about the relations between intuitive *proofs* in informal arithmetic and the formal *derivations* used for metamathematical purposes in all the preceding sections.

0.2. We will take as a representative of (i), Jervell's normal form for first-order arithmetic, which is the same as in Troelstra (1973) (the latter with some variations in the proof and many more details both in the proof and in the applications). It seems to us, in fact, that this is a very successful representative, since it allows, perhaps the most general results possible from a Gentzen-style treatment, including Gentzen's original *Hauptsatz*.

As a representative of (ii) we will take the infinitary system given in Prawitz (1971), pp. 266–267. Actually, only some hints are given there, but our choice is motivated by the fact that such a system is in a natural deduction setting, so that a certain uniform terminology is allowed in considering comparatively (i) and (ii).

The *Hauptsatz* then takes the form of a normal form theorem. The fact that the very successful attempt in (i) uses normalization in a natural deduction calculus, instead of cut elimination in a sequent calculus, deserves some consideration. It was often asserted that natural deduction calculi

137

Maria Luisa Dalla Chiara (ed.), Italian Studies in the Philosophy of Science, 137–152.

were the closest to the informal proofs and gave a good analysis of proofs, but they were not convenient for proof-theoretic investigation: for such a purpose they had to be transformed into sequent calculi. (In fact, this opinion originated with Gentzen himself, who modified his original natural deduction calculus in order to be able to formulate the cut elimination theorem.) It seems that the situation is now changed, after Raggio (1965), and Prawitz (1965), where they were able to formulate the normal form theorem for first-order predicate calculus. Since then, cut elimination could be replaced by normalization in more and more systems, so that sequent calculi cannot be considered, at the present time, the only systems convenient for proof-theoretic investigations; on the contrary, it seems that they are better used to analyze the concept of truth, rather than the concept of derivability (see Prawitz, 1975).

However, the main lines of (i) and (ii) are the same for both natural deduction and sequent calculi, so that we can extend our considerations also to corresponding sequent calculi, with appropriate changes in the terminology.

Knowledge of natural deduction techniques is supposed throughout this paper.

0.3. In this paper we will consider systems of intuitionist elementary arithmetic (Heyting's arithmetic, HA), based on intuitionistic predicate calculus. The reason is that the treatment of intuitionist arithmetic allows the full generality of the proof-theoretic results for elementary number theory, in the sense that results about intuitionist arithmetic extend to classical arithmetic (Peano arithmetic, PA), by means of one of the well-known embeddings of PA into the negative fragment of HA. Some additional complications arise in considering HA instead of PA: it seems, however, that Friedman (1975) may justify an interest in HA for proof-theoretic purposes, beyond reasons of personal taste.

1. THE WEAK NORMAL FORM FOR A FORMAL SYSTEM HA

1.1. *The Formal System HA*

The following HA is given in Jervell (1971), Prawitz (1971), and, using a more abstract form for the basic rules, in Troelstra (1973). HA is the (formal) intuitionist arithmetic, based on first order intuitionist *predicate calculus* in a natural deduction setting (with rules for introducing and eliminating $\wedge$, $\vee$, $\rightarrow$, $\exists$, $\forall$, plus the intuitionist absurdity rule). Free variables (parameters) are notationally different from bound variables. In addition, HA has the symbols and the *basic rules* for 0, =, +, ·, ′ (for zero, equality, addition, multiplication and successor) and the following *induction rule*

$$\text{IND} \qquad \frac{A0 \quad \overset{[Aa]}{Aa'}}{At}$$

where a is a parameter subject to the usual restrictions and t (the induction term) any term.

1.2. *The Weak Normal Form of HA*

1.2.1. The following discussion is contained, with minor variations, in Prawitz (1971).

Generally speaking, a system can be put in normal form if all the rules for not-atomic formulae can be taken, for each logical constant (connectives and quantifiers) in pairs of introduction and elimination rules, such that each pair satisfies the *inversion principle*. We illustrate such a principle with an example: take the pair $(\wedge I, \wedge E)$, for introducing and eliminating $\wedge$: if, in a derivation, the connective $\wedge$ is first introduced by $\wedge I$, deriving $A \wedge B$ from A and B and then (immediately after, or after some consecutive occurrences) $A \wedge B$ is eliminated deriving either A or B by $\wedge E$, the derivation makes a detour: the conclusion of $\wedge E$ was already derived as a premiss of $A \wedge B$. If the inversion principle holds for a logical constant, then applications of *reduction figures* gives a derivation where the detour is eliminated. When all such detours are eliminated, then the derivation is said to be *normal* or *in normal form*.

1.2.2. The induction rule IND in Section 1.1 falls clearly outside the above pairs of rules, hence for HA one cannot show the normal form. The formulation of IND with induction term is motivated by the fact that, taking IND with conclusion as a universally quantified formula (hence as a rule for introducing ∀) the inversion principle for ∀ does not hold, hence not all maximal formulae with ∀ as the main logical symbol can be eliminated (see Prawitz 1971, pp. 262–263).

1.2.3. Having such a disturbing IND with t induction term, one tries to eliminate IND as far as possible.

When t is closed (i.e. $t = \bar{n}$ and $\bar{n}$ is a numeral) then it is possible to derive At without using IND:

$$\sum_0 \quad \frac{\begin{matrix} [Aa] \\ \sum(a) \\ A0 \quad Aa' \end{matrix}}{A\bar{n}} \quad \text{is replaced by} \quad \begin{matrix} \sum_0 \\ [A0] \\ \sum(\bar{0}) \\ [A1] \\ \sum(\bar{1}) \\ \vdots \\ \sum(\overline{n-1}) \\ A\bar{n} \end{matrix} \quad \text{(stop when } A\bar{n} \text{ is met).}$$

When t contains a parameter, then IND is not eliminable. (We will see in Section 1.3 how to handle such a case.)

Clearly, IND being a rule for not-atomic sentences, the definition of normal form given in Section 1.2.1 cannot be satisfied by HA. However, the following definition of *weak normal form* can be given: a derivation of HA is in weak normal form if no introduction rule is followed by the corresponding inverse elimination rule *and*, in applications of IND, t is not a closed term.

The *weak normal form theorem* for HA proves that any given derivation of HA can be transformed into an equivalent one (i.e. with the same conclusion and no more open assumptions) which is in weak normal form.

1.3. *The Proofs of the Weak Normal Form Theorem*

1.3.1. To show the theorem it is necessary to generate an infinite structure from HA. The domain of such a structure consists of derivations in HA, and the set R of the defining relations is given both by the reduction figures and by some other extra relations to be applied in a suitable order.

One of these extra relations (which allows one also to take care of applications of IND with open induction term) is the following ω-reduction:

$$\omega\text{-red)} \quad \sum(a) \text{ reduces to } \sum(\bar{n}), \text{ (any } \bar{n})$$

when a is not proper in $\sum(a)$.

So, denumerably many $\sum(\bar{n})$ are in R-relation with $\sum(a)$. Combined applications of some other reductions with the ω-reduction produce the following facts:

$$\text{(i)} \quad \frac{\begin{matrix} \sum(a) \\ Aa \end{matrix}}{\forall x Ax} \quad \text{is in } R\text{-relation with} \quad \begin{matrix} \sum(\bar{n}) \\ A\bar{n} \end{matrix} \text{ (any } \bar{n})$$

(ii)
$$\dfrac{\begin{array}{cc} & Aa \\ \Sigma_0 & \Sigma(a) \\ A0 & Aa' \end{array}}{At} \quad \text{is in } R\text{-relation with} \quad \begin{array}{c} \Sigma_0 \\ [A0] \\ \Sigma(0) \\ A\bar{1} \\ \Sigma(\bar{1}) \\ \vdots \\ \Sigma(\overline{n-1}) \\ A\bar{n} \end{array} \quad (\text{any } \bar{n})$$

(iii)
$$\dfrac{\begin{array}{cc} & [Aa] \\ & \Sigma(a) \\ \exists x Ax & B \end{array}}{B} \quad \text{is in } R\text{-relation with} \quad \begin{array}{c} [A\bar{n}] \\ \Sigma(\bar{n}) \\ B \end{array} \quad (\text{any } \bar{n}).$$

It is now evident in what sense the structure generated from HA is infinite: not only the tree representing R is infinitely generated (cases (i), (ii) and (iii)), but also there is no finite upper bound for the length of the branches (case (ii)). What is then shown is that the R-relation is well founded, so that bar induction on R can be applied.

The proof of the weak normal form for any Π of HA, follows then from the normal form of the derivations in R-relation with Π, using bar induction as a principle of proof.

1.3.2. We note here, in view of some succeeding considerations, that some derivations in R-relation with any Π of HA are obtained from subderivations of Π: we may then look at the bar induction on R as a transfinite induction over the length of an infinite derivation generated from Π using the derivations on the right of (i), (ii) and (iii), which are in R-relation with Π. We can see immediately that the denumerably many derivations in (i), (ii) and (iii) may be used as ω-rules defining an infinitary system.

1.4. *Applications of the Weak Normal Form*
Notwithstanding the fact that HA is not normalizable in a full sense, many proof-theoretic properties (beyond the consistency) can be derived from the form of the weakly normal derivations. We quote the $\exists$-property and the $\lor$-property for derivations from closed Harrop formulae, the independence-of-the-premiss rule (without parameters), Markov's rule MR_{PR}, some instances of the reflection principle and some other properties (see Troelstra, 1973).

The subformula property for HA is clearly not derivable from the weak normal form: for such a property the full normal form is needed. This is the reason why important applications, as the uniform reflection principle for HA is out of the range of the weak normal form.

2. THE NORMAL FORM FOR AN INFINITARY SYSTEM HA^∞

2.0. A complete Schütte's style treatment for a natural deduction system of intuitionistic arithmetic is not available in the literature. Its main lines can be taken from Schütte's original work, which is a classical system of arithmetic with $\lor$, $\neg$ and $\forall$ as only logical symbols (see Mendelson (1964), which follows rather closely Schütte (1951)). In our schematical presentation we will spread out the hints given in Prawitz (1971).

2.1. *The System HA^∞*

HA^∞ is an infinitary (with ω-rules) system of first-order arithmetic based on first-order intuitionistic predicate calculus. Everything is as in Section 1.1, except that all terms are closed; $\forall I$ and IND are substituted by $\omega\forall I$; and $\exists E$ is substituted by $\omega\exists E$:

$$\omega\forall I \qquad \frac{A\bar{0},\, A\bar{1},\, A\bar{2},\, ...,\, A\bar{n},\, ... \,(\text{all } n)}{\forall x Ax}$$

$$\omega\exists E \qquad \frac{\exists x Ax \quad \overset{[A\bar{0}]}{B},\quad \overset{[A\bar{1}]}{B}, ..., \overset{[A\bar{n}]}{B}, ... \,(\text{all } n)}{B}$$

At application of $\omega\exists E$, all $A\bar{n}$ are discharged. By definition, a prooftree is well founded, and, for any given prooftree there is a finite upper bound for the degree of the maximal formulae.

No attempt is made in making the ω-rules effective.

2.2. *The Normal Form for HA^∞*

It is now immediately clear that the ω-rules of HA^∞ satisfy the inversion principle; in fact, by the new rules, the following reductions can be given (in addition to those given for HA):

$$\omega\forall\text{red} \qquad \frac{\overset{\Sigma_0}{A\bar{0}},\, ...,\, \overset{\Sigma_n}{A\bar{n}},\, ...}{\dfrac{\forall x Ax}{A\bar{n}}} \qquad \text{reduces to} \qquad \overset{\Sigma_\omega}{A\bar{n}}$$

$$\omega\exists\text{red}\quad \frac{\dfrac{\sum'}{A\bar{n}}}{\quad}\frac{\begin{array}{c}\\ \sum_0(0)\\ B\end{array}}{\begin{array}{c}[A0]\\ \end{array}}\quad\cdots\quad\frac{\begin{array}{c}[A\bar{n}]\\ \sum_n(\bar{n})\\ B\end{array}}{}\quad\cdots\cdots \quad\text{red. to}\quad \frac{\begin{array}{c}\sum'\\ [A\bar{n}]\\ \sum_n(n)\\ B\end{array}}{}$$

The figure $\omega\exists$red:

$$\omega\exists\text{red}\qquad \frac{\dfrac{\sum'}{A\bar{n}}\big/BxAx \qquad \dfrac{[A0]}{\sum_0(0)}\;B \quad\cdots\quad \dfrac{[A\bar{n}]}{\sum_n(\bar{n})}\;B \qquad\cdots\cdots}{B}\qquad \text{red. to}\qquad \frac{\sum'}{[A\bar{n}]}\;\;\frac{\sum_n(n)}{B}$$

$$\omega\exists E\text{ red}\qquad \frac{\exists xAx \qquad \dfrac{\begin{array}{cc}[A0] & [A\bar{n}]\\ \sum_0(0) & \sum_n(\bar{n})\\ B & B\end{array}\;\cdots}{B}\qquad\Sigma}{F}\qquad \text{red. to}\qquad \frac{\exists xAx \qquad \dfrac{\begin{array}{cc}[A0] & [A\bar{n}]\\ \sum_0(0) & \sum_n(n)\\ B\;\cdots & B\;\cdots\;\Sigma\end{array}}{F}}{F}$$

HA^∞ satisfying the inversion principle, the normal form for it can be defined as in first-order predicate calculus: a derivation of HA^∞ is normal if no formula in it (nor sequences of immediate occurrences of the same formula) is simultaneously introduced by an introduction rule and eliminated by the inverse elimination rule.

2.3. *The Proofs of the Normal Form*

2.3.1. Let Π be a derivation of HA^∞ such that the highest logical degree of the maximal formulae is d. By induction over the (transfinite) length of Π one shows, first, that, by applications of the reduction figures, there is a well-founded equivalent Π' such that its length depends on the length of Π and d is decreased. Then, using at most d applications of the above result, one shows that all maximal formulae are eliminated.

2.3.2. *Infinite structures and infinite derivations*

In comparison with Section 1.3 the proof of the normal form is very simple: no complicated relation has to be studied, the only reductions being those given in Section 2.2. The complications in handling the R-relation are avoided because infinite derivations are allowed: in fact we see that the difficulties there were originated by defining some specific subderivations to be in R-relation to a given derivation and by ω-substitution: if one has infinite derivations, the first difficulty is absorbed in proofs by transfinite induction on the length of derivations, and the second one disappears completely being the ω-substitution absorbed in the definition of the infinite derivations.

2.3.3. We note here, also in view of succeeding developments, that the infinite derivations allowed in HA^∞ are more than the infinitely many subderivations in R-relation with some Π given in Section 1.3.1: the latter are generated from Π in a specific way, the former are more generally defined. This fact will be used in the sequel: to investigate derivability in HA, the latter are more useful than the former.

2.4. *Applications of the Normal Form for HA$^\infty$*

2.4.1. By imbedding HA into HA$^\infty$ (i.e. by showing that to every given Π of HA, an equivalent Π' of HA$^\infty$ can be given), and by the properties of the normal form for HA$^\infty$, one has the following:
  (i) if $\bar{0} = \bar{1}$ is derivable in HA, then it is derivable in HA$^\infty$;
  (ii) $\bar{0} = \bar{1}$ is not derivable in HA$^\infty$.
By (ii') and the contraposition of (i), one has the consistency of HA:
  (iii) $\bar{0} = \bar{1}$ is not derivable in HA.

2.4.2. The other proof theoretical properties for HA given in Section 1.4. are not possible from the normal form of HA$^\infty$. In fact they hold for HA$^\infty$, but HA$^\infty$ is stronger than HA, hence they cannot be extended to HA.
  For example: the normal form of HA$^\infty$ yields the $\exists$-property for HA$^\infty$; but the $\exists$-property for HA cannot be shown. If, instead, HA$^\infty$ would have been equivalent to HA, we could do the following: suppose $\exists x A x$ to be derivable in HA; by equivalence it is also derivable in HA$^\infty$. By the $\exists$-property for HA$^\infty$, $A\bar{n}$ is derivable in HA$^\infty$. Apply again equivalence and get $A\bar{n}$ in HA.

2.4.3. The subformula property clearly holds for HA$^\infty$. But it is not possible to get the application of the uniform reflection principle for HA: with such ineffective ω-rules even the formulation of the principle is not possible: what needs to be done is to make the infinitary system constructive in some way. This can be done by giving, by definition, the ω-rules recursively describable; however, we may also show that the constructivity requirement is satisfied in an informal way. We shall see that the last way is more profitable, since we can have an infinitary system equivalent to HA (while recursive ω-rules yield stronger systems).

3. THE NORMAL FORM FOR ω-HA IN HA

3.0. Finding the weak normal form for HA is rather complicated; on the other hand, interesting applications are possible (except the reflection principle). Finding the normal form for an infinitary system is much more simple; on the other hand, the applications are very poor, unless such a system is eqivalent to HA.

3.0.1. Attempts to weaken HA$^\infty$ by recursive and primitive recursive ω-rules are better used for other purposes than having a system equivalent to HA (in fact they cannot succeed in that).

3.0.2. If we consider Section 2.4.1. we see that a solution is already at hand: in imbedding HA into HA$^\infty$, every derivation Π of HA is translated into an equivalent on Π' of HA by means of *some instances* of the ω-rules, whose premisses are generated from subderivations of Π. If we restrict our consideration to such Π', we have immediately an infinitary subsystem of HA$^\infty$, which is equivalent to HA. We call it ω-HA.

From the (easy) normal form of HA$^\infty$ we can then have the normalizability of all the derivations in ω-HA, and, hence, all the applications which in Section 2.4.2 were not possible. As to the application of Section 2.4.3, we remark that we use only some constructive instances of the ω-rules: to the purpose of formulating the reflection principle the recursive encoding can then follow later.

3.0.3. The above approach is exploited in Tosi (A). In the following we will show how ω-HA is obtained and stress some basic facts.

3.1. *The Infinitary System ω-HA*

3.1.1. As we have seen in Section 3.0.2, the system ω-HA is obtained with reference to HA.

ω-HA is the system whose language and rules are as in HA, except the following:

(i) we do not have free variables (parameters): the terms are only the numerals;

(ii) Instead of $\forall I$ and IND we have $\omega\forall I$ in Section 3.1.2;

(iii) instead of $\exists E$ we have $\omega\exists E$ in 3.1.3. In Section 3.1.2. and 3.1.3 below, when a derivation $\dfrac{\Pi(a)}{Aa}$ is referred to, we suppose that above the conclusion Aa all the suitable ω-rules are yet substituted for all the applications of $\exists E$, $\forall I$, IND above Aa, so that $\dfrac{\Pi(\bar{n})}{A\bar{n}}$ belongs to ω-HA. $\Pi(a/\bar{n})$ is the result of substituting $\bar{n}$ for a in all the occurrences of a.

3.1.2. *The ω-Rules for the Universal Quantifier in ω-HA*
Suppose the following denumerable sequence of derivations $\{\Pi_n\}(n = \bar{0}, \bar{1}, \bar{2}, \ldots)$ be given in ω-HA. Then we can conclude $\forall x Ax$ if and only if one of the two following cases holds:

3.1.2.1. For all n, $\Pi_n = \dfrac{\Pi(a/\bar{n})}{A(a/\bar{n})}$, where Aa is the premiss of an application of $\forall I$ in HA.

3.1.2.2. For $n = 0$, $\Pi_0 = \dfrac{\Pi}{A0}$, where $A0$ is the the zero-premiss of an application of IND in HA;
for $n > 0$,

$$\Pi_{n+1} = \dfrac{\begin{matrix}\Pi_n\\ [A\bar{n}]\end{matrix}}{A\dfrac{\Pi'(a/\bar{n})}{n+1}}$$

where $\Pi'(a)$ is the derivation of the induction-premiss of IND in HA.

$$\dfrac{[Aa]}{\dfrac{}{Aa'}}$$

3.1.3. *The ω-Rule for the Existential Quantifier in ω-HA*

Suppose there are given in ω-HA $\dfrac{\Pi}{\exists x A x}$ and the denumerable sequence of derivations

$$\left\{ \begin{pmatrix} [A\bar{n}]\\ \Pi(\bar{n})\\ B \end{pmatrix}_n \right\} (n = 0, 1, 2, \ldots),$$

then we can conclude B not depending on $A\bar{n}$ (all n) if and only if

$$\begin{pmatrix} [A\bar{n}]\\ (\Pi\bar{n})\\ B \end{pmatrix}_n = \begin{matrix} [A(a/\bar{n})]\\ \Pi(a/\bar{n})\\ B \end{matrix},$$

where B is the minor premiss of $\exists E$ in HA.

3.2. *The Normal Form for $\omega + HA$ in HA^∞*

3.2.1. *Having infinite derivations it is easy to normalize by applications of the reduction figures given Section 2.2.*
What is to be noted here is that the normal derivations, corresponding to given (not normal) derivations of ω-HA, may not belong to ω-HA. In fact the ω-rules of ω-HA are defined by reference to their form, and such form is not stable under some of the reduction figures used in the normalization process. But, for the applications we are looking for, this is not necessary: it is sufficient

(i) to have, for every Π in ω-HA, an equivalent Π' which is normal (it does not matter whether Π' belongs to ω-HA or to HA$^\infty$); we call this: *normal form of ω-HA in HA$^\infty$*.

(ii) in view of the reflection principle, that the constructivity of the infinite derivation is preserved under the reduction figures, which is the case; in fact a reduced derivation is either a subderivation of a given constructive derivation or it is obtained by composition of given subderivations.

3.2.2. The fact that the normal derivations may not belong to ω-HA, explains why the subformula property, which holds for the normal derivations, do not extend to HA: in fact, every derivation Π of ω-HA has, by definition, a back translation Π' in HA, and a normal Π would yield a normal Π'.

3.2.3. In Tosi (A) it is shown how to find a normal derivation corresponding to every derivation of ω-HA. This can be seen as the normalization of a proper subsystem of HA$^\infty$, and this fact makes the proof easier. In fact all the derivations to be normalized have a finite bound for the length of the maximal segments and, hence, the strategy given there to eliminate in a finite number of steps the maximal segments with common root, can be developed. Actually this fact is not strictly necessary; a normalization of the full HA$^\infty$ is also possible by a different strategy; but this would probably require an assignment of ordinals less simple than ours (which is as simple as possible), and a rather complicated series of reductions.

3.3. *Applications*

3.3.1. Recall Section 2.4.2: having now normal derivations, we can show for them the usual closure properties as in Section 1.4 and, by equivalence, extend them to HA.

3.3.2. Besides the applications from Jervell's normal form, we can show also the uniform reflection principle for HA (for short, RFN). RFN is equivalent to transfinite induction up to ε_0 in this sense:

$$\text{HA} + \text{RFN} \equiv \text{HA} + \text{TI}_{\varepsilon_0}$$

That HA $+$ RFN $\vdash$ TI$_{\varepsilon_0}$ is shown in Kreisel-Levy (1968). That HA $+$ TI$_{\varepsilon_0}$ $\vdash$ RFN was first shown in Lopez Escobar (1976). Thanks to our construction of ω-HA, a more refined use of RFN is made possible: in fact the requirement of Kreisel-Mints-Simpson (1975, p. 100, lines 1–3) is satisfied by our proof of RFN in Tosi (A). Specifically, if one pays attention to the logical complexity of the predicate to which $\text{TI}_{\varepsilon_0}$ is applied, one finds that the proofs of (ii) and (iii) of Theorem 7.1 do not require $\text{TI}_{\varepsilon_0}$ beyond $d_A + 1$, where d_A is the logical degree of A; and, if $d_A = 0$ (i.e., A is primitive recursive) one needs $\text{TI}_{\varepsilon_0}$ applied to a primitive recursive predicate only.

This gives significance to the consistency proof of HA by $\text{TI}_{\varepsilon_0}$ (for a discussion see Kreisel (1971)).

3.4. *The ω-Rules of ω-HA Are Not Unnecessary Detours*

Looking comparatively at the normal forms in Sections 1 and 3, it may seem that (leaving aside RFN) the latter makes some detours with respect to the former: in fact, in the latter we start from HA, generate ω-HA, establish the normal form, and go back to HA for applications; while in the former we start from HA, find a normal form in it, and give applications (without passing through ω-HA and its normal form). It it not so; in fact, as we have seen in Section 1.3, to establish the weak normal form of HA, it is necessary to argue by induction on the infinite R-relation, and this is equivalent to arguing by transfinite induction on the length of infinite derivations generated by specific ω-rules (see Section 1.3.2). Hence we may conclude that ω-HA is not a detour: it is simply a different way of looking at the same object. On the contrary, we may say, looking again comparatively at Sections 1 and 3, that, *in some sense* the weak normal form for HA is a detour. If we look closely at the R-structure, we see that it is not necessary to go back to the normal form of HA to get the applications: they are already possible from the normal form of the infinite structure (as we have done).

However (explaining why we said above that the weak normal form is *in some sense* a detour) it is to be remarked that the closure properties shown by means of the weak normal form for HA are of a different character than those shown by means of the normal form for ω-HA: the former refer to derivations, the latter to the consequence relation. For example, take the $\exists$-property: by the former we can *show* a derivation of $A\bar{n}$ in HA; by the latter we can say that *there is* such a derivation (as already remarked in 3.2.2, there is no standard translation from a normal $\mathit{\Pi}$ in HA^∞ back to a normal $\mathit{\Pi}'$ in HA).

So, with respect to closure properties, the normal form of ω-HA is not really an alternative to the weak normal form of HA (it is an alternative with respect to RFN). It is really an alternative, we think, to the normal form of the full HA^∞: closure properties are out of its range.

4. INFORMAL PROOFS ON NATURAL NUMBERS AND FIRST-ORDER RULES

4.0. In this section we will briefly comment on some of the results and methods of the preceding sections, trying to make explicit some assumptions which, clearly, did not need to be explicitly noted in a formal treatment.

First of all we will write some obvious considerations (to be used more than once) about *proofs* (in intuitive arithmetic) and *justifications* (both of proofs and of formal derivations). Then we will draw some conclusions.

4.0.1. *Informal Proofs and Justifications*

Consider the following informal proofs of arithmetic and their justifications:

(i) If we have a *uniform* proof of Ai (i arbitrary natural number), then we have a proof of An for all n. The justification is the following: we have a proof of $A0$, a proof of $A1$, and so on.

(ii) If we have a 'first proof of $A0$ and a second uniform proof of $Ai+1$ under the assumption Ai (i arbitrary natural number), then we have a proof of An for all n, not depending on that assumption. The justification is the following: a proof of $A0$ is given by the first proof; a proof of $A1$ is given by the proof of $A0$ plus the second proof, where $i = 0$; a proof of $A2$ is given by the proof of $A1$ plus the second proof, where $i = 1$; and so on. Such a proof we call a proof by *composition* of proofs; the length of a proof of An depends on n, since it requires n compositions.

(iii) If we have a first proof of An for some n, and a second uniform proof of B from Ai (i arbitrary), then we can conclude B not depending on Ai. The justification is the following: the first proof insures that, for some i, Ai is proved: hence the second proof derives B as desired.

The proofs in (i), (ii) and (iii) are finite. Their justifications, however, refer to an infinite number of proofs (one for each natural number) and to the way in which such proofs are obtained.

4.0.2. *The Formal Rules and Their Justification as Valid Rules for Arithmetic*

Formal rules express finitely (i.e. in a finite number of symbols) finite informal proofs. $\forall I$, IND and $\exists E$, as rules for arithmetic, express, respectively, the proofs of (i), (ii) and (iii) Section 4.0.1. Their justification is exactly the same, once, instead of natural numbers, one speaks of numerals (as names for natural numbers). In other words: their justification is given via the ω-substitution and the same process of composition which justifies the arithmetical induction, when, in addition, one knows that the numerals are interpreted over all the natural numbers.

4.0.3. *The ω-Rules for ω-HA*

The ω-rules for ω-HA express finite proofs *together* with their justification (which is that of Section 4.0.1): they are infinite just because such a justification requires infinitely many steps.

4.1. Considering, informally, the generation of ω-HA one may ask: why such a generation yields a system which is as strong as HA? We think that the answer has something to do with the considerations above (which were, indeed, heuristical guides); in fact the ω-rules of ω-HA express the same proofs which are expressed by the formal rules of HA. That such a statement is not as trivial as it may seem can by seen by considering the unrestricted ω-rules of HA^{∞} or the recursively or primitive recursively restricted ω-rules: they always yield systems stronger than HA; informally seen: they express more proofs than those expressed by the formal rules of HA. Considering, in this way, that the unrestricted and recursively restricted ω-rules are of less use in investigating derivability in HA, we may conclude that the derivations most useful for being investigated in metamathematical investigations are those which express just the intuitive proofs about natural numbers; considering that the ω-rules of ω-HA are in some way necessary (see Section 3.4) we may conclude that the derivations used in metamathematical investigations are those which express the intuitive proofs about natural numbers together with their justifications.

4.2. *Standard Arithmetic and Proofs on Normal Form*

The ω-rules of ω-HA are not satisfiable over a non-standard model: in fact, as we said in Section 4.0.3, they contain their own justification, and such a justification is standard, in the sense that it cannot be referred to a non-standard model. Consider, for example, the uniform application of $\omega\forall I$, and let $\bar{c}$ in $A\bar{c}$ be a term interpretable over a non-standard element (in a language enlarged by a new constant $\bar{c}$ satisfying $\bar{c} > n$ for all n, usually introduced to obtain a non-standard model): then the infinite sequence of the premisses $A\bar{0}$, $A\bar{1}$, $A\bar{2}$, ... would never contain $A\bar{c}$; hence the conclusion $\forall xAx$ would be completely unjustified.

Let us now consider the formal rules of HA: they do not include the standard justification: they are open for non-standard justifications. Considering again the example above, the proper parameter a in the premiss of $\forall I$ can be interpreted over a non-standard element.

4.2.1. Considering the relation above between the ω-rules of ω-HA and the standard model, we may say that the normal form of Section 3 is *in some sense* semantical, the standard model entering the definition of ω-HA (cf. Section 4.2.2. for an explication of 'in some sense').

If, on the other hand, we consider the weak normal form in Section, 1 we cannot say that it is semantical on the grounds of the same argument: as seen above, the formal rules of HA do not require necessarily reference of the standard model. However, we may conclude that the proof is semantical (in exactly the same sense) on the grounds of the arguments in Section 1.3: the ω-substitution and the compositions of derivations which enter into the definition of the R-relation are exactly those given (in Section 4.0.2) for justifying the formal rules as valid rules for the standard model.

In other words: the same *arguments* that justify the formal rules as valid rules for the arithmetic, play a central role also at a metamathematical level.

4.2.2 In *what sense* have we said that the proofs of the normal form are semantical? Certainly not in a usual sense: the justifications of the rules of a system are given for semantical purposes: but in a syntactical (or: metamathematical) investigation, as that establishing the normal form, they may be forgotten, the formal system having an autonomous life.

However:

(i) in Section 3, we generate ω-HA from HA in such a way that it is not interpretable over a non-standard model;

(ii) in Section 1, the infinitely many derivations which are generated from HA by the R-relation are derivations of formulae interpretable over standard propositions and not over non-standard propositions.

This seems to suggest that the standard model (entering the justifications of the rules) has not only a heuristical influence in proving the normal form, but is also privileged among the arithmetical models in the proofs of the normal form. Perhaps the situation may be made more clear by the analogy with the semantic tableaux systems: the tableaux are syntactical objects, but are motivated on semantical grounds.

(Of course: the above 'semantic' proof is quite a different thing from the usual semantical proof of the consistency of arithmetic, where HA is interpreted over the structure N: the former proof is constructive, while the second is very ineffective).

REFERENCES

Friedman, H. (1975), *Notices AMS* **22**, 476.

Gentzen, G. (1935), 'Investigations into logical deduction', in M.E. Szabo (ed.) *The Collected Papers of G. Gentzen*. Amsterdam North Holland, 1969.

Gentzen, G. (1938), 'New version of the consistency proof for elementary number theory', in Szabo (ed) 1969.

Jervell, H.R. (1971), 'A normal form in first-order arithmetic', in *Proceedings of the Second Scandinavian Logic Symposium*, I.E. Fenstad (ed.). Amsterdam, North-Holland.

Kreisel, G. (1971), 'A survey of proof theory II' in *Proceedings of the Second Scandinavian Logic Symposium*, I. E. Fenstad (ed.). Amsterdam, North-Holland.

Kreisel G. and Levy, A. (1968), 'Reflection principles and their use for establishing the complexity of axiomatic systems', *Zeitschrift für mathematische Logik und Grundlagen der Mathematik* **14**, 97–142.

Kreisel, G., Mints G.E., and Simpson, S.G. (1975), 'The use of abstract languages in elementary mathematics: some pedagogic examples', in *Logic Colloquium*. Berlin, Springer. Springer Lecture Notes **453**, 38–129.

Lopez-Escobar, E.G.K. (1976), 'On an extremely restricted ω-rule', *Fundamenta Mathematicae* **90**, 159–172.

Mendelson E. (1964), *Introduction to Mathematical Logic*. Princeton, Van Nostrand.

Prawitz, D. (1971), 'Ideas and results in proof theory', in *Proceedings of the Second Scandinavian Logic Symposium*, I. E. Fenstad (ed.). Amsterdam, North-Holland.

Prawitz, D. (1975), 'Comments on Gentzen-type procedures and the classical notion of truth', in *Proof Theory Symposium, Kiel 1974*. Berlin, Springer. Springer Lecture Notes **500**.

Raggio A. (1965), 'Gentzen's Hauptsatz for the systems NI and NK', *Logique et analyse* **30**, 91–100.

Scarpellini, B. (1969), 'Some applications of Gentzen's second consistency proof', *Mathematische Annalen* **181**, 325–344.

Schütte K. (1951), 'Beweistheoretische Erfassung der unendlichen Induktion in der Zahlentheorie', *Math. Annalen* **122**, 369–389.

Schütte, K. (1960), *Beweistheorie*. Berlin, Springer.

Tosi P. (A) 'Normal derivability and first order arithmetic', to appear in *Notre Dame Journal of Formal Logic*.

Troelstra A. S. (ed.) (1973), *Metamathematical Investigation of Intuitionistic Arithmetic and Analysis*. Berlin, Springer. *Springer Lecture Notes* **344**.

MAURIZIO NEGRI

CONSTRUCTIVE SEQUENT REDUCTION
IN GENTZEN'S FIRST CONSISTENCY PROOF
FOR ARITHMETIC*

1. The purpose of this paper is to discuss Gentzen's first consistency proof for arithmetic, the 'galley proof' (published for the first time in the *Collected Papers*, Amsterdam, 1969), by analyzing its methods of proof. Opposing positions on this subject have been taken by Bernays (1970) and Kreisel (1971). According to Kreisel's analysis, which uses a result of Tait, it turns out that the 'galley proof' does not involve methods of proof exceeding transfinite induction up to ε_0.

The galley proof is reformulated here in the following manner. Any constructive proof that every derivable sequent is reducible furnishes a constructive proof of the consistency of arithmetic. One defines a function that associates in an effective manner each derivable sequent with a reduction rule, that is, with a sequence of operations which – when applied to the sequent – transform it into a normal form. Since the proof is intended to show the validity of the formal system of arithmetic with respect to the property of 'being reducible' or 'posssessing a reduction rule', it will be necessary first to associate the initial points of any derivation with reduction rules (which will be represented by reduction functions), and then to define, for each inference rule of the formal system, a functional ϕ that carries any reduction function of the premises to a reduction function of the conclusion. In defining this functional for the case of the cut-rule, we are faced with the necessity of using non-finitary but still constructive means. In fact, in this case ϕ is defined by recursion on the 'levels' of reduction functions, where these levels are expressed by ordinal numbers less than ε_0.

Just as the assignment of ordinals to derivations in the 1936 proof constitutes a measure of their complexity, so in this version of the 'galley proof' the level is a measure of the complexity of the reduction process represented by the reduction rules. This makes it possible to reason inductively on the complexity of the reduction of a sequent, as occurs at Lemma 14.6 of the 'galley proof'; and it allows us to define ϕ by induction on the complexity of the reduction.

---

*I wish to thank Prof. Warren Goldfarb of Harvard University for his invaluable assistance in revising this translation.—Tr.

153

Maria Luisa Dalla Chiara (ed.), Italian Studies in the Philosophy of Science, 153–168.
Copyright © 1980 *by D. Reidel Publishing Company.*

The reduction process is represented by means of reduction trees, which play an essential role in assigning ordinals as levels. This assignment is done in such a way that, if α is the level of the function which reduces sequent S, and β is that of the function which reduces sequent S', where S' is obtained from S by means of a reduction step, then $\beta < \alpha$.

In the 'galley proof', the reduction of sequents is obtained directly: the problem is to prove that one is dealing with a finite process for any possible choice. In the 1936 proof, sequent reduction is obtained indirectly, by reducing the derivations: the problem is to prove that a reduced form of the sequent is obtained after a finite number of steps carried out on the derivation for any possible choice. Induction up to ε_0 then enters as the method of proof necessary to assure that the reduction of any derivation stops after a finite number of steps; this only indirectly assures that the reduction of sequents stops after a finite number of steps.

The 'galley proof' is easy to follow because there is no mediation of the syntactic plane: it is a result more in the field of constructive semantics than in proof theory. For Gentzen "the statability of a reduction rule . . . [is] the formal replacement of the informal concept of truth; it provides us with a special finitist interpretation of propositions . . . " In our analysis of the 'galley proof' the problem is not to prove constructively that each sequence of reductions has an end: in fact it cannot be otherwise, given that the reduction-tree assigned to each derivable sequent, which represents its reduction, always has a finite number of levels – in other words, each branch is a finite sequence of points (reduction steps). Rather the problem will be to define the functional ϕ which associates every derivable sequent with its reduction rule. Hence the non-finitary part of the process is not so much the reduction of sequents as the association of each sequent, on the basis of its derivation, with its reduction rule. Every derivable sequent is reducible by finitary methods. Consequently it might seem possible to prove the consistency of arithmetic also by finitary means; but to know whether "each derivable sequent is reducible by finitary means", that is, to associate each derivable sequent with its reduction, induction up to ε_0 is necessary. The proof of consistency is at hand only when not only the means for reducing the sequents but also the information necessary to ascertain that they are sufficient has been furnished.

2. Let us consider the formal system in which arithmetic is formalized in Bernays (1970). It is a calculus of sequents having the form $\Gamma \rightarrow A$ where Γ is a finite sequence of formulas and A is a single formula. The logical

symbols $\vee$ (or), Σ (there is), and $\Rightarrow$ (implies) are defined in terms of $\&$ (and), $\sim$ (not) and Π (for all). As Bernays writes:

(1) Logical initial sequents, i.e. sequents of one of the forms

$$A \,\&\, B \to A; \quad A \,\&\, B \to B; \quad A, B \to A \,\&\, B;$$
$$A, \sim A \to 1 = 2; \quad \sim \sim A \to A; \quad (\Pi x)F(x) \to F(t),$$

where x is a bound variable, and t a term.

(2) Arithmetical initial sequents, i.e. sequents whose formulas are equations and which have the property that by replacing each free variable by a numeral (of course equal variables by equal numerals) and by computing the function values, either the succedent formula gets the truth value 'true' or one of the antecedent formulas gets the truth value 'false', according to the usual valuation of numerical equations. [Bear in mind that the numerical functions involved are all computable: hence in the final analysis it is a question of verifying numerical equations.]

The rules of inference are:

(a) Rules of structural change in a sequent, permitting one,

–to interchange the order of the formulas of the antecedent,

–to add an arbitrary formula to the antecedent,

–to delete a repetition of a formula in the antecedent,

–to change a bound variable of a universal quantifer, everywhere in its scope, into another bound variable.

(b) Logical inference schemata

$$\frac{\Gamma \to A \quad A, \Delta \to B}{\Gamma, \Delta \to B} \qquad \text{(cut)}$$

$$\frac{\Gamma, A \to 1 = 2}{\Gamma \to \sim A} \qquad \text{(negation introduction)}$$

$$\frac{\Gamma \to F(a)}{\Gamma \to (\Pi x)F(x)} \qquad$$ (where a is a free variable not occurring in any formula of Γ nor in $F(x)$, and where x is a bound variable) (universality introduction)

(c) Induction

$$\frac{\Gamma \to F(1) \quad F(a), \Delta \to F(a+1)}{\Gamma, \Delta \to F(t)}$$

where t is a term and a a free variable not occurring in Γ, Δ, $F(1)$ or $F(t)$.

In this calculus from any two formulas A, $\sim A$ we can derive, using the initial sequent $A, \sim A \to 1 = 2$ and cut, the sequent $\to 1 = 2$; hence, in order to prove consistency of the considered formal system, it is sufficient to show that the sequent $\to 1 = 2$ is not derivable. [Bernays (1970), pp. 410–11.–Tr.]

In order to prove the system's consistency Gentzen formulates the property of reducibility which the sequents can possess, and which intuitively amounts to a constructive concept of truth. A sequent is said to be reducible if it can be brought to a reduced form by means of a sequence of reduction operations (which we shall list immediately below): here a sequent in reduced form is a sequent either whose succedent is a true atomic formula or else whose succedent is a false atomic formula and whose antecedent contains a true atomic formula. Given that the predicates and arithmetical functions which are used are decidable, a sequent in reduced form can be considered true in a constructive sense. If there exists a sequence of operations which lead the sequent to a reduced form – that is, a reduction rule – we shall say that the sequent is reducible.

There are eight types of reduction operations or steps, and they have a preferential order in the sense that (a_1) has preference over all other steps, and (a_1) and (a_2) have preference over the other reduction steps. As listed in Bernays, they are:

(a_1) Replacing a free variable, wherever it occurs in the sequent, by the same numeral, which can be arbitrarily chosen.

(a_2) Replacing a function symbol all of whose arguments are constants by its value.

(b_1) when the succedent has the form $(\Pi x)F(x)$ replacing it by $F(k)$, where k is an arbitrarily chosen numeral.

(b_2) When the succedent has the form $A \,\&\, B$, replacing it by A or by B, according to an arbitrary choice.

(b_3) When the sequent has the form $\Gamma \to\, \sim A$, replacing it by $A, \Gamma \to 1 = 2$.

(c) When the succedent is a false numerical equation:

(c_1) Replacing an antecedent formula $(\Pi x)F(x)$ by $F(k)$, or adding $F(k)$ to it in the antecedent, where k is a numeral. [As we shall see, in this case choices are not possible with respect either to the numeral k or to retaining or not the formula in the antecedent.]

(c_2) Replacing an antecedent formula $A \,\&\, B$ by one of the formulas, A, B or adding one of these in the antecedent [without the possibility of choice as before].

(c_3) If an antecedent formula $\sim A$ occurs, replacing the succedent formula by A and possibly cancelling the formula $\sim A$ in the antecedent [also without the possibility of choice]. [Bernays (1970), p. 411.–Tr.]

A reduction process for a sequent consists in a series of transformations of the sequent's form, that is in a manipulation of signs. In instances where such a process gives rise to arbitrary choices (steps (a_1), (b_1) and (b_2)), it is understood that the sequent is reducible if a reduced form is reached by every possible choice. The fact that in certain cases infinitely many different choices are possible is the point at which an infinitary element is introduced.

3. Corresponding to each reduction step we now define a function which, when applied to a sequent to be reduced, furnishes as a value the result of reducing that sequent by the reduction step. While a reduction rule was first considered to be a series of steps which lead to a reduced form, it will

now be a function defined in terms of the functions which express the reduction steps. The aim of the consistency proof is to assign to each derivable sequent a reduction function analogous to the reduction rules of Gentzen's proof. We shall indicate the functions corresponding to the reduction steps with the letters $f_{a_1}, f_{a_2}, f_{b_1}, f_{b_2}, f_{b_3}, f_{c_1}, f_{c_2}, f_{c_3}$.

Of these, the definitions of f_{a_2}, f_{b_3} and f_{c_i} ($i = 1, 2, 3$) present no difficulty, given that application of the corresponding reduction steps leads to a unique value. In the case of f_{a_2}, when applied to a sequent in which there is a term constituted by a function symbol applied only to constants, the value is obtained from that sequent by replacing the term in question with the result of the computation of its value. In the case of f_{b_3}, when applied to a sequent of the form $\Gamma \to {\sim} A$, the value is the sequent A, $\Gamma \to 1 = 2$. As for f_{c_i}, where it seems that arbitrary choices are entailed, the result is determined univocally in regard to whether or not the formula in the antecedent that is to be reduced is preserved or not. Whether or not it is to be preserved is decided according to whether or not the sequent to be reduced has been obtained by means of a contraction applied to two formulas like the one to be reduced. If it has been, the formula will be preserved.

As for the functions f_{a_1}, f_{b_1}, f_{b_2}, corresponding to reduction steps in which arbitrary choices are made, they are applied to a pair constituted by the sequent to be reduced and a natural number. This allows one to individuate univocally each of their values, in other words, to individuate each sequent obtained from the given sequent by application of the reduction step. The first of these, f_{a_1}, when applied to a sequent containing a free variable a and to a natural number n gives as value the result of substituting the numeral corresponding to n for all occurrences of a in that sequent. The function f_{b_1} when applied to the sequent $\Gamma \to (\Pi x)A(x)$ and to the number n gives as value the sequent $\Gamma \to A(n)$. The function f_{b_2} when applied to the sequent $\Gamma \to A \& B$ and to the number $2n$, gives as value the sequent $\Gamma \to A$, while, when applied to the same sequent and to the number $2n + 1$, it gives $\Gamma \to B$ as value.

It is assumed that it is possible to operate on the functions $f_{a_i}, f_{b_i}, f_{c_i}$ by means of composition. The reduction function of a sequent must furnish a reduced form of the sequent once it is applied to the sequent to be reduced and, if the function involves n components of the type $f_{a_1}, f_{b_1}, f_{b_2}$, to an n-tuple of natural numbers.

4. This section consists in a transcription of Gentzen's proof in which the

reduction functions of the preceding section are substituted for the reduction steps, except for the case of the cut-rule, which will be the subject of
Sections 6 ff.

In order to prove that every derivable sequent is reducible, we begin by
proving that it is possible to associate every initial sequent with a reduction function. Now the reduction of initial arithmetical sequents involves
only steps of types a_1 and a_2. The reduction function will be obtained from
f_{a_1} and f_{a_2} by simple composition.

The reduction of initial logical sequents is obtainable from that of sequents of the form $A \to A$, where A is any formula. The associated functional will be constituted in part like the preceding one, whenever A contains free variables and non-computed terms. We then pass to the reduction of the formula A in the consequent, which will occur by means of
functions f_{b_i} according to the principal connective of the formula to be
reduced. Here too any components f_{b_1} and f_{b_2} of the function to be defined
are defined in relation to a natural number, and thus in the end the reduction function associated with the initial sequent will be defined in relation
to n-tuples of natural numbers, where n is the number of its components
of types $f_{a_1}, f_{b_1}, f_{b_2}$. If the reduction of A ends with a true atomic formula
for each of the possible choices, then components of the types f_{c_i} operating
on the antecedent are unnecessary. In cases where, for some n-tuple, the
reduction of A ends with a false atomic formula, it is necessary to com-
plicate the construction of the reduction function slightly by specifying
some steps to be executed on the antecedent, in other words, by adding
components of the types f_{c_i}. The reduction which takes place on the antecedent involves no arbitrary choices, that is, it occurs without extending the
n-tuple necessary for reducing the consequent. In fact, the choice of A or
B in the reduction of $A \& B$ or that of $A(n)$ in the reduction of $(\Pi x)A(x)$,
are already specified in the information contained in the part of the reduction which leads A to a false atomic formula, and this information is
encoded in a specific n-tuple of natural numbers. It is sufficient to specify
the steps of the types c_i, that is, to specify the values of the functions f_{c_i},
in correspondence to the steps carried out according to such an n-tuple in
order to be certain of finally reaching a false atomic formula in the antecedent, and thus a reduced form of $A \to A$. Hence the reduction function
$A \to A$ is defined in correspondence to sequents and n-tuples of natural
numbers if, in the reduction, the functions $f_{a_1}, f_{b_1}, f_{b_2}$ have been applied
n times.

It remains to prove that if it is possible to associate a reduction function with each premise of an inference rule, then it is possible to obtain from the functions of the premises a function which reduces the conclusion of the application of the rule. Let us examine first of all the cases of structural rules, that of $\sim$ introduction, and that of $\mathit{\Pi}$ introduction. The case of the induction rule can be handled in terms of that of the cut-rule (see Bernays (1970, p. 412) or Gentzen (1969, pp. 205, 206). The structural rules and the two introduction rules present no difficulties, unlike the cut-rule. For the case of the cut-rule involves the complexities of all the elimination rules which can be defined from it by using the initial logical sequents, and above all the reduction of the induction rule.

As for the exchange rules, addition in the antecedent, and change of bound variable, it is evident that if there is a reduction function for their premises, the same function can reduce the consequence. The case of the contraction rule requires the intervention of a reduction of the antecedent in which the reduced formula is preserved. Suppose $A, \Gamma \to B$ is inferred from $A, A, \Gamma \to B$ by means of contraction, and suppose that there is a reduction function for $A, A, \Gamma \to B$. If such a function provides a reduced form by reducing B to a true atomic formula or by reducing some formula of Γ to a false atomic formula without involving A, then the same function also allows the reduction of the conclusion of the rule. In the opposite case the reduction involves the formula A. Thus in the function there will be a f_{c_i} component which effects the reduction of $A, A, \Gamma^* \to B^*$ to $A, A', \Gamma^* \to B^*$ (where the asterisks indicate the execution of reduction steps). Let us now complete an analogous reduction step on the antecedent of the sequent $A, \Gamma^* \to B^*$ obtained by contraction, but here preserving formula A. We see immediately that the result of this operation can be reduced with the same sequence of operations that reduced the premise of the rule. In point of fact the reduction of the consequence of the rule is obtained by substituting the first component which operates in the antecedent on A with an analogous component which preserves formula A. As for the $\sim$ introduction rule, it will be sufficient to add a component f_{b_3} to the function which reduces the premise $\Gamma, A \to 1 = 2$, so that if f is the function which reduces the premise, then the function for the conclusion will be $f(f_{b_3})$. As for the $\mathit{\Pi}$ introduction rule, if the reduction function of the premise $\Gamma \to F(a)$ is $f(f_{a_1})$ (given that the sequent contains the free variable a), then the reduction function of the consequence will be $f(f_{b_1})$.

5. Let us now consider the case of the cut-rule in order to define a func-

tional ϕ which carries reduction functions f and f' of the premises $\Gamma \to A$ and $A, \Gamma \to B$ of the rule, respectively, to a function that reduces the conclusion $\Gamma, \Delta \to B$. Suppose that the cut-formula A is atomic, that is, of degree zero. In such a case the reduction of the conclusion is easily obtained by applying successively the functions f' and f that reduce the two premises. The function that reduces $\Gamma, \Delta \to B$ will thus be indicated by $f(f')$. Suppose that A is in fact true and that f' reduces $A, \Delta \to B$. Then it may happen that B is reduced either to a true or a false atomic formula, but that some Δ-formula is reduced to a false atomic formula. In such a case the steps carried out by f are superfluous. On the other hand should the reduction of $A, \Delta \to B$ depend on A, which must be a false atomic formula, then the reduction of $\Gamma \to A$ brought about by f must occur by reducing some Γ-formula to a false atomic formula. Thus, if – after having applied f' to $\Gamma, \Delta \to B$ – a reduced form has still not been reached (because the reduced form of $A, \Delta \to B$ depends on A), it is sufficient to apply the function f in order to reach a reduced form. This furnishes the basis of a definition of ϕ by induction on the degree of A.

Suppose now that we are able to obtain the function that reduces the conclusion from functions that reduce the premises of a cut-rule, whenever the cut-formula has degree less than n. Let us suppose therefore that the cut-rule in question has a cut-formula of degree n. As in the preceding case let us apply f' that reduces the premise $A, \Delta \to B$. If the reduction furnished by f' does not involve the cut-formula, then the reduction of $\Gamma, \Delta \to B$ is immediately obtained from f'. If instead the reduction of $A, \Delta \to B$ involves A, transforming it at a certain point to A' by means of a step of type f_{c_i}, and *if the reduction effected in this way does not involve the preservation of A*, one proceeds as follows. The application of f' to $\Gamma, \Delta \to B$ will reduce it to a sequent $\Gamma, \Delta^* \to B^*$, not in reduced form, where B^* is a false atomic formula and no Δ^*-formula is a false atomic formula. In this case a function which completes the reduction of $\Gamma, \Delta^* \to B^*$ is obtainable by the inductive hypothesis. In fact, by the inductive hypothesis it is possible to obtain, from functions g and g' that reduce premises $\Gamma \to A'$ and $A', \Delta^* \to B^*$ of a cut-rule, a function g'' that reduces the conclusion $\Gamma, \Delta^* \to B^*$ of this rule: for the cut-formula A' here has degree less than n (given that A' contains one connective fewer than A). Having obtained g'', the sequent $\Gamma, \Delta \to B$ will be reduced by $g''(f')$. (The functions g' and g are obtainable from f' and f, respectively by eliminating from these last the steps leading from $A, \Delta \to B$ to A', $\Delta^* \to B^*$, and from $\Gamma \to A$ to $\Gamma \to A'$ respectively; in other words, by eliminating everything up to the first step involving A.)

In the case where no reductions are carried out on the cut-formula in the antecedent that involve the preservation of the formula, we can define the functional ϕ in the following way, by induction on the degree of the cut-formula:

$$\phi(f, f', 0) = f(f')$$
$$\phi(f, f', n) = (\phi(g, g', (n - 1)))(f')$$

The most interesting part of the proof occurs in the case that the reduction of the premise $A, \Delta \to B$ contains at least one reduction step carried out on the antecedent which involves A and in which the formula is preserved, passing to the sequent $A', A, \Delta^* \to B$.

Let us look briefly at the lines of Gentzen's proof. He assumes he has already proven the possibility of reducing the conclusion of a cut-rule having the following premises: $\Gamma \to A$ and $A, A', \Delta^* \to B^*$. As can be seen, the degree of the cut-formula is constant while the second sequent has undergone at least one reduction step. It is in fact a partial result of the reduction process, which by hypothesis exists, of the premises $A, \Delta \to B$. To sum up, for the reduction of the application of the cut-rule

$$\frac{\Gamma \to A \qquad\qquad A, \Delta \to B}{\Gamma, \Delta \to B}$$

we substitute the reduction of two applications of the cut-rule

$$\frac{\Gamma \to A' \qquad \dfrac{\Gamma \to A \qquad\qquad A, A', \Delta^* \to B^*}{A', \Gamma, \Delta^* \to B^*}}{\Gamma, \Delta^* \to B^*} \qquad\begin{matrix}(1)\\(2)\end{matrix}$$

of which (1) still has A as cut-formula and one of the two premises has undergone at least one reduction step, while (2) has cut-formula of degree less than A and thus the inductive hypothesis can be applied to it.

Naturally to suppose that it is possible to reduce (1) means having presupposed something very similar to the theorem which is intended to be proved (Lemma 14.6 of the 'galley proof'), save for the fact that at least one of the two premises is now reduced by at least one step. The procedure avoids circularity only if one reasons by induction on the length of the reduction process. In fact, if, in order to prove the theorem, we must suppose we proved it for a cut-rule in which a premise has undergone a reduction, we must subsequently prove this hypothesis, which in its turn refers to an analogous theorem in which such a premise has been further reduced, and so on. Thus a succession of theorems must be proved in which the premise in question advances toward its reduced form, which exists by hypothesis. According to Gentzen, that this process ends in a finite number of steps depends on the finitude of the reduction process. Let us see what complications are involved in this case in the definition of functional ϕ which furnishes the reduction of the conclusion of the rule.

6.1. In order to discuss the case of the cut-rule it is necessary to define ϕ by induction on a characteristic of the reduction functions, the level, which represents in a certain sense a measure of the complexity of the reduction carried out by such a function. So we shall assign an ordinal number representing the level to each derivable sequent and to the corresponding reduction function. The guiding principle in assigning an ordinal number as the level of a sequent is that each sequent obtained by applying reduction steps for the given sequent is assigned a smaller ordinal than the given sequent. It is possible to obtain this result if a reduction tree is associated with each derivable sequent in such a way that each sequent obtained from S by reduction steps has as its associated reduction tree a subtree of the tree associated with S. Ordinal numbers will then be assigned to the reduction trees in such a way that the ordinal of a tree is always larger than that of each of its subtrees. In this way each derivable sequent S, by means of its reduction tree T, is associated with an ordinal α such that any sequent S' obtained by reduction from S has ordinal $\beta < \alpha$, its reduction tree T' being a subtree of T.

6.2. We now define the concept of reduction tree. By reduction tree is meant a finitely or non-finitely generated tree of finite level. A tree is said to be finitely generated if each point has a finite number of immediate successors, otherwise it is said to be non-finitely generated. The level of each point is given by a natural number assigned to the point by a function f defined as follows. If O is the origin of the tree, then $f(O) = 0$; in other words, zero is the level of O. If point x is an immediate successor of point y and $f(y) = n$, then $f(x) = n + 1$. The level of a tree is the maximum of the levels of its points; and the tree has a finite level if the maximum is given by a natural number. If a tree has finite level, the maximum length of the branches is finite, where by the length of a branch is meant the number of points which constitute it, excluding the origin.

A reduction tree lends itself to the representation of the reduction process of a sequent. It is sufficient to place the sequent in the origin, then in the successive level, as succeeding points, to place the sequents obtained from such a sequent by means of a reduction step, and so on. The terminal points are reduced forms of the sequent placed in the origin. On the basis of the possible reduction steps, the number of successors of a point can be 1, 2, or denumerable. Nevertheless, each branch will be finite, corresponding to the fact that each reduction must end after a finite number of steps.

6.3. Let us now assign an ordinal number to reduction trees in the following manner. Starting from the origin, the choice which is exercised in passing to the successive level is encoded by a natural number: the fact that infinitely many different choices can be made implies the use of the totality of natural numbers. In the end, each branch of the tree will appear as a finite sequence of natural numbers and the information contained in such a sequence is sufficient to lead from the origin of the tree to the selected terminal point, in other words, from the given sequent to its selected reduced form. The ordering of these sequences is the ordinal number of the given tree. If the tree has level n, in other words, if the maximum length of the branches is n, each branch can be viewed as a sequence of natural numbers of length n. For the branches which stop at a level $n\text{-}m$, it is sufficient to add m zeroes. If the tree is finitely generated, each branch is a sequence of length n of natural numbers less than q for a certain q; for the ordering of the branches q^n is sufficient, and this is an ordinal less than ω. If the tree is not finitely generated, each branch is a sequence of length n of natural numbers. Now the ordinal ω^n is constituted by all sequences of length n of natural numbers; thus ω^n is the ordinal which provides the ordering of the branches of the given tree. Since it is not possible to establish a finite limit to the number of reduction steps necessary for a sequent, that is, to the level of the tree, ω^ω becomes the limit of the necessary ordinals, according to this method of assigning natural numbers to the points of the tree.

6.4. The assignment of ordinal numbers to derivable sequents occurs first by associating ordinals with initial sequents and then by determining the ordinal of each sequent obtained by means of an inference rule on the basis of the ordinal number assigned to the premises.

Ordinal numbers less than ω^ω are associated with the initial sequents: this reflects the structure of their reduction trees. Reduction rules, apart from the cut-rule, do not involve significant modifications in the transition from the ordinal number of the premises to that of the consequence. In fact, the reduction tree of the consequence does not have a very different structure since at the most it has a larger number of levels. In order for the subtrees to have a smaller ordinal, it is sufficient to assign an ordinal smaller than ω^ω to the tree of the conclusion. In the case of the cut-rule the situation changes drastically since the reduction tree of the consequence is much more complicated than those of the premises. We distinguish two cases.

Case 1

How does the reduction tree for $\Gamma, \Delta \to B$ appear once ones for $\Gamma \to A$ and for $A, \Delta \to B$ are known? First of all it will consist of a section of the tree for $A, \Delta \to B$, where, by section of a tree of level n is meant the tree obtained by eliminating the points of the level larger than m, $0 < m < n$. To characterize this section it is sufficient to eliminate all the points obtained by acting on the formula A in the antecedent and all the points dependent on them. In parallel fashion Gentzen's proof prescribes that the same reduction used for $A, \Delta \to B$ be carried out on $\Gamma, \Delta \to B$, up to the first reduction step carried out on A. In this way a tree is obtained where the terminal points not yet in reduced form correspond to the sequent $\Gamma, \Delta^* \to B^*$. (In Gentzen's proof $\Gamma, \Delta^* \to B^*$ is precisely the result of the application of the existing reduction of $A, \Delta \to B$ to $\Gamma, \Delta \to B$.) Now Gentzen's proof prescribes that the reduction of $\Gamma, \Delta^* \to B^*$ be carried out; this reduction is obtained from the application of the cut-rule with premises $\Gamma \to A'$ and $A', \Delta^* \to B^*$: by the inductive hypothesis this is reducible. Similarly, the reduction tree for $\Gamma, \Delta^* \to B^*$ is to be added below those points of the tree that are not in reduced form, obtaining the former tree from trees for $\Gamma \to A'$ and for $A', \Delta^* \to B^*$.

These last are subtrees of those for $\Gamma \to A$ and for $A, \Delta \to B$, which exist by hypothesis. Consequently it is not possible to determine the tree for $\Gamma, \Delta^* \to B^*$ directly, but, by carrying out the same procedure used to obtain the tree of $\Gamma, \Delta \to B$ from those of the premises of the cut-rule, we will be able to say that it certainly begins and develops like the tree for $A', \Delta^* \to B^*$ up to a certain level, that is up to the point where the existing reduction for $A', \Delta^* \to B^*$ involves the formula A'. In Gentzen's proof in parallel fashion the sequent $\Gamma, \Delta^* \to B^*$ is reduced to the sequent $\Gamma, \Delta^{**} \to B^{**}$ by using the steps of the reduction of $A', \Delta^* \to B^*$ until they involve A'. At this point we can add to the partial tree obtained for $\Gamma, \Delta \to B$ the part which we know of the tree for $\Gamma, \Delta^* \to B^*$: the points not in reduced form of the tree so obtained will correspond to sequents $\Gamma, \Delta^{**} \to B^{**}$. In this way, with successive additions of trees below the points not yet in reduced form, we will finally complete the tree for Γ, $\Delta \to B$. That the process does end is ensured by the diminution of the degree of the cut-formula A.

The assignment of an ordinal must take into account the structure of the tree which has been delineated. Since the tree has been obtained by means of successive additions of trees below the terminal points of a tree, each branch can be considered as a finite sequence of finite sequences of natural numbers, in other words, as a sequence of branches of the component trees. The length of the sequence equals the degree n of the cut-formula (the number of times which the preceding procedure is iterated). The length of the sequences that are the elements of the sequence of length n has as its upper limit the maximum length of the branches of the trees of the premises. If α is the maximum of the ordinals of the premises, α^n will be the ordinal to be associated with the tree of the conclusion. Every subtree will clearly have a smaller ordinal. The ordinal α^n will be smaller than ω^ω which remains the limit of the necessary ordinals.

Case 2

In this case it is also possible to extract a procedure for constructing the tree for $\Gamma, \Delta \to B$ from Gentzen's proof, but here we are dealing with a more complicated process than the preceding one. In fact, in the latter, one proceeded by induction on the degree of the cut-formula, whereas now there are two inductions: one on the degree of the cut-formula and one on the length of the reduction of the premises of the cut-rule. In fact, once $\Gamma, \Delta \to B$ has been reduced to $\Gamma, \Delta^* \to B^*$ by means of the reduction of $A, \Delta \to B$, the reduction of $\Gamma, \Delta^* \to B^*$ results from the following two applications of the cut-rule:

$$\frac{\Gamma \to A' \quad \dfrac{\Gamma \to A \quad A, A', \Delta^* \to B^*}{A', \Gamma, \Delta^* \to B^*}}{\Gamma, \Delta^* \to B^*}$$
$$\tag{1}$$
$$\tag{2}$$

The inductive process which furnishes us with the tree for $\Gamma, \Delta^* \to B^*$ on the basis of that for $\Gamma \to A'$ and that for $A', \Gamma, \Delta^* \to B^*$, in relation to the cut-rule (2), is completely similar to that dealt with in Case 1. Nevertheless the tree for $A', \Gamma, \Delta^* \to B^*$ in its turn results from an inductive process on the length of the reduction of the premises $\Gamma \to A$ and $A, A', \Delta^* \to B^*$ which is completely similar to the process by which the tree for Γ, $\Delta \to B$ is constructed from those for $\Gamma \to A$ and A, $\Delta \to B$.

The same way that Gentzen assumes he has proven his Lemma 14.6 in the case where the premises have undergone some reductions, so we assume we have already attained the tree for A', Γ, $\Delta^* \to B^*$. What distinguishes Case 1 from Case 2 is precisely the fact that while in the first the reduction of A', $\Delta^* \to B^*$ is contained in that of A, $\Delta \to B$, and hence is already given, in Case 2 that for A', Γ, $\Delta^* \to B^*$ is still to be determined. In any case, assuming that we do have the reduction, the tree for Γ, $\Delta^* \to B^*$ can be determined, as in the first case, by means of the addition of a series of trees below the points not in reduced form, an addition to be carried out a finite number of times. Now in the tree so obtained we isolate the subtrees which correspond to the added trees, and the tree as a whole is considered as a tree of trees. This corresponds to the fact that the reduction of A', Γ, $\Delta^* \to B^*$ is still to be determined, and in Gentzen's proof it is equivalent to the assumption of having already proved the lemma in the case in which one of the premises has been reduced. In their turn the added trees turn out to be trees of trees and the same holds true for their constituents. This process comes to an end after a finite number of steps because each time the process is repeated the trees from which one starts are always smaller, in other words, of a smaller level, given that they are subtrees of those from which one began in the preceding step of the process.

The reduction tree has a structure similar to that of Case 1, yet the trees from which it is constituted are still to be determined. Thus, as in Case 1, each branch of the tree is presented as a sequence of length n, but the length of the sequences of which it is composed cannot be determined directly. Each of these component sequences appears as a sequence of length n of sequences whose length is still to be specified. If the maximum of the ordinals of the premises is an ordinal $\alpha < \omega^\omega$, then each branch of the tree for the conclusion is presented as a sequence of length n of sequences of length n of ... sequences of length n of natural numbers, where 'sequence of' occurs α times. The ordinal which is needed for the tree is thus $((\omega^n)^n)^{\cdots}$ (α times) $= \omega^{n \cdot n \cdots}($$\alpha$ times$) = \omega^{(n^\alpha)}$. Still $n^\alpha = \alpha$, since $\alpha = \omega^q$ for a certain q. The limit of the necessary ordinals is thus shifted to ω^{ω^ω}, always maintaining as a guiding principle that each subtree has a smaller ordinal. Still, it is now possible to have ordinals α, $\omega^\omega < \alpha < \omega^{\omega^\omega}$ in the premises. Correspondingly the limit of the necessary ordinals will be shifted. It is clear that as long as ω^α is larger than α, α cannot constitute a limit for the necessary ordinals. This type of ordinal assignment imposes as a limit the first ordinal α such that $\omega^\alpha = \alpha$: in other words, ε_0.

6.5. Once an ordinal representing the level has been assigned to the derivable sequents and to each sequent obtainable from them by reduction, the same ordinal is associated with the corresponding reduction function. It is now possible, by recursion on the level of the reduction functions, to define the functional ϕ which associates a reduction function for the conclusion with the reduction functions for the premises of the cut-rule in the case where the cut-formula is preserved. Given that the level-ordinals have ε_0 as a limit, a recursion on the level goes beyond the bounds of the finitary.

In order to underline the analogies between Gentzen's proof and the definition of ϕ, the index of the reduction functions, which should be constituted by an ordinal representing the level, is given by the sequent that the function reduces. The lowering of the ordinal is represented by an advancement in the reduction process of such a sequent.

Let $f_{\Gamma \to A}$ and $f_{A, \Delta \to B}$ be functions that reduce the premises $\Gamma \to A$ and $A, \Delta \to B$. The function $f_{\Gamma, \Delta \to B}$ that reduces the conclusion $\Gamma, \Delta \to B$ of the rule is obtained by means of the functional ϕ which is defined by means of the following nested recursion.

$$\phi(f_{\Gamma \to A}, f_{A, \Delta \to B}) = f_{A, \Delta \to B} \equiv_{df} f_{\Gamma, \Delta \to B} \tag{1}$$

if $\Gamma \to A$ is in reduced form. In such a case, A is atomic; if it is true, the reduction of $A, \Delta \to B$ does not depend on A, thus by operating on B and on Δ one arrives at a reduced form in the case of $A, \Delta \to B$ as well as in that of $\Gamma, \Delta \to B$. If it is false, the reduction of $\Gamma \to A$ derives from the presence of a false atomic formula in Γ. A reduced form of $\Gamma, \Delta \to B$ is thus guaranteed in both cases by the reduction of $A, \Delta \to B$.

$$\phi(f_{\Gamma \to A}, f_{A, \Delta \to B}) = f_{\Gamma \to A} \equiv_{df} f_{\Gamma, \Delta \to B} \tag{2}$$

if $A, \Delta \to B$ is in reduced form. In such a case, in fact, if its reduced form does not depend on A, which is a true atomic formula, $\Gamma, \Delta \to B$ is also in reduced form and thus has no need of transformations. If instead its reduced form derives from A, which is a false atomic formula, the reduction of $\Gamma \to A$ must take place by falsifying a formula of the antecedent Γ, hence the same transformation also leads $\Gamma, \Delta \to B$ to a reduced form.

Once we have established the bases of the recursions on the two variables, we come to the decisive point of the definition of ϕ, the point where Gentzen's method of proof is reflected:

$$\phi(f_{\Gamma \to A}, f_{A, \Delta \to B}) = \phi(f_{\Gamma \to A'}, \phi(f_{\Gamma \to A}, f_{A, A', \Delta^* \to B^*}))(f_{A, \Delta \to B}) \qquad (3)$$

where $\Gamma \to A'$ and $A, A', \Delta^* \to B^*$ are sequents of lower level than $\Gamma \to A$ and $A, \Delta \to B$, respectively, given that they are obtained by applying at least one reduction step.

7. It is possible to conclude that: (1) Gentzen's nonfinitary methods of proof reduce to induction up to ε_0; (2) the non-finitary point resides not so much in the reduction process for the sequents as in the method by which we effectively generate the reduction procedure to be correlated with each derivable sequent, in ascertaining that such a reduction is effective and ends in a finite number of steps.

With regard to the fact that the ordinals called for by the proof have ε_0 as a limit, it can be noted that this derives both from the principle adopted, by which each reduction operation must lower the ordinal (in other words, lead to a subtree), and from the type of transformation to which trees of the premises are subjected in the case of the cut-rule in order to obtain a tree of the consequence.

There probably are connections between the fact that, in the case of the cut-rule in which the cut-formula is preserved in the antecedent, the structure of the reduction tree leads to such high assignments of ordinals, and the fact that ϕ is in this case defined by means of a nested recursion (see Tait, 1961).

BIBLIOGRAPHY

Bernays, P. (1970), 'On the original Gentzen consistency proof', in *Intuitionism and Proof Theory*. Amsterdam, North Holland. pp. 409–17.

Gentzen, G. (1969), *Collected Papers*, ed. by M.E. Szabo. Amsterdam, North Holland.

Kreisel, G. (1971), 'Book review of Gentzen's *Collected Papers*,' *Journal of Philosophy*, **68**, 238–65.

Kreisel, G. (1976), 'Wie die Beweistheorie zu ihren Ordinalzahlen kam und kommt', *Jahresbericht der Deutschen Mathematiker-Vereinigung* **78**, 177–223.

Tait, W.W. (1961), 'Nested recursion,' *Mathematische Annalen* **143**, 236–50.

Tait, W.W. (1965), 'Functionals defined by transfinite recursion', *Journal of Symbolic Logic* **30**, 155–192.

INDUCTIVE LOGIC AND INDUCTIVE STATISTICS

1. The title of the present article needs an explanation which, in one sense, can also be intended as a premise: that is, the statistics with which we intend to deal is that generally known as mathematical.

It is well known that there are two parts to the theory of statistics. The first is descriptive, the second is generally called mathematical. To use J.M. Keynes' words, the first one "devises numerical and diagrammatic methods by which certain salient characteristics of large groups of phenomena can be briefly described; and . . . provides formulae by the aid of which we can measure or summarise the variations in some particular character which we have observed over a long series of events or instances" [1]. The second, which Keynes named inductive, "seeks to extend its description of certain characteristics of observed events to the corresponding characteristics of other events which have not been observed" [2]. But then the use of the term 'mathematical' is, at best, inadequate for the individuation of the second part of the theory of statistics. It can also be misleading, in that it suggests that this part of the theory of statistics is deductive.

In this article we want to deal with the second part of the theory of statistics which, according to what we have just said, shall henceforth be called *inductive statistics*.

2. It is well known that the fundamental thesis of the neo-Bayesian (Bayesian for brevity) approach to inductive statistics is that inductive statistics is based on Bayes' theorem: that is, every statistical inference starts with an assignment of initial probabilities to be changed into the final probabilities after the observations are taken into account. It is a widespread conviction that to accept the Bayesian approach also implies accepting a subjectivistic philosophy of probability and then a certain kind of idealism. In our opinion this conviction is deeply wrong, and one of the purposes of this article is to show that the Bayesian approach to inductive statistics suggests in a very natural way a materialistic philosophy of probability. In doing so, we shall also try to answer I. Hacking, who rightly says:

169

Maria Luisa Dalla Chiara (ed.), Italian Studies in the Philosophy of Science, 169–183.

It is curious how papers sympathetic to inductive logic ignore Ramsey's fundamental challenge. The first obligation of the realist who does not like Berkeley or phenomenalism is to kick a stone or hold up a hand and insist that whatever be the external world, at least there is one. Inductive logicians have not honored this first obligation, of providing a *prima facie* case that there are *any* interpersonal relations of credibility that satisfy quantitative or even comparative probability axioms [3].

For the present, we leave this philosophical question and return to inductive statistics. The subjectivists maintain that after the acceptance of the Bayesian approach, all problems are, at least in principle, solved. The probabilities involved in Bayes' theorem represent the subjective opinions of the man who makes the inference. These opinions can be quantified in many ways, but cannot be rationally explained. On the contrary, our opinion is that the acceptance of the Bayesian approach to inductive statistics has opened up many foundational problems of great importance. The theory that tries to solve these problems is inductive logic.

3. It is clear, even if not explicitly stated, that, in his last works, R. Carnap [4] intends inductive logic to be the rational reconstruction of the methods of determining probability values used in inductive statistics. This reconstruction is worked out axiomatically. In the axiomatic systems for inductive logic, probability is a primitive notion. Hence in these systems probability values are determined without ever having to go back to the properties of probability other than the ones explicitly stated as axioms.

We believe that Carnap's decision to restrict his work to the rational reconstruction of statistical inferences – hence to the rational reconstruction of the methods for determining probability values used in inductive statistics – is quite suitable. This is not because we think that inductive logic must deal only with this type of inference while neglecting universal inferences. However, it is undeniable that the known ways of assigning probability values to inductive generalizations are either non-scientific or, when they are scientific, completely unrealistic [5]. The sole scientific and realistic methods for determining probability values known at present are those used in inductive statistics. Hence in order to accomplish useful work, inductive logic must be concerned with these methods, leaving the analysis of universal inferences to the future whenever possible.

From this point of view, the task of inductive logicians is to isolate the methods for determining probability values used in inductive statistics (possibly a small number of them), to axiomatize them and then to develop the consequences of these axioms deductively. They must not try to impose the axioms of inductive logic and therefore the methods of inference on the statisticians. The statisticians must have the final choice of these methods. The work of inductive logicians must give statisticians the knowledge necessary for carrying out a rational choice.

We believe that this is the task of inductive logic at least in the present stage of its development. The reason for that is twofold. First, as we have just now said, the purely scientific and realistic inductive inferences known at present are those used in inductive statistics. Hence, these inferences must be taken into account by those intent on analyzing inductive inferences on a logical level. Second, only in this way, that is, only through the rational reconstruction of these inferences can inductive logic refine its conceptual tools. We strongly believe that in the future inductive logic will be able to elaborate methods from itself for determining probability values. Then its task will be not only the rational reconstruction of the methods used by statisticians but also the search for new methods of inductive inference.

4. Applying Carnap's results [6] we shall try to give an example of such a work relative to a simple statistical problem.

Suppose [7] that a population of N individuals is composed of N_1 members of type P_1 and of N_2 members of type P_2 (in short, $N_1 P_1$ and $N_2 P_2$) such that $N_1 + N_2 = N$. A sample of n individuals is drawn in such a way that any set of n individuals in the population is equally likely to be taken. Of the sample, $n_1 P_1$ and $n_2 P_2$ are such that $n_1 + n_2 = n$. We want to infer something about N_1, given N, n_1 and n_2.

The number of possible samples is $\binom{N}{n}$ and the number of them with $n_1 P_1$ and $n_2 P_2$ is $\binom{N_1}{n_1} \binom{N-N_1}{n-n_1}$. Any two samples exclude each other and we have supposed that they are equally probable. Moreover some samples with n individuals must occur. Hence the probability that any particular sample will occur is $1/\binom{N}{n}$ and the probability of the observed sample when of N individuals N_1 are P_1 will be

$$\frac{\binom{N_1}{n_1}\binom{N-N_1}{n_2}}{\binom{N}{n}}. \tag{1}$$

We have no information to say that one value of N_1 is more probable than another. Hence we take all their *a priori* probabilities as equal, that is we put these probabilities equal to

$$1/(N + 1). \tag{2}$$

Finally applying Bayes' theorem we obtain the probability of N_1, given N, n_1 and n_2

$$\frac{\binom{N_1}{n_1}\binom{N - N_1}{n_2}}{\sum_{i=0}^{N}\binom{i}{n}\binom{N - i}{n_2}}.$$

Thus our problem is solved. But what have we really assumed with the two hypotheses we have made? To summarize. We have assumed:

(I) any set of n individuals in the population is equally probable to be taken;

(II) any possible value of N_1 is *a priori* equally probable. What is the real meaning of these hypotheses?

5. To try to answer this question we take a language $\mathscr{L}$ with a family of two monadic predicates P_1 and P_2 and N individual constants a_1, ..., a_N. In order to explain (I), we suppose that in the population, N_1 individuals are of type P_1. Then a possible composition of the population will be

$$U_i = P_i a_1 \cap \cdots \cap P_i a_N$$

where N_1 $P_i a_i$, are $P_1 a_i$ and N_2 $P_i a_i$ are $P_2 a_i$. There are $\binom{N}{N_1} U_i$ and let $\{U_i\}_{i \in I}$ with a suitable I, be the set of all U_i. The hypothesis for the population $N_1 P_1$ and $N_2 P_2$ is then $\bigcup_{i \in I} U_i = U_{N_1}$.

If we adopt the hypothesis of symmetry with respect to individuals [8] we have

$$\mathscr{C}(U_i/U_{N_1}) = \mathscr{C}(U_j/U_{N_1});$$

but $U_i \cap U_j = \emptyset$ if $i \neq j$, then

$$1 = \mathscr{C}(\bigcup_{i \in I} U_i/U_{N_1}) = \sum_{i \in I} \mathscr{C}(U_i/U_{N_1})$$

$$= \binom{N}{N_1}\mathscr{C}(U_i/U_{N_1}) \text{ with } i \in I;$$

that is, if $i \in I$

$$\mathscr{C}(U_i/U_{N_1}) = 1/\binom{N}{N_1}.$$

A sample with $n_1 P_1$ and $n_2 P_2$ will be

$$E_j = P_j a_1 \cap \cdots \cap P_j a_n$$

where n_1 $P_j a_i$ are $P_1 a_i$, and n_2 $P_j a_i$ are $P_2 a_i$. There are $\binom{n}{n_1}$ E_j and let $\{E_j\}_{j \in J}$ with a suitable J, the set of all E_j. The hypothesis that in the sample $n_1 P_1$ and $n_2 P_2$ is then $\bigcup_{j \in J} E_j = E_{n_1}$.

There are only

$$\binom{n}{n_1}\binom{N-n}{N_1-n_1} U_i \in \{U_i\}_{i \in I}$$

such that there is a $E_m \in \{E_j\}_{j \in J}$ with $E_m \subset U_i$; let $\{U_i\}_{i \in K}$ with a suitable K, the set of all this U_i. Then the probability to choose the sample E_{n_1} is the probability to choose a population of $\{U_i\}_{i \in K}$. That is

$$\mathscr{C}(E_{n_1}/U_{N_1}) = \mathscr{C}(\bigcup_{i \in K} U_i/U_{N_1})$$

$$= \sum_{i \in K} \mathscr{C}(U_i/U_{N_1}) = \binom{n}{n_1}\binom{N-n}{N_1-n_1} \mathscr{C}(U_i/U_{N_1})$$

with $i \in K$, and finally

$$\mathscr{C}(E_{n_1}/U_{N_1}) = \frac{\binom{n}{n_1}\binom{N-n}{N_1-n_1}}{\binom{N}{N_1}} = \frac{\binom{N_1}{n_1}\binom{N-N_1}{n_2}}{\binom{N}{n}}.$$

Thus we have the likelihood (1) and we have also shown that the very meaning of (I) is the symmetry with respect to individuals.

6. In order to partially explain (II) we observe that this supposition amounts to the *a priori* equiprobability of the U_{N_1} with $0 \leqslant N_1 \leqslant N$. The application of the multiplicative axiom to the *a priori* probability of

$$U_i = P_1 a_1 \cap \cdots \cap P_1 a_{N_1} \cap P_2 a_{N_1} a_{N_1+1} \cap \cdots \cap P_2 a_N$$

gives [9]

$$\mathscr{C}(U_i/Z) = \mathscr{C}(P_1 a_1/Z) \times \mathscr{C}(P_1 a_2/P_1 a_1) \times \cdots$$
$$\times \mathscr{C}(P_1 a_{N_1}/P_1 a_1 \cap \cdots \cap P_1 a_{N_1-1}) \times \mathscr{C}(P_2 a_{N_1+1}/P_1 a_1 \cap \cdots \cap P_1 a_{N_1})$$
$$\times \mathscr{C}(P_2 a_{N_1+2}/P_1 a_1 \cap \cdots \cap P_1 a_{N_1} \cap P_2 a_{N_1+1}) \times \cdots$$
$$\times \mathscr{C}(P_2 a_N/P_1 a_1 \cap \cdots \cap P_2 a_{N-1}).$$

But because of the symmetry with respect to individuals, $\mathscr{C}$ is a symmetric function, i.e. there are numerical functions C_i with $1 \leqslant i \leqslant 2$ such that

$$\mathscr{C}(U_i/\mathbf{Z}) = C_1(0, 0) \times C_1(1, 0) \times \cdots \times C_1(N_1 - 1, 0)$$
$$\times C_2(N_1, 0) \times C_2(N_1, 1) \times \cdots \times C_2(N_1, N_2 - 1).$$

Now we suppose that $\mathscr{C}$ is regular and F has γ-equality with respect to $\mathscr{C}$, *i.e.*, if H and E are any propositions on $\mathscr{L}$, molecular, and such that $E \cap H \neq \varnothing$, then $\mathscr{C}(H/E) > 0$, and $\mathscr{C}(P_1 a_i/\mathbf{Z}) = \mathscr{C}(P_2 a_i/\mathbf{Z})$ or equivalently $C_1(0, 0) = C_2(0, 0)$, then F is a symmetric family, that is, if $h, k \in N$, then $C_1(h, k) = C_2(k, h)$, in particular $C_1(0, 1) = C_2(1, 0)$. Following Carnap we put $C_1(0, 1) = C_2(1, 0) = G_i^1(0)$ with $1 \leqslant i \leqslant 2$.

It is well known that if we suppose that $C_i(h, k)$ is a linear function of the ith argument, then

$$C_i(h, k) = \frac{h - (2h - 1)G_i^1(0)}{h + k - 2(h + k - 1)G_i^1(0)}.$$

Let

$$G_i^1(0) = \lambda/2(\lambda + 1) \text{ with } 0 \leqslant \lambda \leqslant \infty, \text{ [10]} \tag{3}$$

then $C_i(h, k) = (h + \lambda/2)/(h + k + \lambda)$, and thus

$$\mathscr{C}(U_i/\mathbf{Z})$$
$$= \frac{\lambda/2(1+\lambda/2)\ldots(N_1-1+\lambda/2)/2(1+\lambda/2)\ldots(N_2-1+\lambda/2)}{\lambda(1+\lambda)\ldots(N-1+\lambda)}.$$

We have already seen that in $\{U_i\}_{i \in I}$ there are $(N!/N_1!N_2!)U_i$, then

$$\mathscr{C}(U_{N_1}/\mathbf{Z}) = \frac{N!}{N_1!N_2!}$$
$$\frac{\lambda/2(1+\lambda/2)\ldots(N_1-1+\lambda/2)/2(1+\lambda/2)\ldots(N_2-1+\lambda/2)}{\lambda(1+\lambda)\ldots(N-1+\lambda)}$$

To obtain $\mathscr{C}(U_{N_1}/\mathbf{Z}) = 1/N + 1$, and thus to achieve (2), we must put $\lambda = 2$. This assumption (see (3)) means that $G_i^1(0) = 1/3$. This has at least two meanings:

(a) $G_i^1(0) > 0$;

(b) $G_i^1(0) < 1/2$.

(a), that we have already implicitly assumed with $\lambda > 0$, is assured from the regularity we have supposed. (b), which is equivalent to $\lambda < \infty$, follows from the principle of positive instantial relevance. Note that (b) holds because we have assumed the γ-condition, otherwise $G_i^1(0)$ depends also from $C_i(0, 0)$.

We have thus shown what are the hypotheses that are behind (I) and

(II). That is, the hypotheses that adopting (I) and (II) we really suppose are, at least, the following: (i) regularity; (ii) symmetry with respect to individuals; (iii) γ-equality; (iv) linearity of $C_i(h,\ k)$ relatively to h; (v) principle of positive instantial relevance. In other words: every time we suppose that the sample of the population satisfies (I), we suppose that (ii) holds. Every time we suppose that the possible values of N_1 in the population satisfy (II), we suppose, at least, that (i)-(v) hold. Clearly, (II) depends also on other hypotheses that at present we cannot explicitly state, i.e. the hypotheses that allow us to choose the value 1/3 for $G_i^1(0)$ in the interval (0, 1/2).

7. In this section, we want give further examples of rational reconstruction of statistical inferences. We start with a very famous inference i.e., the rule of succession of Laplace. In this inference we suppose that the population is composed of countably many individuals and that it holds that

$$\lim N_1/N = p \in [0,\ 1] \text{ when } N \to \infty. \tag{4}$$

Moreover the likelihood is

$$\mathscr{C}(n_1,\ n/p) = \binom{n}{n_1} p^{n_1}(1 - p)^{n_2}, \tag{5}$$

the *a priori* distribution is

$$\mathscr{C}(dp) = dp, \tag{6}$$

the final distribution is then

$$\mathscr{C}(dp/n_1,\ n) = \frac{(n + 1)!}{n_1!n_2!}\ p^{n_1}(1 - p)^{n_2}\ dp.$$

With a language with individual constants we can repeat what we have said in Section 5 and then in order to obtain (5) it is sufficient to impose (4). Also (6) can be obtained as in Section 5 imposing finally (4). The rational reconstruction of (6) is again partial but now there is a new problem to be solved behind the choice of a precise value for $G_i^1(0)$. We mean the validity or at least the justification of the use of (4) in calculating (6). With (4) we suppose that the relative frequency of p_1 has a limit when the population tends to infinity. This assumption can be made for calculating the likelihood, but seems problematic for calculating the *a priori* distribution. Anyway (4) cannot be simply accepted but needs further investigation in order to clarify its logical status completely.

The second example is the inference in which we use the normal distribution both for the likelihood and the initial distribution. The calculation of the likelihood can be made in the way we have just suggested, supposing that also the number of individuals in the sample tends to infinity, that is, as suggested from the theorem of de Moivre–Laplace. The use of this theorem can also help us to calculate the initial probability, but in this case we need new hypotheses besides those whose validity we have supposed for the likelihood [11]. It then follows that the use of the same distribution for the likelihood and the initial distribution are based on completely different hypotheses.

The third example is that it is also possible to give a rational reconstruction in terms of Carnap's results for the Beta distribution, as initial distribution, when its parameters are natural numbers and thus also for the related Gamma and Exponential distributions [12]. The possibility of such a reconstruction for the Beta distribution when its parameters are real numbers is an open problem.

8. We now come back to the problem of the philosophy of probability which we faced in Section 2, with the aim of justifying the point of view we then expressed. In order to do this, however, it is necessary to look at the history of the foundations of probability in a new way.

The first problems faced and the first results obtained by the use of probabilistic notions were related to games of chance. Even in the eighteenth century, the great majority of applications of probability theory were related to the problem of gambling. Even when this was not the case, games of chance were the privileged model, viz. one attempted to reduce every problem to a problem of gambling.

In games of chance the problem of determining probability values is conceptually simple; and this is because the set-up used in these situations (coins, dice, *roulettes*, cards) is not fixed. The set-up thus leads to equally likely elementary cases. Therefore, it is only a question of determining them. Sometimes it is possible to do this in a very simple way; sometimes this is not the case. In any case the determination is always possible. The same holds for favorable cases. The determination of probability values is therefore always possible by determining the ratio of favorable cases to the total number of equally likely ones.

What we have just seen is a method for determining probability values. This method is based on the knowledge of the set-up by which games of chance are carried out. This knowledge leads to the conclusion that events of a given set are equally probable. Classical theory, after having discovered this method, has preferred it and adopted it as the definition of probability [13].

Laplace's work made the theory of probability a completely scientific theory. This greatly enlarged its field of application. The effect of this success of probability theory was two-fold. On one hand, it supported the classical definition as the very root of probability theory. On the other hand, the inadequacy of the classical definition in the new fields of application became more pronounced. Therefore it became clearer and clearer that the classical definition reflected only the way of determining probability values adopted in one of the possible fields of application of the theory.

Through the classical definition one points out a method of determining probability values, but other methods are also important and these are completely neglected by the classical definition.

The concept of probability was widely used in the seventeenth and eighteenth centuries even for problems related to insurance against risks: that is, given the frequency with which a certain event occurred under given conditions in a series of past occasions, it was a matter of determining the probability of its occurrence on a future occasion when those conditions were again fulfilled.

After the scientific reformulation of probability theory by Laplace, the natural sciences became a new field for the application of probabilistic notions. Laplace himself had shown the great advantages of applying these notions to astronomy, to geodesy, to the study of tides and to meteorology. Thanks to F. Galton's use of probability theory in his research on heredity and the contributions of K. Pearson, W.S. Gosset and R.A. Fisher, probability theory became an indispensable tool in scientific experimentation; that is, since the second half of last century the theory of statistics has taken the place of games of chance as the privileged model of probability theory.

The concept of probability that statisticians claimed to use was a sort of ratio of the number of individuals with certain characteristics to the total number of individuals in a 'hypothetical infinite' population. The meaning of this last term has never been clarified by statisticians. Because of this, the concept of probability that they used was ambiguous. On the other hand, their method of calculating probability values was not ambiguous. This method, based on the knowledge of relative frequencies, could in no way be reduced to the ratio of favorable cases to the equally likely ones.

This latent contrast between the notion of probability used in the theory of probability and in the theory of statistics already stressed by L. Ellis and J. Venn was exploded by R. von Mises. He attacked the classical definition for being completely inadequate for the use that was made of the notion of probability within the more impotant sphere of the application of probabilistic notions. In fact, von Mises says

According to a certain insurance table, the probability that a man forty years old will die within the next year is 0.011. Where are the 'equally likely cases' in this example? While are the 'favourable' ones? Are there 1000 different possibilities, eleven of which are 'favourable' to the occurrence of death, or are there 3000 possibilities and thirty-three 'favourable' ones? [14]

Clearly the central point of von Mises' criticism is as follows: in calculating the probability of an individual's death, the method made explicit by the classical definition cannot be used. In general terms, it is impossible to reduce the methods of calculating probability values used in inductive statistics to that used in games of chance. Therefore the contrast that von Mises denounces is not so much a contrast between definitions of probability as it is a contrast between methods of calculating probability values.

Von Mises' definition is an attempt to make the concept of probability used in the theory of probability adequate to the method of calculating probability values used in inductive statistics at the beginning of our century. The very reason for the success of the frequentistic point of view must be sought in the fact that this definition gave a probabilistic base to the method already widely used by statisticians to calculate probability values.

However, what had already occurred concerning the classical definition occurred again for the frequentist definition. Through their definition of probability the frequentists favored one method of calculating probability values and disregarded all others. They even went so far as to refuse to accept the existence of other methods, particularly that expressed by the classical definition. Furthermore, they dogmatically denied the possibility of individuating new methods of calculating probability values. For the frequentists a probabilistic statement is meaningful when it is based on repeated experimental observations, otherwise it is meaningless.

However, methods of determining probability values for unrepeatable events, or more generally, for hypotheses had also been elaborated by classical inductive statistics, as it were: for example, methods of estimation and the theory of testing statistical hypotheses. These methods are based on transforming the logical relations connecting hypothesis to experimental data into numerical values. As Fisher writes, the logical basis of these methods (tests of significance) "was the elementary one of excluding, at an assigned level of significance, hypotheses, or views of the casual background, which could only by more or less implausible coincidence have led to what had been observed" [15]. And it is still Fisher who supports the validity of these methods: "Such inferences we recognize to be *uncertain* inferences, but it does not follow from this that they are not mathematically rigorous inferences" [16].

On the other hand, the growth of the application of probabilistic notions made it clear that the limitation imposed by frequentists on the concept of probability was unwarranted. From the thirties on, these notions were used more and more by industrial management: quality control in industrial mass production is one example of this. In order to make any decision in these cases, one must evaluate the probability of events of which one does not have much prior experience. The frequentistic point of view therefore contrasted obviously with this use of the notion of probability. To quote B. de Finetti: "It would not be difficult to admit that the subjectivistic explication is the only one applicable in the case of practical predictions (sporting results, meteorological facts, political events, etc.) which is not ordinarily placed in the framework of the theory of probability, even in its broadest interpretation" [17]. The methods of calculating probability values in the cases mentioned by de Finetti must be different from the one based on relative frequency.

In general, we have seen that statisticians had elaborated methods of calculating probability values even for events that are not repeatable. These methods were not based on knowledge of relative frequency, at least not solely relative frequency. The range of events for which methods of calculating probability values exist was wider than that indicated by frequentists. And if the statisticians were to reject the notion of the probability of a hypothesis, as H. Jeffreys says, they would have, "deprived themselves of any way of saying precisely what they mean when they decide between hypotheses" [18].

The definitions of probability proposed by logicists and subjectivists represent an attempt to reconcile the definition of probability with these modern uses of probabilistic notions: that is, it is an attempt, using the definition of probability, to reconstruct the methods used in modern inductive statistics to determine the probability values of hypotheses for which a great mass of past experiences is not available.

The logicistic and subjectivistic definitions of probability have the same aim but they differ from one another because they refer to different statistical uses of the notion of probability. The logicistic attempt refers to the methods elaborated by statisticians for determining probability values of hypotheses. This attempt comes closest to the way of intending the determination of probability values of inductive logic as expressed in Section 3.

Unlike the other point of view, logicism has been developing for a fairly long time. Just consider that Keynes had Ellis' and Venn's frequentism as a reference point; when Carnap published his first great work on the foundations of probability, the frequentistic point of view was in regression at least as far as the foundations of probability are concerned.

The consequence of this is the existence of many logicistic definitions of probability, at times contrary to one another. However, at present this does not interest us. What does. interest us is that these definitions are once again attempts to reconstruct methods of calculating probability values used in inductive statistics. Two examples of this are sufficient. The definition of Carnap's c^* amounts to the reformulation of methods of calculating probability values founded on the assumption of the equiprobability of statistical distributions. The definition of probability and the work of Jeffreys are so strictly related to the statistical methods that they represent the first and, until now, one of the best attempts of a Bayesian reconstruction of inductive statistics.

The subjectivists attempt to reconstruct the methods of determining probability values used in decision theory. But they stress the decision itself which the decision-maker makes. In other words, they merely take note of the decisions and refuse to enlarge their analysis to include the arguments on which the decisions are founded, i.e., to the relation between the hypothesis and the information available to the decision-maker. And this is because they maintain that the analysis of these relations cannot be performed at the level of the foundations of probability, but must be left to psychology.

Once again the definition of probability proposed by the subjectivists is an attempt to make probability theory adequate to the methods used in the modern theory of statistics by means of the definition of probability. To do this, it will suffice to consider the definition of probability proposed by F.P. Ramsey, based on partial belief and its measure in terms of goods, and the one proposed by de Finetti based on bets.

To conclude, the various definitions of probability proposed since the times of Laplace, before being attempts to discover the 'true' nature of probability, were methods for determining probability values. Or perhaps it is better to say, that they were methods for determining probability values improperly adopted as explications of the concept of probability.

9. The task of inductive logic and the history of the foundations of probability as we have outlined them in Sections 3 and 8 suggest two considerations. First, the problem of the nature of probability is not a scientific problem but a philosophical one. Moreover the problem concerning the nature of probability was mainly faced dogmatically [19]. Second, the history of the foundations of probability is the history of the discovery of new methods for determining probability values and then of new ways of making inductive inferences. This discovery has been determined by the enlargement of the field of application of probabilistic notions.

The second consideration is particularly important in order to justify our philosophical point of view. It shows that the various definitions of probability which have been proposed since Laplace are attempts to adapt the concept of probability to the methods of determining probability values used in the inductive statistics of that period. The various definitions of probability proposed from time to time were merely attempts to prefer one method, disregarding the others. The preferred method was then dogmatically proposed as the 'true' nature of probability.

It follows that inductive logic recovers the valid core of this research, that is, it is the continuation of this research, but from an antidogmatic point of view. In fact, dogmatism, understood as the attempt to prefer one of the methods of determining probability values while disregarding the others, is incompatible with inductive logic. Otherwise it would be impossible to accomplish the rational reconstruction of all the methods of determining probability values used in inductive statistics.

Because of its task, inductive logic does not presuppose any explicit definition of probability. But this does not mean that it is unable to give any valid indication about the nature of probability. In our opinion, at least in one sense, these indications are sufficiently precise and the philosophy of probability cannot neglect it. Nevertheless, the sole indication we intend to stress now is the one most clearly suggested by the history sketched in the preceding section as well as by the task of inductive logic outlined in Section 3. This indication is a materialistic one. The objective nature of probability is shown by the discovery of new methods of determining probability values. Every new method is a new step toward the knowledge of the nature of probability because it shows us a new characteristic of probability. Constant progress in this direction is undoubtedly the strongest argument in favor of the objectivity of probability. In fact, this progress deepens the knowledge of something that seems to exist independently of the mind of the observers who have discovered the methods for determining probability values, and this would still exist if we were not available to make inductive inferences. In other words, this progress indicates probability as an objective reality whose characteristics are reflected by the methods for determining probability values. These methods, elaborated by inductive statistics, are rationally reconstructed by inductive logic.

Someone might now say that we are trying to propose a sort of frequentism or propensionalism. We shall answer that this is not the case and that in our opinion these interpretations represent the metaphysical aspect of the materialistic philosophy of probability. And this is because we believe that a materialistic philosophy of probability, closely tied to a method of determining probability values (in this case to that based on relative frequency), places itself outside history, and we cannot accept it. On the contrary, what we propose is a non-metaphysically materialist philosophy of probability based primarily on the historical development we have described in the preceding section.

NOTES AND REFERENCES

[1] J.M. Keynes, *A Treatise on Probability* (1921). New York, Harper, 1962, p. 327.
[2] *Ibid.*
[3] I. Hacking, 'Propensities, statistics and inductive logic', in *Studies in Inductive Logic and Probability*. 2 vols. Berkley, Univ. of California Press. Part I, vol. I. (R. Carnap and R. C. Jeffrey eds.), 1971, pp. 33–165; Part II, vol. II (R. C. Jeffrey ed.), 1980, pp. 7–155.
[4] R. Carnap, 'A basic system of inductive logic', Part I, in *Studies in Inductive Logic and Probability*, Vol. I, ed. by R. Carnap and R. Jeffrey. Berkeley, University of California Press, 1971, pp. 33–165; 'A basic system of inductive logic', Part II (forthcoming).
[5] Toraldo di Francia, in a valuable analysis of induction in physics, concludes the section in which he discusses the relations between physics and inductive logic with these words: "This authorizes us to conclude that there cannot be at present any interaction between inductive logic and physics, and that no such interaction can be expected in a reasonable future". See G. Toraldo di Francia, 'Induction in Physics', *Rivista del Nuovo Cimento* 4, 144–165 (1974).
[6] In this paper we make use from time to time of the results stated in the latest works of Carnap cited above, without referring to them explicitly.
[7] In the exposition of this problem we closely follow Jeffreys. See H. Jeffreys, *Theory of Probability*. 3rd edition. Oxford, Oxford University Press, 1961, pp. 59 and 125.
[8] Exchangeability, in 'de Finettian' terms.

[9] Z is the set of all models of $\mathscr{L}$; the possibility space in probabilistic terms.

[10] $\lambda = 0$ is excluded from the regularity we have assumed. $\lambda = \infty$ means of course taking the limit of $\lambda/2(\lambda + 1)$ for $\lambda \to \infty$.

[11] See D. Costantini, 'Sul differente uso della distribuzione normale: probabilitá iniziali e verosimiglianze', *I fondamenti dell'inferenza statistica*. Florence, Dipartimento statistico, 1978.

[12] See D. Costantini, 'Considerazioni sulle ipotesi profonde della distribuzione Beta', *Statistica* **38** 313–322, (1978).

[13] A similar thesis was also expressed by Gini and Renyi. See C. Gini, 'Concept et mesure de la probabilité', *Dialectica* **3**, 1949; A. Renyi, *Probability Theory*. Amsterdam, North-Holland, 1970, p. 41.

[14] R. von Mises, *Probability, Statistics and Truth* (1928); 2nd revised edition, London, Allen and Unwin, 1961, p. 69.

[15] R.A. Fisher, *Statistical Methods and Scientific Inference*. Edinburgh, Oliver and Boyd, 1956. p. 4.

[16] R.A. Fisher, 'The Logic of Inductive Inferences', *J. Royal Statistical Society* **98** 39 (1935).

[17] B. de Finetti, 'Foresight: its Logical Law, its Subjective Sources' (1937) in *Studies in Subjective Probability*, ed. by H.E. Kyburg, Jr., and H.E. Smokler. New York, Wiley, 1964.

[18] H. Jeffreys, *The Theory of Probability*, from the Preface to the first edition, p. ix.

[19] A valuable exception is Carnap's first approach, based on the two concepts of probability, i.e., Probability 1 and Probability 2.

PART II

FOUNDATIONS OF EMPIRICAL SCIENCES

MARIA LUISA DALLA CHIARA

IS THERE A LOGIC OF EMPIRICAL SCIENCES?*

Some years ago, in the fifth volume of the *Boston Studies in the Philosophy of Science*, Hilary Putnam published a paper entitled 'Is Logic Empirical?'.** In my opinion, this paper represents a very important turning point in the philosophical discussion concerning the relations between logic and the empirical sciences. At first sight, Putnam's thesis (according to which logic may be *dependent* on experience) appears to be quite *revolutionary*, in the sense that it goes against a deeply-rooted philosophical opinion, which has been systematically defended in the history of modern logic from Leibniz to Carnap and which asserts that a characteristic feature of formal logic is its absolute independence from any content. Historically this view has often been connected to a kind of faith in the uniqueness of logic.

However, if we look at the actual development of logical theories in our century, we can recognize that such a faith only represents a kind of 'inertial prejudice'. As a matter of fact, logic is today, without any doubt, *pluralistic*; and an interesting result is that we are now able to solve, in a formal way, the following philosophical problems. Let us suppose that A and B represent two intelligent beings using two different logics, L_A and L_B respectively, (for instance, A might be classically minded and B intuitionistically minded). Let us put the following questions:

(1) Can A and B communicate?

(2) Can A recognize that B uses a different logic from his own (and vice-versa)?

(3) Can A *describe* and *justify* the logic used by B (and vice-versa)?

In the case of a number of pairs of logics $\langle L_A, L_B \rangle$ we know that our three questions admit a positive answer. In order to see this technically, it is sufficient to use some of the well-known formal interpretations (or, possibly, metatheoretical descriptions) of L_A in L_B and of L_B in L_A. For instance, let L_A represent classical logic (CL) and L_B represent intuitionistic logic (IL). The intelligent being B could describe the logical behavior of A by means of the so-called 'Gödel–Glivenko interpretation' of CL in IL. As a consequence he would say that A wrongly identifies two different ideas of a disjunction (α or β, and not (not α and not β)) and two different ideas of an existential quantifier (there exists at least one such that α, and

Maria Luisa Dalla Chiara (ed.), Italian Studies in the Philosophy of Science, 187–196.
Copyright © *1980 by D. Reidel Publishing Company.*

not for all not α), which have to be sharply distinguished from one another. Further he wrongly identifies any sentence with its double negation. At the same time, A could describe the logical behavior of B, for instance, by means of a *Kripkian* semantics for IL, or even by means of a *modal interpretation* of IL.

These formal relations between L_A and L_B seem to confirm at the same time that one and the same intelligent being may use different logics in different situations and preserve, in spite of this, a consistent rational behavior.

As a consequence, if logic is not unique and if such plurality of logics prevents neither consistency nor communication, there naturally arises the following question: can experience influence the *choice* of the 'suitable logic' in different situations? In particular, may some physical theories need some special logics which may be different from the logics of a number of standard mathematical theories and even from the logics of other physical theories?

Actually, the problems concerning the so-called 'logics of the empirical sciences' are split into two different classes of questions (and they must be sharply distinguished from one another):

(1) *Metalogical* questions (as is well known in the literature, this field of research has often been called 'formal methodology of empirical sciences'[1];

(2) *Object-logical* questions that give rise to *new* logical calculi.

As to our first class of questions, I want to sketch briefly only one particular problem which has been studied in some papers by G. Toraldo di Francia and myself.[2] This problem concerns the foundations of a kind of 'model theory for physical theories'. As we will see, such an 'empirical model theory' seems to be more suitable than the standard model theory for the formal description of some important relations, which obtain between concrete physical theories.

The starting point of this approach is represented by a formal characterization of the intuitive concept of *physical theory*. From an abstract point of view, a (formalized) physical theory T can be described as a pair consisting of a *formal system* FS and of the class K of all the *physical models* of FS, i.e. the *physical structures*, where the axioms of FS are *true* and where the truth is preserved by the rules of inference of FS.

By physical structure we mean a structure

$$\mathcal{M} = \langle \mathcal{M}_0, S, Q_1, ..., Q_n, \rho \rangle$$

where

(1) $\mathcal{M}_0$ represents the *mathematical part* of $\mathcal{M}$ (ordinarily the standard model of a mathematical theory).

(2) $\langle S, Q_1, ..., Q_n \rangle$ represents the *operational part* of $\mathcal{M}$: S is a set of *physical situations* consisting of physical systems in specified states, $Q_1, ..., Q_n$ are *operationally defined quantities*.[3]

(3) ρ is a function which associates a mathematical interpretation in $\mathcal{M}_0$ with any term of the operational part of $\mathcal{M}$.

We suppose that, generally, in one and the same physical situation, $s \in S$, one may carry out, for a quantity Q_i, more than one measurement (for instance, one might have to measure different masses or different times). We denote these different results for Q_i measured in s by $Q_{i_1}(s)$, $Q_{i_2}(s)$, ... Now, in classical macrophysics a result $Q_{i_l}(s)$ may be represented by an interval I_{i_l} of real numbers (of course, the length of I_{i_l} depends on the precisions of the instruments used in our measurement). But more generally (both in macro- and microphysics) one may obtain a probability distribution as a result. Of course this probability distribution is experimentally determined only up to a double 'error', depending on both the resolving power ε^p of the apparatus used and the accuracy ε^p with which we measure probability: as a consequence, its graphical representation will not consist of a curve but of a curved strip Σ_{i_l}, whose magnitude depends both on ε^A and ε^p. We define an *ideal value* of $Q_{i_l}(s)$ either as a real number belonging to the interval I_{i_l} (if $Q_{i_l}(s) = I_{i_l}$), or as a function representing a probability distribution which is comprised of the strip Σ_{i_l} (if $Q_{i_l}(s) = \Sigma_{i_l}$). In these two different cases we will speak of *deterministic* and *probabilistic* results (ideal values) respectively.

As to the formal language L which is associated to $\mathcal{M}$, we suppose that L contains *special* (*physical*) *variables* d_{i_l} and p_{i_l}, ranging over the deterministic and probabilistic ideal values respectively. A *sentence* of the language L will have generally the form $\alpha(d_{i_l}, q_{j_k})$, where each $d_{i_l}(p_{j_k})$ stands for a (possibly empty) sequence of physical variables, and where at most the physical variables may occur free. Such a sentence may be either defined in a physical situation s, or not defined. Intuitively, α is defined in s, when what α asserts is reasonably interpretable in s. This requires that for each physical variable occurring in α, the corresponding physical quantity is operationally defined with respect to s.

Now we are able to define the concept of truth of a sentence α in a physical situation s and in a physical structure: $\alpha(d_{i_l}, p_{j_k})$ is said to be *true* in s ($\underset{s}{\models} \alpha(d_{i_l}, p_{j_k})$) iff

(a) α is defined in s:

(b) there exists an ideal value d_{i_l} of $Q_{i_l}(s)$ and there exists an ideal value p_{j_k} of $Q_{j_k}(s)$ such that the mathematical sentence $\alpha(d_{i_l}, p_{j_k})$ is true in the mathematical part $\mathcal{M}_0$ of $\mathcal{M}$ ($\underset{\mathcal{M}_0}{\models} \alpha(d_{i_l}, p_{j_k})$).[4]

We will say that α is *true* in the physical structure $\mathcal{M}$ ($\underset{\mathcal{M}}{\models} \alpha$) iff

(a) α is defined for at least one s of $\mathcal{M}$;

(b) for any s of $\mathcal{M}$ for which α is defined, $\underset{s}{\models} \alpha$.

A *model* of a (physical) formal system FS is a physical structure $\mathcal{M}$ such that:

(a) the mathematical part $\mathcal{M}_0$ of $\mathcal{M}$ is the standard model of the mathematical subsystem of FS;

(b) $\mathcal{M}$ verifies all the axioms of FS, and the truth in $\mathcal{M}$ is preserved by all the rules of inference of FS.

A sentence α of a physical theory $T = \langle \text{FS}, K \rangle$ will be called a *valid* sentence of T ($\underset{T}{\models} \alpha$) iff α is true in every model $\mathcal{M}$ of $T(\mathcal{M} \in K)$.

By a *partial model* of a theory T we will understand a physical structure $\mathcal{M}$, verifying all the axioms of T that are defined in at least one physical situation s of $\mathcal{M}$, and whose concept of truth is preserved by all the rules of inference of T.

Clearly the fundamental difference between this definition of truth and the standard one, is represented by the fact that, according to our definition, the truth of a sentence *depends* essentially on the precision of our instruments.

In order to understand better the intuitive meaning of this definition of truth, which represents a sort of *empirical truth*, let us consider the following example. Let us ask: what does it mean that the second law of dynamics ($f = ma$) is true in a physical structure $\mathcal{M}$? According to our definition, we will have:

$$\underset{\mathcal{M}}{\models} f = ma \text{ iff}$$

(a) $f = ma$ is defined for at least one physical situation s of $\mathcal{M}$;

(b) for any s for which $f = ma$ is defined (i.e for any s for which one can measure the three quantities, force (F), mass (M) and acceleration (A) and obtain as a result three intervals $F(s)$, $M(s)$, $A(s)$), it holds that, for at least one $a \in F(s)$, for at least one $b \in M(s)$, for at least one $c \in A(s)$, $a = bc$ is mathematically true.

As a particular example of an application of this kind of 'physical model theory', I want to discuss a problem which represents a much debated question of modern physics.

As is well known, after the advent of quantum mechanics and of the Copenhagen interpretation, it has been customary for many physicists to think that: (1) nature is *indeterministic*; (2) however, that part of nature which is termed macroscopic behaves in a *deterministic* way (though exceptions are known). There is an apparent contradiction between the two statements. Namely: how can a deterministic theory, such as classical mechanics, be in some sense a subtheory of an indeterministic theory, such as quantum theory? Actually, within the framework of standard model theory (if we refer to the ordinary metatheoretical relation of *subtheory*) this problem can hardly be solved in a satisfactory and rigorous way. But in the new approach this apparent contradiction can be easily resolved. Let us first try to answer the question: when can a physical theory T_1 be called physically weaker than another theory T_2?

From an intuitive point of view, a physical theory T_1 is considered weaker than a theory T_2, when the *domain of validity* of T_1 is included in the domain of validity of T_2. In other words, whenever a class of physical phenomena is explained by T_1, it must be explained also by T_2. Now, in our empirical model theory, a class of physical phenomena is represented by an operational structure $\langle S, Q_1, ..., Q_n \rangle$ within a physical model of the theory under investigation. Of course, the mathematics of a physically stronger theory T_2 may be mathematically stronger than the mathematics of the physically weaker theory T_1. These heuristic considerations naturally suggest the following formal definition: T_1 is *physically weaker* than T_2 $(T_1 \preccurlyeq T_2)$ iff

(a) the mathematical subtheory of T_1 is *mathematically weaker*[5] than the mathematical subtheory of T_2;

(b) any physical model of T_1 can be transformed into a partial model of T_2. In other words, for any physical model $\mathcal{M} = \langle \mathcal{M}_0, S, Q_1, ..., Q_n, \rho \rangle$ of T_1 there exists a partial model $\mathcal{M}' = \langle \mathcal{M}'_0, S', Q'_1, ..., Q'_n, \rho' \rangle$ of T_2, whose operational part coincides with the operational part of $\mathcal{M}$ (apart from a possible change of the order of the quantities).

Now we want to define the concepts of *deterministic* (*indeterministic*) theory. From an intuitive point of view it seems reasonable to claim that: a theory is deterministic when all the uncertainties of its predictions fall within the resolving power of the instruments. Roughly speaking, the 'fault' of any uncertainty is due to the instruments and not to the theory itself. This suggests the following formal definition; let us first define when a theory T is *deterministic* with respect to one of its models $\mathcal{M}$: T is called deterministic in a model $\mathcal{M}$ iff any probabilistic result measured in a physical situation s of $\mathcal{M}$, i.e. any probability distribution p obtained as a result of a measurement in s (up to the double error $\langle \varepsilon^A, \varepsilon^p \rangle$) is such that: (a) the standard deviation of p is of the order of the resolving power of the instrument A; (b) in the wings p differs from the normal error distribution of the instrument by less than $\varepsilon^p/2$.

Finally, we say that T is *deterministic* if it is deterministic in all its models, *indeterministic* otherwise.

According to these definitions, we may consistently have: classical mechanics (CM) is deterministic; quantum mechanics (QM) is indeterministic (because it is deterministic in nearly all its macroscopic models but indeterministic in its microscopic models); CM $\leqslant$ QM (indeed, any model of CM can be transformed in a partial model of QM).

As a consequence of this analysis we are now in a position to better classify some puzzling cases that arise when a microscopic and unpredictable event can cause a macroscopic event (we may recall, for instance, Schrödinger's cat or Bridgman's battleship that might be made to sink by the unpredictable passing of a particle through a slit).

Of course this kind of 'physical model theory' does not represent, at first sight, a real 'change of logic' but only an enrichment of the traditional *metamathematics* (perhaps a kind of 'metaphysics', if we are not afraid of using this 'bad word'!)

Nevertheless, one can easily see that this change of the metatheoretical foundations may have some consequences for even the logic of the object-theories. Indeed our metatheoretical analysis gives rise to the following logical situation:

(1) our semantics is bivalent and all classical logical laws turn out to be valid (if the mathematical part of any model of the theory under investigation is a classical model);

(2) nevertheless the sentential connectives turn out to be non-truth functional. Indeed, as a consequence of our truth definition we have: the truth of a negated sentence is not generally equivalent to the truth of the positive sentence; similarly for the other connectives.

If we want to have at our disposal also truth functional connectives, we may extend our logical language and introduce, by trivial semantic definitions, the usual truth functional connectives (which we must distinguish from our previous connectives). We will have for instance $\neg$, $\wedge$, $\vee$, $\rightarrow$ (non-truth functional); and $\neg$, $\sqcap$, $\sqcup$, $\rightarrow$ (truth functional). In this way, we may obtain a kind of *multiple logical situation*, which (through a convenient combination of the two classes of connectives) admits of certain kinds of contradictions and of violations of *tertium non datur*.

But I will not describe here this multiple empirical logic,[6] since I would like to devote the final part of my paper to a short discussion of a general philosophical problem concerning the 'most lucky candidate' for the role of a logical calculus for physical theories; this is represented by *quantum logic*.[7]

As is well known, the literature on quantum logic today is quite extensive and even somewhat confusing (some years ago Van Fraassen very aptly spoke in this connection of the 'Labyrinth of Quantum Logics'[8]. Today, one may propose a number of different theses on quantum logic in one of these two senses: (1) different syntactical and semantic characterizations of QL as logic; (2) different hypotheses on the role of QL in the logical construction of QM. I would like to discuss here only the following question: can we *experimentally* determine which is the most *suitable* logic for a rigorous description of the micro-world? In particular, can we refer to a physical experiment, which seems to force us to choose a special form of QL as the 'right' logic of microphysics?

As is well known, some authors have indicated, for instance, the two-slit experiment as a kind of *experimentum crucis* which could prove that the micro-world logically behaves according to QL instead of CL. Actually, the two-slit experiment represents quite an exciting physical experiment for a logician. Indeed it gives rise to a logical situation, which, from an abstract point of view, can be described as follows:[9] we are interested in a sentence α of the theory T that represents QM (α assert that the detected distribution of the particles involved in the two-slit experiment represents an 'additive pattern'). Now our experimental evidence forces us to conclude that: not $\models_T \alpha$. At the same time we can logically prove that $\models_T \alpha$. Of course this is paradoxical! Since physicists generally trust experience more than formal proofs one must discover what is wrong in the proof (say $\mathscr{P}$) of α. Of course the proof $\mathscr{P}$ makes use of some assumptions. If we analyze the formal version of $\mathscr{P}$, we see that the different assumptions involved in $\mathscr{P}$ may be classified as follows: (1) arithmetical laws (they turn out to be all quite trivial); (2) probabilistic laws; (3) logical laws; (4) 'hidden hypotheses' implicit in the language of the metatheory (where we carry out the proof $\mathscr{P}$). The most important hidden hypothesis regards the *logical individuality* of the particles which occur in the physical situation under consideration. These particles are supposed to be legitimate *logical subjects* and thus normal *supporters of predicates*. Properties which are expressible in the language of QM are correctly predicable of them. Now, the 'orthodox' way out of the paradox may be described as essentially founded on the refusal of our hiddeen hypotheses. As a consequence, the proof $\mathscr{P}$ breaks down, since it is linguistically incorrect.[10] From an intuitive point of view, this kind of solution leads to the conclusion that the concept of 'micro-object' does not generally have all and the same properties that hold for the concept of 'macro-object'.

On the contrary the quantum logical solution preserves the hidden hypotheses and rejects only one logical law which is used in the proof: namely the distributivity law of conjunction with respect to disjunction (which is a law of CL but not of QL). Indeed, in the proof $\mathscr{P}$ we apply at a certain point the following instance of the distributivity law:

$$Xp \wedge (Ap \vee Bp) \leftrightarrow (Xp \wedge Ap) \vee (Xp \wedge Bp)$$

(a particle p has arrived at a certain region X of the screen with detectors and has gone through either slit A or slit B iff either p has arrived at X having gone through A or has arrived at X having gone through B).

Although, at first sight, the orthodox solution and the quantum logical solution appear to be very far from each other, I will try and justify the claim that they are compatible in a strong sense. For, whereas the orthodox physicist claims 'the particles are *anomalous individuals*', the quantum logician can reply: 'let us suppose that a classically minded person persists and describes the particles as 'normal individuals', then he must admit that this kind of individual has logically anomalous behavior. The reason why these special individual p's fail to satisfy the distributivity laws can be justified in a simple way. The p's are not normal *supporters* of predicates, in the sense that a predicate of the language may be strongly indetermined for a p. Further it may happen that a disjunctive predicate is determined for a p, although both members of the disjunction are indetermined; thus it may happen that a sentence like $Xp \wedge (Ap \vee Bp)$ turns out to be determined and true although $Xp \wedge Ap$ and $Xp \wedge Bp$ are both non-true. In other words, the quantum logical solution represents a possible explication of the sense in which the microparticles are anomalous individuals. The asserted anomaly simply means that a lot of physical properties can be semantically indetermined for the elements of the quantum-universe.

In my opinion, a rigorous analysis of the two-slit experiment (as well of other examples that one could recall) confirms the following conclusion: QL *may be* a suitable logic for the description of particular situations that arise within QM. QL is not *the* logic of QM in the strong sense that the whole of QM should be formalized in a quantum logical calculus. Indeed, so far, QL seems to govern only particular sublanguages of QM and its metatheory. Generally it may be applicable in a number of different theoretical situations where, for different reasons, it is convenient to *modalize* the classical logical constants in some way.[11]

Let us now turn to our general question: is there a logic of the empirical sciences? As a consequence of our analysis, the most natural reply seems to be the following: the existence of some *experimenta crucis* (for instance the two-slit experiment), which should force us to choose a particular logic for the description of certain classes of physical phenomena, appears to be quite improbable. Nevertheless today we have at our disposal a number of non-classical logical calculi and metalogical theories, which in many situations turn out to be suitable formal tools for a rigorous description of the physical world.

Of course there naturally arises here the question: how can we choose among all these different logical theories? I am aware that at first glance a most convenient solution to this question is represented by a kind of *conventionalistic attitude*: logics should be regarded as languages, and the choice of the suitable logic should be determined, for instance, by practical reasons. However, there are some strong objections against such an attitude. The fundamental objection is, in my opinion, the following: logics do not really behave as languages. The reason is because scientific theories are not invariant with respect to a translation of logic, whereas they generally turn out to be invariant with respect to a translation of language.

Perhaps we will be able to apply, some day, to logics, what Reichenbach some years ago maintained to be true of geometries: we cannot solve the problem, 'what is the right geometry of the real world?'. However, we are able to decide which is the most *natural* geometry with respect to a given physical theory'. So far, unfortunately the problems concerning logics seem to be, in this connection, much more complicated than the corresponding problems concerning geometries.

NOTES

[1] See for instance [1].
[2] See [2–4]. A similar approach has been proposed independently by Wójcicki [5].
[3] These concepts admit of a formal description in the metatheory of T (see [4]).
[4] $\models_{\mathcal{M}_0} \alpha$ represents the standard model-theoretical notion of truth.

[5] This concept can be formally defined by means of the standard relation of *theoretical inclusion*, or, more generally, by means of the relation of *relative interpretability* between theories.

[6] This has been studied in [6].

[7] For quantum logic, see the papers of Beltrametti and Cassinelli, and of Guccione in this volume.

[8] See [7].

[9] More details about this argument can be found in [8]. See also the paper of Guccione in this volume.

[10] Formally, such an attitude may recall Russell's theory of type-solution to the paradoxes of set theory.

[11] A modal interpretation of quantum logic has been studied in [9, 10].

REFERENCES

Proceedings of the Boston Colloquium for the Philosophy of Science 1966/68, ed. by R.S. Cohen and M. Wartofsky. Dordrecht, Reidel, 1969, pp. 216–241.

**This is a slightly modified version of a paper read at the Boston Colloquium for the Philosophy of Science, September, 1975.

[1] M. Przełecki, K. Szaniawski, and R. Wójcicki (eds.), *Formal Methods in the Methodology of Empirical Sciences*. Warsaw, Ossolineum, 1976.

[2] M.L. Dalla Chiara and G. Toraldo di Francia, 'A logical analysis of physical theories', *Rivista del Nuovo Cimento* 3, 1–20 (1973).

[3] M.L. Dalla Chiara and G. Toraldo di Francia, 'The logical dividing line between deterministic and indeterministic theories', *Studia Logica* 35, 1–5 (1976).

[4] M.L. Dalla Chiara and G. Toraldo di Francia, 'Formal analysis of physical theories', Lecture Notes for the 'Enrico Fermi' Summer School on 'Problems in the Foundations of Physics', Varenna, 1977. Amsterdam North-Holland, 1979.

[5] R. Wójcicki, 'Set Theoretic Representations of Empirical Phenomena', *Journal of Philosophical Logic* 3, 337–343 (1974).

[6] M.L. Dalla Chiara, 'A multiple sentential logic for empirical theories', in [1].

[7] B.C. Van Fraassen, 'The labyrinth of quantum logics', in R. S. Cohen and M. Wartofsky (eds), *Logical and Epistemological Studies in Contemporary Physics*, Dordrecht, Reidel 1974.

[8] M.L. Dalla Chiara, 'A general approach to non-distributive logics', *Studia Logica* 35, 23–36 (1972).

[9] R.I. Goldblatt, 'Semantic analysis of orthologic', *Journal of Philosophical Logic* 3, 19–35 (1974).

[10] M.L. Dalla Chiara, 'Quantum logic and physical modalities', *Journal of Philosophical Logic* 6, 331–347 (1977).

ALDO BRESSAN

ON PHYSICAL POSSIBILITY

1. INTRODUCTION

Two notions of physical possibility are to be considered: phys. $Poss_1$ which can be substantially characterized as a certain limit of technical possibility for physical assertions (i.e., assertions that involve only physical and logico-mathematical terms), and phys. $Poss_2$ defined as compatibility with physical laws in a suitable sense, it being understood that these laws are considered as expressed in an extensional language.[1]

At least in the last decades, phys. $Poss_2$ (or some variant of it) has been considered substantially as a definable substitute or a definitional explicatum of the (intuitive) ordinary notion of physical possibility, possibly referred to the real world.[2] Among other things, this point of view is not suited to the use of modal logic, and is likely to have slowed up its development.[3]

Some of my works dealing with the axiomatization of classical particle mechanics, following Mach and Painlevé and/or (non-quantum) relativistic physics (cf. Bibliography, items [1], [2]), are based only on phys. $Poss_1$. The use of phys. $Poss_2$ would have been circular in that the notion of physical possibility substantially occurs in the axioms.

Now, some philosophers of science agree in criticizing the use of phys. $Poss_2$ as an explicandum of physical possibility primarily because, in short, in order to be sure that an assertion p is physically possible in a satisfactory sense – such as the intuitive one used by Hutten (cf. Note 2) – it is not sufficient to realize its compatibility with physical laws should p involve some physical constants: one must also realize that these constants have real values.

In the present article – which extends some considerations made by me in [5a], p. 256 (cf. Note 3) – I shall try to study physical possibility in a detailed way. This notion is related to a theory, supposedly based on extensional logic: let *r-possible* mean being compatible with the axioms of the theory r (which may be non-physical). It is important to distinguish the cases where r is purely physical from those where r is, so to speak, physico-astronomical. In the former case, r-possibility is the usual notion of physical possibility according to the definition of Hutten,

197

Maria Luisa Dalla Chiara (ed.), Italian Studies in the Philosophy of Science, 197–210.

for example. In the latter case *r*-possibility will be called (an instance of) phys. Poss$_3$. It is natural to consider this third notion of physical possibility in order to attempt to cope with the criticism above. We claim the following theses (cf. Note 3).

(1) *Phys. Poss$_2$ is unsatisfactory as an explicatum of physical possibility* (*it is inadequate for this purpose*).

(2) *Phys. Poss$_3$* (*though more adequate than phys. Poss$_2$*) *cannot be taken as a* (*definitional*) *explicatum of physical possibility in a non-circular way.* (*It is based on the complex notion of constitutive equations, which is implicitly based in turn on phys. Poss$_1$.*)

I conclude that the use of phys. Poss$_2$ made by Hutten for example is illusory, as well as a similar use of phys. Poss$_3$. Therefore I strongly prefer to use the simpler notion of phys. Poss$_1$, e.g. in its versions developed in my [1], [3], [5, §§2.7], or [6, §2], as an explicatum for physical possibility (presented explicitly as a primitive notion). This notion is closely connected with the idea of reproducing phenomena and hence with the notion of time. In order to give an explanation of phys. Poss$_1$ it is natural to use the ordinary intuitive notion of time. In [5] and [6] the same explanation is given by referring to classical time, or to a time notion relative to a given inertial space within special relativity (or general relativity, provided something such as Fock's conjecture is accepted); and this is done in order to arrive at a natural (modally) absolute concept of event point (which allows us to grasp easily a natural modally absolute concept of inertial spaces: cf. Definition 18.8 in [3] or Definition 5.3 in [5].)

In order to characterize phys. Poss$_1$, the use of inertial space-time frames, for example, though practically convenient, is by no means necessary. Incidentally, this is important especially in connection with axiomatic theories such as the modal non-quantum theory of continuous media presented in [2] or [4, Section 83] within general relativity (without anything such as Fock's conjecture).

Let me add that I agree with the main thesis of much North American philosophy which holds that science can be adequately described in purely extensional terms, provided this thesis is meant in a weak way (cf. [5a], p. 254, 1.2, b). However, this is compatible with my thesis that modal notions are useful for improving the foundations of natural sciences, in particular for reducing primitive notions and giving a better and deeper insight into the basic laws. For instance the objections to circularity raised against Newton's theory of mechanics do not hold for the foundations of mechanics according to Mach and Painlevé; and apart from these objections, the latter theory is more profound. Some objections to the latter theory are due mainly to two facts: modal logic is not yet widely known and many accounts of its foundations have defects; in particular, possibility postulates are lacking. Instead, in these foundations, for example, possibility postulates are as important as existence postulates in geometry.

The theses presented above, in particular (2), are supported in this paper from a general point of view, and also in a particular way by taking into account constitutive equations, whose mathematical theory has been developed considerably in the last decades within both classical physics and (general) relativity (cf. [4] and [11]).

The considerations regarding physical possibility developed in this article can be extended to causal possibility in a natural way.

More particularly, in Section 2, phys. $Poss_1$ is characterized in terms of technical possibility, perhaps in a more elaborate way than usual. In Section 3 some examples involving some physical constants (masses) show that phys. $Poss_2$ – i.e., the definitional explicatum for physical possibility according to Hutten, for example (cf. Note 2) – is unsatisfactory even as a non-definitional explicatum and in particular has a certain metaphysical character. Furthermore the use of the explicatum phys. $Poss_2$ is shown to be incompatible with the following irrefutable principle:

If some technically possible initial conditions determine a (physically possible) process on the bases of physical laws, this is also technically possible.

In Sections 4 and 5, the distinction between universal laws and constitutive non-universal laws, and the distinction between purely physical theories and physico-astronomical theories is taken into account in a natural attempt to avoid the defects of phys. $Poss_2$: this possibility notion is replaced by phys. $Poss_3$.

Some advantage is obtained, however, in Section 6 where thesis (2) is proved; furthermore the defects of phys. $Poss_2$ are emphasized so that they are maintained in connection with general constitutive equations to an even greater extent than is the case in connection with physical constants.

2. ON CAUSAL AND PHYSICAL POSSIBILITIES IN THE FIRST SENSE

One must often decide to bring into being one out of several mutually exclusive future situations, any one of which can be produced. For instance, men are generally (in particular, Adam is) able to render the following sentence true:

(1)   Adam's car will be in the garage tomorrow.

Therefore (1) is said to be *technically possible*. In addition men are capable
of rendering (1) false. Therefore, this sentence is said to be technically
contingent. In this century a rocket trip from the Earth to the Moon is
technically possible, while in the past century it was not so. However, this
depends on the fact that the technicians of the past century were not skillful enough. Therefore one can say that this trip (or its happening in the
past century) was causally possible.

More generally the following intuitive characterization of causal possibility in terms of technical possibility seems sufficiently clear. Let the
sentence p assert that a certain phenomenon takes place in the bounded
space-time region $\mathscr{R}_4$. We say that p is *causally possible*$_1$ – where the index
1 is to be read as 'in the first sense' – if p would be technically possible
should some technicians exist who have the capabilities of some (suitable)
future technicians. (For the sake of completeness let us add that the technicians who realize p must do so by working only outside $\mathscr{R}_4$.)

If p expresses a phenomenon that occupies an unbounded space-time
region $\mathscr{R}_4$, p will be called *causally possible*$_1$ if its restriction to every
bounded part of $\mathscr{R}_4$ is causally possible$_1$.

Perhaps some readers prefer replacing the last definition with the following (more complex) one, because it requires fewer experimental capabilities: The above assertion p is said to be *causally possible*$_1$ if $\mathscr{R}_4$ can be
covered by some *open* bounded regions such that the restriction of p to any
of them is causally possible$_1$.

If p is causally possible, and it is a physical assertion in that it concerns
only logical (mathematical) and physical notions, then p is said to be
phys. Poss$_1$ (cf. Note 1).

In case assertion p is physical, its physical possibility$_1$ can be characterized intuitively without using counter-factual conditionals. To do this,
let us consider the *translate* of p by the time τ. For instance, if p is P_{t_1,t_2},
i.e. 'this stone is in this hole in the time interval $t_1 \mapsto t_2$, then the translate
of p by the time τ is $P_{t_1+\tau,t_2+\tau}$, i.e. 'this stone is in this hole in $t_1 + \tau \mapsto
t_2 + \tau$.

Now consider a physical assertion, p, connected with a bounded space-
time region $\mathscr{R}_4$. We can say that p is *phys. Poss*$_1$ if its translate by some
time τ is technically possible. The general case is reduced to this as before.

Let us note that the time notion referred to above is the ordinary one, so
that it is regarded as absolute (observer independent) by classical physi-

cists and it can be regarded as one of the infinitely many time notions admitted, for example, by most physicists working in general relativity. At least when Fock's conjecture about general relativity is disregarded, the last notions are not modally absolute, which has no bearing on our first explanation of phys. $Poss_1$. Instead, in connection with the translate of an assertion p_t by the time τ, it is useful to state eplicitly that in the absence of any (modally) absolute time-notion, our second explanation of phys. $Poss_1$ has this sense: p_t is phys. $Poss_1$ if its translate $P_{t+\tau}$ by some time τ in connection with an admissible time notion that coincides with the real one up to the instant t, is technically possible.

Let us add incidentally that in some cases, for example in connection with a bounded space-time region, $\mathscr{R}_4$, with an exactly specified space-time metric, we must be content with the capability of reproducing it within an arbitrarily preassigned degree of approximation and disregarding its orientation. Of course, this difficulty concerning technical possibility is closely connected with our difficulty (or inability) to check or observe whether or not a reproduction, for example, of $\mathscr{R}_4$ really has the exact given space-time metric.

Any of the above explanations for phys. $Poss_1$, and e.g. the axiom system for modal logic in [3], for example, render phys. $Poss_1$ an explicatum for physical possibility.

3. PHYSICAL POSSIBILITY IN A SECOND SENSE

Note that knowing the meaning of 'being physically possible' does not imply knowing which assertions *are* physically possible. In order to decide whether or not a particular assertion p is physically possible, physical laws (and the results of some experimental observations) are often taken into account. This induces us to consider a second notion of physical possibility:

We say that p is *phys. $Poss_2$* if p is compatible with physical laws in some suitable sense similar to that of Hutten who means that these laws and p imply no contradictions (cf. Note 2).

Similarly *causal possibility₂* can be defined as being compatible with physical, chemical, and biological laws.

Various general physical theories have been constructed at different times, which describe physical phenomena with increasing accuracy, with better and better approximation, e.g., classical physics, relativity theories, classical and relativistic quantum theories.[4] All these theories are mutually incompatible. Hence we must refer phys. $Poss_2$ to the laws of a given r, among them. Then we speak of r-possibility, (as was said in Section 1). It is useful to identify r with classical physics for this purpose, because this theory is the simplest and the closest to our everyday way of looking at the physical world. In this physics we can speak of absolute time, the physical space is Euclidean, and material particles can be thought of as having well-defined positions and velocities at every instant.

Suppose now, for a moment, that the laws of physics are finitely many and all known. Then the physical possibility of p can be tested in a purely logical and hence precise way. Phys. $Poss_1$ is directly based on nature, so that it is a refined (or specified) version of the intuitive notion of physical possibility. Generally phys. $Poss_1$ has not been used as an explicatum for it, probably because the laws governing its use were not known precisely, or the known systems of these laws were not considered sufficiently practical. Instead phys. $Poss_2$ was considered as a useful equivalent to phys. $Poss_1$. The relation between the two notions is somewhat similar to the one between deduction in the ordinary sense and that defined in theories of mathematical logic such as [9]. For example, in [9] the adequacy of the formal notions defined in it is asserted.

A formal notion is adequate if it holds whenever the corresponding intuitive notion holds (with the possible exception of some marginal situations).

As was said in Section 1, especially in the past decades the thesis that phys. $Poss_2$ is adequate (or is an explicandum of physical possibility) was widely asserted (cf. Note 2). Roughly speaking this implies that phys. $Poss_2$ assertions are phys. $Poss_1$ assertions. To support criticism of the above view, already hinted at, let us now show by example that, generally, the phys. $Poss_2$ motions of a mechanical system are *more* than its motions which are phys. $Poss_1$.

Let m_0, m_1, ... be the real values for the masses of the sun $\mathscr{P}_0$ and the planets $\mathscr{P}_1$, ... of the solar system $S = (\mathscr{P}_0, \mathscr{P}_1, ...)$; and for the sake of simplicity let us consider S as a particle system.

The positions P_0, P_1, ... and velocities v_0, v_1, ... of $\mathscr{P}_0$, $\mathscr{P}_1$, ... respectively at the instant t_0 can be chosen in a (very) arbitrary way in that, if we fix the points P_0, P_1, ... and vectors v_0, v_1, ... , then the following is phys. $Poss_1$ and phys. $Poss_2$:

Initial conditions (at t_0): At the instant t_0, $\mathscr{P}_i$ is in P_i with velocity v_i: $(i = 0, 1 ...)$.

The subsequent motion of S (assumed to be isolated) is determined by these (initial) conditions and the above (real) values m_0, m_1, ..., For every choice of P_0, P_1, ... and v_0, v_1, ... there is exactly one corresponding motion $\mathcal{M}$ of S that is phys. Poss_1. Instead, for the same choice, infinitely many motions of S are phys. Poss_2, because physical laws are compatible with the assertion for example, that $\mathcal{P}_r$ has the mass $(r + 1)m_r$ $(r = 0, 1 \ldots)$. More precisely, corresponding to the arbitrary positive values m'_0, m'_1 ... for the masses of $\mathcal{P}_0$, $\mathcal{P}_1$, ... with $m'_0 = m_0$ and $m'_i \neq m_i$ for some i, there is a motion, say $\mathcal{M}'$, that is phys. Poss_2, but not phys. Poss_1, for S.

This conclusion can be restated as follows: phys. Poss_2 is unsatisfactory because there are (infinitely) many sentences that are phys. Poss_2 but no experimenters, however skillful, can render these sentences true. Thus physical experiments (alone) are unable to check the phys. Poss_2 of many assertions that are phys. Poss_1. This constitutes in some sense a metaphysical feature of phys. Poss_2.

Remembering the foregoing argument, we can briefly make some remarks that perhaps add stronger support to the thesis that phys. Poss_2 is inadequate. To this end we consider the following irrefutable principle relating technical possibility to any satisfactory notion of physical possibility (it has obvious generalizations).

If a motion $\mathcal{M}$ of an isolated mechanical system is physically possible in the time interval $t_0 \mapsto t_1$ and its initial conditions (i.e. those of positions and velocities) at the instant t_0 are technically possible, then $\mathcal{M}$ is technically possible.

Now let P_i and v_i be the real position and velocity of $\mathcal{P}_i$ respectively at the instant t_0 $(i = 0, 1, \ldots)$. Then the above-mentioned initial conditions can be carried out experimentally (with vanishing work). Hence they are technically possible. Hence the above motion $\mathcal{M}'$ of the solar system (which can be reasonably regarded as isolated) has technically possible initial conditions, is technically impossible (because it is not phys. Poss_1), and in spite of this, is phys. Poss_2. Hence phys. Poss_2 is an unsatisfactory notion of physical possibility.

4. ON PHYSICAL LAWS

The following considerations about kinds of physical laws constitute a preliminary for an attempt at improving phys. Poss_2.

Some physical laws are universal, i.e. they hold alike for all material bodies (or particles); for example, such are the fundamental law of dynamics and Newton's law of gravitational forces. Incidentally the latter can be regarded as a universal constitutive law.

To give an example of a non-universal physical law let us consider a homogeneous gas (gaseous body) $\mathscr{G}$. If this is simple enough, there is a function $f(T, V)$ that fulfills the condition.

(C) If (*at the instant* t) $\mathscr{G}$ *occupies* (*homogeneously*) *a region of volume V and has* (*uniformly*) *the temperature T and pressure p, then*

$$p = f(T, \ V). \tag{2}$$

Equality (2) is called the *state equation* of $\mathscr{G}$. It is a particular *constitutive equation* of $\mathscr{G}$; and f is called a *constitutive function* of $\mathscr{G}$. Condition (C) can be called a *constitutive law* of $\mathscr{G}$. Clearly it is not a universal law; and it is known only if the mathematical function f is explicitly expressed.

Incidentally, at least when condition (C) is to be used as a definition of the state equation of $\mathscr{G}$, the 'if . . . , then . . . ' occurring in it can be meant as a causal or physical implication (compulsorily) in the first sense, but not as a material implication[5] or a physical implication in the second sense.

The gaseous body $\mathscr{G}$ also has a constitutive equation for its internal energy E. For the sake of simplicity let this equation have the form

$$E = g(T), \tag{3}$$

which holds for perfect gases. The corresponding constitutive law reads as follows

If (*at* t) *T is the* (*uniform*) *temperature of* $\mathscr{G}$ *and E is its internal energy, then* (3) *holds.*

Let m_0 be the (rest) mass of $\mathscr{G}$ when $\mathscr{G}$ has a vanishing temperature, and hence the internal energy $g(0)$. By the relativistic equivalence principle for mass and energy, the increment $\Delta E = g(T) - g(0)$ for the internal energy of $\mathscr{G}$ is associated with the increment Δm for the mass of $\mathscr{G}$, given by

$$\Delta m = c^{-2}\Delta E, \qquad [\Delta E = g(T) - g(0)], \tag{4}$$

where c is the speed of light *in vacuo*. Hence when $\mathscr{G}$ has the temperature T, its mass m is $m_0 + \Delta m$, so that, according to relativistic physics $\mathscr{G}$'s mass has the constitutive equation

$$m = m_0 + c^{-2}[g(T) - g(0)]. \tag{5}$$

In ordinary cases Δm is so small that it cannot be measured, so that (5) can be replaced by

$$m = m_0, \tag{5'}$$

which therefore can be regarded naturally as a constitutive equation. Thus, on the one hand, that the earth has a certain mass can be regarded as true on the basis of a physical law (favoring Hutten's definition of phys. Poss as compatibility with physical laws), but on the other hand this law is not universal; furthermore the unsatisfactory features pointed out about mass in Section 3 hold for all non-universal constitutive laws.

5. PHYS. POSS$_3$ AS AN ALTERNATIVE TO PHYS. POSS$_2$

The notion of phys. Poss$_2$ is based on some physical theory r; it is r-possibility. Not all choices of r are suitable bases for it: indeed (as was said) in classical mechanics following Painlevé (cf. [10]), physical possibility is primarily used in certain essential counterfactual conditionals (see also [1, 3, 5, 6]. Incidentally, if r is extensional, r-Poss defined as compatibility with r's axiom (in some suitable sense) has been accepted as a characterization of the intuitive notion of phys. Poss; no one would do the same in the opposite case.

At this point it is natural to search for a suitable theory r, possibly not purely physical, or as yet unknown, such that r-possibility may be satisfactorily substituted for phys. Poss$_2$ (cf. Section 2).

Taking into account that Hutten, for example, considers physically possible assertions involving real bodies such as the moon (cf. Note 2), we suppose that r contains laws which can be regarded as more astronomical (or geographical) than physical.

As a preliminary, the state equation (2) can be referred to a material point M of $\mathcal{G}$, provided by V we mean the specific volume of $\mathcal{G}$ at M.

We are interested in very general properties of the theory r, so that for our purposes we can surely use the following assumptions which afford a very much simplified picture of the world.

(a) *The whole universe can be regarded as a (non-homogeneous) fluid $\mathcal{F}$.*

(b) *Electromagnetic and diffusion phenomena as well as irregular motions (such as slidings or fractures) do not occur or are negligible.*

(c) *Chemical reactions are negligible.*

Having fixed a reference instant t^* and a Galilean frame G, the arbitrary material point M of $\mathscr{F}$ can be determined by means of its coordinates in G at t^*. Let

$$p = f(T, V, M) \tag{6}$$

be the state equation of the universe (relative to t^*): *if at the material point M of $\mathscr{F}$ and at the instant t the pressure, temperature, and specific volume of $\mathscr{F}$ have the values p, T, and V, respectively, then (6) holds.*

Note that the 'if . . ., then . . .' above can also be considered as a causal or physical implication (in the first sense) but not as a material implication.

Let our analogue of (3) for $\mathscr{F}$ be

$$\varepsilon = g(T, V, M), \tag{7}$$

where ε is the density of the internal energy of $\mathscr{F}$ at M.

Let us add the other (non-universal) constitutive equations, among others those for mass density and the coefficient of thermic conductivity, all in a form such as (7). These equations together with dynamic equations and the first (and second) principles of thermodynamics (which are universal laws) constitute a finite number of physico-astronomical laws. Let r_1 be the theory based on them, so that r_1-possibility is (an instance of) phys. Poss$_3$. Given the above simplifying assumptions concerning the world, *a physical assertion is r_1-possible iff it is phys. Poss$_1$*.

Thus under assumptions (a) to (c), the notion of phys. Poss$_3$ relative to the physico-astronomical theory r_1 appears much more acceptable than phys. Poss$_2$, for example, in a version similar to Hutten's [Section 2].

Furthermore r_1-possibility can be improved. We can drop assumption (a) by referring to a suitable theory r_2, of the same kind as r_1, but based on a modern general theory of materials (including solids and materials with memory). At least in part, we can also drop assumption (b) by using a physico-astronomical theory r_3 based on already known classical or relativistic theories. Furthermore it is not unreasonable to conjecture that on the basis of general (relativistic) physical and chemical theories, not yet completely developed, a physico-chemical astronomical theory r_4 can be constructed by which assumption (c) can also be dispensed with.

6. SERIOUS DEFECTS OF PHYS. POSS$_3$

Even r_4-possibility has serious defects. For the sake of simplicity we show them by taking into account the preceding, simplifying hypotheses (a) to (c) in Section 5, and r_1-possibility.

In purely physical theories, constitutive equations are studied from a general point of view, in that the physical properties, for example, that are consequences of their form are investigated. In the physico-astronomical theory r_i constitutive equations such as (6) and (7) are regarded as determined mathematically in connection with every real material point M. Therefore they shall never be known completely. Hence r_i-possibility $(i = 1, ..., 4)$ will always remain unknown.

It is important to note that, since we were unable to specify the constitutive equations (6) and (7) mathematically, we characterized them by an essential use of causal possibility: precisely phys. Poss_1. Actually, theory r_1 can be presented (as far as it is correct) only by using primitive functional constants $\bar{f}, \bar{g}, ...$ for 'f' and 'g', ... in (6) and (7), and by interpreting them just on the basis of the general notion of constitutive equations (because we can specify them only partially mathematically). To this end we need modal notions—in particular phys. Poss_1. We conclude that the following holds.

OBJECTION 1. *It would be circular to define physical possibility as r_1-possibility, or as phys. Poss_3 with respect to some physico-astronomical theory.*

Let us compare the preceding situation of r_i-possibility $(i = 1, ..., 4)$ with the analogue for a purely physical theory r (cf. Section 2). We have shown that r-possibility is not adequate in that it is not equivalent to phys. Poss_1; however it can be defined in an extensional metalanguage (r and r_1 to r_4 are obviously supposed to be extensional).

On the the other hand, for $i = 1, ..., 4$, r_i-possibility is adequate but because of our incomplete knowledge of physico-astronomical theories such as r_i, the metalanguage in which we define r_i-possibility must be modal; hence we can regard this notion as a special kind of physical possibility but (cf. Objection 1) it would be circular to consider it as a definitional explicatum of the general notion of physical possibility, or in particular as a definition of physical possibility, of the kind hoped for, for example, by Hutten (cf. [8]).

Let us emphasize that the defects of physical possibility$_2$ shown in Section 2 in connection with physical constants (masses) reappear to a greater extent by considering constitutive equations. Indeed a purely physical theory of these equations tells us that if a material has certain constitutive equations, then certain processes are physically possible for it. However, infinitely many mathematical determinations of the constitutive functions of the material are compatible with all physical laws (including the constitutive ones). Hence infinitely many processes for the material are phys. $Poss_2$, and not phys. $Poss_1$.

Moreover it is easy to realize that there are processes for the universe that are phys. $Poss_2$ and have technically possible initial conditions, but are not technically possible. Incidentally the situation with phys. $Poss_2$ is worse in connection with constitutive equations than with physical constants, because in the typical case, physical constants (i.e. masses) can be determined by knowledge of an arbitrarily short part of the real process, whereas this is by no means the case with constitutive equations, especially for materials with memory. This implies that the processes that are phys. $Poss_2$ and not phys. $Poss_1$ are much more numerous in the second case.

Now let us accept the admittedly hardly realistic assumption that one can mathematically specify the constitutive equations of the universe, such as (6) and (7). In spite of this the following holds.

OBJECTION 2. Assumption (c) in Section 5, considered for the sake of simplicity, is not true. By means of chemical processes, new chemical substances (various scores per year) are constructed and, even disregarding this fact, chemical processes occur. Hence there are physical bodies existing only after the reference instant t^*, so that their constitutive equations are not included in equations such as (6) or (7), or in the analogues of these equations for r_2 or r_3. Hence the theory r_4 which takes chemical processes into account must be referred to. It can be concluded that *the physical possibility$_3$ of a physical assertion is to be referred not only to physical laws but also to chemical laws.*

Of course it is not wrong to assume r_4-possibility, for example, as an explicatum of physical possibility. However, this involves the use of the general notion of constitutive equations as an auxiliary primitive notion. Furthermore this notion is modal, hence it involves a notion of physical possibility such as phys. $Poss_1$; and it is very complex (in modern mathematical physics and engineering, more and more general kinds of constitutive equations are considered). Hence the simpler notion of phys. $Poss_1$ seems to me a much better non-definitional explicatum for physical possibility than phys. $Poss_3$. The last notion has been considered in this paper as a refinement of phys. $Poss_2$. The serious defects of phys. $Poss_2$ have apparently not been known by many people and this explains in part its rather wide acceptance as a definitional explicatum for physical possibility.

NOTES

[1] We shall use phys. Poss_r (phys. Poss_r) for physical possibility (physically possible) in the rth sense.

[2] Hutten uses phys. Poss_2 for physical possibility; and he does so not (only) in connection with a particular physical theory r (by saying, for example, that a process expressible in r is physically possible iff it is logically compatible with the axioms of r), but (also) in connection with the real world: a trip to the moon, says Hutten in 1956, is technically impossible but physically possible – cf. [8], p. 52.

[3] At the congress of 'Logic and Ontology' held at Salzburg in 1973, the various conferences were followed by discussions. Some of these discussions have been published, with some additions, in [5a] and contain some of my remarks against considering phys. Poss_2 as an explicatum of physical possibility. In the present paper my theses, briefly asserted in 1973, are much more elaborated and supported by examples and considerations that are in part connected with constitutive equations.

[4] Actual relativistic theories of quantum mechanics are generally considered unsatisfactory.

[5] For the sake of simplicity assume that body $\mathscr{G}$, which has the state equation (2), always occupies the same region of volume V_0. Then, for $V \neq V_0$ and for every instant t the following assertion is false.

(C_1) *At the instant t $\mathscr{G}$ occupies a region of volume V and has the temperature T and the pressure p (uniformly).*

Hence for $V \neq V_0$ (C_1) implies materially every assertion and in particular the equality

$$p = f(T, V) + a(V - V_0)^2 \tag{2'}$$

where a is a positive constant.

In addition, since (C) and the above assumption for $\mathscr{G}$ and V_0 hold, for $V = V_0$ assertion (C_1) implies materially (2) and hence $(2')$ again. Thus (C_1) implies materially $(2')$ in every case. Hence, if (C) is meant as a material implication and is used to define the state equation of $\mathscr{G}$, then both (2) and $(2')$ turn out to be such equations, which is obviously unacceptable from the physical point of view.

[6] We are referring to the 'presupposed interpretation' following Carnap (cf. [7], p. 172).

REFERENCES

[1] Bressan, A., 'Metodo di assiomatizzazione in sense stretto della Mectanica classica. Applicazione di esso ad alcuni problemi di assiomatizzazione non ancora completamente risolti', *Rend. Sem. Mat. Univ. di Padova* 32, 55–212 (1962).

[2] Bressan, A., 'Una teoria di relatività generale includente, oltre all'elettromagnetismo e alla termodinamica, le equazioni costitutive dei materiali ereditari. Sistemazione assiomatica', *Rend. Sem. Mat. Univ. di Padova* 34, 74–109 (1964).

[3] Bressan, A., *A General Interpreted Modal Calculus*. New Haven, Yale University Press, 1972.

[4] Bressan, A., *Relativistic Theories of Materials*. Berlin, New York, Springer, 1978, pp. 286.

[5] Bressan, A., 'On the usefulness of modal logic in axiomatizations of physics' in *PSA 1972* (*Boston Studies in the Philosophy of Science*, Vol. 20) ed. by K.F. Schaffner and R.S. Cohen. Dordrecht and Boston, Reidel, 1974; reprinted (pp. 5–28), with discussion (pp. 251–258) in [5a].

[5a] *Problems in Logic and Ontology*, ed. by E. Morscher, J. Czermak, and P. Weingartner. Graz, Akademische Druck-u. Verlagsanstalt, 1977.

[6] Bressan, A., 'On the semantics for the language ML^ν based on a type system, and those for the type-free language ML^∞', *J. Phil. Logic* 3, 171–194 (1974).

[7] Carnap, R., *Introduction to Symbolic Logic and its Applications*. New York, Dover Publications, 1958.

[8] Hutten, E. H., *The Language of Modern Physics*. London, George Allen and Unwin; New York, Macmillan, 1956.

[9] Mendelson, E., *Introduction to Mathematical Logic*. New York, Van Nostrand Reinhold Co., 1964.

[10] Painlevé, P., *Les axiomes de la mécanique*. Paris, Gauthier-Villars, 1922.

[11] Truesdell, C., and Noll, W., 'The non-linear field theories of mechanics', *Handbuch der Physik*, Vol. III/3. Berlin, Springer, 1965.

A MUCH USED LOGICAL NOTION OF PHYSICAL POSSIBILITY AND GÖDEL'S UNDECIDABILITY THEOREM*

A. INTRODUCTION

In agreement with many other authors, Hutten writes:

(a) "so long as no known physical law is violated we are entitled to say that it is physically possible to observe a given thing or event" [8, p. 52]. This is often reworded into the following sentence:

(b) *If A is logically compatible with known physical laws, A can be said to be physically possible.*

In Section 3 (of my article 'On physical possibility', briefly 'On phys. Poss.') a not completely specified notion of phys. Poss. similar to the above logical one – cf. (b) – is called phys. $Poss_2$. In preparing Section 3, I at first identified phys. $Poss_2$ with the latter notion. However, very recently I noted that the first version of phys. $Poss_2$ has very unacceptable features. Therefore in writing the final version of Section 3 'logical compatibility' was replaced, in the definition of phys. $Poss_2$, with 'compatibility in a suitable sense similar to that of Hutten'; but this sense was not specified.

In section B the above-mentioned unacceptable features are pointed out and emphasized. In Sections C and D two not purely logical versions of phys. $Poss_2$ are presented, to be considered as mathematical and physical, respectively. They lack the above unacceptable features, at least in connection with several significant physical theories. However, they also lack certain positive features of the first version of phys. $Poss_2$ and thus they are more unsatisfactory than the logical version of phys. $Poss_2$ appeared to be, on the basis of 'On phys. Poss.', before its unacceptable features, pointed out in this supplement, were known. Hence new evidence has been produced in favor of a thesis set forth in that article; it is convenient to replace phys. $Poss_2$ with phys. $Poss_1$ in interpreting physical theories.

B. THE LOGICAL NOTION OF PHYSICAL POSSIBILITY, PHYS. $POSS_2$, AND GÖDEL'S UNDECIDABILITY THEOREM

Let r be a theory of (classical or relativistic non-quantistic) physics that, contains all known physical laws expressible in it and, is framed in a sublanguage $\mathscr{L}$ of the extensional language L used by us. The assertion $\mathscr{A}$

211

Maria Luisa Dalla Chiara (ed.), Italian Studies in the Philosophy of Science, 211–214.
Copyright © 1980 *by D. Reidel Publishing Company.*

of $\mathcal{L}$ is said to be *r-possible$_1$*, briefly $\Diamond_{1r}\mathcal{A}$ – where the index 1 stands for 'in the first sense' – if $\mathcal{A}$ is logically compatible with r's axioms.

Through mathematical analysis r embodies a theory of natural numbers (arithmetic). In order to consider the question from a general point of view, let r be framed within a logical theory that embodies Zermelo – Fränkel set theory and includes quantification over properties, such as the NBG theory (derived especially from von Neumann, Bernays, and Gödel) presented in [9, Chapter 4]. In addition we know r's axioms. Hence r must be regarded as a recursively axiomatic extension of NBG. Since NBG is (essentially recursively undecidable and) essentially incomplete – cf. [9, p. 193] – , r has (infinitely many) undecidable sentences. Let $\mathcal{U}$ be a true one among them. Then $\Diamond_{1r}\sim\mathcal{U}$ is a true sentence of L. Thus $\sim\mathcal{U}$ *is r-possible$_1$ in spite of $\mathcal{U}$ being a true mathematical sentence.*

Since r-possibility$_1$ is a direct version of Hutten's phys. Poss. – cf. (a) and (b) in (A) – , $\sim\mathcal{U}$ is phys. Poss. in Hutten's sense. This can also be checked directly by noting that $\sim\mathcal{U}$ violates no physical law. The same can be said of $\sim\mathcal{U}\wedge\mathcal{V}(1)$ where (c) $\mathcal{V}(r)$ says (in $\mathcal{L}$) that in a certain space-time region some electromagnetic energy is travelling with the speed rc, where c is the speed of light *in vacuo*.

The sentence $\sim\mathcal{U}_1\wedge\mathcal{V}_{(1)}$ has been considered because it is contingent, so that it can be regarded as (expressing) an event which, on the one hand, is phys. possible in Hutten's sense – cf. (a) in (A) – but on the other hand is false (by the conjunct $\sim\mathcal{U}$). A better event having these two properties is (expressed by) sentence $\mathcal{W}$ defined by $\mathcal{W}\equiv_D\mathcal{V}(1 + C_{\mathcal{U}})$, where $C_{\mathcal{U}}$ is the characteristic function of $\mathcal{U}$, so that it equals 1 [0] for $\mathcal{U}$ true [false].

Indeed $\Diamond_{1r}\sim\mathcal{U}$ holds (where r is special relativity) and $\sim\mathcal{U}$ logically implies $C_{\mathcal{U}} = 0$ and hence $\mathcal{V}(1 + C_{\mathcal{U}}) \equiv \mathcal{V}(1)$. Then we have $\Diamond_{1r}\mathcal{W}$ (and $\mathcal{W}$ violates no physical law). Thus, on the one hand, $\mathcal{W}$ is phys. Poss. in Hutten's sense, and on the other hand, $\mathcal{W}$ says (in L) in a way not known by (all) the users of r, that some electromagnetic energy is travelling with the speed $2c$.

The set of known physical laws is increasing, so that the corresponding set of phys. Poss. assertions in Hutten's sense is decreasing. Let us remark that on the one hand the phys. Poss. for energy to travel with the speed $2c$ appears to be denied by actually known laws and on the other hand *this possibility will always be asserted by $\mathcal{W}$ for a suitable choice of $\mathcal{U}$* – in spite of the increase of the set of known arithmetical truths.

The troublesome phys. Poss. sentences $\mathcal{U}$ to $\mathcal{W}$ above contain an unusual arithmetic sentence. We now show that this property is not essential for the troublesome sentences that interest us. It suffices to note that every acceptable notion of phys. Poss. fulfills the following

REQUIREMENT. *Phys. Poss. consequences of true mathematical statements are phys. Poss.*

Indeed let $\mathscr{A}$ be any wff (well formula) of $\mathscr{L}$. Since $\sim\mathscr{U}$ logically implies $\mathscr{U} \supset \mathscr{A}$ and $\Diamond_{1r} \sim \mathscr{U}$ holds, we have $\Diamond_{1r}(\mathscr{U} \supset \mathscr{A})$. By the requirement above, this yields $\Diamond_{1r}\mathscr{A}$ if we identify phys. Poss with the $\Diamond_{1r}$, which substantially is Hutten's logical notion of phys. Poss. We conclude that *under the last assumption every sentence of $\mathscr{L}$ is phys. Poss.*

C. A 'MATHEMATICAL' VERSION OF PHYS. POSS$_2$

By the preceding considerations it is natural to replace $\Diamond_{1r}$, i.e., logical compatibility, with r's axioms, with mathematical compatibility with those axioms ($\Diamond_{2r}$):

We say that $\mathscr{A}$ is *mathematically compatible* with r's axioms, or briefly that $\mathscr{A}$ is r-possible$_2$, if $\mathscr{A}$ is r'-possible$_1$, where r' is the extension of r obtained by adding r's axioms with all true mathematical assertions (relative to the standard model of the purely mathematical part r_M of r).

From the semantic point of view note that the sentence $\mathscr{A}$ of $\mathscr{L}$ is r-possible$_2$ iff it is true in all models of r that induce the standard model on r_M (i.e., that are also models of r').

The 'suitable' compatibility notion used in Section 3 can be identified with r-possibility$_2$. The analogue holds for phys. Poss$_3$ [cf. Section 5].

I think, for example, that the new version of phys. Poss$_2$ avoids the trouble of the first version, at least in connection with some significant choices of the physical theory r. In this case the new version has the unsatisfactory features pointed out in 'On phys. Poss.', but can be used as an auxiliary notion in dealing with phys. Poss$_1$.

On the other hand, let us note that, since the above theory r' is not recursively axiomatizable, [cf. 9, Chapt. 3, Sect. 6], r-possibility$_2$ lacks the positive properties of r-possibility$_1$ due to r-possibility$_1$ being based on a finite set of axioms (Hutten speaks of known laws). Just because r-possibility$_2$ is based in the same way on (infinitely many) unknown (arithmetical) sentences (which are axioms of r') the new version of phys. Poss$_2$ appears even more unsatisfactory than the first one did, on the basis of 'On phys. Poss', before its defects connected with Gödel's undecidability theorem were noted.

We identify r-possibility$_r$ for $r = 1$ [2] with logical [mathematical] compatibility with r's axioms. On the other hand (abstract) mathematics is often said to be reduced to logic. Hence, it is now natural to stress that this reduction is restricted to notions and does not hold for assertions. Essential mathematical axioms assert the existence of some large sets. This is substantially done, in particular, by one of Peano's axioms.

Let us add that mathematics is independent of logic even more than the above facts show. Indeed the proofs of Gödel's undecidability theorem and its extensions put in evidence a denumerable and logically independent set of arithmetic sentences of $\mathscr{L}$ that are undecidable (in r) and can be recognized (in L) to be true; furthermore these sentences differ from the above mathematical axioms in that they somehow assert their own unprovabilities in (suitable) extensions of r.

D. A 'PHYSICAL' VERSION OF PHYS. POSS$_2$

In the definition of r-possibility$_2$ the physical theory r is arbitrary. Then r may contain a (physical) structure similar to the one of natural numbers, so that r-possibility$_2$ may be affected (for some choices of r) by the same trouble that arithmetic gives to possibility$_1$ – cf. Section B. Therefore it is useful to strengthen the semantic characterization of possibility$_2$ into a definition of the following kind: The sentence $\mathscr{A}$ of $\mathscr{L}$ is said to be r-possible$_3$($\Diamond_{3r}\mathscr{A}$), if it is true in some model belonging to a certain class K_r of models of r, which in particular include the standard model of r_M. The class K_r is supposed to have other restrictions. If, for example *mass-point* (MP), *event-point* (EP), and *worldLine* (wL) are primitive terms of r, then the models of K_r associate MP and EP with their ordinary interpretations and they associate wL with a function from mass-points to suitable subsets of EP.

Since the models of K_r associate physical [mathematical] constants with physical [mathematical] entities, r-possibility$_3$ can be said to be physically compatible with r's axioms.

Let us remark that the class K_r has to be specified case by case. Furthermore, r-possibility$_3$ is worth being considered, in spite of its dependence on K_r, because it is the closest to a notion of phys. Poss. used sometimes by physicists, among the notions of phys. Poss. of Hutten's type considered here.

NOTE

* This work was done mainly during the author's visit at the Logic Group, Australian National University, Canberra, September–December 1979.

E.G. BELTRAMETTI AND G. CASSINELLI

PROBLEMS OF THE PROPOSITION-STATE STRUCTURE OF QUANTUM MECHANICS

1. In this paper we shall be mainly concerned with two issues which are of paramount importance in the foundations of quantum mechanics. The first is the superposition principle. The second is the possibility of performing the measurement of any observable of a given physical system by a so-called ideal, first-kind measurement, which means, in particular, that the repetition of the same observation would give with certainty the result initially found, without further changing the state of the physical system. The purpose is to examine these two issues beyond the usual Hilbert space formulation of quantum mechanics. Hence it is natural to adopt the simplifying restrictions which are common in studies of quantum logic, in particular the restriction to dichotomic observables, or yes-no experiments, i.e., to those observables whose outcome reduces to a binary choice (the marks 'yes' and 'no'). The states of the physical system will be thought of as the different preparations of the system. Each state will assign a definite probability to the 'yes' outcome of any yes-no experiment. Two yes-no experiments will be considered as equivalent if there is no state which assigns to their 'yes' outcomes different probabilities. A class of equivalent yes-no experiments is called a proposition on the system; the set of all the propositions will be denoted by $\mathscr{L}$, while single propositions will be denoted by the letters $a, b, \ldots$. We shall use the Greek letters $\alpha, \beta, \ldots$ to denote states, while $\mathscr{S}$ will stand for the set of them.

States and propositions will be used, in what follows, as primitive undefined quantities, the comments above providing only a physical interpretation. In the following sections we shall be engaged in different aspects and, in each case, the precise assumptions needed for $\mathscr{L}$ and $\mathscr{S}$ will be specified. We leave aside the somewhat controversial problem of whether some economy can be made by adopting, for instance, propositions as primitive and states as secondary defined objects: the last is Piron's well known point of view (Piron, 1976).

2. We shall first examine where the superposition principle finds its place within the logical structure of quantum mechanics. For this, we make the following assumption:

215

Maria Luisa Dalla Chiara (ed.), Italian Studies in the Philosophy of Science, 215–235.
Copyright © 1980 by D. Reidel Publishing Company.

ASSUMPTION 2.1. $\mathcal{L}$ is an orthomodular poset, $\mathcal{S}$ is a set of probability measures on $\mathcal{L}$, order determining and σ-convex.

In fact, these hypotheses on $\mathcal{L}$ and $\mathcal{S}$ are supported on very general grounds, as Mackey (1963), in particular, has shown, in any probabilistic description of a physical system. Recall that $\mathcal{S}$ is said to be order determining on $\mathcal{L}$ when $\alpha(a) \le \alpha(b)$, for every $\alpha \in \mathcal{S}$ gives $a \le b$ in $\mathcal{L}$. We leave open, at the moment, whether $\mathcal{S}$ is also strongly ordering on $\mathcal{L}$, i.e., whether $\mathcal{S}_1(a) \subseteq \mathcal{S}_1(b)$, with $\mathcal{S}_1(a) = \{\alpha \in \mathcal{S} : \alpha(a) = 1\}$, gives $a \le b$ in $\mathcal{L}$.

Thus, in order to fit the quantum mechanical superposition principle within this background, we consider a notion of superposition of states which is a slight modification of the one proposed by Varadarajan (1968).

Given $\alpha \in \mathcal{S}$ let us denote by $\mathcal{L}_1(\alpha)$ the set of the propositions whose outcome is certainly yes for the state α:

$$\mathcal{L}_1(\alpha) = \{a \in \mathcal{L} : \alpha(a) = 1\}.$$

Let S be a subset of $\mathcal{S}$; then we say that $\beta \in \mathcal{S}$ is a superposition of the states of S if

$$(1) \qquad \mathcal{L}_1(\beta) \supseteq \bigcap_{\alpha \in S} \mathcal{L}_1(\alpha)$$

i.e., if all the propositions which give with certainty the yes outcome for all the states of S do the same for the state β. We then define the closure $\bar{S}$ of a set S of states as the set containing all the states of S and all their superpositions:

$$(2) \qquad \bar{S} = \{\beta \in \mathcal{S} : \mathcal{L}_1(\beta) \supseteq \bigcap_{\alpha \in \mathcal{S}} \mathcal{L}_1(\alpha)\}$$

Of course, for every $S \subseteq \mathcal{S}$

$$(3) \qquad \bar{S} \supseteq S.$$

A set S is called closed with respect to superposition whenever $S = \bar{S}$. The set

$$\mathcal{M} = \{S \subseteq \mathcal{S} : S = \bar{S}\},$$

i.e. the set of all the closed subsets of $\mathcal{S}$, will be of relevance in what follows. We have:

THEOREM 2.1. *$\mathcal{M}$ is a complete lattice with respect to the ordering defined by set inclusion: the meet $(\wedge)$ and join $(\vee)$ are given by*

$$S_1 \wedge S_2 = S_1 \cap S_2, \qquad S_1 \vee S_2 = \overline{S_1 \cup S_2}$$

($\cap$ and $\cup$ are set intersection and union), the zero and unit elements are the empty set and the whole $\mathcal{S}$, respectively.

Proof. First remark that, for arbitrary S, $S_i \subseteq \mathcal{S}$:

(i) $\bigcap\limits_{\alpha \in S} \mathcal{L}_1(\alpha) = \bigcap\limits_{\alpha \in \bar{S}} \mathcal{L}_1(\alpha),$

(ii) $\bar{S} = \bar{\bar{S}},$

(iii) $S_1 \supseteq S_2 \Rightarrow \bar{S}_1 \supseteq \bar{S}_2,$

(iv) $\bigcap\limits_{i \in \mathcal{I}} \bar{S}_i = \overline{\bigcap\limits_{i \in \mathcal{I}} \bar{S}_i},$ $\quad \mathcal{I}$ arbitrary index set.

The equality (i) follows, after remarking that, owing to (3), the right-hand side contains the left-hand side, but the converse is also true because, if $\beta \in \bar{\bar{S}}$ then $\mathcal{L}_1(\beta) \supseteq \bigcap\limits_{\alpha \in S} \mathcal{L}_1(\alpha)$. The equality (ii) follows after remarking that, owing to (3), $\bar{\bar{S}} \supseteq \bar{S}$, but if $\bar{S}$ were strictly contained in $\bar{\bar{S}}$ then choose $\beta \in S$ and, by use of (i),

$$\mathcal{L}_1(\beta) \supseteq \bigcap\limits_{\alpha \in \bar{S}} \mathcal{L}_1(\alpha) = \bigcap\limits_{\alpha \in S} \mathcal{L}_1(\alpha) \text{ i.e., } \beta \in \bar{S}, \text{ hence } \bar{\bar{S}} \subseteq \bar{S}.$$

The statement (iii) is the obvious consequence of (2). The equality (iv) follows after remarking that (3) gives $\bigcap \bar{S}_i \subseteq \overline{\bigcap \bar{S}_i}$; on the other hand, if $\beta \in \overline{\bigcap \bar{S}_i}$, then

$$\mathcal{L}_1(\beta) \supseteq \bigcap\limits_{\alpha \in \cap \bar{S}_i} \mathcal{L}_1(\alpha),$$

hence

$$\mathcal{L}_1(\beta) \supseteq \bigcap\limits_{\alpha \in \bar{S}_j} \mathcal{L}_1(\alpha)$$

for every $j \in \mathcal{I}$, therefore $\beta \in \bar{\bar{S}}_j$, and, by use of (ii), $\beta \in \bar{S}_j$, for every $j \in \mathcal{I}$, whence $\beta \in \bigcap \bar{S}_i$. The proof of the theorem only requires us to note that, by (iv), the intersection of closed sets is closed and, for any $S_0 \in \mathcal{M}$ such that $S_0 \subseteq S_i$ for every $i \in \mathcal{I}$ we get $S_0 \subseteq \bigcap S_i$, so that the meet is given by set intersection; moreover, by (ii), the closure of the set union is closed and, for any $S_0 \in \mathcal{M}$ such that $S_0 \supseteq S_i$ for every $i \in \mathcal{I}$ we get $S_0 \supseteq \bigcup S_i$, hence, by (iii), $S_0 \supseteq \overline{\bigcup S_i}$. $\qquad \square$

The result expressed by this theorem holds true independently of the explicit form of the superposition of states. In fact, the essential steps are the property (3) and the points (ii), (iii) occurring in the proof of the theorem: they ensure that $S \mapsto \bar{S}$ is a closure operation in the sense of Moore (Birkhoff, 1967): it is then a general theorem that the set of closed subsets is a complete lattice with meet and join as defined in the theorem independently of the explicit form of the closure operation.

We stress that the adopted notion of superposition of states and the quoted properties of $\mathcal{M}$ do not require specifying whether pure states exist in $\mathcal{S}$. On the other hand, as will become clear in what follows, the possibility of superposing pure states to get pure states is peculiar to quantum systems. It is thus worth noticing that, restricting $\mathcal{S}$ to the (non empty) subset $\mathcal{P}$ of its pure states and restricting corresondingly $\mathcal{M}$ to the set $\mathcal{M}_{\mathcal{P}}$ of all the subset of $\mathcal{P}$ which are closed under superposition, one still obtains for $\mathcal{M}_{\mathcal{P}}$ a lattice structure; such a result is due to Gudder (1970) and, as a matter of fact, the previous Theorem 2.1 paraphrases and generalizes Gudder's proof.

It is also worth noticing that the adopted definition of superposition of states (1) includes, as a particular case, the mixtures of states, the only superpositions occurring in classical mechanics.

In this case the closure $\bar{S}$ of S takes the form

$$\bar{S} = \left(\sum_{\alpha_i \in S} t_i \alpha_i \, ; \, t_i \geq 0, \, \sum t_i = 1 \right).$$

It is then easy to check (Beltrametti and Cassinelli, 1976) that $\mathcal{M}$ is a distributive lattice, and that $\mathcal{L}$ and $\mathcal{M}$ are isomorphic Boolean algebras (Varadarajan, 1968).

According to Varadarajan the superposition principle of quantum mechanics amounts to saying that $\mathcal{M}$ and $\mathcal{L}$ are isomorphic as lattices, though they are no longer Boolean algebras.

A first hint in favor of this relation between $\mathcal{L}$ and $\mathcal{M}$ is provided by remarking that $\mathcal{S}_1(a)$ belongs to $\mathcal{M}$ for every $a \in \mathcal{L}$ (recall that we have defined $\mathcal{S}_1(a) = \{\alpha \in \mathcal{S} : \alpha(a) = 1\}$). In fact, if $\beta \in \overline{\mathcal{S}_1(a)}$, we get, by (2), $a \in \mathcal{L}_1(\beta)$, whence $\beta \in \mathcal{S}_1(a)$ and $\overline{\mathcal{S}_1(a)} \subseteq \mathcal{S}_1(a)$, hence, by (3), $\mathcal{S}_1(a)$ is closed. After this remark we can reduce the superposition principle to the

ASSUMPTION 2.2. $\mathcal{S}$ is strongly ordering on $\mathcal{L}$ and, for every $S \in \mathcal{M}$ there exists an element a_S of $\mathcal{L}$ such that $S = \mathcal{S}_1(a_S)$.

In fact, $\mathcal{S}$ being strongly ordering, the mapping $a \mapsto \mathcal{S}_1(a)$ is an order preserving injection of $\mathcal{L}$ into $\mathcal{S}$; moreover, due to the possibility of associating to any $S \in \mathcal{M}$ a proposition a_S such that $S = \mathcal{S}_1(a_S)$, (its unicity is a consequence of the strongly ordering hypothesis), the mapping $a \mapsto$

$\mathscr{S}_1(a)$ provides an isomorphism between $\mathscr{M}$ and $\mathscr{L}$. Thus $\mathscr{L}$ inherits from $\mathscr{M}$ the structure of complete lattice, while $\mathscr{M}$ inherits from $\mathscr{L}$ the ortho-complementation

$$S \mapsto S^{\perp} \quad \text{with} \quad S^{\perp} = \mathscr{S}_1(a_S^{\perp}).$$

Summing up, $\mathscr{L}$ and $\mathscr{M}$ are, at this stage, equipped with isomorphic structures of complete orthomodular lattices.

In order to visualize in more familiar terms the role of the superposition principle we need to enrich the structure of $\mathscr{L}$ (hence of $\mathscr{M}$) with further properties demanded by the description of physical quantum systems. In particular, we refer to atomicity and the so-called covering property.

Consider first atomicity, which finds its counterpart, in the Hilbert space formulation of quantum mechanics, in the existence of maximal, or complete, sets of compatible observables. Owing to the isomorphism between $\mathscr{L}$ and $\mathscr{M}$, focus attention on how atomicity sets up in $\mathscr{M}$. Since the ordering relation in $\mathscr{M}$ is just set inclusion, the atoms of $\mathscr{M}$ are the closed sets whose non-empty subsets are not closed. Remark that an element of $\mathscr{M}$ can contain a finite number of pure states, but there are infinitely many non-pure states if the number of pure states is greater than one (the possibility of an infinite choice of weights for mixtures of states is here assumed). Therefore, since we are looking for atoms of $\mathscr{M}$, it is plausible to suppose that pure states exist and that the sets formed by only one pure state belong to $\mathscr{M}$: this is the only chance for $\mathscr{M}$ to possess finite elements, and, moreover, the elements formed by only one state are obviously atoms. We formalize this point by means of

ASSUMPTION 2.3. $\mathscr{S}$ contains a non-empty subset $\mathscr{P}$ of pure states; the sets formed by only one pure state belong to $\mathscr{M}$ and exhaust the atoms of $\mathscr{M}$.

This condition ensures that there are no superpositions of a single pure state, so that the set $\{\alpha\}$, with $\alpha \in \mathscr{P}$, is closed. In other words, if $\alpha \in \mathscr{P}$, there is no $\beta \in \mathscr{S}$ such that $\mathscr{S}_1(\beta) \supseteq \mathscr{S}_1(\alpha)$. This is the form in which atomicity is introduced by Gudder (1970).

Having classified the atoms of $\mathscr{M}$, we get, by the isomorphism between $\mathscr{L}$ the $\mathscr{M}$, all the atoms of $\mathscr{L}$: they are those propositions p such that $\mathscr{S}_1(p)$ is a set formed by just one pure state. Let $\mathscr{A}$ be the set of the atoms of $\mathscr{L}$: the isomorphism between $\mathscr{L}$ and $\mathscr{M}$ determines, by restriction, a one-to-one correspondence between $\mathscr{A}$ and $\mathscr{P}$.

The existence of atoms in $\mathscr{M}$ is not sufficient to ensure that $\mathscr{M}$ is atomic:

we further require that every element of $\mathcal{M}$ contains at least one atom, hence that every closed set of states contains at least one pure state. On physical grounds we guarantee this property by means of the further atomicity condition:

ASSUMPTION 2.4. Every not pure state can be expressed as a mixture of pure states.

A closed set containing a mixture certainly contains all pure states occurring in the mixture; thus $\mathcal{M}$ is atomic. Actually we have even more. Let $S \in \mathcal{M}$ and let $P \subseteq \mathcal{P}$ be the set of the pure states belonging to S: on account of the last two atomicity assumptions and taking into account the form of the join in $\mathcal{M}$, we get

$$(4) \qquad S = \bigvee_{\alpha \in P} \{\alpha\}, \qquad S \in \mathcal{M}$$

so that every element of $\mathcal{M}$ is the join of its atoms. By the isomorphism, the same property is shared by $\mathcal{L}$. For every $a \in \mathcal{L}$ let $\mathcal{A}_a$ be the set of atoms contained in a:

$$(5) \qquad \mathcal{A}_a = \{p \in \mathcal{A} : p \leq a\}.$$

Then

$$(6) \qquad a = \bigvee_{p \in \mathcal{A}_a} p, \qquad a \in \mathcal{L}.$$

In the standard terminology of lattice theory the properties (5) and (6) are referred to by saying that $\mathcal{M}$ and $\mathcal{L}$ are not only atomic but even atomistic. Let us mention (Maeda and Maeda, 1970, p. 134) that it is always possible to pick up from $\mathcal{A}_a$ a countable sequence $\{p_i\}$ of mutually orthogonal atoms such that $a = \bigvee p_i$. The countability requirement finds its analog in classical mechanics in the separability of the Boolean σ-algebra, while in standard quantum mechanics it corresponds to the separability of the Hilbert space.

We add a useful definition. Let $\alpha \in \mathcal{S}$; the proposition $\sigma(\alpha)$ defined by

$$(7) \qquad \mathcal{L}_1(\alpha) = \{a \in \mathcal{L} : a \geq \sigma(\alpha)\},$$

is called the support of α (Pool, 68). If α is a pure state, then $\sigma(\alpha)$ is proved to be the atom of $\mathcal{L}$ associated to α by the isomorphism between $\mathcal{M}$ and $\mathcal{L}$ (Beltrametti and Cassinelli, 1973). If α is a mixture we write it in the form

$$\alpha = \sum t_i \alpha_i, \qquad \alpha_i \in \mathscr{P}$$

and $\sigma(\alpha)$ is the join of the atoms of $\mathscr{L}$ associated to the pure states α_i (Beltrametti and Cassinelli, 1973)

$$\sigma(\alpha) = \bigvee \sigma(\alpha_i).$$

By use of the notion of support we can easily rephrase the definition (1) of superposition by saying that $\beta \in \mathscr{S}$ is a super-position of the states belonging to $S \, (\subset \mathscr{S})$ if

$$(8) \qquad \sigma(\beta) \leq \bigvee_{\alpha \in S} \sigma(\alpha).$$

Denoting by μ the mapping which isomorphically maps $\mathscr{M}$ onto $\mathscr{L}$, by π, its restriction to the pure states (the atoms of $\mathscr{M}$), by ϕ, the mapping $\mathscr{S} \to \mathscr{M}$ which maps the state $\alpha = \sum t_i \alpha_i{}^{'}$ $\alpha_i \in \mathscr{P}$, into the closed set $\{\alpha_1, \alpha_2, \ldots\}$, we can summarize the various correspondences defined so far by Figure 1. where $i_{\mathscr{A}}$ and $i_{\mathscr{P}}$ are the canonical inclusions, respectively, of $\mathscr{A}$ in $\mathscr{L}$ and, of $\mathscr{P}$ in $\mathscr{M}$. From this diagram we can extract, with standard algebraic notations, the two exact and commutative diagrams shown as Figures 2 and 3 (where O stands for the zero element of $\mathscr{L}$).

We have to remark that the structure of $\mathscr{L}$ is not yet rich enough to ensure that, given any two pure states α_1, α_2, there exists a pure state which is superposition of them. To do that we need to come to the covering property of $\mathscr{L}$.

Recall that, for $a, b \in \mathscr{L}$, b is said to cover a if b is larger than a and if

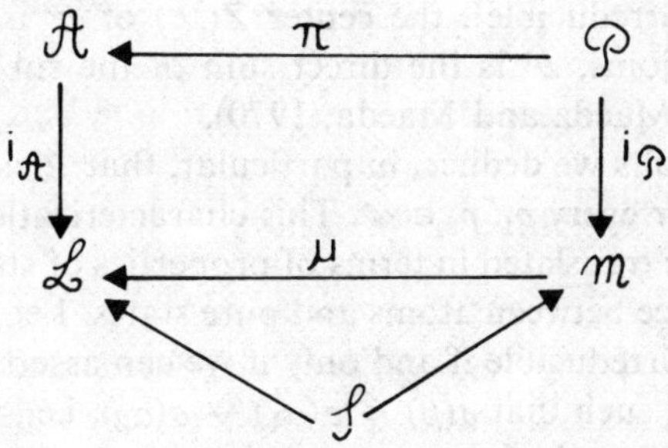

Fig. 1

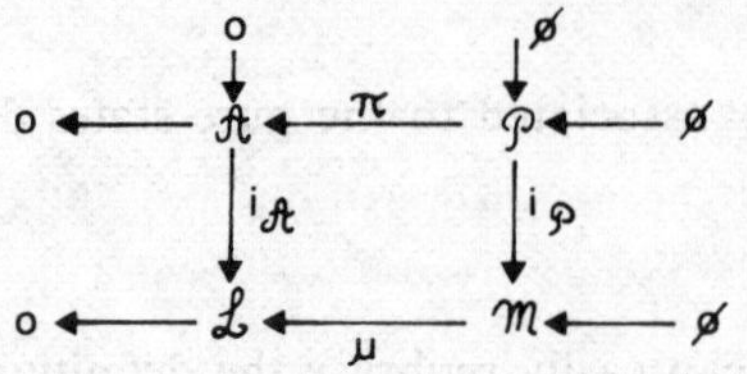

Fig. 2 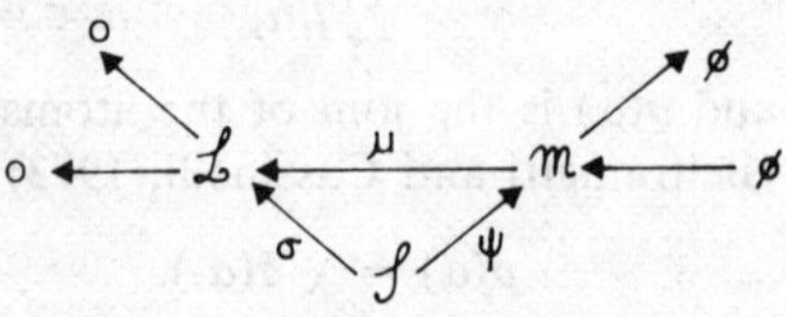   Fig. 3

there is no other element in between them: technically if $a \leq x \leq b$ it implies either $x = a$ or $x = b$. Then $\mathscr{L}$ has the covering property if, given $a \in \mathscr{L}$ and any atom p such that $p \not< a$, $a \vee p$ covers a. $\mathscr{L}$ being an ortho-modular atomistic lattice, the covering property can be equivalently for-mulated saying that $(p \vee a^\perp) \wedge a$ is an atom for every $a \in \mathscr{L}$ and for any atom p such that $p \not< a^\perp$.

Then suppose:

ASSUMPTION 2.5. $\mathscr{L}$ has the covering property.

It may be shown (Maeda and Maeda, 1970) that the relation $\approx$ defined below is an equivalence relation. We say that $p_1 \approx p_2$, $p_1, p_2 \in \mathscr{A}$, if either $p_1 = p_2$ or there exists a third atom $q \neq p_1, p_2$ such that $q < p_1 \vee p_2$. Then consider the partition of $\mathscr{A}$ defined by $\approx$ and denote by $\mathscr{J}$ the index set which labels the equivalence classes. Let z_i, $i \in \mathscr{J}$, be the join of all the atoms which belong to the ith equivalence class. The sublattices (segments) of $\mathscr{L}$

$$\mathscr{L}[0, z_i] = \{a \in \mathscr{L} : a \leq z_i\}, \qquad i \in \mathscr{J}$$

are shown to be irreducible[1], the center $Z(\mathscr{L})$ of $\mathscr{L}$ is exhausted by the z_i's and by their joins, $\mathscr{L}$ is the direct sum of the sublattices $\mathscr{L}[0, z_i]$, i ranging over $\mathscr{J}$ (Maeda and Maeda, 1970).

From these results we deduce, in particular, that $\mathscr{L}$ is irreducible if and only if $p_1 \approx p_2$ for every $p_1, p_2 \in \mathscr{A}$. This characterization of the irreduci-bility of $\mathscr{L}$ can be translated in terms of properties of states making use of the correspondence between atoms and pure states. Let α_1, α_2 be two pure states: then $\mathscr{L}$ is irreducible if and only if we can assert the existence of a third pure state β such that $\sigma(\beta) < \sigma(\alpha_1) \vee \sigma(\alpha_2)$, i.e., on account of (8), of a third pure state which is superposition of α_1 and α_2. The unlimited possibility of superposing pure states to get new pure states corresponds, according to the terminology of physics, to a quantum system not admitting 'superselection rules'.

3. We consider now a formalism such that, besides propositions and states, another set, which we call the set of operations, is taken into account. The mathematical structure we shall suppose in this approach for the sets of propositions and states is very poor and has an immediate physical meaning; we shall show how the existence of the operations can enrich this structure, and we shall, in what follows, investigate the intertwinings between the properties of the propositions and states and those of the operations. The basic idea which lies at the foundations of the concept of operation is the transformation induced in the state of a physical system by the action of the experimental apparata used to verify a proposition.

This idea has been advanced and worked out, by means of the mathematical techniques of involutive semigroups, by Pool (1968): what follows is an adaptation, to slightly different hypotheses, of his work.

We denote, as usual, by $\mathscr{L}$ the set of all propositions; we do not assume any structure for $\mathscr{L}$, which is simply a set. A state, α, on $\mathscr{L}$, is a function from $\mathscr{L}$ to the unit real interval $[0, 1]$, and $\alpha(a)$ is interpreted as the probability of the yes outcome of a when the system is in the state α; the set of all states will be denoted by $\mathscr{S}$. Our hypothesis on the pair $(\mathscr{L}, \mathscr{S})$ is the following:

ASSUMPTION 3.1. In $\mathscr{L}$ there exist two distinguished elements O and I such that $\alpha(O) = 0$ and $\alpha(I) = 1$ for every $\alpha \in \mathscr{S}$. Moreover for any $a \in \mathscr{L}$ there exists in $\mathscr{L}$ another unique element, that we shall denote as $a^{\perp}$, such that

$$\{\alpha \in \mathscr{S} : \alpha(a) = 0\} = \{\alpha \in \mathscr{S} : \alpha(a^{\perp}) = 1\};$$
$$\{\alpha \in \mathscr{S} : \alpha(a^{\perp}) \neq 0\} = \{\alpha \in \mathscr{S} : \alpha(a) \neq 1\}.$$

We now ask whether we can define a map of $\mathscr{S}$ into $\mathscr{S}$, whcih represents the transformation of the state of a system when a proposition is verified by means of an ideal, first-kind measurement (in the sense of quantum theory of measurement), consequently we shall call such a map an ideal first kind mapping.

Let us, first of all, settle some notation: for every mapping f of $\mathscr{S}$ into $\mathscr{S}$ we shall denote by $\mathscr{D}[f]$ the subset of $\mathscr{S}$ on which f is defined, and by $\mathscr{R}[f]$ its range, that is the image of $\mathscr{S}$ under f, $\mathscr{R}[f] = \{\alpha \in \mathscr{S} : \alpha = f(\beta)$

for some $\beta \in \mathscr{D}[f]\}$. Given any $a \in \mathscr{L}$, $a \neq O, I$, an ideal first-kind map of $\mathscr{S}$ into $\mathscr{S}$, to be denoted by Ω_a, is characterized by the following properties:

$$\text{(9)} \qquad \mathscr{D}[\Omega_a] = \{\alpha \in \mathscr{S}: \alpha(a) \neq 0\},$$

$$\text{(10)} \qquad \text{if } \alpha \in \mathscr{D}[\Omega_a] \text{ and } \alpha(a) = 1 \text{ then } \Omega_a\alpha = \alpha,$$

$$\text{(11)} \qquad \Omega_a\alpha(a) = 1 \text{ for every } \alpha \in \mathscr{D}[\Omega_a].$$

Points (9), (10), (11) characterize first kind maps; from points (10) and (11) it is easy to show that $\mathscr{R}[\Omega_a] = \{\alpha \in \mathscr{S}: \alpha(a) = 1\}$. With the aid of this identification we can restate the properties of the mapping $a \mapsto a^\perp$, in the following way:

$$\mathscr{R}[\Omega_{a\perp}] = \mathscr{C}\mathscr{D}[\Omega_a] \text{ and } \mathscr{D}[\Omega_{a\perp}] = \mathscr{C}\mathscr{R}[\Omega_a]$$

(where $\mathscr{C}$ denotes set-theoretic complement relative to $\mathscr{S}$). The elements O and I have the properties that $\mathscr{D}[\Omega_O] = \varnothing$, $\mathscr{D}[\Omega_I] = \mathscr{R}[\Omega_I] = \mathscr{S}$, $\Omega_I\alpha = \alpha$ for every $\alpha \in \mathscr{S}$; moreover $\mathscr{C}\mathscr{R}[\Omega_I] = \varnothing = \mathscr{D}[\Omega_O]$ and $\mathscr{C}\mathscr{D}[\Omega_O] = \mathscr{S} = \mathscr{R}[\Omega_I]$, so that $O^\perp = I$ and $I^\perp = O$.

We say that two maps, f and g, of $\mathscr{S}$ into $\mathscr{S}$ commute when $f \circ g$ and $g \circ f$ have the same not empty domain in $(f \circ g)\alpha = (g \circ f)\alpha$ for every $\alpha \in \mathscr{S}$ (here we denote by $\circ$ the usual composition of mappings). With this definition in mind, we will say that a first-kind map is ideal when

$$\text{(12)} \qquad \alpha(b) = 1 \text{ implies } \Omega_a\alpha(b) = 1 \text{ whenever } \Omega_a \text{ and } \Omega_b \text{ commute.}$$

Our fundamental hypothesis about ideal first kind maps is:

ASSUMPTION 3.2. There exists a one-to-one correspondence between $\mathscr{L}$ and the set $\{\Omega_a: a \in \mathscr{L}\}$, given by $a \mapsto \Omega_a$.

Ω_a is thus interpreted as the state transformation caused by an ideal, first kind measurement of a. For a more detailed discussion of the properties of the Ω_a's and of Assumption 3.2 we refer to Beltrametti and Cassinelli (1976).

Let us now consider the set T_Ω, of all mappings of $\mathscr{S}$ into $\mathscr{S}$, defined by

$$T_\Omega = \{\Omega_{a_1} \circ \cdots \circ \Omega_{a_n}: a_1, \ldots, a_n \in \mathscr{L}, n \text{ finite}\}.$$

The elements of T_Ω will be called operations.

We want to stress that it may happen (and it can be shown by examples) that two (or more) sequences of propositions, $a_1, \ldots a_n, b_1, \ldots, b_m$, can exist such that the mappings $\Omega_{a_1} \circ \cdots \circ \Omega_{a_n}$ and $\Omega_{b_1} \circ \cdots \circ \Omega_{b_m}$ have the same domain and

$$(\Omega_{b_1} \circ \cdots \circ \Omega_{b_m})\alpha = (\Omega_{a_1} \circ \cdots \circ \Omega_{a_n})\alpha$$

for every α of the domain. In this case the two sequences $a_1, \ldots, a_n$, $b_1, \ldots, b_m$ provide two realizations of the same mapping

$$\Omega_{a_1} \circ \cdots \circ \Omega_{a_n} = \Omega_{b_1} \circ \cdots \circ \Omega_{b_m}.$$

If there exists a sequence $a_1 \ldots a_n$ such that $\Omega_{a_1} \circ \cdots \circ \Omega_{a_n}$ has an empty domain, the mapping $\Omega_{a_1} \circ \cdots \circ \Omega_{a_n}$ will be identified with the null map Ω_0. Moreover it can be shown by counterexamples that, in general, a composition of ideal first-kind maps is not ideal first-kind.

T_Ω has a natural structure of semigroup, with respect to composition of maps, and it possesses a zero and unit element, Ω_0 and Ω_I, respectively; indeed

$$\Omega_0 \circ x = x \circ \Omega_0 = \Omega_0 \text{ and } \Omega_I \circ x = x \circ \Omega_I = x \text{ for all } x \in T_\Omega.$$

Now suppose that for the set of operation the following holds:

ASSUMPTION 3.3. If $a_1, \ldots, a_n, b_1, \ldots, b_m$ are such that

$$\Omega_{a_1} \circ \cdots \circ \Omega_{a_n} = \Omega_{b_1} \circ \cdots \circ \Omega_{b_m},$$

then we have also

$$\Omega_{b_m} \circ \cdots \circ \Omega_{b_1} = \Omega_{a_n} \circ \cdots \circ \Omega_{a_1}.$$

By means of this assumption we can give T_Ω the structure of an involution semigroup; in fact we can define a mapping $x \mapsto x^*$ of T_Ω onto itself in the following way: if $\Omega_{a_1} \circ \cdots \circ \Omega_{a_n}$ is a realization of $x \in T_\Omega$ then $x^* = \Omega_{a_n} \circ \cdots \circ \Omega_{a_1}$. The mapping $x \mapsto x^*$ is well defined and is an involutive anti-automorphism of T_Ω, that is, $x^{**} = x$ and $(x \circ y)^* = y^* \circ x^*$ for all $x, y \in T_\Omega$. The set $P(T_\Omega)$ of the projections of T_Ω is defined by

$$P(T_\Omega) = \{e \in T_\Omega : e \circ e = e = e^*\}.$$

The elements of T_Ω of the form Ω_a, for some $a \in \mathscr{L}$, are projections but in general there are projections that are not of the form Ω_a. $P(T_\Omega)$ has some properties which are common to the set of projections of involution semigroups (Maeda and Maeda, 1970; Foulis, 1960): there exists a natural partial ordering in $P(T_\Omega)$ defined as follows

$$e \leq f \text{ when } e \circ f = e = f \circ e; \qquad e, f \in P(T_\Omega).$$

The mappings Ω_0 and Ω_I are the least and greatest elements, respectively of the partially ordered set $P(T_\Omega)$; in fact it is easily seen that

$$\Omega_0 \leq e; \, e \leq \Omega_I \text{ for all } e \in P(T_\Omega).$$

The ordering relation in $P(T_\Omega)$ induces, by restriction, an ordering relation in the set $\{\Omega_a : a \in \mathscr{L}\}$:

$$\Omega_a \leq \Omega_b \text{ when } \Omega_a \circ \Omega_b = \Omega_a = \Omega_b \circ \Omega_a; \, a, b \in \mathscr{L}.$$

It is possible to characterize this ordering relation in another equivalent way.

LEMMA 3.1. $\Omega_b \circ \Omega_a = \Omega_a$ *if and only if* $\mathscr{S}_1(a) \subseteq \mathscr{S}_1(b)$
Proof. Suppose first that $\Omega_b \circ \Omega_a = \Omega_a$, then $\mathscr{R}[\Omega_b \circ \Omega_a] = \mathscr{R}[\Omega_a]$; we have already seen that $\mathscr{R}[\Omega_a] = \mathscr{S}_1(a)$, moreover $\mathscr{R}[\Omega_b \circ \Omega_a] \subseteq \mathscr{R}[\Omega_b] = \mathscr{S}_1(b)$ for the range of a composition of maps is certainly contained in the range of its first left factor, so $\mathscr{S}_1(a) \subseteq \mathscr{S}_1(b)$. Conversely, suppose $\mathscr{S}_1(a) \subseteq \mathscr{S}_1(b)$: notice that for every $a \in \mathscr{S}$

$$\mathscr{S}_1(a) = \mathscr{R}[\Omega_a] \subseteq \mathscr{D}[\Omega_a] = \{\alpha \in \mathscr{S} : \alpha(a) \neq 0\},$$

hence

$$\mathscr{R}[\Omega_a] = \mathscr{S}_1(a) \subseteq \mathscr{S}_1(b) = \mathscr{R}[\Omega_b] \subseteq \mathscr{D}[\Omega_b].$$

By this fact

$$\mathscr{D}[\Omega_b \circ \Omega_a] = \{\alpha \in \mathscr{D}[\Omega_a] : \Omega_a \alpha \in \mathscr{D}[\Omega_b]\} = \mathscr{D}[\Omega_a],$$

moreover

$$\Omega_a \alpha \in \mathscr{R}[\Omega_a] = \mathscr{S}_1(a) \subseteq \mathscr{S}_1(b) \text{ for all } \alpha \in \mathscr{D}[\Omega_a];$$

then

$\Omega_a \alpha(b) = 1$ and $\Omega_b(\Omega_a \alpha) = \Omega_a \alpha$ (by 3.2). We have thus shown that $\Omega_b \circ \Omega_a = \Omega_a$. Concluding $\Omega_b \circ \Omega_a = \Omega_a$ is equivalent to the relation $\mathscr{S}_1(a) \subseteq \mathscr{S}_1(b)$. $\qquad\square$

Owing to the one-to-one correspondence between the set $\{\Omega_a : a \in \mathscr{L}\}$ and $\mathscr{L}$, the ordering relation defined above can be transferred in $\mathscr{L}$, so we have

$$a \leq b \text{ when } \mathscr{S}_1(a) \subseteq \mathscr{S}_1(b); \, a, b \in \mathscr{L}.$$

In particular $a = b$ when $\mathscr{S}_1(a) = \mathscr{S}_1(b)$. By this remark we can show that the mapping $a \mapsto a^\perp$, introduced in Assumption 3.1, is an involutive one, that is $a^{\perp\perp} = a$. In fact we have

$$\mathscr{S}_1(a^{\perp\perp}) = \{\alpha \in \mathscr{S}: \alpha(a^{\perp}) = 0\}$$
$$= \mathscr{C}\{\alpha \in \mathscr{S}: \alpha(a) \neq 0\}$$
$$= \{\alpha \in \mathscr{S}: \alpha(a) = 1\} = \mathscr{S}_1(a).$$

Moreover if $\mathscr{D}[\Omega_a] = \mathscr{D}[\Omega_b]$ we have $\mathscr{S}_1(a^{\perp}) = \mathscr{S}_1(b^{\perp})$ hence $a^{\perp} = b^{\perp}$, and, by the involutive property of $^{\perp}$, $a = b$. Then we can write

$$(13) \qquad a = b \Leftrightarrow \mathscr{D}[\Omega_a] = \mathscr{D}[\Omega_b].$$

Among involutive semigroups of particular importance for our purposes are the so-called Baer *-semigroups (Maeda and Maeda, 1970; Foulis, 1960) because of their relation with orthomodular lattices. A Baer *-semigroup is an involutive semigroup T with a zero element, equipped with a mapping $x \mapsto x'$ of T into the set $P(T)$ of its projections, such that

$$\{y \in T: x \circ y = 0\} = x'T, \qquad \forall x \in T.$$

Thus we ask which additional condition T_Ω must fulfill in order to become a Baer *-semigroup. Let us assume the following

ASSUMPTION 3.4. For every $x \in T_\Omega$ there exists $a_x \in \mathscr{L}$ such that

$$\mathscr{D}[x] = \mathscr{D}[\Omega_{a_x}].$$

The unicity of a_x follows from (13).

By means of Assumption 3.4, T_Ω can be equipped with the structure of a Baer *-semigroup, in the following way. Define a map of T_Ω into $P(T_\Omega)$ by

$$x \mapsto x' = \Omega_{a_x^{\perp}}, \qquad x \in T_\Omega;$$

LEMMA 3.2. *T_Ω, equipped with the mapping defined above, is a Baer *-semigroup.*
Proof. Let us first suppose that $y \in T_\Omega$ is such that $x \circ y = \Omega_{O'}$, this means

$$\mathscr{R}[y] \subseteq \mathscr{C}\mathscr{D}[x] = \mathscr{C}\mathscr{D}[\Omega_{a_x}] = \mathscr{R}[\Omega_{a_x^{\perp}}],$$

hence $\Omega_{a_x^{\perp}} \circ y = y$, so that

$$\{y \in T_\Omega: x \circ y = \Omega_O\} \subseteq x' \circ T_\Omega.$$

Conversely, suppose that $y \in x' \circ T_\Omega$ then

$$\mathscr{R}[y] \subseteq \mathscr{R}[\Omega_{a_x^{\perp}}] = \mathscr{C}\mathscr{D}[\Omega_{a_x}],$$

whence $x \circ y = \Omega_0$, so that

$$\{y \in T_\Omega : x \circ y = \Omega_0\} \supseteq x' \circ T_\Omega. \qquad \square$$

Let us remark that in case $x = \Omega_a$, for some a, $x' = \Omega_a$. In a Baer *-semigroup it is known (Maeda and Maeda, 1970; Foulis, 1960) that a projection e has the property $e'' = e$ (and it is called closed) if and only if $e = x'$ for some element x of the semigroup, then in our case the mapping $x \mapsto \Omega_{a_x^\perp}$ ensures that closed projection have the form Ω_a for some $a \in \mathscr{L}$. On the other hand, being $\Omega'_a = \Omega_{a^\perp}$, the involutive property of the mapping $\perp$ in $\mathscr{L}$ ensures us that $\Omega''_a = \Omega_a$ for every $a \in \mathscr{L}$. Hence the set $P'(T_\Omega)$ of all closed projections of T_Ω coincides with $\{\Omega_a : a \in \mathscr{L}\}$

$$P'(T_\Omega) = \{\Omega_a : a \in \mathscr{L}\}.$$

The set of the closed projections in a Baer *-semigroup is known to possess a number of relevant properties (Maeda and Maeda, 1970; Foulis, 1960). In the sequel we shall examine how these properties can be translated into $\mathscr{L}$ by means of the one-to-one correspondence between $P'(T_\Omega)$ and $\mathscr{L}$ given by $a \mapsto \Omega_a$.

(1) $P'(T_\Omega)$ is a complete lattice, i.e. the join and meet of elements of any subset of $P'(T_\Omega)$ exist in $P'(T_\Omega)$. Then $\mathscr{L}$ is also a complete lattice.

(2) The restriction to $P'(T_\Omega)$ of the mapping $x \mapsto x'$ is an orthocomplementation in $P'(T_\Omega)$. Due to the fact that $\Omega'_a = \Omega_{a^\perp}$ it is recognized that the map $a \mapsto a^\perp$ is an orthocomplementation in $\mathscr{L}$ and that the orthocomplement of a in $\mathscr{L}$ defined by the one-to-one correspondence between $\mathscr{L}$ and $P'(T_\Omega)$), is just $a^\perp$.

(3) $P'(T_\Omega)$ is an orthomodular poset, the property inherited by $\mathscr{L}$ reads

$$a \leq b \Rightarrow b = a \vee (b \wedge a^\perp), \; a, b \in \mathscr{L}.$$

(4) The meet in $P'(T_\Omega)$ is explicitly given in terms of the operation of the semigroup by

$$(14) \qquad \Omega_a \wedge \Omega_b = \Omega_a \circ (\Omega'_b \circ \Omega_a)'.$$

This formula cannot be translated in $\mathscr{L}$ because in general $\Omega'_b \circ \Omega_a$ is not an element of $P'(T_\Omega)$; all we can say is that $a \wedge b$ in $\mathscr{L}$ is the element which is in one-to-one correspondence with the element $\Omega_a \circ (\Omega'_b \circ \Omega_a)'$ in $P'(T_\Omega)$.

(5) Two elements Ω_a and Ω_b of $P'(T_\Omega)$ commute as maps (that is $\Omega_a \circ \Omega_b = \Omega_b \circ \Omega_a$) if and only if a and b commute in the sense of orthomodular lattices. In this case the formula for the meet of two projections simplifies and we get

$$\Omega_a \wedge \Omega_b = \Omega_a \circ \Omega_b \quad \text{when} \quad (a, b)C \quad \text{(or, equivalently} \quad \Omega_a \circ \Omega_b = \Omega_b \circ \Omega_a).$$

Then Assumptions 3.1–3.4 allow us to conclude that the set $\mathscr{L}$ of propositions is a complete orthomodular lattice and that commutativity in $\mathscr{L}$ is equivalent to commutativity of ideal first-kind maps, a fact which has a precise physical meaning. Moreover the ordering relation in $\mathscr{L}$ is related to the properties of the set $\mathscr{S}$ of states on $\mathscr{L}$ so that

$$a \leq b \quad \text{when} \quad \mathscr{S}_1(a) \subseteq \mathscr{S}_1(b).$$

It is not possible, in the scheme outlined in this section, to identify the elements of $\mathscr{S}$ with probability measures on $\mathscr{L}$, that is, to the extent of our knowledge, it is not possible to show that the elements of $\mathscr{S}$ have the property

$$\alpha(\vee a_i) = \Sigma \alpha(a_i) \quad \text{when } a_i \text{ is a disjoint sequence of elements of } \mathscr{L}.$$

As a last remark we want to point out that with the definition and properties of $\mathscr{S}$ adopted in the present section it is possible to define the closure of a set of states and the set $\mathscr{M}$ of all closed subsets of $\mathscr{S}$ exactly in the same way of Section 2. To obtain the isomorphism between $\mathscr{L}$ and $\mathscr{M}$ we had to adopt Assumption 2.2; in the framework of this section this is not the case, in fact we have the following:

THEOREM 3.1. *For every* $S \in \mathscr{M}$, *there exists* $a \in \mathscr{L}$ *such that* $S = \mathscr{S}_1(a)$.
Proof. We first remark that one easily gets, by use of (14),

$$\mathscr{R}[\Omega_a \wedge \Omega_b] = \mathscr{R}[\Omega_a] \cap \mathscr{R}[\Omega_b], \qquad \forall a, b \in \mathscr{L}.$$

Given any $S \in \mathscr{M}$, we then consider the subset of $\mathscr{L}$

$$\mathscr{L}_S = \{a \in \mathscr{L} : \mathscr{R}[\Omega_a] \supseteq S\},$$

and we have

$$\mathscr{R}\left[\bigwedge_{b \in \mathscr{L}_S} \Omega_b\right] = \bigcap_{b \in \mathscr{L}_S} \mathscr{R}[\Omega_b] \supseteq S.$$

However, strict inclusion cannot hold, for this would imply the existence of some

$$\beta \in \bigcap_{b \in \mathscr{L}_S} \mathscr{R}[\Omega_b], \qquad \beta \notin S,$$

so that

$$\mathscr{L}_1(\beta) \supseteq \bigcap_{a \in S} \mathscr{L}_1(\alpha)$$

hence $\bar{S} \supset S$, contrary to the assumption $S \in \mathscr{M}$. $\qquad\qquad\square$

In this way the superposition principle follows from the assumptions of this Section.

Before closing this Section let us remark that, by use of the properties of the Ω_a's, one may prove, (Beltrametti and Cassinelli, 1976) the identity

$$\mathscr{S}_1(a^\perp) \cup \{\alpha \in \mathscr{D}[\Omega_a] : \Omega_a\alpha(b) = 1\} = \mathscr{S}_1(a^\perp \vee (a \wedge b)).$$

The left-hand side represents the set of states which either give with certainty the 'no' outcome of a, or are transformed by the measurement (with 'yes' outcome) of a into states which give with certainty the 'yes' outcome of b. We can regard these states as those which make true the sentence 'a implies b': here we are dealing with an implication often considered in (quantum) logic and sometimes called counter-factual (Stalnaker and Thomason, 1970; van Fraassen, 1973; Hardegree, 1974, 1975). The previous identity thus shows that the proposition $a^\perp \vee (a \wedge b)$ represents that implication: for orthomodular lattices it plays a role similar to the one played, for Boolean algebras, by the usual material implication $a^\perp \vee b$.

Of course we do not claim to solve, by this remark, the not obvious problem of determining the significant implication(s) holding in quantum logic: for an analysis of this problem within a general approach to non-distributive logics we refer to Dalla Chiara (1976).

4. In the previous section we have seen that, if Assumptions 3.2–3.4 on existence and properties of ideal first-kind maps hold, it is possible to show that $\mathscr{L}$ is a complete orthomudular lattice, that $\mathscr{S}$ is strongly ordering on $\mathscr{L}$, and that for the pair $(\mathscr{L}, \mathscr{S})$ the superposition principle holds. In this section we want to reverse the problem and show that there exist pairs $(\mathscr{L}, \mathscr{S})$ for which the existence of ideal first-kind maps can be proved. The $(\mathscr{L}, \mathscr{S})$ pairs we shall examine here are not the most general (from a mathematical point of view) one can consider (Cassinelli and Beltrametti, 1975) but allow one to avoid some mathematical technicalities. Our hypotheses will be the following.

ASSUMPTION 4.1. $\mathscr{L}$ and $\mathscr{S}$ are such that for them Assumptions 2.1–2.4 of Section 2 hold.

We recall some consequences of this assumption: $\mathscr{L}$ becomes complete atomic and there exists a bijection between the set $\mathscr{P}$ of pure states of $\mathscr{L}$ and the set $\mathscr{A}$ of atoms of $\mathscr{L}$. This bijection is the restriction to $\mathscr{A}$ of the bijection μ between $\mathscr{L}$ and $\mathscr{M}$ (we have denoted it by π); then, if $\alpha \in \mathscr{P}$, $\pi(\alpha) = \mu(\{\alpha\}) \in \mathscr{A}$. Moreover we have seen in Section 2 that the support of the pure state $\alpha \in \mathscr{P}$ is the atom $\pi(\alpha)$, and that $p \in \mathscr{A}$ is the support of the unique pure state $\pi^{-1}(p)$. As far as not pure states are concerned, we have seen that our hypotheses are sufficient to ensure us that every state has a support and every $a \in \mathscr{L}$, $a \neq O, I$, is the support of some state; the explicit form and properties of the mapping σ (support) have been exploited in Section 2. Then we adopt again

ASSUMPTION 4.2. $\mathscr{L}$ has the covering property.

For every $a \in \mathscr{L}$, $p \in \mathscr{A}$, let $\varphi_a(p) = (p \vee a^{\perp}) \wedge a$ denote the so-called Sasaki projection; it is known that $\varphi_a(p) \in \mathscr{A}$ and that $\varphi_a(p) = O$ if and only if $a \perp p$ (for these and other basic properties of Sasaki projections the reader is referred to Maeda and Maeda (1970). For every $a \in \mathscr{L}$ $(a \neq O, I)$, let us consider the mapping of $\mathscr{P}$ into $\mathscr{P}$ defined by $\pi \circ \varphi_a \circ \pi^{-1}$. As π is a bijection between $\mathscr{A}$ and $\mathscr{P}$ and φ_a maps $\mathscr{A}$ into $\mathscr{A}$, this map is well defined, and it is easy to see that it is the unique mapping, for every fixed a $(a \neq O, I)$, such that the diagram shown as Figure 4 is commutative. Let us denote $\pi \circ \varphi_a \circ \pi^{-1}$ by Ω_a. The following theorem is a justification for this notation.

THEOREM 4.1. (1) *The domain* $\mathscr{D}[\Omega_a]$ *of* Ω_a *above defined is given by*

$$\mathscr{D}[\Omega_a] = \{\alpha \in \mathscr{P} : \pi(\alpha) \not\perp a\}.$$

(2)   If $\alpha \in \mathscr{D}[\Omega_a]$ and $\alpha(a) = 1$ then $\Omega_a \alpha = \alpha$.

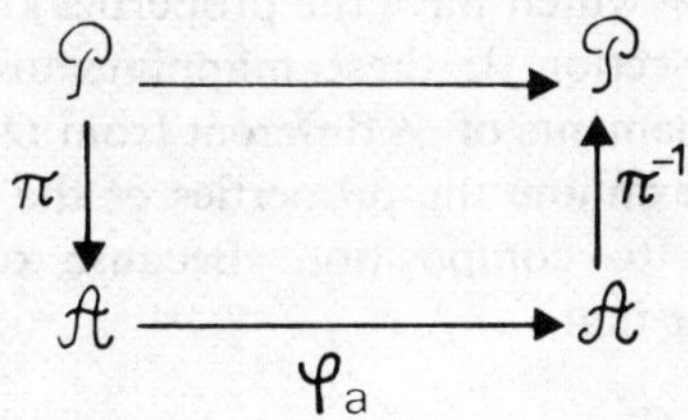

Fig. 4

(3)   If $\alpha \in \mathscr{D}[\Omega_a]$ then $\Omega_a \alpha(a) = 1$.

(4)   If $a, b \in \mathscr{L}$, $(a, b)C$ and $\alpha \in \mathscr{D}[\Omega_a]$ then $(b) = 1$ implies $\Omega_a \alpha(b) = 1$.

Proof. (1) Given $\alpha \in \mathscr{P}$, $\varphi_a(\pi(\alpha)) = O$ if and only if $\pi(\alpha) \perp a$, then $\varphi_a(\pi(\alpha)) \in \mathscr{A}$ if and only if $\pi(\alpha) \not\perp a$. Thus Ω_a is defined on the set $\{\alpha \in \mathscr{P} : \pi(\alpha) \not\perp a\}$.

(2) $\alpha(a) = 1$ implies $\pi(\alpha) \leq a$, hence (Maeda and Maeda 1970) $\pi(\alpha)$, a, $a^\perp$ form a distributive triple, so that

$$\varphi_a(\pi(\alpha)) = (\pi(\alpha) \vee a^\perp) \wedge a$$
$$= (\pi(\alpha) \wedge a) \vee (a^\perp \wedge a) = \pi(\alpha)$$

whence

$$\pi^{-1}(\varphi_a(\pi(\alpha))) = \Omega_a \alpha = \alpha.$$

(3) Owing to the commutativity of Figure 4 above, we have

$$\pi(\Omega_a \alpha) = \varphi_a(\pi(\alpha)), \qquad \forall \alpha \in \mathscr{D}[\Omega_a],$$

hence $\pi(\Omega_a \alpha) \leq a$, for φ_a projects $\mathscr{L}$ onto the sublattice $\mathscr{L}[O, a]$; this implies $\Omega_a \alpha(a) = 1$.

(4) $\alpha(b) = 1$ implies $\pi(\alpha) \leq b$ hence $\varphi_a(\pi(\alpha)) \leq \varphi_a(b)$ since φ_a is monotone. From $(a, b)C$ it follows that a, $a^\perp$, b form a distributive triple, hence

$$\varphi_a(b) = (b \vee a^\perp) \wedge a = (b \wedge a) \vee (a^\perp \wedge a) = b \wedge a.$$

The commutativity of the diagram which defines Ω_a ensures that $\pi(\Omega_a \alpha) = \varphi_a(\pi(\alpha))$. Therefore $\pi(\Omega_a \alpha) \leq \varphi_a(b) = b \wedge a \leq b$, hence $\Omega_a \alpha(b) = 1$. $\square$

This theorem (which can be proved in a more general context (Cassinelli and Beltrametti, 1975)) ensures us that in the pair $(\mathscr{L}, \mathscr{S})$ for which the hypotheses of this Section hold, we can prove the existence of a family of mappings of $\mathscr{P}$ into $\mathscr{P}$ which have the properties of ideal first kind mappings, as stated in Section 3; these mappings are in one-to-one correspondence with the elements of $\mathscr{L}$ different from O and I.

Now we want to examine the properties of the mappings Ω_a's, defined above, with respect to composition. Because of the fact that $\Omega_a = \pi \circ \varphi_a \circ \pi^{-1}$, it is clear that

$$\Omega_{a_1} \circ \cdots \circ \Omega_{a_n} = \pi \circ \varphi_{a_1} \circ \varphi_{a_2} \circ \cdots \circ \varphi_{a_n} \circ \pi^{-1}.$$

THEOREM 4.2. *If there are two sets of elements of* $\mathscr{L}$: $a_1, ..., a_n$ *and*

$b_1, \cdots, b_m$ such that $\Omega_{a_1} \circ \cdots \circ \Omega_{a_n} = \Omega_{b_1} \circ \cdots \circ \Omega_{b_m}$, then we have also $\Omega_{b_m} \circ \cdots \circ \Omega_{b_1} = \Omega_{a_n} \circ \cdots \circ \Omega_{a_1}$.

Proof. We have

$$\Omega_{a_1} \circ \cdots \circ \Omega_{a_n} = \pi \circ \varphi_{a_1} \circ \cdots \circ \varphi_{a_n} \circ \pi^{-1}$$

and

$$\Omega_{b_1} \circ \cdots \circ \Omega_{b_m} = \pi \circ \varphi_{b_1} \circ \cdots \circ \varphi_{b_m} \circ \pi^{-1},$$

hence

$$\varphi_{a_1} \circ \cdots \circ \varphi_{a_n} = \varphi_{b_1} \circ \cdots \circ \varphi_{b_m}.$$

Because φ_{a_i} and φ_{b_i} are projections in the Baer *-semigroup of residuated mappings (or emimorphisms) of $\mathscr{L}$ (Foulis, 1960), we have

$$(\varphi_{a_1} \circ \cdots \circ \varphi_{a_n})^* = (\varphi_{b_1} \circ \cdots \circ \varphi_{b_m})^*$$

so that

$$\varphi_{a_n} \circ \cdots \circ \varphi_{a_1} = \varphi_{b_m} \circ \cdots \circ \varphi_{b_1},$$

and also

$$\Omega_{a_n} \circ \cdots \circ \Omega_{a_1} = \Omega_{b_m} \circ \cdots \circ \Omega_{b_1}. \qquad \square$$

In this way it is shown that the Assumption 3.3 holds for the composition of the Ω_a's defined in this section. The last point to show is the Assumption 3.4; we have:

THEOREM 4.3. *If $\Omega_{a_1} \circ \ldots \circ \Omega_{a_n}$ is such that its domain is not empty, then there exists a unique element $a \in \mathscr{L}$, $a \neq O$, such that*

$$\mathscr{D}[\Omega_{a_1} \circ \cdots \circ \Omega_{a_n}] = \{\alpha \in \mathscr{P} : \alpha(a) \neq 0\} = \mathscr{D}[\Omega_a].$$

Proof. $\alpha \in \mathscr{D}[\Omega_{a_1} \circ \cdots \circ \Omega_{a_n}]$ if and only if $(\varphi_{a_1} \circ \cdots \circ \varphi_{a_n})(\pi(\alpha)) \neq O$, then $\mathscr{D}[\Omega_{a_1} \circ \cdots \circ \Omega_{a_n}] = \{\alpha \in \mathscr{P} : (\varphi_{a_1} \circ \cdots \circ \varphi_{a_n})(\pi(\alpha)) \neq O\}$. The composition $\varphi_{a_1} \circ \cdots \circ \varphi_{a_n}$ of Sasaki projections is an element of the involutive semigroup of residuated mappings (or emimorphisms) of $\mathscr{L}$ (Foulis, 1960); if φ is a residuated mapping of $\mathscr{L}$ we have $\varphi(a) = O$ if and only if $a \perp \varphi^*(I)$ (Foulis, 1960), then

$$(\varphi_{a_1} \circ \cdots \circ \varphi_{a_n})(\pi(\alpha)) = O \text{ if and only if}$$
$$\pi(\alpha) \perp (\varphi_{a_1} \circ \cdots \circ \varphi_{a_n}) * (I).$$

By the definition of support this means

$$(\varphi_{a_1} \circ \cdots \circ \varphi_{a_n})(\pi(\alpha)) \neq O \text{ if and only if}$$
$$\alpha((\varphi_{a_1} \circ \cdots \circ \varphi_{a_n}) * (I)) \neq 0.$$

Then if we define $a = (\varphi_{a_1} \circ \cdots \circ \varphi_{a_n}) * (I)$ we have

$$\mathscr{D}[\Omega_{a_1} \circ \cdots \circ \Omega_{a_n}] = \{\alpha \in \mathscr{P} : \alpha((\varphi_{a_1} \circ \cdots \circ \varphi_{a_n}) * (I)) \neq 0\}.$$

Then if we define $a = (\varphi_{a_1} \circ \cdots \circ \varphi_{a_n}) * (I)$, we have

$$\mathscr{D}[\Omega_{a_1} \circ \cdots \circ \Omega_{a_n}] = \{\alpha \in \mathscr{P} : \alpha((\varphi_{a_1} \circ \ldots \circ \varphi_{a_n}) * (I)) \neq 0\} = \mathscr{D}[\Omega_a].$$

The unicity of a and the fact that $a \neq O$ are obvious.   □

Thus we have seen that, adopting Assumptions 4.1 and 4.2, it is possible to prove the existence of a family of maps of $\mathscr{P}$ into $\mathscr{P}$ which has all the properties stated in Section 3 for ideal first-kind mappings.

NOTE

[1] In an orthomodular lattice we say that a and b commute, and write $(a, b)C$, when $a = (a \wedge b) \vee (a \wedge b^{\perp})$. The center of the lattice is the set of the elements which commute with all the elements. The lattice is irreducible if its center contains the zero and the unit elements alone.

REFERENCES

Beltrametti, E.G. and Cassinelli, G. (1973), *Z. Naturforsch.* **28a**, 1516.

Beltrametti, E.G., and Cassinelli, G. (1976), *La Rivista del Nuovo Cimento* **6**, 321.

Birkhoff, G. (1967), *Lattice Theory*. Third edition. Amer. Math. Soc. Colloq. Publ., New York, 1967.

Cassinelli, G., and Beltrametti, E.G. (1975), *Commun. Math. Phys.* **40**, 7.

Dalla Chiara, M.L. (1976), 'A General approach to non-distributive logics', *Studia Logica* **30**, 23–36.

Foulis, D.J. (1960), *Proc. Amer. Math. Soc.* **11**, 648.

van Fraassen, B.C. (1973), in *Contemporary Research in the Foundations and Philosophy of Quantum Theory*. Dordrecht, Reidel.

Gudder, S.P. (1970), *J. Math. Phys.* **11**, 1037.

Hardegree, G.M. (1974), *Synthese* **29**, 63.

Hardegree, G.M. (1975), *J. Phil. Logic* **4**, 399–421.

Mackey, G.W. (1963), *The Mathematical Foundations of Quantum Mechanics*. New York, W.A. Benjamin.

Maeda, F., and Maeda, S. (1970), *Theory of Symmetric Lattices*. Berlin, Springer.

Piron, C. (1976), *Foundations of Quantum Physics*. Reading, W.A. Benjamin.

Pool, J.C.T. (1968), *Commun. Math. Phys.* **9**, 118.
Stalnaker, R., and Thomason, R.H.: *Theoria* **36**, 23.
Varadarajan, V.S.: (1968), *Geometry of Quantum Theory*. Vol. I. Princeton, Van
  Nostrand.

QUANTUM LOGIC AND THE TWO-SLIT EXPERIMENT

1. As is well known, the two-slit experiment is related, within Quantum Mechanics (Q.M.), to the dualistic character of radiation (for instance light), whose behavior is correctly represented as a wave-like phenomenon during its passage from the source, through the slits, up to the screen; and by a particle model (for instance, photons) when it impinges on the screen causing the extraction of electrons.

The two-slit experiment is, indeed, a diffraction experiment as schematically depicted in Figure 1. The source is in O, D is a screen with two parallel slits and S is a photosensitive screen where the diffraction pattern is observed. A similar experiment can be run using *material* particles (for instance, electrons) instead of radiation. In this case a crystal replaces the slits and the diffraction pattern is observed in a Wilson-chamber by means of the distribution of particle-traces. Reverting now, for the sake of convenience, to the case of optical radiation, one finds oneself, as is well known, in very peculiar difficulties, whenever one tries to represent the phenomenon in its entirety only in particle terms (photons). For instance, if one closes slit F_2 (see Figure 2) a characteristic diffraction pattern will be observed around point P_1.

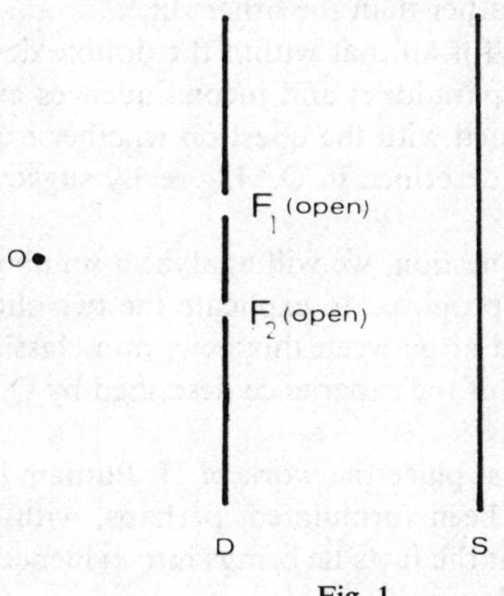

Fig. 1

Maria Luisa Dalla Chiara (ed.), Italian Studies in the Philosophy of Science, 237–247.
Copyright © 1980 D. Reidel Publishing Company.

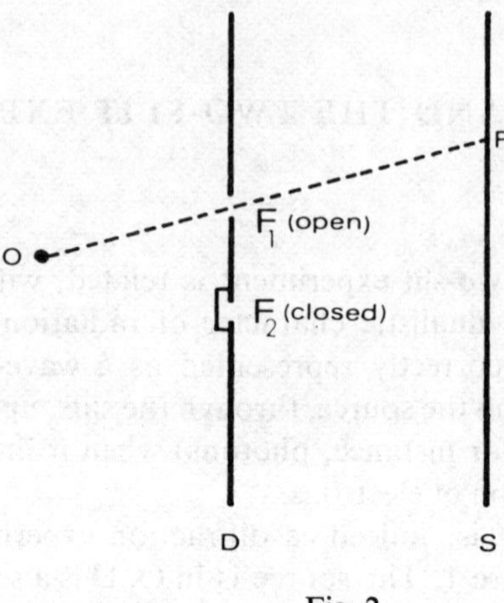

Fig. 2

However, if both slits, F_1 and F_2, are open, the intensity of the diffraction pattern, related, as is well known, to the probability that a photon impinges on the screen, will not be the simple superposition of the diffraction patterns singly obtained through F_1 or F_2 alone.

So, within the 'particle only' outlook, an argument of the following general type has often been presented. 'Since a photon *must* have gone through *either* F_1 *or* F_2, one finds oneself in the awkward situation of suspecting that the photon *knows* whether one of the slits is closed or even *prefers* to go through one rather than the other slit.'

While stressing once and for all that within the double description, i.e. wave-particle, outlook, all paradoxes and inconsequences evaporate, the present note will be concerned with the question whether experiments on physical microsystems – as described in Q.M. – really suggest a different, non-classical, logic.

In order to answer this question, we will analyze a number of different positions which have been proposed to explicate the two-slit experiment, and relations will be hinted at between this new, non-classical logic and the essential indeterminism of the experience described by Q.M..

2. Let us consider in the first place the work of H. Putnam [1]. Although his arguments might have been formulated, perhaps, with a trifle more clarity, I have no doubt that the facts he brings into evidence may be substantially formulated as follows.

Let A_1 be the statement 'the photon goes through slit n.1' and A_2 the statement 'the photon goes through slit n.2'. Let us further denote with P (A_i, R) the probability that the photon impinges on region R under assumption A_i. "These probabilities", Putnam asserts, "may be computed from quantum mechanics or classical mechanics (they are the same, as long as only one slit is open), and checked experimentally by closing one slit, and leaving only A_1 (respectively A_2) open in the case of P (A_I, R) (respectively, $P(A_2, R)$)" [1, pp. 222–223]. The problem arises in case both slits are open. Indeed, in such a case one has:

(i) P_Q is the probability that the photon impinges on R as predicted by Q.M.

(ii) P_Q accords with the experiment.

(iii) P_C is the probability that the photon impinges on R, as predicted by Classical Mechanics (C.M.).

(iv) P_C is arrived at following classical logic.

(v) $P_C \neq P_Q$.

Now items (i), (ii) and (v) together assert that photons do not behave classically (*and this is corroborated by the experimental evidence*), while asserting only (iii) and (iv) *in isolation* is equivalent to asserting the correctness of using classical logic in dealing with the behavior of photons.

Let A now be the statement 'photons behave as *classical particles*' and B the statement 'it is correct to use classical logic in dealing with the behavior of photons'.

It is then possible to state Putnam's thesis (considering item (v)) as:

$$\sim(\sim A \wedge B)$$

that is to state that there is a *de facto* incompatibility in asserting both $\sim A$ and B. However, it will be necessary to discuss in more detail items (iii) and (iv) of my rendition of Putnam's arguments.

The probability, P_C, that the photon (with both slits open) impinges on R, as predicted by C.M., could be written as

$$\tfrac{1}{2} P(A_1, R) + \tfrac{1}{2} P(A_2, R)$$

as a function of $P(A_1, R)$ and $P(A_2, R)$ previously defined.

In fact Putnam argues that, according to the classical viewpoint, in the first place the probability that the photon goes through slit 1 is equal to the probability that it goes through slit 2 and furthermore the disjunction $A_1 \vee A_2$ is *de facto* true because one is taking into account only experiments in which the photon *does* go through the first screen. Therefore one has:

$$"P(A_1 \vee A_2, R) = P((A_1 \vee A_2) \wedge R)/P(A_1 \vee A_2)$$
$$= P(A_1 \wedge R \vee A_2 \wedge R)/P(A_1 \vee A_2)$$
$$= P(A_1 \wedge R)/P(A_1 \vee A_2) + P(A_2 \wedge R)/P(A_1 \vee A_2)$$

(here $P((A_1 \vee A_2) \wedge R)$ means the probability of $(A_1 \vee A_2) \wedge R$, and similarly for $P(A_1 \vee A_2)$, etc.). Since $P(A_1) = P(A_2)$, we have $P(A_1 \vee A_2)$ $= 2P(A_1) = 2P(A_2)$. Thus $P(A_1 \wedge R)/P(A_1 \vee A_2) = P(A_1 \wedge R)/2P(A_1)$ $= \frac{1}{2} P(A_1, R)$ and similarly $P(A_2 \wedge R)/P(A_1 \vee A_2) = P(A_2 \wedge R)/2P(A_2)$ $= \frac{1}{2} P(A_2, R)$. Substituting these expressions in the above equation yields:

$$P(A_1 \vee A_2, R) = \frac{1}{2} P(A_1, R) + \frac{1}{2} P(A_2, R)$$

Now a crucial step in this derivation was the expansion of $(A_1 \vee A_2) \wedge R$ into $A_1 \wedge R \vee A_2 \wedge R$. This expansion is *fallacious* in quantum logic; thus the above derivation also fails" [1, p. 223].

Putnam thus concludes by saying that whoever believes in the validity of Quantum Logic (Q.L.) (whose logical connectives corresponding to the classical *et* and *vel* do not satisfy a distributive law) "would see no reason to predict $P(A_1 \vee A_2, R) = \frac{1}{2} P(A_1, R) + \frac{1}{2} P(A_2, R)$" [1, p. 223].

3. Heelan [2] radically criticizes Putnam's viewpoint of the two-slit experiment as well as the alleged necessity of introducing a new logic, different from the classic one, and suggested by the kind of experience described by Q.M. Heelan's arguments about the two-slit experiment are as follows.

Let A_1 be the statement 'the photons go through slit 1', A_2 the statement 'the photons go through slit 2' and D the statement 'the distribution of photons on the screen is the arithmetic sum of the distributions which obtain when a single slit is open'.

Since D is factually false, let us consider the argument:

> Major: $(A_1 \vee A_2) \supset D$
> Minor: $\sim D$
> Conclusion: $\sim(A_1 \vee A_2)$

Here, the inference scheme used is the *modus tollens* and as no supporter of Q.L. considers it invalid, Heelan feels justified in using it.

On the other hand the conclusion cannot be accepted and it remains for us to find out how to avoid it. To this end the major or the minor can be negated. But the minor is a statement of experimental fact and therefore the only remaining possibility is that of declaring the major false. Now, the major is formed of the three statements A_1, A_2 and D which are formulated in the language of classical physics which in turn presupposes – as Heelan implies – the validity of its conceptual framework.

These statements are bound with the logical connectives '$\vee$' and '$\supset$', which connectives, I wish to stress for the sake of the observations at the end of this paper, are those of classical logic. Thus Heelan asserts "the most natural solution which suggests itself is that the language framework and concepts of classical physics do not apply to electrons in the two-slit experiment. (Heelan is considering electrons instead of photons, but this is immaterial.) Underlying the belief that $(A_1 \vee A_2) \supset D$ is the set of classical assumptions that each particle is an independent kinematical unit, that each travels along a uniquely determined trajectory from the source to the screen, that each particle consequently passes through one and only one slit. In this model, the set of particles arriving at any area of the screen would be the set-theoretic union of two sets of particles, of those that pass through slit 1 and those that pass through slit 2. In rejecting the validity of classical descriptive assumptions for the 2-slit experiment, one can falsify the major without impugning the validity of classical logic" [2, p. 321].

4. Gardner's critique [3] of Putnam's interpretation of the two-slit experiment proceeds in the name of a general principle, maintained for instance by Quine, which states that in the face of a choice between different alternative revisions of a theory in order to take unexpected observations into account, the similarity of a new theory with the old one is strong grounds for preferring it. He then goes on to argue that reasonable options can be found within Q.M. which would allow for no substitution for classical logic.

Let us take a closer look at Gardner's argument. According to him, the statement, that the intensity in R (with both slits open) is $\frac{1}{2} P(A_1, R) + \frac{1}{2} P(A_2, R)$, follows, by using the usual logic and probability theory, from the assumption that every photon has a definite trajectory which goes through one or the other slit. Gardner then maintains that Putnam's argument goes as follows. Since the statement that the intensity in R is $\frac{1}{2} P(A_1, R) + \frac{1}{2} P(A_2, R)$ is false, a revision of the logic used to derive it is necessary. But "the two-slit experiment itself determines the total probability $P(A_1 \vee A_2, R)$ and not the separate probabilities $P(A_i, R)$, $i = 1, 2$. Apparently, the only way we can determine the results of passage through one slit is by closing the other. But how do we know $P(A_1, R)$ is the same whether or not slit 2 is open? Do not the experimental results constitute a refutation of this very assumption?" [3, p. 526].

Hence, "though initially puzzling, this effect is explainable, in the sense that it follows (by standard logic) from the axioms of quantum theory together with certain initial conditions" [3, p. 526].

5. If my interpretation of Putnam's arguments about the two-slit experiment is correct, it follows that what he really proves is

$$\sim(\sim A \wedge B)$$

where, let us recall it again, A is the statement 'photons behave as classical particles' and B is the statement 'it is correct to use classical logic in dealing with behavior of photons'. It can then be immediately inferred that Q.M. requires a logic different from classical logic. Indeed, it only needs to observe that

$$\sim(\sim A \wedge B) \supset (\sim A \supset \sim B)$$

is a tautology whose antecedent is considered valid on empirical grounds.

Therefore the consequent can also be considered valid with at least the same degree of credibility, and its meaning is that when one deals with physical objects which do not behave classically, but with physical objects with a quantum behavior, then the logic suggested by experience is not classical logic. One could object that the kind of inference here employed still belongs to classical logic [2], but this is perfectly legitimate for one who assumes a viewpoint favorable to a plurality of logics, each suggested by a different kind of experience [4].

Reconsidering now Heelan's critiques of Putnam's position, we may notice that his central observation consists in surmising that Putnam incorrectly uses

$$A_1 \vee A_2 \supset D \tag{1}$$

because he is dealing with quantum particles Here of course A_1, A_2 and D are as defined in Section 3 while '$\vee$' and '$\supset$' are used classically.

Now of course Heelan is perfectly right in asserting that (1) does not apply for non-classical particles! But using (1) to describe the behavior of non-classical particles is a mistake at a *logical level* because it is just the connectives '$\vee$' and '$\supset$' of classical logic which fail in connecting elementary statements describing quantum behaviors.

Indeed, it is not the case that photons are dealt with as classical particles (which Putnam implicitly denies when he states that Q.M. predicts the correct probability), but rather that *if* one wants to use classical logic, *then* photons are dealt with as classical particles. There is nothing unlawful, I think, in asserting either A_1 or A_2 *separately*, both classically and quantum mechanically. This is so because neither proposition necessarily implies, in isolation, that each photon be an independent kinematic entity traveling along a path uniquely determined.

It is the statement $A_1 \vee A_2 \supset D$ which raises so many problems just because it is built up with logical connectives modeled upon a kind of experience essentially different from that adequately described by Q.M. [4]. It is this statement which, just because of the classical interpretation of connectives, is characteristic of the language of classical physics.

Considering Gardner's critiques now, we have to recall that in Putnam's proof of

$$P(A_1 \vee A_2, R) = \tfrac{1}{2} P(A_1, R) + \tfrac{1}{2} P(A_2, R),$$

$P(A_1, R)$ is the probability in case slit 2 is closed, and $P(A_2, R)$ in case slit 1 is closed.

But $P(A_1 \vee A_2, R)$ denotes the probability that the photon goes through either slit *when both are open* (i.e. the probability that, in this case, the photon goes through the first screen and impinges on R) and nothing guarantees that, letting, $P_\vee(A_i, R)$ be the probability of a photon going through slit i when both are open, the equality

$$P_\vee (A_i, R) = P(A_i, R) \tag{2}$$

obtains.

Now I would like to observe that, although it is a fact that *if* we consider classical particles (and therefore trajectories uniquely determined), *then* $P_\vee(A_i, R) = P(A_i, R)$ holds, it does not follow from $P_\vee(A_i, R) = P(A_i, R)$ that the particles are classically behaved.

It remains therefore to establish whether assuming (2) as valid implies a contradiction within the whole of Q.M.. In any case if (2) does not hold, then Gardner's objections are well founded. In this case items (iii) and (iv) of Putnam's argument as summarized in Section 2 might become inessential and the following argument could be proposed instead. All phenomena involved *could* be adequately described if the $P_\vee(A_i, R)$ could be substituted for the $P(A_i, R)$ in the expansion of $P(A_1 \vee A_2, R)$. The use of the $P_\vee(A_i, R)$ would then be implicitly equivalent to assert the non classical nature of the particles involved and the logic used *could* still be classical logic. In this case $P(A_1 \vee A_2, R)$ would not be the P_C of Section 2, but should (or could?) be that experimentally measured P_Q whose expansion in terms of the $P_\vee(A_i, R)$ would only need *the sole use of classical logic.*

But we are then in the presence, I think, of a devastating problem. If $P_\vee(A_i, R) \neq P(A_i, R)$, $i = 1, 2$, *is it then possible to perform a measurement* of $P_\vee(A_i, R)$, $i = 1, 2$? If this is impossible (and I do not see how it could be done), then it will always be possible to express $P(A_1 \vee A_2, R)$ as a formula composed of two sub-formulas themselves without logical connectives, one referring to slit 1, the other to slit 2, bound together with some classical logical connective, but from the viewpoint of the empirical assessment of its truth value it would be considered only as an elementary formula because the two statements

    (a) the value of $P_\vee(A_1, R)$ is $\bar{x}$ with $0 \leqslant \bar{x} \leqslant 1$,
    (b) the value of $P_\vee(A_2, R)$ is $\bar{y}$ with $0 \leqslant \bar{y} \leqslant 1$,

in general will be neither verifiable nor falsifiable. In any event it does not seem to contradict the principles of Q.M. to posit $P_\vee(A_i, R) = P(A_i, R)$, $i = 1, 2$.

Now, if this assumption of equality fails, then the two-slit experiment does not seem to be an argument either for or against the thesis that logic is empirical (even if it is not beyond doubt that Q.L. holds at the level of *physical qualities* [2, 4, 5]) while, in case the assumption holds, then the thesis will be strongly corroborated.

Presently a question arises that to accept the thesis that logic is empirical is to accept that different kinds of experience may suggest or force different logics upon us! [4]. In my opinion, the problem of discriminating between "suggesting' and 'forcing' is ill-posed; the true problem is that of the possibility of interpreting one logic within another (consider, for instance, the interpretation by Gödel and Glivenko of classical logic within intuitionism). A positive answer to this problem is, I think, a matter of fact: it may be and then it may not, even if it seems to attract a high degree of credibility (or perhaps hope). Still an important question remains. Dalla Chiara [6] puts it in the following form: "Is it admissible that the same intelligent being *will use* different logics in different situations, while maintaining a rational, consistent behavior?"

Well, in the first place I must admit to a substantial agreement with

Dalla Chiara's answer to the above question, i.e.: "the answer would be obviously negative if one insisted that a rational being could be identified in a single formal system. However, many arguments tend to assimilate the rational activity of an intelligent being to a multiple system of partial formal systems. . . . The reference to a multiple system of formal systems with *no last element* seems unavoidable. And, for the previous arguments it is possible that the elements of this multiple system use different logics without necessarily generating situations of incompatibility or of loss of communication." [6, p. 349–350]

However, I would wish to underline the fact that, as formulated, the above question may appear somewhat ambiguous, because the phrases appearing in it such as 'intelligent being' and 'rational consistent behavior', if used as unanalyzed terms with respect to the question of the plurality of logics, and with a strong appeal to common sense, are themselves rather vague terms.

Perhaps, without entering into a most complex argument, and referring explicitly only to the problem at hand, one could maintain that a *necessary* condition for attributing the character of intelligent beings to a set of subjects is that they be able to *build* and *use* at least one logic.

The behavior of one of these beings could be then defined as 'rational consistent behavior' (consistent within *that* logic) on the basis of the use that subject makes of such logic in the sum total of his behavior.

How then is it possible to define the rational consistent behavior of a member (or of a sub-society) of the set of subjects if the logics built and used (according to the different empirical situations) are more than one? Probably the correct answer still lies in the ability of the subject to interpret its own logics within each other.

I would like to hint at a last question of fundamental importance, i.e. the relationship between Q.L. and the indeterminism of Q.M. From many quarters it has been maintained that adopting a non-distributive logical structure would be a sufficient measure for avoiding all the so-called *anomalies* of Q.M. Finkelstein, for instance, asserts that "all the anomalies of quantum mechanics, all things that make it so hard to understand complementarity, interference, etc., are instances of non-distributivity" [7, p. 208]. Putnam himself says: "The only laws of classical logic that are given up in quantum logic are distributive laws, . . . and every single anomaly vanishes once we give these up" [1, p. 226]. Now, if one accepts this point of view, I think that Haack is right when she says: "Neither Finkelstein nor Putnam offers any general proof of this claim. . . .

These arguments are unfortunately inconclusive, since though it is true that the distributive laws are used in Putnam's derivation of the paradoxes, it doesn't follow that the paradoxes cannot be derived without them" [8, p. 162].

The above thesis of the removal of *supposed* paradoxes has sometimes been carried to extremes (although this is not the case for Putnam) by positing a sort of opposition between Q.L. and indeterminism. Audi, for instance, who is definitely opposed to any Q.L., asserts in his interesting book [9], that the main motivation behind the attempts to replace classical logic with some quantum logic is rooted in a desire of explain away the so-called causal anomalies of Q.M. and he sustains this assertion by quoting Reichenbach's argument [10] concerning his own attempt to build a three-valued logic for Q.M. Reichenbach (whose thesis on three-valued logic will not be considered here; for a critical discussion see the well-known book by Born [11] and van Fraassen [12], Putnam [13] and Zinov'ev [14]) indeed seems to support the more radical point of view. However, once ascertained that in Q.M. it is the case that indeterminism rather than a-causality obtains [11], I think that the problem of the opposition between indeterminism and Q.L. is in fact ill-posed [4].

It is not that Q.L. solves in a deterministic sense the *so-called* causal anomalies of Q.M. (since "determinism, far from being logically necessary or empirically established even in the classical domain, is actually very difficult to maintain in microphysics" [15, p. 23]) but rather Q.L. appears to be the logical structure suggested by a realm of experience as clearly and fundamentally indeterministic as that adequately described by Q.M..

REFERENCES

[1] Putnam, H., 'Is Logic Empirical?', *Boston Studies in the Philosophy of Science*, vol. V. Dordrecht and Boston, Reidel, 1969, pp. 216–241.

[2] Heelan, P., 'Quantum and Classical Logic: Their Respective Roles', *Boston Studies in the Philosophy of Science*, vol. XIII. Dordrecht and Boston, Reidel, 1974, pp. 318–349.

[3] Gardner, M., 'Is Quantum Logic Really Logic?', *Philosophy of Science* **38**, 508–529 (1971).

[4] Guccione, S., 'Some Remarks about Semantics of Quantum Logic'. Unpublished paper.

[5] Birkhoff, G., and J. von Neumann, 'The Logic of Quantum Mechanics', *Annals of Mathematics* **37**, 823–843 (1936).

[6] Dalla Chiara, M.L., 'Il problema della pluralità delle logiche', *Atti del XXIV Congresso Nazionale di Filosofia*, L'Aquila, 1974. Rome, Società filosofica italiana, 1974. vol. II, part 1, pp. 345–351.

[7] Finkelstein, D., 'Matter, Space and Logic', *Boston Studies in the Philosophy of Science*, vol. V. Dordrecht and Boston, Reidel, 1969. pp. 199–215.

[8] Haack, S., *Deviant Logic*. Cambridge, Cambridge University Press, 1974.

[9] Audi, M., *The Interpretation of Quantum Mechanics*. Chicago, University of Chicago Press, 1973.

[10] Reichenbach, H., *Philosophic Foundations of Quantum Mechanics*. Berkeley, University of California Press, 1944.

[11] Born, Max, *Natural Philosophy of Cause and Change*. Oxford, Clarendon Press, 1949.

[12] van Fraassen, B.C., 'The Labyrinth of Quantum Logics', *Boston Studies in the Philosophy of Science*, vol. XIII. Dordrecht and Boston, Reidel, 1974, pp. 224–254.

[13] Putnam, H., 'Three-Valued Logic', *Phil. Studies* **8**, 73–80 (1957).

[14] Zinov'ev, A.A., *Philosophical Problems of Many-Valued Logic*. Dordrecht and Boston, Reidel, 1963.

[15] Jauch, J.M. 'Determinism in Classical and Quantal Physics', *Dialectica* **27**, 13–26 (1973).

P. CALDIROLA AND E. RECAMI

CAUSALITY AND TACHYONS IN RELATIVITY*

CONTENTS

249

ABSTRACT. In the first part of this paper we consider standard (Special) Relativity and show that a 'Third Postulate' (the Reinterpretation Principle) is necessary to avoid information transmissions into the past. Such a Third Postulate allows—at the same time—to predict the existence of antiparticles within a purely relativistic context.

In the second part of the paper, we take due account also of tachyons. The theory of tachyons already has a long story, but the causality problems connected with it seem to have not yet been generally understood. We therefore perform a thorough analysis of the whole problem. We show in what sense the previous Third Postulate is enough to enforce the law of (retarded) causality even when in presence of tachyons. A careful, physical analysis is made of the definitions of causal connections, laws, descriptions, sources and detectors, etc. The kinematical problems related to usual macro-objects interacting via (micro and macro) tachyons are studied.

As examples of the clarifying power of 'extended relativity' even with regard to usual physics, the topics of advanced solutions, of the CPT theorem, of crossing relations, and so on, are investigated.

Other problems (e.g. about virtual particles, tachyon localization, vacuum decays, etc.) are briefly investigated, which are relevant to the completeness and self-contained-ness of the present analysis of causality and of tachyons.

PART I

1. FOREWORD AND HISTORICAL REMARKS

We shall consider the theory of Special Relativity [1] as the typical framework for considering the problem of *causality* in modern physics. We are going to claim that the 'principle of reinterpretation', or equivalent principles (see the following), has necessarily to be assumed as the Third Postulate of Special Relativity (SR) in order to *avoid* information transmission into the past. Incidentally, that 'principle' will allow us to understand the connection between matter and antimatter, since it allows the prediction of the existence of antiparticles. Furthermore, we shall show that the very Third Postulate itself (or rather *the same* set of postulates) is quite adequate for deriving a fully *causal* theory even in presence of faster-than-light objects (FTLO). The FTLO's have been given the name 'Tachyons' (T) in Reference 2, from the Greek word equivalent to 'swift'. We shall call 'Bradyons' (B) the usual, slower-than-light objects [3, 4], from the Greek word equivalent to 'slow'. And last, we shall call 'Luxons' ($\mathcal{L}$) the objects, like photons, travelling exactly at the speed of light [5].

The 'Principle of Reinterpretation' was first introduced by Stückelberg [6] and by Feynman [7], and then also used in the case of tachyons by Sudarshan and co-workers [5].

As regards tachyons, as far as we know, the first author mentioning FTLO's was Lucretius [8], as outlined by Corben [9] who quoted lines 161–165 in book 2 of *De Rerum Natura*. Instead, let us here explicitly quote *another* passage, i.e. lines 201–203 in book 4: [8]

Quone vides citius debere et longius ire
multiplexque loci spatium transcurrere eodem
tempore quo Solis pervolgant lumina caelum? [10]

After Lucretius, we do not know at present of any other progress until Thomson's [11], Heaviside's [12], Des Coudres' and particularly Sommerfeld's [12] works. In 1905, however, together with relativity [1], the conviction unfortunately spread that light-speed in vacuum was the upper limit of *any* speed, the early twentieth century physicists being misled by the evidence that normal particles cannot overcome that speed. They behaved like Sudarshan's imaginary demographer studying the population patterns of the Indian subcontinent [13]:

Suppose a demographer calmly asserts that there are no people north of the Himalayas since none could climb over the mountain ranges! That would be an absurd conclusion. People of central Asia are born there and live there: They did not have to be born in India and cross the mountain ranges. So with FTLO's.

Moreover, Tolman [14] *believed* he had shown, in his old 'paradox', that the existence of Superluminal particles allowed information transmission into the past (*anti-telephone*).

Therefore one had to wait almost until the sixties [15] before seeing the tachyon problem re-examined by the French Arzeliès [16], the German Schmidt [16], the Japanese Tanaka, the Soviet Terletsky, and the Indian Sudarshan and co-workers [5]. After Reference 5, a number of people started studying the subject, among whom were Feinberg [2] (in the USA) and Recami and colleagues [17] (in Europe). For other historical details, see Reference 18. In the following, let us forget about tachyons, until they are explicitly mentioned.

2. THE POSTULATES OF SPECIAL RELATIVITY REVISITED

2.1. *The Three Postulates of SR*

Even today, the best 'background' for analyzing the essential aspects of time and causality is still that of SR, in which the framework is a four-dimensional, *pseudo-Euclidean* space-time. Let us remember that a suitable choice of postulates for the theory of SR [1] is the following [19, 20]:

(1) *Principle of Relativity*: Physical laws of Electromagnetism and of Mechanics are covariant (= invariant in form) when going from an inertial observer to another inertial observer.

(2) *Space-time is homogeneous and space is isotropic.*

Notice that the postulate of light-speed invariance is *not* strictly necessary, since it can be derived [21] from the above Postulates (1) and (2). Moreover, if we want, as we do, to *avoid* information transmission into the past, a *Third Postulate* is necessary:

(3) *Principle of Retarded Causality*: For every observer, *causes* chronologically precede their own *effects* (for the definition of 'causes' and 'effects' see the following).

This 'Third Postulate' can also be called the *Principle of Reinterpretation* for reasons that we shall see, and it will be shown to be equivalent to assuming that: 'Negative-energy objects or particles travelling forward in time *do not exist*; and physical signals are transported only by objects that appear as carrying positive energy' (such a form being clear within information theory).

An important point, as already mentioned in Section 1, is that from 'Postulate (3)' existence of anti-matter will be inferred.

Let us add that the above postulates imply that the observations from an observer O are transformed into the observations from another observer O', in uniform rectilinear relative-motion, by means of the *Lorentz Transformations*, whose geometrical meaning is depicted in Figure 1.

Postulate (2) is justified by the fact that from it the conservation laws of energy, momentum and angular-momentum follow, which laws are well verified by experience, at least in our 'local' space-time region.

Postulate (1) is inspired by the observation that *all* the inertial frames, in uniform straight relative-motion, should be *equivalent*[1], since we usually do not find any reasons for considering one of them as 'privileged'.

2.2. *The Problem of the Absolute Frame*

In recent times, however, it seemed possible to assume as *privileged* the *frame* in which the cosmic radiations – coming from the whole cosmos –

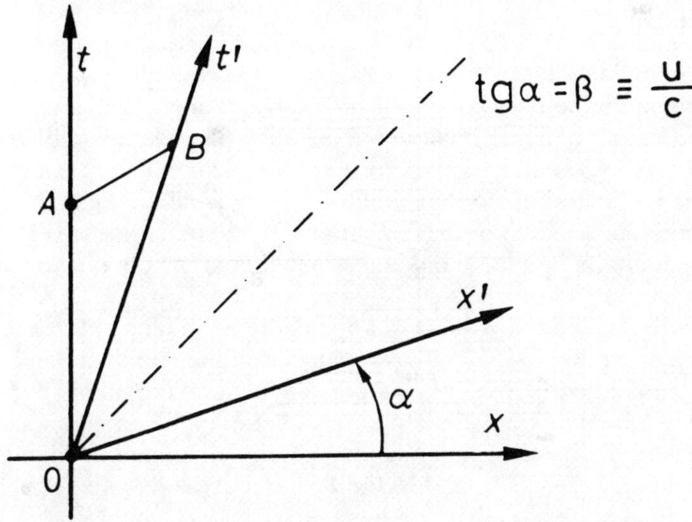

Fig. 1

happen to be *isotropically* distributed. For instance, the 'absolute frame' can be defined as the one in which the so-called 3 K 'fossil radiation' (possibly the remnant of the *big-bang*, i.e. of the initial 'explosion' of our cosmos) comes isotropically from all space directions. It seems that the Earth's absolute speed might be [22] of the order of 600 Km/s.

Nevertheless, the theory of SR – if we still maintain the three postulates above – does not change in an essential way, as shown, for example, in Reference 23. We should merely notice that the problem of the 'ether' (or, if you prefer, of the 'space' or of the 'vacuum') is still quite present in contemporary physics.

3. REINTERPRETATION PRINCIPLE, ADVANCED CAUSALITY (AND ANTIMATTER)

3.1. *The Third Postulate* (RIP)

In order to go back to our 'Third Postulate', let us now consider Figure 2, where, for simplicity, a two-dimensional space-time is depicted. When we are in the position $x = 0$ at time $t = 0$, we usually incline to consider as 'existing' all the x-axis events. However, if another inertial observer, O', moving along the positive x-axis, overtakes us at the origin-event, then at *the same* time $t = t' = 0$ he will tend to consider as 'existing' all the x'-axis events. Therefore, if we want to be able to start discussing and ex-

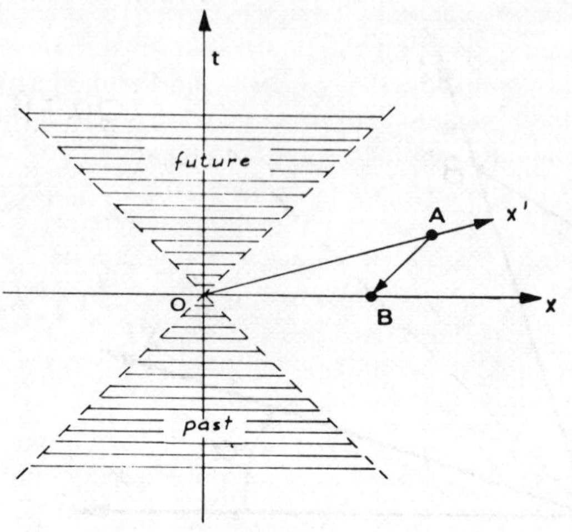

Fig. 2

changing information with him, we must first be prepared to consider that *all* chronotopical events 'exist'[2] (at least the ones *outside* the past-future zone of the light-cone). Then, nothing *a priori* prevents event *A* from influencing event *B* (see Figure 2). It is precisely to forbid such a possibility that we introduce the 'Third Postulate' (or 'RIP' = Reinterpretation Principle). Our point is that, since we 'explore' the Minkowski space-time going *forward in time* (along the direction determined by thermodynamics and by the cosmological evolution) [24], any observer will see the event *B* of Figure 2 as the first one and the event *A* as the last one. Moreover, it has been shown in Reference 18 that *an object going backwards in time* (Figure 2) *corresponds* in the space *dual of* the chronotopical one, i.e. *in the four-momentum space* (see Figure 3 (a)), *to an object carrying negative energy*. And, vice-versa, changing the energy-sign in one

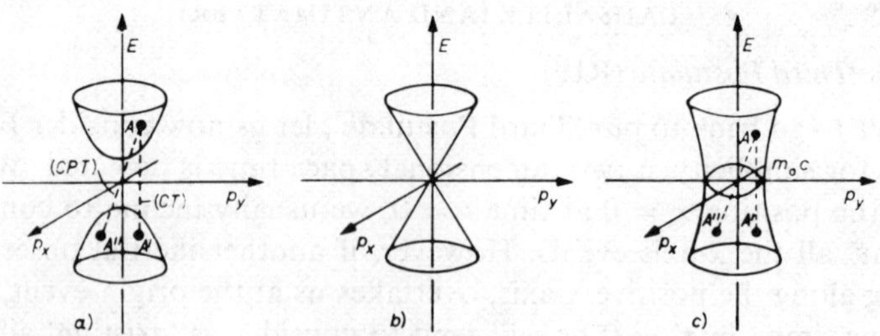

Fig. 3

space corresponds to changing the sign of time in the other (dual) space [18].

We can easily understand this, starting from the *safe consideration* of something that we already, *surely* know from common experience: a positive-energy object going forward in time. If we now want to apply to it an operation turning its motion backwards in time, then Postulates (1) and (2) oblige us to use a *non-orthochronous* Lorentz transformation. But *any* Lorentz transformation changing the sign of the fourth-component of the chronotopical 4-vector (i.e. of *time*) will change also the sign of the fourth-component of the four-momentum vector (i.e. of *energy*) and of any other 4-vector associated to the same observed object.

This is true also in Quantum Field Theory (QFT), i.e. in relativistic quantum mechanics: for example, if

$$f(\boldsymbol{p}, E) = \frac{1}{(2\pi)^2} \int \tilde{f}(\boldsymbol{x}, t) \exp\left[i\boldsymbol{p} \cdot \boldsymbol{x} - iEt\right] d^4x,$$

then [18]:

$$(1) \qquad f(\boldsymbol{p}, -E) = \frac{1}{(2\pi)^2} \int \tilde{f}(\boldsymbol{x}, -t) \exp\left[i\boldsymbol{p} \cdot \boldsymbol{x} - iEt\right] d^4x.$$

Then, it is easy to convince ourselves that those two paradoxical occurrences (negative energy and motion backwards in time) will be *reinterpreted* in a quite *orthodox* way by any observer, when they are – as they actually are – simultaneous.

Namely, let us suppose (Figure 4) that a particle P, with *negative* energy

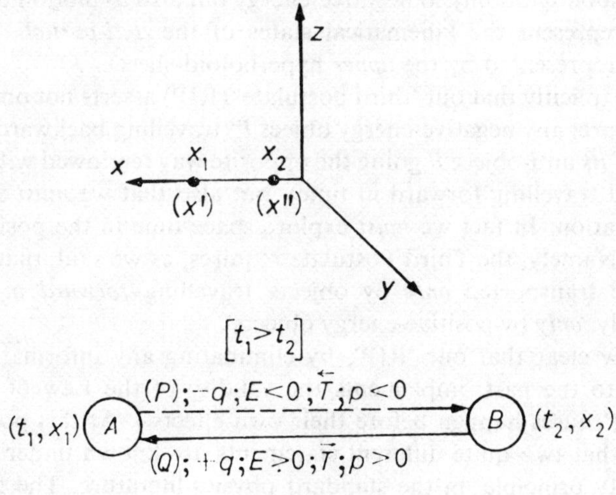

Fig. 4

and e.g. *charge*[3] $-e$, travelling *backwards* in time, is emitted by A at time t_1 and absorbed by B at time $t_2 < t_1$. Therefore, at time t_1, object A 'loses' negative energy and charge $-e$, i.e. *gains* positive energy and charge $+e$. And, at time $t_2 < t_1$, object B 'gains' negative energy and charge $-e$, i.e. *loses* positive energy and charge $+e$. In fact, emission of a negative quantity is equivalent to absorption of a positive quantity, and vice-versa. The physical phenomenon here depicted will of course appear to be nothing but the exchange *from B to A* of a (normal) particle Q, with positive energy, charge $+e$, and travelling *forward* in time.

We have however seen that Q has the *charge opposite* to P; this means that our 'reinterpretation procedure' operates – among other things – a *charge conjugation*, C. A closer inspection [18, 25, 26] of the 'RIP' tells us that indeed Q will appear as the ANTIPARTICLE[4] of P:

$$(2) \qquad Q = \bar{P}.$$

We mean that *the concept of anti-matter is a purely relativistic one*; and that, on the basis of the double sign (Figure 3(a))

$$(3) \qquad E = \pm \sqrt{p^2 + m_0^2}, \qquad [c \equiv 1],$$

existence of antiparticles could have been predicted since 1905 – with exactly the properties they actually showed when later discovered – provided that recourse had been made to the 'reinterpretation principle'. We therefore mean that the points of the *lower* hyperboloid-sheet in Figure 3(a), – since they correspond not only to negative-energy but also to motion backwards in time – represent the kinematical states of the *anti-particle* $\bar{P}$ (of the particle P represented by the *upper* hyperboloid-sheet).

Notice explicitly that our 'third postulate' (RIP) asserts not only that we *can* reinterpret any negative-energy object P (travelling backwards in time) in terms of *its* anti-object $\bar{P}$ going the opposite way (endowed with positive energy and travelling forward in time), but also that we *must* apply that reinterpretation. In fact we *must* explore space-time in the positive time-direction. Namely, the Third Postulate requires, as we said, that 'physical signals are transported *only* by objects travelling *forward in time*, or, equivalently, *only* by positive-energy objects'.

It is now clear that our 'RIP', by eliminating any information-transmission into the past, implements the validity of the Law of Retarded Causality ('causes happen before their own effects'). At this point, let us underline that two quite different statements are known under the name of 'causality principle' in the standard physics literature. The first statement bears that name very improperly, since it merely requires non-existence of faster-than-light signals: we shall give up such an assumption. The second statement asserts that *causes* must chronologically precede their own *effects*. This second statement is adopted [25] by us as the definition of *causality*[5] (or rather of *retarded causality* [27]).

Let us also, *explicitly* observe that the reinterpretation procedure exchanges the roles of *source* and *detector*, and that – with reference to Figure 2 – every observer will deem B to be the source and A the detector of the (*reinterpreted*) antiobject $\bar{P}$.

Here we want to anticipate that our Third Postulate seems to allow solving even the paradoxes connected with the fact that many physical problems admit, besides standard, 'retarded' solutions, also *advanced solutions*: such 'advanced solutions' merely appear to represent *antiparticles travelling the opposite way* [18, 25, 26]. For instance, if Maxwell's equations admit solutions in terms of *outgoing* (polarized) photons of helicity $\lambda = +1$, then they will admit also solutions in terms of *incoming* (polarized) photons[6] of helicity $\lambda = -1$.

4. EXTENDED RELATIVITY. CASE OF TACHYONS. DESCRIPTION AND LAWS

4.1. *Extended Relativity*

All the considerations of Section 3 assume a more compact form when we allow room also for *Superluminal* (= faster-than-light) frames of reference and for tachyons [18], so to take account of *all* space-time rotations (for $0 < \alpha < 2\pi$: see Figures 1 and 5) as *generalized* Lorentz transformations.

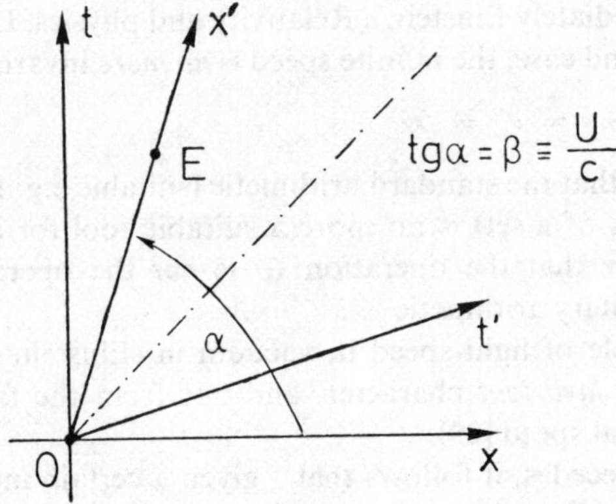

Fig. 5

In this connection, let us explicitly emphasize that it is possible to *extend* [28] Special Relativity so as to consider Superluminal frames and objects *without violating the principle of retarded causality* [26]. Namely, it is enough to start from the above mentioned three Postulates (Section 1), without assuming *a priori* $|v| < c$. The reinterpretation principle (Third Postulate) will be shown to be sufficient for solving *all* causal paradoxes [18]. Notice that the Principle of Relativity says nothing with respect to the value of the relative speed between the inertial frames considered, and that from the Postulates (1) and (2) one merely derives [20, 21] that an *invariant* speed (*not* a maximal speed) must exist.

Namely, from Postulates (1) and (2) it follows [20, 21] that a quantity w^2 (having the physical dimensions of the square of a speed) must exist, which has *the same value* according to all the inertial frames:

$$(4) \qquad w^2 = \text{invariant.}$$

If we assumed $w = \infty$, as done in Galilean relativity, then we would get classical (Galilean – Newtonian) physics. In such a case, the invariant speed is the infinite one; and – if we symbolically indicate by $\oplus$ the operation of *speed composition* – we can write

$$(5) \qquad \infty \oplus v = \infty.$$

But experience has shown us that the invariant speed is *finite* (and real), namely that it is the *speed c of light* in vacuum. In this case, the invariant speed is c:

$$(6) \qquad c \oplus v = c,$$

and we get immediately Einstein's Relativity and physics. Let us emphasize that, in this second case, the infinite speed is *no more* invariant:

$$(6 \; bis) \qquad \infty \oplus v = V' \neq \infty.$$

It simply means that the standard arithmetic (suitable e.g. for counting the discrete elements of a set) is no more a suitable tool for adding physical speeds; *or rather* that the operation $\oplus$ is *not* the operation $+$ of the standard, elementary arithmetic.

The special role of light-speed in vacuum in (Einstein's) Relativity [1] follows from its *invariant* character, and not from the fact that it is *or is not* the maximal speed [29].

From what precedes, it follows that – given a certain inertial reference-frame – the class $\{I\}$ of the *inertial* reference-frames *a priori* consists of all the frames f moving with constant relative-velocity u, where $-\infty < |u| < +\infty$.

When extrapolating usual Lorentz Transformation (LT) for angles $|\alpha| > 45°$ (see Figure 5, where for simplicity we consider the 2-dimensional case), i.e. considering also tachyonic reference-frames, we are led [18] to a new group, G, of 'Generalized Lorentz Transformations' (GLT) which is constituted by *all* the 'rotations' in Minkowski space-time for $0 \le \alpha \le 360°$. The essential point for getting that result is the following.

4.2. *Duality Principle*

Let us choose the particular inertial frame s_0. The lightspeed c — because of its invariant-quantity character — allows an exhaustive partition of frames $f \in \{I\}$ into two subclasses $\{s\}$, $\{S\}$ of frames having speeds $u < c$ and $U > c$ *relative* to s_0, respectively. For simplicity, in the following we shall consider ourselves as 'the observer s_0'. Frames $s \in \{s\}$ will be called *subluminal* (= slower-than-light), and frames $S \in \{S\}$ *Superluminal*. The relative speed of two frames s_1, s_2 (or S_1, S_2) will always be smaller than c; and the relative speed between two frames s, S will be always larger than c. The important point is that the above, exhaustive partition is invariant when s_0 is made to vary inside $\{s\}$ (or inside $\{S\}$); on the contrary, when we pass from $s_0 \in \{s\}$ to a frame $S_0 \in \{S\}$, the subclasses $\{s\}$, $\{S\}$ are interchanged with each other (cf. References 30, 31). At the present time, we neglect *luminal* frames ($u = U = c$) as 'unphysical', even if mathematical use of 'infinite-momentum frames' has spread out recently in physics.

One can immediately deduce a 'Duality Principle' [31, 32], which may be briefly put in the form 'the terms B, T, s, S do not have an absolute meaning, but only a *relative* one'. Let us notice that the *opposite* assumption, that the bradyonic/tachyonic character was absolute, would lead immediately to the impossibility of defining Superluminal frames [30].

4.3. *Bradyons and Tachyons. Subluminal and Superluminal Lorentz Transformations*

We shall neglect space-time translations, i.e. consider only the so-called *restricted* Lorentz transformations. All frames are supposed to have the same event as their origin. Let us also remember that in Minkowski space bradyons are characterized by time-like world-lines, luxons by light-like world lines, and tachyons by space-like world lines.

Now, the transformations L, effecting transition between two inertial frames $f_1, f_2 \in \{I\}$, and satisfying Postulates (1) and (2), must be linear and must preserve the four-vector magnitudes, *apart from the sign* [31, 32]. This point is proved e.g. in Reference 33, as a consequence of light speed invariance. Therefore, transformations L between two inertial frames f_1, f_2 must be such that

$$(7) \qquad x_0'^2 - \mathbf{x}'^2 = \pm (x_0^2 - \mathbf{x}^2)$$

for every four-vector $x \equiv (x_0, \mathbf{x})$, where x means either 4-position, or 4-momentum, or 4-velocity, or 4-current, and so on. We choose throughout this work the metric-signature $(+ - - -)$; natural units $(c = 1)$ will be adopted when convenient.

In the particular case of chronotopical vectors, Equation (7) will read

$$(8) \qquad c^2 t'^2 - \mathbf{x}'^2 = \pm (c^2 t^2 - \mathbf{x}^2),$$

or rather (by using $g_{\mu\nu} = \delta_{\mu\nu}$ and Einstein's notations):

$$(9) \qquad c^2 t'^2 + (i\mathbf{x}')^2 = \pm [c^2 t^2 + (i\mathbf{x})^2].$$

In the following we shall always avoid explicit use of a metric tensor [34] – as well as in Equation (9) – by writing the generic chronotopical vector as $x \equiv (x_0, x_1, x_2, x_3) \equiv (ct, ix, iy, iz)$.

It is easy to convince ourselves that the sign *plus* in Equations (7)–(9) refers to the usual case of subluminal relative speeds, whilst the sign *minus* has to be chosen for Superluminal relative speeds [34].

Postulates (1) and (2) allow considering frames s and S on an *equivalent* footing (cf. the following); therefore, even Superluminal observers S must be supposed to be able to fill space with meter-sticks and (synchronized) clocks, all at rest relative to S: that is to say to build up their 'lattice-work of meter-sticks and clocks' [34].

From the requirement that Superluminal frames are physical [30], it follows of course that objects must exist which are *at rest* relative to S and *tachyons* relative to frames s. From the further fact that luxons $\mathscr{L}$ show the same velocity to any observer s or S, it can be deduced that a bradyon $B(S)$ relative to an S will be a tachyon $T(s)$ relative to any s, and vice-versa:

$$(10) \qquad B(S) = T(s); \quad T(S) = B(s); \quad \mathscr{L}(S) = \mathscr{L}(s).$$

This accords [31, 32] with the Duality Principle, that we are going to complete by *adding* that [31] 'frames S are supposed to have at their disposal exactly the same physical objects as frames s have, and vice-versa.'

In conclusion, when frames s, S observe the same event, 'time-like' vectors transform into 'space-like' vectors, and vice-versa, in going from s to S or from S to s. On the contrary, it is well-known that usual LT's, from s_1 to s_2, or from S_1 to S_2, preserve the four-vector *type*. One is therefore allowed to say that (subluminal) LT's are expected to be such that:

$$(11a) \qquad c^2 t'^2 + (ix')^2 = + [c^2 t^2 + (ix)^2], \qquad [\beta^2 < 1],$$

while '*Superluminal Lorentz Transformations*' (SLT), from s to S or from S to s, are expected to be such that [$\beta \equiv u/c$]:

$$(11b) \qquad c^2 t'^2 + (ix')^2 = - [c^2 t^2 + (ix)^2], \qquad [\beta^2 > 1].$$

Of course, also tachyons will possess *real* rest-masses (since they are just usual particles with respect to their own rest-frames f, where f are Superluminal frames to us). From Equation (11b), applied to 4-momentum vector, one can immediately derive for tachyons the relation

$$(12) \qquad E^2 - p^2 = -m_0^2 < 0, \qquad [m_0 \text{ real}].$$

Therefore, one has:

$$(13a) \qquad p^2 = m_0^2 > 0 \qquad \text{for bradyons (case I, or } time\text{-}like),$$
$$(13b) \qquad p^2 = 0 \qquad\qquad \text{for luxons (case II, or } light\text{-}like),$$
$$(13c) \qquad p^2 = -m_0^2 < 0 \quad \text{for tachyons (case III, or } space\text{-}like).$$

In four-momentum space (see Fig. 3), equations (13) represent respectively: (i) for B's, a two-sheeted hyperboloid of rotation around the E-axis; (ii) for $\mathscr{L}$'s, a double indefinite cone, having E as axis; (iii) for T's, a single-sheeted rotation hyperboloid. In all cases m_0 is real, and we have $|v| = |p/E|$. For obvious reasons, in Figure 3 only the 3-dimensional 'space' $p_z = 0$ has been depicted. Remember that any SLT maps the 'interior' of the light-cone $p^2 = 0$ into its 'exterior', and vice-versa (as one can show e.g. within the mathematical 'theory of catastrophes') [35a], even if such a mapping is one-to-one only *almost everywhere*. [Note: $v = $ velocity.]

It may be noted that: (a) the speed c preserves of course its character of *limit* kinematical-parameter of our four-dimensional cosmos [24, 35b, 18] (even if we know that such a limit has two 'sides'); (b) tachyons will slow down when energy increases and accelerate when their energy decreases.

In particular, divergent energies are needed to slow down the tachyon speed towards the (lower) limit c. On the contrary, when tachyon's speed tends to infinity, its energy tends to zero; this prevents violation of the common postulate that 'energy can be transmitted only at *finite speed*', since a tachyon shows zero energy to the *same* observers to whom it presents divergent speed. Notice that a bradyon may have zero momentum (and minimal energy m_0c^2), and a tachyon may have zero energy (and minimal momentum magnitude m_0c); however bradyons B (Figure 3a) cannot exist at zero energy, as tachyons T (Figure 3c) cannot exist at zero momentum – with respect to the observers to whom they appear as tachyons! It is immediately seen that infinite speed belongs only to tachyons corresponding to the intersection of the hyperboloid in Figure 3c with the hyperplane $E = 0$.

Incidentally, since transcendent tachyons do transport momentum, they allow getting the *rigid-body* behavior even in SR. As a consequence, in elementary-particle physics, tachyons might a *priori* result as useful for interpreting diffractive scatterings, or the so-called pomeron-exchange reactions, and elastic scatterings [35c].

4.4. *The Generalized Lorentz Transformations*

Let us here skip the problem of explicitly finding the GLT's, which are discussed elsewhere [35d]. Let us only specify that the new group G of GLT's will be (if we represent the transformations by 4×4 matrices):

$$
\begin{aligned}
G &= \{+\Lambda_<\} \cup \{-\Lambda_<\} \cup \{-i\Lambda_>\} \cup \{+i\Lambda_>\}, \\
\Lambda_< &\equiv \Lambda(\beta^2 < 1); \quad \Lambda_> \equiv \Lambda(\beta^2 > 1); \qquad \beta \equiv u/c,
\end{aligned}
\tag{14}
$$

so that, if $L \in G$, then also $-L \in G$, $\forall L \in G$. In Equation (14), the set $\{+\Lambda_<\}$ is the one of the usual proper, orthochronous, subluminal LT's; the set $\{-\Lambda_<\}$ the one of the corresponding *non*-orthochronous LT's; and the sets $\{\pm i\Lambda_>\}$ the ones of the SLT's, where the $\Lambda_>$'s are matrices formally identical to the $\Lambda_<$'s but containing values of β in the range $\beta^2 > 1$.

Notice that:

$$
\det L = +1, \quad \forall L \in G;
$$

in fact all GLT's are space-time rotations (cf. Figures 1 and 5). In short, SLT's are obtained from usual LT's by multiplying the latter by the imaginary unit i *and simultaneously* by changing β into $1/\beta$ (cf. Reference 18). More precisely, our symbol i in four dimensions represents (rather than the ordinary 'imaginary unit') a quantity, e.g. a 4×4 matrix, whose square is -1. For its physical interpretations, see [18].

4.5. *Equivalence of Bradyonic and Tachyonic Inertial Frames. Descriptions and Laws.*

Before going on, let us show with some rigour that the usual definition of *equivalence* can actually be extended also to Superluminal inertial frames. Let us choose [35e] a set $\mathscr{R}$ of certain well-defined reference-frames r, the set $\mathscr{P}$ of the phenomena p of mechanics and electromagnetism, and the set $\mathscr{D}$ of the descriptions d (of phenomena $p \in \mathscr{P}$ from frames $r \in \mathscr{R}$). All observers r are supposed to possess the same instruments, both physico-experimental and mathematico-theoretical (*i.e. the same theory too*). Strictly speaking, one has to deal with the 'triads' dpr, elements of the set $\mathscr{DPR} \equiv \mathscr{D} \times \mathscr{P} \times \mathscr{R}$, the Cartesian product of the three sets considered. As depicted in Figure 6 (which has only the task of supporting intuition), to the same p there correspond two descriptions d_1, d_2 in two different frames r_1, r_2, and so on: given any *two elements* out of d, p, r, the third one must be univocally fixed, in our assumptions. We may write [35e]

$$d_1 p r_1 \xrightarrow{r_1 \to r_2} d_2 p r_2, \qquad \text{i.e.} \quad \{r_1 \to r_2, p \to p \Rightarrow d_1 \to d_2\},$$

$$d p_1 r_1 \xrightarrow{r_1 \to r_2} d p_2 r_2, \qquad \text{i.e.} \quad \{r_1 \to r_2, d \to d \Rightarrow p_1 \to p_2\}.$$

Let us define the subsets $\Delta_r \in \mathscr{D}$:

(15) $d \in \Delta_r \Leftrightarrow dpr \in \mathscr{DPR} \equiv \mathscr{D} \times \mathscr{P} \times \mathscr{R}.$

Then, following References 35e, we shall say that two frames r_1, r_2 are *equivalent* ($\doteq$) if Δ_r is *mapped onto itself* when passing from r_1 to r_2:

(16) $r_1 \doteq r_2 \Leftrightarrow \Delta_{r_1} = \Delta_{r_2} \Leftrightarrow \forall d \in \Delta_{r_1} \Rightarrow d \in \Delta_{r_2} \text{ and } \forall d' \in \Delta_{r_1} \Rightarrow d' \in \Delta_{r_1}.$

Such a condition operates an exhaustive partition [35e] of set $\mathscr{R}$ into sub-

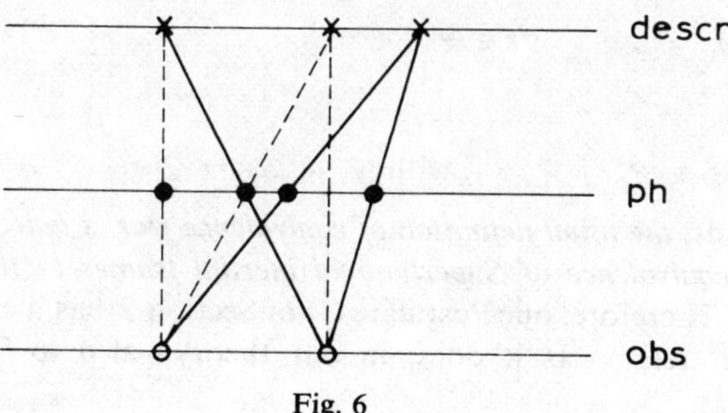

Fig. 6

sets of equivalent frames. Conversely, given a frame r and a set $\mathscr{P}$ of phenomena, it is possible to build up the set $\mathscr{R}$ of frames equivalent to r. It is well known that, given an inertial frame $r \equiv s_0$ and the set $\mathscr{P}_<$ of usual mechanical and electromagnetic phenomena, a class of equivalent frames is the one $\mathscr{R}_s$ of the usual (subluminal) inertial frames s, where $\mathscr{R}_s \equiv \{s\}$. It means that (speaking loosely) we can write, given the set $\mathscr{P}_<$:

$$\Delta_r = \Delta \equiv \Delta_<, \qquad \forall r \in \mathscr{R}_s.$$

Then, due to the Relativity and Duality principles, the set $\Delta \equiv \Delta_<$ of descriptions will correspond – according to any Superluminal inertial frame $S_0 \in \mathscr{R}_S$, where $\mathscr{R}_S \equiv \{S\}$, – to another, new set $\mathscr{P}_>$ of mechanical and electromagnetic phenomena; and, given any frame $S_0 \in \mathscr{R}_S$, all the other frames S of the set $\mathscr{R}_S$ will be equivalent to it (with respect to the phenomena $\mathscr{P}_>$).

Now, according to SR, let us *assume* that the usual inertial frames $s \in \mathscr{R}_s$ are equivalent *also* when considering *all* the mechanical and electromagnetic phenomena $p \in \mathscr{P}$ associated to both subluminal and Superluminal bodies. In particular, $\mathscr{P}$ will contain $\mathscr{P}_< \cup \mathscr{P}_>$, plus other phenomena (i.e. the phenomena referred to both bradyonic and tachyonic sources and detectors; or, more generally, the phenomena to which both bradyons and tachyons participate, besides photons). In other words, let us assume that, given the set $\mathscr{P}$:

$$(17) \qquad \Delta'_r = \mathscr{D}, \qquad \forall r \in \mathscr{R}_s.$$

At this point, we can say that also frames $S \in \mathscr{R}_S$ are equivalent to the frame s_0 (and to the other frames $s \in \mathscr{R}_s$) *if* the set $\mathscr{D}$ is always mapped onto itself under all the transformations $S \rightleftarrows s$, $S_1 \rightleftarrows S_2$, $s_1 \rightleftarrows s_2$, with respect to the whole set $\mathscr{P}$ of (generalized) mechanical and electromagnetic phenomena; i.e. (loosely writing) *if*, given the set $\mathscr{P}$:

$$(18) \qquad \mathscr{D}_r = \mathscr{D}, \qquad \forall r \in \mathscr{R}$$

where

$$\mathscr{R} \equiv \mathscr{R}_s \cup \mathscr{R}_S \qquad \text{with } \mathscr{R}_s \cap \mathscr{R}_S = \varnothing.$$

In other words, *the usual definition of equivalence works quite well also for defining the equivalence of Superluminal inertial frames to the usual subluminal ones.* Therefore, our Postulate (1) of Section 2 has a clear meaning even when it refers – as it does, in our theory – also to Superluminal inertial frames.

Extended Relativity, being based on Postulates (1)–(3), does of course realize the validity of condition (18). For instance, let us confine ourselves, for simplicity, to the descriptions $\mathscr{D}_<$, $\mathscr{D}_>$ (from a frame $r \equiv s_0$) of the phenomena p belonging to the set $\bar{\mathscr{P}}$ of the motions of single, free (bradyonic or tachyonic) particles, respectively. Then, given the phenomena $p \in \bar{\mathscr{P}}$:

$$(19) \qquad \mathscr{D} = \mathscr{D}_< \cup \mathscr{D}_>, \qquad \forall r \in \mathscr{R},$$

and, when passing from an s to an S, $\mathscr{D}_<$ goes into $\mathscr{D}_>$ and vice-versa, but *the whole $\mathscr{D}$ is mapped onto itself, as required*. (By the way, notice that the boundary $\mathscr{D}_< \cap \mathscr{D}_>$, representing free photons, goes into itself).

Let us now try to define [35e] physical *laws*. Given a phenomenon p, if d_1 and d_2 are its descriptions in the frames r_1, r_2 respectively, and if the transformation L is such that

$$(20) \qquad Lr_1 = r_2,$$

we shall consequently use the convention of writing

$$(20 \; bis) \quad Ld_1 = d_2.$$

Let us suppose, now, that we have a *criterion* $\bar{C}$ for a given description d to belong to the set $\mathscr{D}$ of the descriptions of phenomena $p \in \mathscr{P}$ from the frame $r \in \mathscr{R}$; we then write

$$(21) \qquad \bar{C}(d) \text{ verified} \Leftrightarrow d \in \mathscr{D}.$$

We shall call C a 'good criterion' if it holds for any d:

$$(22) \qquad \forall d \in \mathscr{D} \Rightarrow C(d) \text{ verified} \Leftrightarrow d \in \mathscr{D}.$$

It follows that the 'good' criteria C are *covariant* (in form) under any L:

$$(22') \qquad C(Ld) \text{ verified} \Leftrightarrow Ld \in \mathscr{D}.$$

We shall by definition call C (or better the union of the various, possible good criteria C_1, C_2, ...) the *ensemble of the physical laws* of phenomena $p \in \mathscr{P}$ as seen by frames $r \in \mathscr{R}$. Conversely, a proposition will be considered a physical law if it is a part of C. *In other words*, given $\mathscr{R}$ and $\mathscr{P}$, we define 'physical law' as any proposition regarding a $p \in \mathscr{P}$ which is covariant within $\mathscr{R}$.

Moreover, let us assume that we know, besides the class $\mathscr{A}$ of the usual physical laws (of mechanics and electromagnetism) for bradyons and antibradyons, also the class $\mathscr{B}$ of the physical laws for tachyons and antitachyons. When we pass from a subluminal frame s to a Superluminal frame S, class $\mathscr{A}$ will of course have to transform into class $\mathscr{B}$, and vice-versa. *In this sense*, the totality of physical laws ($\mathscr{A} \cup \mathscr{B}$) will be covariant under the whole group G, i.e. *G-covariant*. And *in this sense inertial frames* (with relative speeds $|u| \gtrless c$) *are all equivalent*.

We have already seen – as better shown elsewhere [18] – that physical laws (of SR) may be written in a (universal) form valid for both B's and T's, a form obviously coinciding with the usual one in the bradyonic case. For example, we have shown in Reference 18 that the G-covariant expression

$$m = \frac{m_0}{\sqrt{|1 - \beta^2|}}, \qquad (\beta^2 \gtrless 1, \ m_0 \text{ real}),$$

in such a form has a 'universal' validity.

From previous considerations a '*Rule of Tachyonization*' immediately follows, extending Parker's principle [32] to the four-dimensional space-time: 'The relativistic laws (of mechanics and electromagnetism, at least) for tachyons follow by applying a SLT – e.g. a 'transcendent transformation' [18] – to the corresponding laws for bradyons'. Such a rule may be also named the '*rule* of extended relativity'.

At last, let us remark incidentally that the consideration of 'Extended Relativity' [28] prompts us to choose a *five-dimensional* [36] space-time (at least) as a better background for *mechanics* theories (see the following).

5. CAUSALITY AND TACHYONS

In the case of tachyons, it is *even clearer* that our 'Third Postulate' (as-serting, e.g., that negative-energy particles, travelling forward in time, do not exist) does easily eliminate any 'information transfer' backwards in time. In fact (cf. Figure 3c), to get transition from a (standard) tachyon A (with positive energy and moving forward in time) to a negative-energy tachyon A', a usual LT [a LT operates a movement *on the same* hyper-boloid *sheet*] is enough. The fact that such a LT will change not only the sign of energy but also of time is easily seen by comparing Figures 3c and 7.

Let us first look at Figure 3c, and consider a frame s_0, and then a con-tinuous succession of reference-frames moving with increasing positive speeds $u < c$ along the x-direction, which observe *the same* free tachyon T. When varying the observer within that succession, the point K, repre-senting the kinematical state of the observed tachyon, moves from its initial position $A \equiv K(s_0)$ – representing, for example, tachyon T travelling along the positive x-direction with speed $V > c$ – towards a final position A'. In order to go from the upper ($E > 0$) region to the lower ($E < 0$) one, the representative point K must cross the 'plane' $E = 0$. In such a position, since $V = p/E$, $[c = 1]$, the point $K(E = 0)$ refers to a *transcendent* tachyon, i.e. to a tachyon T endowed with infinite speed – and minimal momentum $m_0 c$. It is easy to calculate that, with respect to s_0, the *critical* frame s_∞ wherefrom T appears to be transcendent is the one with relative-speed $u = c^2/V < c$. Incidentally, if we confine ourselves for simplicity's sake – as before – to motions along x, then a one-to-one correspondence

$$v \leftrightarrow c^2/v$$

can be set between subluminal frames (or objects) with speed $v < c$ and Superluminal frames (or objects) with speed $c^2/v \equiv V > c$.

Any observer coming after s_∞ in the above succession of frames should therefore see T endowed with a negative energy E (cf. Figure 3c). Now, let us pass to Fig. 7. It is easy to realize that the frame s_∞ will be represented by axes (x_∞, t_∞) rotated with respect to (x, t) by an angle α_∞ such that x_∞ is superimposed to the 'world-*line*' OT of the considered *free* tachyon T. The above frame-succession, in the chronotopical space, is got by increasing α (from *zero*) with continuity: the frames attributing $E > 0$ to T correspond to $\alpha < \alpha_\infty$, and the frames that should attribute $E < 0$ to T are rotated by $\alpha > \alpha_\infty$. But inspection of Figure 7 immediately confirms that the latter ones should *also* see tachyon T moving backwards in time (*besides* having negative energy)!

It is therefore straightforward to realize that (cf. Figure 7), since point A' should represent a negative-energy tachyon T *travelling backwards in time, then* (owing to the 'RIP') it actually represents nothing but an anti-tachyon $\bar{\text{T}}$ (travelling, with positive energy, forward in time).

We are left with *no* motion backwards in time.

6. MATTER AND ANTIMATTER

We have seen that the 'RIP' *eliminates* any information transmission into the past (even by tachyons); and simultaneously it allows us to predict – merely from Relativity – the existence of *antimatter*.

In fact, in our theory, antiparticles $\bar{P}$ are nothing but particles P *in the*

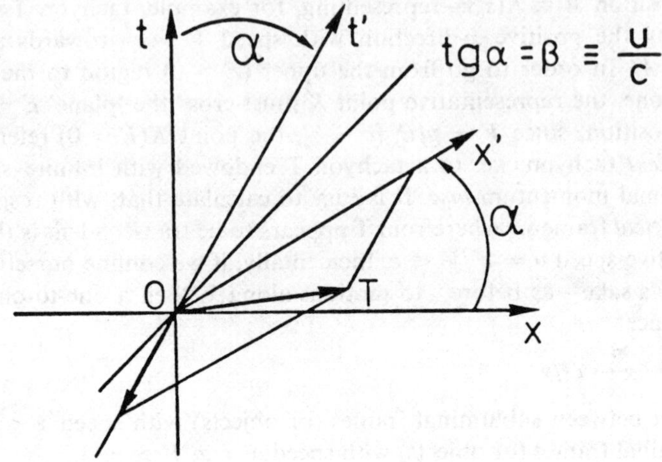

Fig. 7

state with 'negative energy and motion backwards in time'; 'particles' P *in that state* will indeed appear to us as antiparticles $\bar{P}$ (endowed with *positive* energy and motion *forward* in time) since we must explore space-time in a unique time-direction (that we called positive by definition). Our Third Postulate assumes that negative-energy particles moving forward in time (and then positive-energy particles moving backwards in time) *do not exist*.

Namely, in our theory we showed that, given a tachyon T, a *usual* LT can transform it into an object $\bar{T}$ expected to have *exactly* all the properties that antiparticles actually showed in the experiments.

In the case of bradyons B, however (cf. Figure 3a), by means of a usual LT one *cannot* leave his hyperboloid sheet. That is to say, a usual LT cannot bring an upper-hyperboloid point (representing a particle B) into a lower-hyperboloid point (representing the antiparticle $\bar{B}$).

It follows that – in the case when we *confine* ourselves to usual LT's – then the '*matter*' or '*antimatter*' character is invariant for bradyons, but is relative to the observer for tachyons.

However, when eliminating the previous restriction, then – by means of GLT's, e.g. by means of two SLT's – we can indeed pass from particles to antiparticles even in the case of bradyons. Thus, in *Extended* Relativity, the matter/anti-matter character is relative to the observer also for usual particles and objects.

Namely – let us repeat it – a particle P in the kinematical state corresponding to a point of the lower hyperboloid (Figure 3a) has been shown to appear as the *antiparticle* $\bar{P}$ of P, *in the usual sense*. The fact is interesting that, once the notion of *particle* is introduced (as is usually done in SR), merely from SR itself the concept of *antiparticle* follows (cf. Equation (3)).

At last, let us confine ourselves for simplicity's sake to *boosts* along x, i.e. to collinear Lorentz transformations along the x-direction: then, the four subsets in Equation (14) of GLT's describe transitions, *from* the initial frame s_0 (e.g. with *right-handed* space-axes), *not only to* all frames f^R moving along x with all possible speeds u, where $-\infty < u < +\infty$, *but also to all left-handed* frames f^L, moving as well with all possible speeds u along x (cf. Figures. 2–6 and 11 of Reference 18).

In fact, when overcoming the *transcendent frame* (relative to s_0) $f(\infty) = f(U = \infty)$, we pass[7] from frames f^R (e.g., with a *right-handed* spatial frame) to totally inverted frames $f^L = (\text{PT})f^R$, with a *left*-handed spatial frame, a reversed time-axis [8], etc. This could have been expected, since the *total inversion*, or strong reflection, PT (where P is the space-parity operation and T the time-reversal) is nothing but a particular 'rotation' in 4-dimensional space-time; and we saw that GLT's coincide with all space-time 'rotations' *when* we do not restrict ourselves to the usual, proper, orthochronous LT's, so that[9] PT $\in G$ (see the following).

Loosely speaking, we can say that[10] – after applying the 'RIP' – if an *ideal* frame f could undergo a trip along the axis (*circle*) of the speeds, then it would come back (after having bypassed $f(\infty)$) with inverted spatial axes (space *parity*) and *with particles transformed into anti-particles*.

7. SOLVING CAUSAL PARADOXES FOR TACHYONS

We have seen that in the case of tachyons it is even *clearer* that our Third Postulate does eliminate any causality violation. However, this success is obtained at the price of *abandoning the old conviction that judgment about what is 'cause' and what is 'effect' is independent of the observer*. In fact, in the case examined above (Figs. 3c and 4) the initial observer (s_0) will judge the event at A as *causing* the event at B. Conversely, any observer s', which interprets the same phenomenon as exchange of an antitachyon $\bar{T}$ from B to A, will judge *the event at B as the cause of the event at A* [37].

Nevertheless, *all observers will always see the cause to precede chronologically its own effect* [18].

Once again, the *Law* of 'Retarded Causality' is relativistically covariant, and holds for all inertial observers, both subluminal and Superluminal. On the other hand, we have *not* to expect covariance of the phenomenon-description and of the 'description details' at all: in the present case, of the assignment of 'cause' and 'effect' names [38]. Moreover, we shall see [39] that *two events 'causally connected' according to an observer* s_1 (e.g. via a tachyonic exchange) *may even appear as totally uncorrelated to other observers* s_2; so that not even the very existence of a causal correlation is relativistically covariant.

Relativity of judgment about cause and effect, and even more of existence of a 'causal correlation', led to a series of apparent 'causal paradoxes' [40–41] that—even if easily solvable in microphysics [39, 42]—gave rise to some perplexities.

We shall state and resolve here only a causal paradox which seems to be one of the most sophisticated. It was proposed by Pirani [41] in 1970 and *substantially* solved by Parmentola and Yee [37] in 1971, on the basis the ideas expressed in References 39 and 42.

Let us consider four observers A, B, C, D having given [41] velocities in the plane (x, y) with respect to a fifth observer s_0. Let us suppose that the four observers are given in advance the instruction to emit a tachyon as soon as they receive a tachyon from another observer, so that the following *chain of events* takes place (see Figure 8). Observer A initiates the

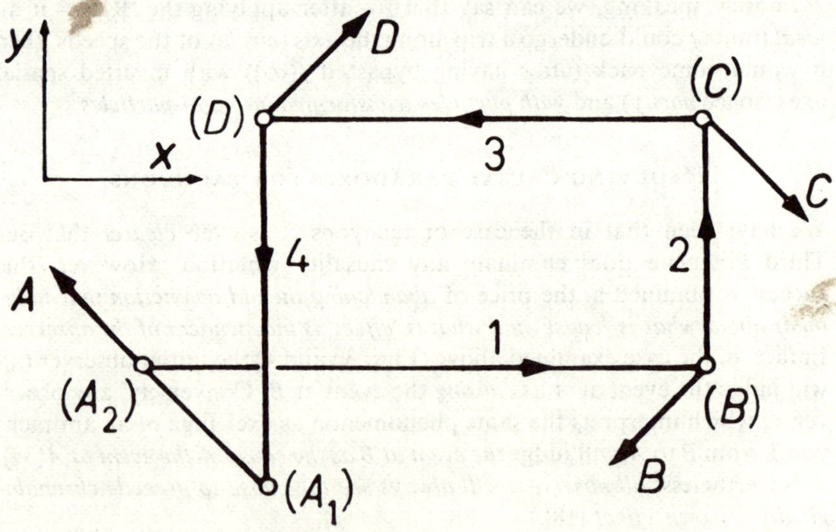

Fig. 8

experiment by sending tachyon 1 to B; observer B immediately emits tachyon 2 towards C; observer C sends tachyon 3 to D, and observer D sends tachyon 4 back to A, with the result [41] that A receives tachyon 4 (event A_1) *before having initiated the experiment* by emitting tachyon 1 (event A_2). The sketch of this '*Gedanken Experiment*' is in Figure 8, where oblique vectors represent observer velocities relative to s_0, and lines parallel to the Cartesian axes represent the tachyon paths.

It is important to notice that Figure 8 does *not* represent the process's actual description by any observer [25]! In fact, the arrow of each tachyonic line simply denotes its motion direction with respect to the observer that emitted that particular tachyon. [By the way, tachyons and observers' velocities on the contrary can be chosen [41] in such a way that all tachyons effectively appear to observer s_0 to move in directions opposite to the ones indicated in Figure 8.]

The paradoxical situation above does actually arise *by mixing together observations by four different observers* [25, 43, 44]. On the contrary, it is necessary to investigate how each observer describes the event chain.

Following Reference 43, let us pass, to this end, to Minkowski space and study the space-time description given, e.g., by observer A. From a dynamical viewpoint, the other observers may be replaced by external force-fields that scatter the tachyons (or by atoms, able to absorb and emit tachyons).

In Figure 9 it is clearly shown that the absorption of 4 happens *before*

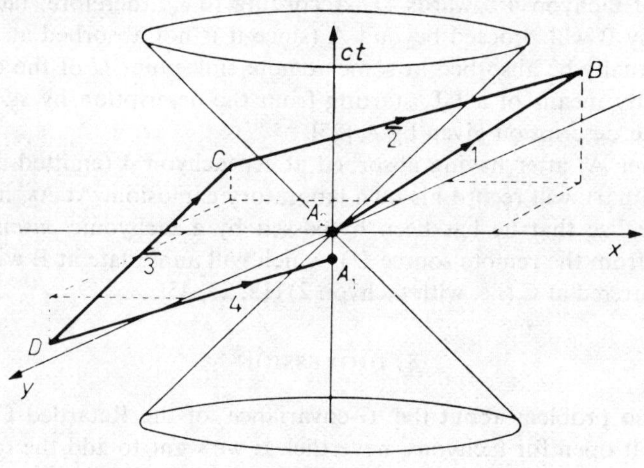

Fig. 9

the emission of 1. It *might* seem that one can send signals into the past of A. However, observer A will effectively see the sequence of events as follows: event D consists in the creation of pair $\bar{3}$ and 4 by the external field; tachyon 4 is then absorbed at A_1, while $\bar{3}$ is scattered at C (transforming into tachyon $\bar{2}$); event A_2 is the emission, by A itself, of tachyon 1 that annihilates at B with tachyon $\bar{2}$. Therefore, according to A, one has essentially an initial pair-creation at D, and a final pair-annihilation at B; and tachyons 1, 4 do *not* appear causally correlated at all [18, 25, 43]. In other words, according to A the emission of 1 does *not* initiate any chain of events that brings to the absorption of 4, and we are *not* in the presence of any effect preceding its own cause.

Analogously, *orthodox* descriptions (i.e. the descriptions put forth by the remaining observers) may be obtained by Lorentz-transforming the description above by A.

Let us now formulate the same 'paradox' in its *strong* version [18, 25, 43]. Let us suppose that tachyon 4, when absorbed at A_1 by A, *blows up* the whole laboratory of A, eliminating the physical possibility that tachyon 1 (believed to be the sequence starter) is subsequently emitted (at A_2), and thus originating a supposedly contradictory state of things – according to the paradox terms. Following Root and Trefil [45], we can see, on the contrary, how, for example, observers s_0 and A will really describe the phenomenon.

In particular, s_0 will observe the laboratory of A blown up after emission (at A_1) of tachyon $\bar{4}$ towards D. According to s_0, therefore, tachyon $\bar{1}$ emitted by B will proceed beyond A (since it is not absorbed at A_2) and will eventually be absorbed at some remote sink point U of the universe [25, 45]. By means of a LT, starting from the description by s_0, we can obtain the description given by A [25].

Observer A, after having absorbed at A_1 tachyon 4 (emitted at D together with $\bar{3}$), will record his own laboratory explosion. At A_2, however, A will realize that he has been by-passed by a tachyonic *cosmic* ray 1 (coming from the remote source U), which will annihilate at B with tachyon $\bar{3}$ scattered at C (i.e. with tachyon $\bar{2}$) [18, 25, 45].

8. DIGRESSION

Even if no problem about the 'G-covariance' of the Retarded Causality Law is left open for tachyons, nevertheless we want to add the following 'digression', *without any reference to what precedes.*

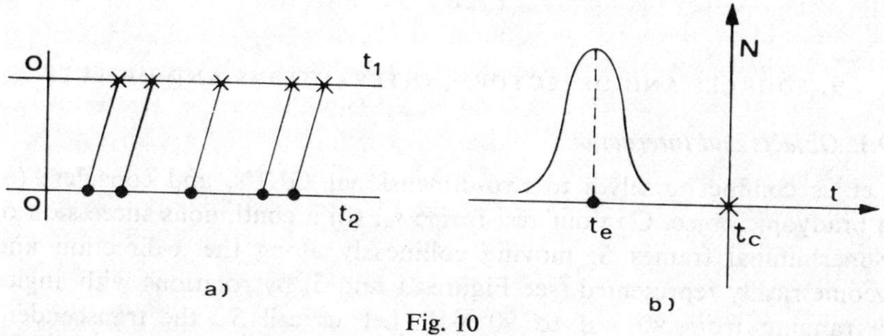

Fig. 10

Let us notice that Postulate (3) is a fundamental hypothesis of ours (in full accord with statistical thermodynamics and with information theory), but *a priori* is *not* logically *necessary* [46]. In fact: (i) Let us suppose that a statistical correlation exists between *two* series of events, in the sense that e.g. each second-series event happens about 1 second *before* a first series of events (see Figure 10). Such a statistical correlation will be called a '*causal connection*'; (ii) Let us now suppose that first-series events are the '*independent*' ones, in the sense that we make them occur e.g. at instants chosen by consulting random-values' tables (maybe produced by a remote computer, having no reasonable relation with the events considered). Such events will be called the *causes*; (iii) The second-series events will be then called the 'dependent' ones in the causal correlation defined at point (i). They will be said to be the '*effects*'; (iv) One may therefore conclude, from the definitions above, that in this case effects *do* chronologically precede their own causes (Figure 10). To conclude the present digression, let us shed some light on the possible nature of our difficulties in conceiving effects chronologically preceding their causes, by reporting the following anecdote [46], which does not involve contemporary prejudices. For ancient Egyptians [18, 46], who knew only the Nile and its tributaries, which all flow from South to North, the meaning of the word 'South' coincided with the one of 'up-stream', and the meaning of the word 'North' coincided with the one of 'down-stream'. When Egyptians discovered the Euphrates, which unfortunately happens to flow from North to South, they passed through such a *crisis* that it is mentioned in the stele of Tuthmosis I, which tells us about "*that inverted water that goes down-stream* (i.e. towards North) *in going up-stream*" [18, 46].

Before closing, let us add that the present 'digression', Section 8, has also *nothing to do* with what follows.

PART II

9. SOURCES AND DETECTORS; INTERACTIONS AND OBJECTS

9.1. *Objects and Interactions*

Let us confine ourselves to two-dimensional GLT's, and consider: (a) a bradyonic object C in our rest-frame s_0; (b) a continuous succession of Superluminal frames S, moving collinearly along the x-direction and geometrically represented (see Figures 1 and 5) by rotations with angles α ranging from $90° - \Delta$ to $90° + \Delta$. Let us call S_∞ the transcendent frame S ($\alpha = 90°$), in which C becomes an infinite-speed *tachyon*. Due to 'RIP', as we *by-pass* the transcendent frame S_∞ (see Figure 11), the new frames will judge the observed object C as an antitachyon $\bar{C}$ *travelling in the opposite direction*.

Vice-versa, we can put ourselves at rest (with respect to s_0) and observe frames S considered as objects, or rather as *one* (tachyonic) object C that moves with speeds *varying* with continuity. We have already analyzed that situation in Figure 4.

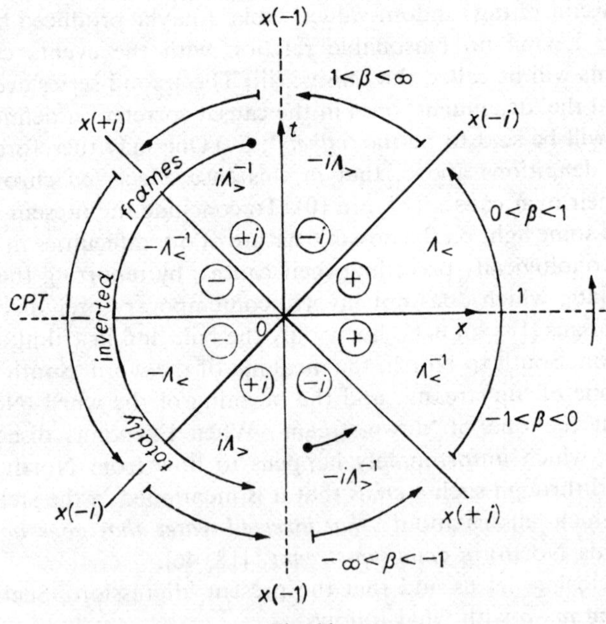

Fig. 11

But close inspection of Figure 4 reveals that the Third Postulate cannot be applied if we (for each bradyon B or tachyon T) do not take account of the proper *sources* and *detectors*. More generally, since we have to deal with exchanges of the emission and absorption roles, the 'reinterpretation procedure' looses its meaning if we cannot refer our B's and T's to some (space-time) *interaction regions*. For example, when a tachyon T overcomes the divergent speed, it passes from being, e.g., a tachyon T *entering* (a certain interaction region) to being an antitachyon T̄ *outgoing* (from that interaction region). In conclusion, the 'RIP' will be *completed* by saying [18]: 'Under a *trans-critical* GLT, when e.g. the roles of emitter and absorber happen to be interchanged, any negative-energy object in the *initial* 'state' physically corresponds to its positive-energy anti-object in the *final* 'state', and vice-versa'.

Of course, the Third Postulate – in order to be used for reinterpreting the effects of GLT's – requires considering processes with both initial and final 'states'.

Therefore, Extended (Special) Relativity imposes that physics must deal with *interactions* rather than with *objects* (in quantum-mechanical language, with 'amplitudes' rather than with 'states').

9.2. *Sources and Detectors*

Now, let us go back again to Figure 4. If the two macro-objects A and B are connected through the exchange of a particle P, then – according to the particular class $\{f_\infty\}$ of observers – object P is endowed with infinite speed: $P \equiv P_\infty$. But, for transcendent particles, the motion-direction along AB (see Figure 4) is *not* defined: in such a limiting case, we can for instance consider P_∞ *either* as a particle P ($v = +\infty$) going from A to B, *or* equivalently as an antiparticle $\bar{P}$ ($v = -\infty$) going from B to A; in the quantum-mechanics (Q.M.) formalism we could write for example:

$$(23) \qquad P_\infty \equiv |P(|v| = \infty) > = a|P(v = +\infty) > + b|\bar{P}(v = -\infty) > ;$$
$$a^2 + b^2 = 1.$$

In order words, if A and B exchange a (bradyonic or tachyonic) object, they are connected via a *symmetrical, instantaneous* interaction. (This perhaps helps justifying the newly increasing use of action-at-a-distance descriptions, as equivalent to the standard ones [47].)

But this also clarifies that *no* source can emit anything if a proper detector is not yet ready to absorb it, somewhere in the universe. *At least* in the case of tachyon exchange, this condition is strictly required by

Relativity (if we confine ourselves only to subluminal LT's). It is not without meaning that, even when using standard SR, Wheeler and Feynman [48] were able to build up *for the limiting case of photons* a theory (equivalent to usual electromagnetism) where sources emit photons just only if their detectors are (already) ready to absorb them.

9.3. *The CPT Theorem*

Let us add here the following. The GLT corresponding to $\alpha = 180°$ (see Figures 1, 5) is the 'strong reflection', or *'total inversion'* operation $PT \equiv -1$, as already mentioned in Section 6. Actually, the product of two SLT's (which one *always* yields a subluminal Lorentz transformation [18]) can yield a transformation *both* 'orthochronous', i.e. of the type $LT = +\Lambda_<$, *and* 'non-orthochronous', i.e. of the type $-LT = (PT)\Lambda_<$; let us *in particular* consider the (non-orthochronous) LT that is usually called PT. Since, in order to reach the value $\alpha = 180°$ (starting from $\alpha = 0°$), we must by-pass the case $\alpha = 90°$ (see Figures 1 and 11), then we have to apply the reinterpretation principle (cf. Figure 4): so that *in a universe with charges* the GLT that we called PT does actually produce the exchange particle $\rightleftarrows$ antiparticle and must be effectively considered a CPT operation [18]:

$$(24) \qquad GLT(\alpha = 180°) = CPT,$$

where here C means inversion of *all* (additive) *'charges'*. Since Equation (24) holds both for B's and for T's, then under the 'strong reflection' $GLT = -1$, particles (B or T) in the initial or final state of an interaction process will be transformed into antiparticles ($\bar{B}$ or $\bar{T}$) in the same state of the same interaction process, and vice-versa [18]. For instance, the two reactions

$$a + b \rightarrow c + d,$$
$$\bar{a} + \bar{b} \rightarrow \bar{c} + \bar{d}$$

are the two different descriptions *of the same phenomenon* as seen by the two different inertial frames s_0 and $-s_0 \equiv (CPT)s_0$, respectively.

It is worthwhile explicitly to notice that, if a GLT acts on a fourvector associated with an object, then it analogously acts also on the *other* fourvectors (as four-momentum, four-current, . . .) associated with that object. In particular, the 'strong reflection' operation, $\bar{P}\bar{T}$, which changes sign to 3-position x and time t, will change sign *also* to the 3-momentum p and to the energy E. Thus we are led, within Extended Relativity, to introduce the new symbols $\bar{P}$ (*strong* Parity) and $\bar{T}$ (*strong* Time-Reversal) for meaning the sign-inversions of the first three components and of the fourth component *of all fourvectors, respectively*. Then, the meaning of Equation (24) is the following:

$$(25) \qquad \bar{P}\bar{T} \stackrel{(RIP)}{=} CPT,$$

where:

$$\bar{P} \equiv P\hat{p} \ldots; \quad \bar{T} \equiv T\hat{E} \ldots,$$

and where C, P, T are the usual discrete symmetries (except for the fact that in our formalism $C \equiv \bar{C}$ means conjugation of *all* the additive 'charges').

It follows that the CPT theorem, requiring covariance of all physical laws under CPT, is a direct consequence of Special Relativity (when we do not confine ourselves to subluminal inertial frames).

Equations (24)–(25) suggest also that, within Relativity, 'the right way of doing PT is doing CPT', and this is the essential teaching of Lee and Yang [49]. In fact, Extended Relativity says that we can safely – i.e. covariantly – reflect space-time (or, in the case when T-covariance is supposed to hold, reflect space) only if we simultaneously apply C, so as to have furthermore particles changed into antiparticles.

Let us close by mentioning that the 'RIP' has been given a new, more elegant formulation in a five-dimensional space (see Reference 50), where the fifth axis is *related* to rest-mass (or rather to proper-time).

10. ONLY LAWS, AND NOT DESCRIPTIONS, ARE COVARIANT

Let us remember that the 'Principle of Relativity' (i.e. our first Postulate) does not require at all that two different observers give the same *description* of the same phenomenon (for instance, due to Doppler effect, they can observe the same object as having *different* colors). It does require only that the two observers must find that phenomenon to be ruled by the same physical *laws* (generally speaking, *conservation laws*).

For instance, let us emphasize that the electric charge – as well as energy, and so on – of an isolated system is *not* required to be a relativistic invariant, but only to be constant (as seen by any observer) during the transformation of the system.

It is instructive to analyze an explicit example (Figure 12). Let us

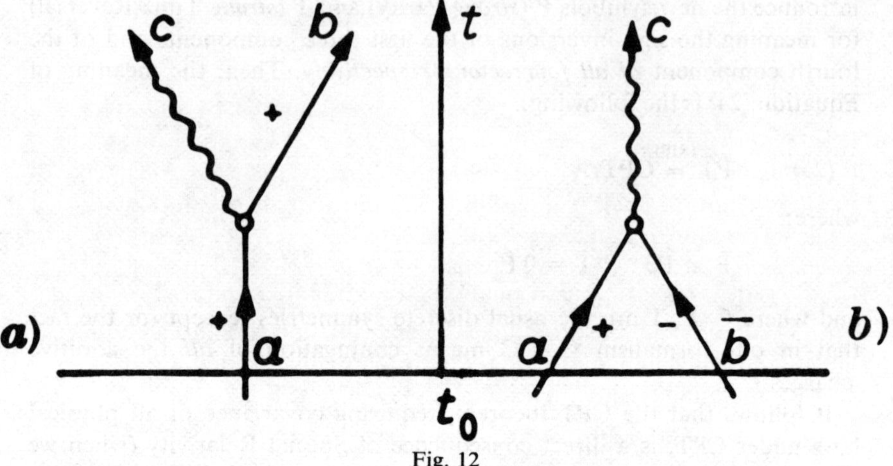

Fig. 12

consider a positively charged particle a that, with respect to a first observer O_1, decays into a neutral particle c and another positively charged particle b having in general different velocity (Figure 12a, where the time-arrow is represented too). It is possible to find another observer O_2, with respect to whom, e.g., the outgoing particle b behaves as an incoming antiparticle b bearing a negative charge. Observer O_2 will judge the process as a 'reasonance' formation or as an 'annihilation' process (Figure 12b). Frame O_1 observes a total electric charge $+1$, while O_2, a total electric charge *zero*. Both observers, however, will agree that the electric-charge conservation-law is verified in the observed process.

Moreover, before interaction O_1 sees *one* particle, while O_2 sees *two* particles.

Therefore, the very number of particles (e.g. *of tachyons*, if we consider only subluminal LT's), at a certain instant, is *not* Lorentz invariant.

However, *the total number of particles* (e.g. of tachyons) *participating in the reaction* (in both initial *and* final states) *is Lorentz invariant*, due to the very features of 'RIP' and of GLT's. Again, we are encouraged to build the physical theories in terms of 'reaction processes' rather than of 'objects'. Such a suggestion has an interesting philosophical meaning.

Before closing this Section. let us underline that the usual proofs of electric-charge invariance hold *only* for subluminal, orthochronous LT's applied to bradyons. We have already shown, on the contrary, that the charge of a particle can change sign under GLT's (cf. Figure 4).

11. MISCELLANEOUS CONSIDERATIONS

11.1 *Again about CPT*

We saw in Section 9 that CPT-covariance is required by the mere SR. It is required, however, *only* for the physical laws of mechanics and electromagnetism (strictly speaking). Extension of the PR (= Principle of Relativity) also to nuclear and subnuclear phenomena (i.e. for strong and weak interactions) should possibly lead to a *new* wider theory (e.g. to *'conformal* relativity'), just as extension from mechanics *to* include electromagnetism led from Galilean–Newtonian theory to Einstein's. Therefore, were, for example, a CPT-covariance violation found in sub-nuclear interactions, it could mean that Lorentz transformations are *no longer* precise enough to be used in building up strong and weak field theory.

11.2. *Crossing Relations*

Besides the CPT theorem, from the mere SR it is possible to derive also the so-called *crossing-relations* (even for *usual* elementary particles). Namely, it has been shown elsewhere [18, 26] that – within Extended Relativity – the *same function* must yield the 'scattering amplitudes' of different processes like [18]:

$$(25) \qquad a + b \to c + d$$

and

$$(25') \qquad a + \bar{c} \to \bar{b} + d$$

in correspondence, of course, to the respective (different) domains of the process-*variables*.

Let us here mention that SLT's could even be considered simply as an 'analytic extrapolation' of the usual ones, which allow us to derive new equations that (as they are expressed in terms of real quantities) admit a clear, new physical meaning. Cf. moreover [18].

In any case, Extended Relativity is useful – even if tachyons should not exist – to understand better also *usual* physics (remember also Section 6).

11.3 *Some Unusual Tachyon-mechanics*

Now, let us go back to tachyons and in particular to Section 4: Namely, let us investigate some unusual consequences of the energy-momentum relation (Equation (12)) for tachyons:

$$(12) \qquad E^2 = p^2 - m_0^2, \qquad (\beta^2 > 1,\ m_0 \text{ real}),$$

where $\beta \equiv V/c$.

For instance, a proton p – when absorbing, e.g., a tachyon t or an antitachyon t̄ (coming from a nearby emitter or from cosmic radiation) – *may a priori* transform *into itself*:

$$(26) \qquad \text{p} + \text{t} \to \text{p}, \qquad \text{p} + \bar{\text{t}} \to \text{p},$$

as it can be kinematically verified in the frame of the initial proton. In the global center-of-mass frame (c.m.f.) it is similarly straightforward to verify that, *on the contrary*, the decay of a proton is *not* kinematically allowed.

However, if we pass from the global c.m.f. to another (subluminal) inertial frame, moving collinearly e.g. with positive speed $u \equiv u_x > c^2/V_x$ (where V_x is the x-component of t or t̄ velocity, and where it is assumed $V_x > c$), we know – from what has been previously seen – that the tachyon t *entering* the first reaction (26) will appear as an *escaping* antitachyon; and the antitachyon t̄ entering the second reaction (26) will appear as an outgoing tachyon. *In the new frame*, therefore, the following reactions are kinematically allowed (in agreement with 'crossing relations'):

$$(27) \qquad \text{p} \to \text{p} + \bar{\text{t}}, \qquad \text{p} \to \text{p} + \text{t} \qquad \text{(only in flight)}.$$

In other words, a proton *in flight* (but not at rest!) can *a priori* be seen to decay into itself plus a tachyon or antitachyon (which will either be absorbed nearby or proceed out to cosmic distances) [51].

More generally, let us notice that *no* body at rest can *emit* any tachyon (in its rest-frame), *unless* it lowers its rest-mass. This can be easily seen, for example, in the case of infinite-speed tachyons. In fact, since transcendent tachyons carry impulse but no energy, they cannot be emitted (nor absorbed[11]) by a body at rest, unless the body lowers its rest-mass, *due to energy conservation*. [Generally speaking, however, a body at rest can *a priori absorb* suitable finite-speed tachyons, both when conserving its rest-mass *and* when changing it].[12] See recent work by Maccarrone and Recami (to appear in Nuovo Cimento A).

11.4. *'Virtual Particles' and Tachyons*

Infinite-speed tachyons, carrying no energy but carrying momentum with magnitude $|\boldsymbol{p}| = m_0 c$, can well be responsible for diffractive scatterings and for other scatterings (and for the so-called 'pomeron-exchange' reactions), as we mentioned before. For instance, let us consider the case when the two bodies *A, B do not* change their rest-masses during the exchange of tachyons, so that *A, B* scatter elastically (in the language of elementary particle physics). Then, *in the c.m.f.*, the two bodies *A, B* happen to exchange only momentum (and no energy), so that they may *naturally* be considered as exchanging infinite-speed tachyons. But this means – *by applying a Lorentz transformation* – that *in general* elastic scatterings can be considered as due to the exchange of *finite-speed tachyons*.

This does agree with the following more general fact. If we briefly invade a field naturally reserved to Q.M. (quantum mechanics), i.e. that one of models for elementary-particle (strong) interactions, and if we confine ourselves to their relativistic (high-energy) behavior, then we find that the so-called *'virtual particles'* generally bear a *negative* four-momentum square:

$$(12') \qquad t \equiv p^2 \equiv E^2 - \boldsymbol{p}^2 < 0$$

just as tachyons do (remember Equation (12)). This fact too suggests that 'virtual particles' (i.e. the objects exchanged between sub-nuclear particles) might be tachyons [53]. (For further details, see References 18, 53 and 54.) Here let us only add that many results that belong to the framework of quantum mechanics seem to be merely derivable from *classical* physics with tachyons (plus suitable *extended* models for elementary particles).

11.5. *Tachyon Localization*

Let us moreover observe that *free bradyons* always admit a particular class of subluminal reference-frames (the rest-frames) whence they appear – in Minkowski space – as *points in space extended in time along a line*. On the contrary, *free tachyons* always admit a particular class of subluminal reference-frames (the 'critical' frames) whence they appear with divergent speed, i.e. as *points in time extended in space along a line* (cf. Figure 1 or 5): see [18].

Considerations of this kind are important for understanding the 'localization' of tachyons, as well – perhaps – as their possible role [53] in hadron structure. Incidentally, in such a respect, an interesting problem would be investigating how a *non-free* bradyon (e.g. a bradyon harmonically oscillating along the *x*-axis) appears to Superluminal frames, in particular to the transcendent one.

Before going on, let us mention that a Superluminal object – when observed by means of its radioemissions (*light*) – will in general appear as occupying *two* positions in space simultaneously. Namely, a tachyon will (in general) be seen as an object that suddenly appears and then splits into two (Superluminal) points travelling in opposite directions. This can be *easily* derived by studying, in Minkowski space-time, the reception by the observer 'at rest' of the light emitted from a Superluminal world-line [18].

11.6. *Vacuum Instabilities*

It is clear that, for each observer, the vacuum can become unstable (via tachyon emission) only by emitting two or more infinite-speed tachyons, in such a way that the total 3-momentum of the emitted set is zero. The emitted total energy will be automatically zero (due to the kinematical properties of transcendent tachyons).

Let us now examine *what* it means to say that a certain observer s sees at the origin, for example, a vacuum decay into a pair of infinite-speed tachyons T, T, travelling along the positive and negative x-axis, respectively, until they are absorbed by two (*moving*) bodies A, B, respectively. Let us suppose that A, B are moving perpendicularly to the x-axis. It will be realized immediately that s can interpret his observations also as due to a pair-creation of infinite-speed tachyons at any point of the x-axis between A and B; or rather at A or B themselves. In other words, he can interpret the observed phenomenon as due to the exchange between A and B of *one* transcendent tachyon: namely, *either* of a transcendent *tachyon* T emited by B and absorbed by A, *or* of a transcendent *antitachyon* $\bar{T}$ emitted by A and absorbed by B. In conclusion, s will see an *elastic interaction* of A and B, due to an infinite-speed transmission of momentum between A and B; and he can choose a description where there is no need to invoke vacuum instabilities in-between.

We should remember that the divergent speed is *not* Lorentz-invariant, the only invariant speed in Extended Relativity being that of light. Therefore, according to new observers s', s'', ..., moving along x, the process will appear as due to the exchange of a tachyon (or antitachyon) *endowed with finite speed*[13].

11.7. *Tachyon Cosmic Flux*

Another possible consideration is the following. Let us consider, for example, a process like the following:

$$(28) \qquad a \rightarrow b + \bar{t},$$

where $\bar{t}$ is an antitachyon. Then, under a suitable LT, a new observer can describe the *same* process as:

(29) $a + t \to b.$

If, in Equation (28), $\bar{t}$ was emitted and then had travelled until absorbed by a (near or *far*) detector, then in Equation (29) t must of course be considered as *emitted* by a (near or *far*) source.

It should then be clear that the tachyon-emission power of matter must be connected with the *tachyon cosmic flux* (as expected also from other, obvious considerations). For instance, if $\Delta\tau$ is the mean-life of particle a for the decay in Equation (28), then:

(30) $[\Delta\tau]^{-1} \propto \phi(t) \cdot \sigma(a, b),$

where $\phi(t)$ is the tachyon (cosmic) *flux*-density and $\sigma(a, b)$ the invariant 'cross-section' of process (29). In other words, $\Delta\tau$ is given by the Lorentz transformation of the average time $\Delta\tau'$ that particle a must spend before being struck by a 'cosmic' tachyon t and transforming into b:

(30') $[\Delta\tau]^{-1} = [L(\Delta\tau')]^{-1}.$

11.8. *Explaining Advanced Solutions*

It has long been known that, in general, relativistic equations admit *advanced* (as well as retarded) solutions. For instance, Maxwell's equations predict both retarded and advanced electromagnetic radiation. Quite naïvely, 'advanced' solutions have been considered as actually associated with motion backwards in time, forgetting the Stückelberg–Feynman 'Reinterpretation Principle' and even the very structure of SR.

Within Extended Relativity, it is easy to explain why *relativistic* (both classical and quantal) equations are expected to admit both retarded and advanced solutions, i.e. solutions *apparently* relative to particles travelling forward and backwards in time, respectively.

In fact, when an equation admits a solution corresponding to (outgoing) particles or photons, *then* a class of suitable GLT's will transform that solution into another one corresponding to (incoming) antiparticles or (anti)photons.

Therefore, if that equation is relativistically covariant (or better *G-covariant*: i.e. it holds for all other inertial observers, both sub- and Superluminal), then *it must also admit solutions relative to incoming antiparticles or photons, whenever* it admits solutions relative to outgoing particles or photons.

We have thus shown that all relativistic equations, covariant under the group G of GLT's *must* admit both retarded and 'advanced' solutions (even if the latter ones are actually *connected to oppositely-directed anti-particles, or photons*, and NOT to objects travelling backwards in time).

The fact that, in general, relativistic equations actually satisfy this requirement – derived from Extended Relativity – is another success [55] of Extended Relativity itself. But it does not mean that all relativistic equations are *already* written (in their present form) in *G*-covariant form. It means, rather, that they can (straightforwardly) be re-written in *G*-covariant form, as we showed elsewhere.

It is obvious to conclude that *no* particle (in particular, *no radiation*) actually travelling backwards in time is either predicted by relativistic equations or expected to be experimentally detectable within the 'world' of SR. We are, however, left with the problem: Why do we usually observe e.g. only the outgoing radiation, and not the incoming (anti)radiation? The whole clue to the answer is in taking into account initial *conditions*. In usual macro-physics some initial conditions are much more probable than other ones. For instance, (mechanical) fluid-dynamics equations *allow* on the see-surface *both* outgoing circular, concentric waves *and* incoming, circular waves tending to a center. However, the initial conditions yielding the first case are much more likely to be observed than the initial conditions yielding the second case.

For a generalization of Maxwell equations for tachyons, see, for example, Reference 18, where also the behavior of tachyons in the gravitational field is discussed.

12. ABOUT CAUSALITY FOR TACHYONS IN MACRO-PHYSICS

12.1. *Macro-objects, Entropy, and Information Transmission by Tachyons*

We want to check now whether our 'Third Postulate' of SR recovers Causality for tachyons not only in the case of elementary processes – as mainly considered, till now – but also *in macroscopic processes*, where entropy considerations are in order.

Let us go back to Figure 4 and *repeat* once more what follows. Let us consider two bodies, A and B (for simplicity, at rest one with respect to the other). Then, let us consider – in their rest-frame s (i.e. in the laboratory) – the exchange of a tachyon T along the positive x-direction, with speed $+V$, from A (source) to B (detector). Now, the frame with speed $u \equiv u_x = c^2/V$ will see the tachyon T travelling at infinite speed, and any frame with speed $u > c^2/V$ would 'see' T as going backwards in time and bearing negative energy: that is to say, it will actually observe an antitachyon $\bar{T}$. In other words, if we consider a (subluminal) boost along the positive x-direction with speed $c^2/V < u < c$, then the new observer s' will *see*, because of 'RIP', an antitachyon $\bar{T}$ emitted by B (source) and absorbed by A (detector), and carrying, of course, positive energy *forward* in time (see Figure 4).

The *point* is that, while s observes e.g. the total energy of A *decreasing* in the process, on the contrary s' observes the *total* energy of A *increasing* in the same process. (We shall soon use these results). However, for each observer the *law* of energy-conservation *is satisfied* in the observed process. Again, only the *description* 'details' are not invariant, in full accord with SR that does *not* require their invariance. Notice that the 'proper energy of A, i.e. the rest-mass of a bradyonic body referred to subluminal frames, is however, Lorentz-*invariant* (as well as its *variations*).

To clarify the situation, let us remember that SR (and Doppler effect) taught us that, for example, the *color* of an object is not Lorentz-invariant (even if nothing prevents any observer, who knows SR, from 'calculating' the 'real' color, i.e. the intrinsic, *proper* color displayed by the object in its rest-frame). The necessary requirement of SR is that each observer puts forward a *self-consistent* description containing the *same* (covariant) physical *laws*.

These simple observations, which also apply to the *previous* case, will help us soon. They teach us that we must not be too surprised if, to s', body A appears to be increasing its total energy, but it appears to be decreasing its total energy to s. Still, all observers can transform their observations into the *proper* frame, where A, B are at-rest, or – more generally – can agree to transform their observations to a (conventionally chosen) 'privileged' frame. *Nevertheless*, all observers – in any case – will put forward a *self-consistent* description, in agreement with SR.

Now, let us ask what happens if, in Figure 4, one may have (instead of *one* tachyon exchange) a tachyon-signal transmission by means of a *modulated tachyon-beam*. Then, *if* entropy-changes and information-exchanges are to be connected, e.g., with the variations of the *proper* internal energies (where 'proper' means measured in the frame at rest with respect to A or B, respectively), they would *not* necessarily behave as the 'total energies'. We shall illustrate the problems which could possibly arise by proposing and considering two paradoxes.

12.2. *Can a Tachyonic-Observer Inform Us about Our Future?*

First, let us consider the case of a subluminal frame (x, t) and of a Super-luminal frame (x', t') boosted along the (positive) x-direction with relative-speed $U > c$ (see Figure 5). Let us choose as common origin O, the event when both frames coincide. To s, the x-axis represents all the events contemporary with O; whilst, to s', all the events contemporary with O are represented by the x'-axis. If information can be carried by (almost) infinite-speed tachyons, then observer s' – when near to O – can receive signals from event E (see Figure 5) and can pass on this information to s. It *seems*, therefore, that s' can communicate (when near O) to s informations on the *future* of s. This problem will be studied elsewhere.

However, it is clear that the 'RIP' teaches us the following: *The Super-luminal signal $E \to O$ will appear to observer s as a slower-than-light signal* (or rather 'antisignal' [56]) *sent by s' from O towards E* (since, with respect to s, it would be a 'negative-energy signal travelling backwards in time' which *must* be reinterpreted according to 'RIP'). It *seems* that, according to s, observer s' is predicting to him nothing but what can *actually* be 'fore-cast' on the basis of our physical operations and of physical laws; i.e. observer s' is merely 'predicting' to s the obvious effects on E of the (anti)-*signals* sent by himself from O towards E, so as to *physically influence* it.

What matters to us here is that even in this case, each observer puts forward a *different description* of the same chain of events, *but* the Law of (Retarded) Causality seems to hold *in all descriptions*, for all observers.

Let us now pass to another example, and spend some more time in investigating the tachyon-mechanics involved in it.

12.3. *Tachyons, Free-Will and Entropy*

Let us now consider a process such as in Figure 4, but with body B moving with respect to body A. Let us, therefore, consider *two* usual, inertial observers (Figure 1) moving *along the x-direction* relative to each other, with speed $u < c$, and let us suppose that (according to the frame $s \equiv (x, t)$) observer s sends a signal to $s' \equiv (x', t')$ by means of a tachyonic beam with speed $V > c^2/u$ along the positive x-direction. According to s', however, the 'tachyon-beam' would actually appear as an antitachyon-beam emitted by s' itself towards s.

Now, we can well imagine that – when overcoming s' at O – observer s told s' about his intention of transmitting to him (i.e. to s') a tachyonic signal (with speed $V > c^2/u$) at time t, whose Lorentz transformed value is time t'. Then, it *seems* that observer s' will be '*compelled*' – at a certain instant $t' - \Delta t'$ – to emit an (anti)-signal[14] towards s, in order . . . to save the validity of the classical theory of tachyons in four-dimensions. But this fact would of course violate the 'free-will' of observer s'; which *can* on the contrary decide not to send any signal (or antisignal) towards s! *A fortiori*, observer s' can be in the impossibility of sending out *that* signal, lacking information on it (for instance, s told s' – at O – that at time t he himself will send to s' one piece of music via a tachyon-beam having $V > c^2/u$; and s' may have no record of that music, so that he *cannot* send out to s any anti-*music*, i.e. any music *by* antitachyon-beams). The situation seems self-contradictory. First of all, let us note that the free-will of observer s' is not actually jeopardized since s' can well decide not to switch on his 'tachyon-radio' B: in such a case nothing will happen [i.e., s will see his signal bypass s' without being absorbed, and s' will see his 'radio B' emitting no (anti)signal at all]. It is interesting, moreover, to analyze more deeply the mechanics related to such emissions and absorptions of tachyons. *For instance*, if we assume the rest-mass of B *not* to change during the tachyon absorption, then (in such a case) the *mechanical laws of tachyons do not allow observer s'* (the moving body B) *to absorb any tachyon with speed $V > c^2/u$.* In other words, s' can get information from s through such a process (tachyon-beam emission by A and tachyon-beam absorption by B), under that assumption, only by means of tachyons having speeds V *smaller* than c^2/u (*in which case no causal problem arises*).

Let us explain that fact. First, let us emphasize again that a body *cannot* emit or absorb any tachyon-momentum, but only *definite* values of it. *For instance*, we already saw that a body A at rest (with respect to an observer s_0) *cannot* emit or absorb tachyons endowed with infinite-speed (with respect to s_0), unless A suitably decreases its rest-mass. This was a trivial consequence of energy-momentum conservation. Here, for simplicity's sake we are considering (moving) bodies B that *do not change* their rest-mass in the process of tachyon absorption. The results we are going to obtain, however, hold also when B increases its rest-mass M_B, or rather when

$$\Delta \equiv M_B'^2 - M_B^2 = 2p_\mu P^\mu - m_0^2 \geq - m_0^2,$$

where m_0, p are tachyon proper-mass and 3-momentum, respectively, and P is the initial 3-momentum of body B; cf. [56], last title.

It must be stressed that we are essentially considering here only tachyon *emission* or *absorption*, i.e. we are investigating only information-transmission via tachyon-beams that are merely *either* emitted (by the 'apparent' source) *or* absorbed (by the 'apparent' detector), but *not* scattered away[15].

Now, then, let us consider a *macro-object B*, with restmass M_B, moving with (subluminal) speed u in the positive x-direction and absorbing tachyons that move (with respect to the same reference frame s_0) along the positive x-axis as well. Let the tachyons be emitted by a body A, having rest-mass M_A, at rest (see the following) in the origin of s_0. Then, in the process of tachyon-absorption by B we shall have [in natural units]:

$$(31) \qquad \sqrt{p^2 - m_0^2} + \sqrt{P^2 + M_B^2} = \sqrt{(p + P)^2 + M_B^2}.$$

One immediately obtains p as a function of m_0 and P,

$$(32) \qquad |p| = \frac{m_0}{2M_B^2} [|P|m_0 + \sqrt{(P^2 + M_B^2)(m_0^2 + 4M_B^2)}],$$

so that, for every value of m_0, the body B can absorb only the tachyons with a definite (*discrete*) magnitude of p.

After having considered the absorption-process (by B), let us turn our attention to the *emission*-process by A. The tachyon-Mechanics is such that body A (at rest in s_0) can *emit* tachyons (of any proper-mass) *only by lowering its rest-mass*, as we know. Thus, in this case, we *cannot* neglect $\Delta M_A \equiv M_A' - M_A$; some trivial kinematics tells us that $[E_T \equiv \sqrt{p^2 - m_0^2}]$:

$$\Delta(M_A^2) \equiv M_A'^2 - M_A^2 = -m_0^2 - 2M_A E_T \leq -p^2 < -m_0^2.$$

If $\Delta(M_A^2)$ is *not* quantized (as we expect when dealing with a macroscopic object A), the energy balance for the emission-process:

$$(33) \qquad M_A = \sqrt{p^2 + M_A'^2} + \sqrt{p^2 - m_0^2}$$

will *not* yield any definite constraint on the value of m_0 (in terms of p and M_A)[16]. Therefore, under the previous assumptions, we are left only with the constraint Equation (32).

From Equation (32), it is straightforward to derive the important relation

$$(34) \qquad V < c^2/u,$$

where we should remember that $u \equiv u_x$ is the relative-speed between A and B, and $V \equiv V_x$ is the speed of the beam-tachyons.

As a conclusion, if A wants to transmit information to B by tachyon-beams, it *cannot* use tachyons with speeds $V > c^2/u$, that is to say it can only use tachyons that will *appear even to B as travelling from A to B*. (under our assumptions).

We can *state that*: Every time that (the macro-object) A transmits information by a modulated tachyon-beam to the (moving) *macro-object*

B, which absorbs the beam (either by conserving or by increasing its rest-mass), then tachyon-dynamics assures us that *also* B will always see the beam as going from A to B, so that in this case *neither* the free-will of observer B will be jeopardized, *nor* we shall meet any contradictory situation (of the kind hypothetically proposed in this Section).

In connection with Equation (34), let us notice that – in four dimensions – the relevant quantity will be, instead of V, the tachyon speed-component along the direction B–A. In connection with Equation (32) let us underline that, *for every* m_0, the body B can absorb only tachyons with *discrete* values of $|p|$ [whenever the quantity $\Delta \equiv M_B'^2 - M_B^2$ is *zero* or assumes only discrete (positive) values: cf. refs. (54, 56)].

12.4. *Again about Laws and Descriptions* (*plus some Science Fiction*).

In the previous sub-section we have been considering *macro-bradyons* essentially absorbing (or emitting) beams made of *micro*-tachyons. Here we want to give a *hint* about how to proceed when dealing also with *macro-tachyons*. At this point, let us take the liberty of using some science fiction. Namely, let us assume that one Reader R_1 sits on a chair and that an 'author' A_1 *switches on*, around him, a (rotating) Kerr or Kerr–Newman *black-hole* K_1 (as big as a building, let us imagine), so that R_1 finds himself inside the ergosphere, near its *internal* boundary-surface [57]. Then, let the chair start receiving angular-momentum, so that R_1 crosses the *external* boundary-surface of the ergosphere, transforming–*let us imagine*–into a tachyon. Afterwards, R_1 travels e.g. along the positive x-direction with speed $V > c$ until he reaches – in very short time – another, far planet, where a pioneer ('author' A_2) has already arrived and switched on a second, analogous Kerr (or Kerr–Newman) black-hole K_2. Then K_2 captures R_1 who, entering the ergosphere of K_2, transforms again into a bradyon. At last, the angular momentum of the R_1-chair is slowed down until R_1 approaches the internal surface of K_2-ergosphere, and *eventually* A_2 switches off K_2 (so that R_1 can leave his chair and walk on the new planet). Of course, K_1 and K_2 are considered practically at rest one with respect to the other (for simplicity).

At this point, if we now consider a second[17] Reader R_2, travelling (along the positive x-direction) with subluminal speed $u > c^2/V$, he will *not* see the reader R_1 going from K_1 to K_2, but on the contrary an 'anti-reader' $\bar{R}_1$ *going from* K_2 *to* K_1. Or, rather, he will observe K_2 *creating the pair* R_1, $\bar{R}_1$ (the fact that in this case the traveller $\bar{R}_1$ is tachyonic and the walking-on-the-new-planet R_1 is bradyonic is not essential, and due to the *internal* action of K_2). We are ready to accept *micro-objects* pair-creation (provided some energy is supplied), but perhaps we are not eager to accept *macro-objects* (reader-antireader) pair-creation. Therefore, if we know (extended) Relativity, a simple 'recipe' can be suggested:

Recipe (for orthodox persons): through the suitable Lorentz transformation, *go* from your description to the description in the K_2 rest-frame: and you shall find a more orthodox description (i.e., reception of the tachyonic R_1, his slowing down, and emission of bradyonic R_1).

Naturally enough, however, we expect that the reader R_2 (and any people that wanted to apply the 'recipe') will operate an analogous Lorentz transformation *whenever* they observe e.g. the dress-color of the (subluminal) reader R_2 as *shifted* owing to the standard 'Doppler effect'!

Fiction aside, let us remember that the descriptions always become *more orthodox* in some suitable frames (e.g. in the detector frame, or in some *rest*-frames, . . .)[18].

13. 'APPEARANCE' AND RELATIVISTIC LAWS

Before concluding, let us further *clarify* the last subsection by adding the following considerations. Recently, Edmonds [58] *believed* he had shown that the controlled creation and absorption of tachyon implies a violation of the Causality Law; in fact, he suggested doing the following.

We build a long rocket sled with a 'tachyon-laser' at the left end and a 'target flower' at the right end. A short lever sticks out of the side of the 'laser'. If we trip the lever, the tachyon laser emits a very sharp, intense burst of tachyons for which we measure the speed of, let us say, V. These tachyons then hit the 'flower' and blast it into pieces. The flower absorbs all the tachyons in the pulse as it explodes, so the tachyons disappear (just as absorbed light disappears).

Now, we accelerate the sled (with 'charged' tachyon-laser and flower attached to it) up to an incoming speed of $-v \equiv -v_x$, relative to our frame, and then turn off its rocket engines.

Moreover, we form a long line of 'astronauts' floating in space along the x-axis (i.e. along the rocket-sled motion line). The sled is aligned along x and moves to the left, in the $-x$ direction. Each astronaut has a 'roulette wheel' in his one hand, and keeps spinning his gambling wheel until he gets, let's say, the number 13. When he happens to do so, he quickly puts out a stick in front of him which *could* beat the trigger on the moving laser. No one in our frame knows when a given astronaut will get 13 to come up. Some astronauts may get 13, but be too far down the line, or find the trigger has already passed them when they get it. But, finally, someone gets the right number, puts out his stick, and finds that the sled is almost at his position and he triggers the laser.

Once the laser fires, the observers travelling with the sled see a burst of tachyons travelling from the laser to the flower. If the sled is moving slowly enough, then we also see the flower blow up at a time *later* than the time at which the laser fires. However, if the sled is fast enough $(v > c^2/V)$, we see a pulse of antitachyons going from the flower to the laser: namely, we *would see* the flower blow up *before* the laser fires. Therefore, the astronaut who triggers the laser sees the laser immediately 'swallowing' a pulse of antitachyons coming from the flower (instead of seeing the laser firing a burst of tachyons towards the flower). In other words, the lucky astronaut will conclude that the flower had to *know in advance* who was going to get 13 (so that it can blow up and create the antitachyon-beam just at the right time, in order for the beam to arrive at the lucky astronaut as he gets the number 13 to come up for him).

We are now going to answer Edmonds' considerations, showing that actually we have *no* paradoxes (and in the meantime clarifying the previous subsection).

Let us first observe that, since 'source' and 'detector' are supposed by Edmonds to be at rest one with respect to the other, according to both laser and flower – i.e. in the laboratory frame – there are *no* problems about the actual flight-direction of the tachyons.

However, if we choose *other* reference frames (as the astronauts), then something *apparently* strange can happen. Namely, on triggering the laser, the lucky astronaut does not see the laser firing tachyons T, but on the contrary absorbing anit-tachyons $\bar{T}$ apparently coming from the flower.

We want to stress that here there are no paradoxes: The astronaut, knowing Relativity (also the Extended one) we hope, can calculate tachyons' behavior in the laser frame, and find out that actually it was the laser which fired out tachyons, even if the very large relative speed between tachyons and astronaut produces a high 'distortion' of the process *observed* by him. In other words, it is not important that it *seems* to the astronaut that the flower precognize the future; it is important, on the contrary, that the flower actually *not*. And in fact, in the flower frame, the tachyons are not emitted but absorbed!

It is possible to *clarify* the whole story *by an* (amusing) *example taken out from usual, standard Special relativity* (let us therefore forget tachyons in what follows).

Let us suppose we are informed about a cosmic fight between two different species of extraterrestrial living-beings, each one driving his own interplanetary rocket, where the rocket-colors are *violet* for the first species and *green* for the second one. Let us suppose moreover that we know that the 'green men' possess an inviolable natural instinct that makes them peaceful, so that they are inhibited from firing their guns; on the contrary, the 'violet men' possess an aggressive, warrior instinct.

When we observe the cosmic battle with our telescope, it can well happen – due to the Doppler effect – that, when a 'violet man' fires his gun and strikes a green rocket, the violet color appears to us as green and *vice-versa* (because of the rocket motions). Then, according to Edmonds, we should deduce that an (inviolable) natural law has been badly violated (the instinctive law of those extraterrestrial beings). On the contrary, we merely *observe – at first sight –* an apparent violation of natural laws; but, if we know physical theories (i.e. Relativity, and the rocket velocities), then we can determine the 'proper colors' of the rockets, in their own rest-frames, and resolve our wondering.

In other words, any observer is capable of understanding the physical happenings in terms of (only) the physics of his own observations *provided he properly uses his knowledge of the theory of relativity.*

This means that we can scientifically *observe* the natural world only if we are endowed with *theoretical* 'instruments', besides the experimental devices.

14. DISCUSSION OF SECTIONS 12, 13

In the previous section 13 (analogously to section 14.4) we suggested passing to the relevant frames in order to obtain more orthodox descriptions, devoid of the distortions introduced by the high (Superluminal) relative speeds. For instance, to decide whether the 'flower' did 'actually' (i.e. *intrinsically*) precognize the future *or not*, one had to go to the flower proper-frame (whence the flower behavior, incidentally, did not appear to violate causality).

That procedure, even if quite natural and generally successful, does however encounter some difficulties:

(i) First of all, when it is applied by observer s to observer s' of the case in section 12.2, it generates obvious perplexities in s; *even if each observer* (s, s') *will see 'causes to precede their effects'*. Such perplexities originate when two observers communicate to each other, i.e. when one and the same observer evaluates and compares different descriptions with his own description.

(ii) Secondly, if 'laser' and 'flower' of section 13 are supposed to be in *relative* motion, that procedure does not eliminate all problems, since we are then left with the same difficulty as discussed in section 12.3.

In section 12.3 we saw the descriptions put forth by 'source' and 'detector' to agree if the relation (cf. page 286)

$$(35a) \qquad \Delta \geq - m_0^2$$

is satisfied. In such a case, *any difficulties essentially vanish*. However, when

$$(35b) \qquad - M_B^2 \leq \Delta < - m_0^2,$$

observer A will see tachyons T emitted by A and *absorbed* by B, while observer B will see a beam of antitachyons $\bar{\text{T}}$ *emitted* by B and absorbed by A. In this case, it seems that some interesting problems remain to be exploited (in tachyon macro-physics), whose discussion apparently requires taking into account different subjects which *may* range from the *peculiar* behavior of tachyon sources and detectors, to the Wheeler-Feynman-type theories, to the spontaneous tachyon-emission properties of matter, to the controllability of tachyon beams (e.g., to the probability that the role of tachyons be confined to micro-physics or to probabilistic phenomena), to information theory and the possibility of transmitting signals by tachyon beams, to the coupling-strength of tachyons to ordinary matter (and even to the question whether Minkowski space-time is enough for allocating the 'free-will' behavior). All these subjects seem to need further, joint investigation in macrophysics.

ACKNOWLEDGMENTS

Part of this paper is based on work done by the authors together with R. Mignani. Other shorter parts are based on work done with V.S. Olkhovsky, M. Pavšič. Thanks are due to A. Agodi, R.S. Cohen, M. Dalla Chiara, G. Toraldo di Francia for discussions, and to S. Leotta for his kind collaboration.

NOTES

* Work partially supported by I.N.F.N., MPI (ex-art. 286 T.U.) and by CNR.

[1] For a definition of 'equivalence', see Section 4.

[2] L. Fantappié was the first to notice that (usual) Special Relativity, when *deprived* of the 'Third Postulate', *would allow* sending signals into the past. See also Reference 44.

[3] Here, and in the following, 'charge' means *any* additive charge, like electric, magnetic, barionic, leptonic, . . . charges. Therefore, our definitions of 'charge conjugation' etc. will be *different* from the usual ones, but in agreement with Reference 18. Cf. References 18, 25, 26.

[4] Actually, the mere 'RIP' yields the *antiparticle* except for the helicity sign. But the *full* result is immediately got when considering the action of the *complete* Lorentz transformation (together with 'RIP').

[5] For definitions of 'causes', 'effects' and 'causal connection' see Section 8.

[6] In fact, antiphotons essentially coincide with photons; see Reference 18.

[7] Since we have to deal with divergent speeds, we may usefully borrow a bit from projective geometry, where e.g. straight lines are considered as large (infinite)-radius circles.

[8] Before applying the 'RIP'! We shall see, in Section 9, that—after application of 'RIP' – our PT operation is essentially the usual CPT. In our formalism, the CPT operation is a *linear* (classical) operator in the pseudo-Euclidean space (since it is an element of G), and a *unitary* (quantum-mechanical) operator when acting on the space of Q.M.-states.

[9] See note 8.

[10] See note 7.

[11] *Remember*, in fact, that transcendent tachyons have no definite direction along their motion line (in the sense that an infinite-speed-tachyon emission is totally equivalent to an infinite-speed-antitachyon absorption, and vice-versa).

[12] Notice that the exchange of finite-speed tachyons *might* be capable of solving e.g. the classical-physics paradoxes [51] connected with pair creation in a constant electric field (see References 52).

[13] Of course, the same description would be given by s', s'', ..., even if observer s chooses to describe his observations in terms of a vacuum-decay of the kind seen above.

[14] An 'anti-signal' is carried by anti-particles just as the 'signal' is carried by (their) particles; but both carry *positive* energy, of course. Cf. also Reference 56.

[15] In fact, we do *not* surely know yet *how* tachyons behave when interacting with matter, even if Extended Relativity can help us guess. (For instance, would the phenomenon corresponding to 'bradyon termalization' be that of having tachyons going approximately at infinite-speed?).

[16] However, we shall get immediately such a constraint when A is a sub-nuclear particle, e.g. when considering the process $\Delta_{33}(1232) \rightarrow p + t$, in the rest-frame of the Δ_{33}-*resonance*. Considerations of this kind are useful when studying the possible role of tachyons in elementary-particle interactions ('*dual theories*', '*strings*', '*Higgs mechanism*', '*instantons*', ..., besides 'virtual particles' and 'pomerons'). Tachyons might even be the strong-field quanta (compare e.g. the electromagnetic decay of an excited atom $A \rightarrow A + \gamma$ with the previous, possible process $\Delta_{33} \rightarrow p + t$).

[17] An ingredient of this subsection is the assumption that the set of Readers contains $n > 2$ elements.

[18] The same also applies, for instance, when body B of Figure 1 has a complicated internal structure, so as to contain *moving* constituent-objects: in this case, B can practically *appear* as absorbing also tachyons with the *prohibited* speeds $V \gtrsim c^2/u$ *in the overall c.m.f. of B* (where u is the speed of B in the c.m.f), even under our hypotheses.

REFERENCES

[1] A. Einstein, *Ann. der Phys.* **17**, 891 (1905).

[2] G. Feinberg, *Phys. Rev.* **159**, 1089 (1976).

[3] E. Recami, *Accad. Naz. Lincei Rendic. Sc.* **49**, 77 (1970).

[4] M. Baldo, G. Fonte and E. Recami, *Lett. Nuovo Cim.* **4**, 241 (1970).

[5] O.M.P. Bilaniuk, V.K. Deshpande and E.C.G. Sudarshan, *Amer. Journ. Phys.* **30**, 718 (1962).

[6] E.C.G. Stückelberg, *Helv. Phys. Acta* **14**, 324, 588 (1941). See also P.A.M. Dirac, *Proc. Roy. Soc.* A **126**, 360 (1930); O. Klein: *Zeit. für Phys.* **53**, 157 (1929).

[7] R.P. Feynman, *Phys. Rev.* **76**, 749, 769 (1949).

[8] T. Lucretius Caro, *De rerum natura*, book **4**, lines 201–203, M.T. Cicero (ed.) (Rome, about 50 B.C.).

[9] H.C. Corben, in *Tachyons, Monopoles and Related Topics*, ed. by E. Recami (Amsterdam, North-Holland, 1978), p.31; *Nuovo Cimento* **29A**, 415 (1975).

[10] *Translation*, "Don't you see how they must go faster and further/And travel a larger interval of space in the same amount of/Time than the Sun's light as it spreads across the sky?".

[11] J.J. Thomson, *Phil. Mag.* **28**, 13 (1889).

[12] A. Sommerfeld, *K. Akad. Wet. Amsterdam Proc.* **8**, 346 (1904); *Nachr. Ges. Wiss. Göttingen*, Feb. 25, (1905), p. 201.

[13] E.C.G. Sudarshan, Report CPT–166 (Austin, Tex., 1972); Report NYO–3399–191/SU–1206–191 (Syracuse, N.Y., 1968).

[14] R.C. Tolman, *The Theory of Relativity of Motion* (Berkeley, Cal., 1917), p. 54. This apparent 'paradox' has been recently proposed again by many physicists, unaware of the already existing literature.

[15] Apart from the purely *mathematical* considerations by E. Majorana: *Nuovo Cim.* **9**, 335 (1932) and *Ann. of Math.* **40**, 149 (1939).

[16] H. Arzelies, *Compt. Rend.* **245**, 2698 (1957); H. Schmidt: *Zeit. Phys.* **151** 365, 408 (1958).

[17] E. Recami, Report IFUM–088/SM (Milano, Aug. 1968); *Giornale di fisica* **10**, 195 (1968); V.S. Olkhovsky and E. Recami. *Nuovo Cimento* **63**, 814 (1969).

[18] E. Recami and R. Mignani, *Rivista Nuovo Cim.* **4**, 209–290, 398 (1974), *and references therein*. See also E. Recami (editor): *Tachyons, Monopoles, and Related Topics*, North-Holland Pub. Co. (Amsterdam, 1978); E. Recami, chapt. 16 in *Albert Einstein 1879–1979: Relativity, Quanta and Cosmology . . .* , ed. by F. De Finis (Johnson Rep. Co., New York, N.Y., 1979), vol. **2**, p. 537–597; E. Recami: chapt. 18 in *Centenario di Einstein: Astrofisica e Cosmologia, Gravitazione, Quanti e Relatività nello sviluppo del pensiero scientifico di A. Einstein*, ed. by M. Pantaleo (Giunti-Barbera, Florence, 1979), p. 1021–1097; G.D. Maccarrone and E. Recami, *Lett. Nuovo Cim. (in press)*; A.O. Barut, G.D. Maccarrone and E. Recami, (to appear).

[19] See e.g. E. Recami and R. Mignani, ref. (18).

[20] See e.g. V. Berzi and V. Gorini, *Journ. Math. Phys.* **10**, 1518 (1969); V. Gorini and A. Zecca, *Journ. Math. Phys.* **11**, 2226 (1970), and references therein; L.A. Lugiato and V. Gorini, *Journ. Math. Phys.* **13**, 665 (1972).

[21] W.V. Ignatowski, *Phys. Zeits.* **21**, 972 (1910); P. Frank and H. Rothe, *Ann. der Phys.* **34**, 825 (1911); E. Hahn, *Arch. Math. Phys.* **21** 1, (1913). Among the most recent works, see e.g. F. Severi in *Cinquant'anni di relativita'*, M. Pantaleo editor (Florence, 1955); V. Gorini and A. Zecca, ref. (10); M. Di Jorio, *Nuovo Cim.* **B22** 70, (1974).

[22] D.W. Sciama, Preprint IC/74/10 (ICTP, Trieste, 1974).

[23] See e.g. C.N. Gordon, *Found. of Phys.* **5**, 173 (1975); S.C. Barrowes, *Found. of Phys.* **7**, 617, (1977).

[24] P. Caldirola and E. Recami, Proceedings of the '1976 *Conference on the Concept of Time*' (S. Margherita L., Italy), *Epistemologia* **1** 263–304 (1978).

[25] E. Recami, in *Enciclopedia EST-Mondadori, Annuario 73* (Milan, 1973), p. 85.

[26] See ref. (19); ref. (25); R. Mignani and E. Recami, *Nuovo Cimento* **24**A, 438 (1974); *Int. Journ. Theor. Phys.* **12**, 299 (1975); *Lett. Nuovo Cim.* **11**, 421 (1974).

[27] P.L. Csonka, *Nucl. Phys.* **21**B, 436 (1970).

[28] See E. Recami and R. Mignani, ref. (18). See also the successive papers A. Yaccarini, *Lett. Nuovo Cim.* **9**, 354 (1974), and (unpublished); R. Mignani and E. Recami, *Lett. Nuovo Cim.* **9**, 357 (1974); **9**, 367 (1974); *Nuovo Cimento* **21**B, 210 (1974); *Lett. Nuovo Cim.* **12**, 263 (1975); H.C. Corben, *Lett. Nuovo Cim.* **11**, 533 (1974); *Int. J. Theor. Phys.* **15**, 703 (1976); R. Mignani and E. Recami, *Lett. Nuovo Cim.* **13**, 589 (1975); *Int. Journ. Theor. Phys.* **12**, 299 (1975); K.T. Shah, *Lett. N. Cim.* **18**, 156 (1977); E. Recami; *Lett. N. Cim.* **22**, 591 (1978); P. Castorina and E. Recami, *Lett. N. Cim.* **22**, 591 (1978). For *different* approaches, see also A.F. Antippa and A.E. Everett, *Phys. Rev.* **8**, 2352 (1973); R. Goldoni, *Nuovo Cimento* **14**A, 501, 527 (1974); *Lett. Nuovo Cim.* **21**, 333 (1978).

[29] E. Recami and E. Modica, *Lett. Nuovo Cimento* **12**, 263 (1975).

[30] E. Recami and R. Mignani, *Lett. Nuovo Cim.* **4**, 144 (1972).

[31] R. Mignani and E. Recami, *Nuovo Cimento* **14**A, 169 (1973); **16**A, 208 (1973); A. Antippa, *Nuovo Cimento* **10**A, 389 (1972); E. Recami and R. Mignani, *Lett. Nuovo Cimento* **8**, 110 (1973); V.S. Olkhovsky and E. Recami, *Lett. Nuovo Cim.* **1**, 165 (1971); R. Mignani, E. Recami and U. Lombardo, *Lett. Nuovo Cim.* **4**, 624 (1972).

[32] L. Parker, *Phys. Rev.* **188**, 2287 (1969).

[33] See e.g. W. Rindler, *Special Relativity* (Edinburgh, 1966), p. 16.

[34] See e.g. L. Landau and E. Lifshitz, *Théorie du champ* (Moscow, 1966); *Mécanique* (Moscow, 1965); E.F. Taylor and J.A. Wheeler, *Space-Time Physics* (San Francisco, Cal., 1966).

[35] (a) K.T. Shah, *Lett. Nuovo Cimento* **18**, 156, (1977), and references therein; and in *Tachyons, Monopoles, and Related Topics*, ed. by E. Recami (Amsterdam, North-Holland, 1978), p. 49; (b) P. Caldirola, M. Pavšič and E. Recami, *Nuovo Cimento* B (1978); *Physics Letters* **A66**, 9 (1978); P. Caldirola and E. Recami, *Lett. Nuovo Cim.* (in press); (c) R. Mignani and E. Recami, *Phys. Letters* **65**B 148 (1976); *Nuovo Cimento* **30**A, 533 (1975); M. Pavšič and E. Recami, *Nuovo Cimento* **A36**, 171 (1976); **A46**, 298 (1978); E. Recami, *Lett. Nuovo Cim.* **21**, 208 P. Castorina and E. Recami, *Lett. N. Cim.* **22**, 195 (1978); (d) E. Recami and R. Mignani, (ref. 19); R. Mignani

and E. Recami, *Lett. Nuovo Cim.* **9**. 357 (1974); **16**, 449 (1976); (e) E. Fabri, *Nuovo Cimento* **14**, 1130 (1959); A. Agodi (unpublished).

[36] E. Recami and C. Spitaleri, *Scientia* **110**, 204 (1975); G. Arcidiacono, *Relatività e Cosmologia* (Veschi Pub., Rome, 1973), p. 233; Th. Kaluza, *Sitzungsber. Preuss. Akad. Wiss., Math. Phys. Kl.* (1921), 966; O. Klein, *Zeits. Phys.* **37**, 895 (1926); A. Einstein and P. Bergmann, *Ann. Math.* **39**, 683 (1938); E. Leibowitz and N. Rosen, *General Relat. Gravit.* **6**, 449 (1973). For the interpretation of the fifth axis, see E. Recami and G. Ziino, *Nuovo Cimento* **33A**, 205 (1976); R.L. Ingraham, *Nuovo Cimento* **12**, 825 (1954); A. Maduemezia. Report IC/68/78 (ICTP, Trieste, 1968); J.D. Edmonds, Jr., *Found. of Phys.* **5**, 239 (1975); P. Caldirola, *Nuovo Cimento* **19**, 25 (1942); *Supplem. Nuovo Cim.* **3**, 297 (1956).

[37] E.C.G. Sudarshan, in *Symposia on Theor. Phys. and Math.*, vol. 10, p. 129 (New York, N.Y., 1970); O.M. Bilaniuk and E.C.G. Sudarshan, *Nature* **223**, 386 (1969); J.A. Parmentola, D.D.H. Yee, *Phys. Rev.* **D4**, 1912 (1971); R.G. Root and J.S. Trefil, *Lett. Nuovo Cim.* **3**, 412 (1970). See also refs. (5, 6, 7, 17).

[38] See ref. (19), and e.g. M. Camenzind, *Gen. Relat. Grav.* **1**, 41 (1970).

[39] E. Recami and R. Mignani, ref. (19); *Lett. Nuovo Cimento* **4**, 144 (1972); **8**, 110 (1973); E. Recami, ref. (25); P.L. Csonka, *Nucl. Phys.* **21B**, 436 (1970); R.G. Root and J.S. Trefil, ref. (37); J.A. Parmentola and D.D.H. Yee, ref. (37); E. Recami, *Found. of Phys.* **8**, 329 (1978).

[40] See e.g. refs. (68) in ref. (18); in particular R. Fox *et al.*, *Proc. Roy. Soc.* **A316**, 515 (1970); *Nature* **223**, 597 (1969); G.A. Benford *et al.*, *Phys. Rev.* **D2**, 263 (1970); J. Strnad, *Fortsch. Phys.* **18**, 237 (1970); W.B. Rolnick, *Phys. Rev.* **183**, 1105 (1969); **D6**, 2300 (1972); D.J. Thoules, *Nature* **224**, 506 (1969); J. Ehlers, private communication to Sudarshan, ref. (37); J. Strnad and A. Kodre, *Lett. Nuovo Cim.* **12**, 261 (1975).

[41] F.A.E. Pirani, *Phys. Rev.* **D1**, 3224 (1970).

[42] E. Recami, ref. (25); E.C.G. Sudarshan, ref. (37); O.M. Bilaniuk and E.C.G. Sudarshan, ref. (37); P.L. Csonka, ref. (39); R.G. Root and J.S. Trefil, ref. (37); J.A. Parmentola & D.D.H. Yee, ref. (37); E. Recami, ref. [39].

[43] J.A. Parmentola and D.D.H. Yee, ref. (37).

[44] E. Recami, *Scientia* **109**, 721 (1974). We take advantage of this opportunity to note that line 17 from the bottom of pag. 723 (. . . "case to introduce the imaginary unit and to write" . . .) is misprinted, and should read: ". . . case to introduce the imaginary unit. On the contrary, we can write[2,19]): . . .".

[45] R.G. Root and J.S. Trefil, ref. (37).

[46] R. Newton, *Phys. Rev.* **162**, 1274 (1967); P.L. Csonka, ref. (39); *Phys. Rev.* **180**, 1266 (1969); E. Recami, ref. (25); E. Recami and R. Mignani, ref. (18).

[47] E.C.G. Sudarshan, Report CPT–55/AEC–14 (Austin, Tex., 1970); A.B. Volkov, *Canad. Journ. Phys.* **49**, 1697 (1971); D. Leiter, *Nuovo Cimento* **2A**, 679 (1971).

[48] J.A. Wheeler and R.P. Feynman, *Rev. Mod. Phys.* **17**, 157 (1945); **21**, 425 (1949).

[49] T.D. Lee and C.N. Yang, *Phys. Rev.* **105**, 1671 (1957).

[50] E. Recami and G. Ziino, *Nuovo Cimento* **33A**, 205 (1976).

[51] See E. Recami and R. Mignani, ref. (18); *Lett. Nuovo Cim.* **8**, 780 (1973); see also ref. (53).

[52] Ya.B. Zeldovich, *Phys. Letters* **52B**, 341 (1974), footnote at pag. 342; and in *Magic*

without magic: *J.A. Wheeler*, edited by J.R. Klauder, pag. 279 (San Francisco, Cal., 1972).

[53] Cf. pag. 265, ref. (18). See also P. Castorina and E. Recami, *Lett. N. Cim.* **22**, 195 (1978); and references therein.

[54] R. Mignani and E. Recami, *Nuovo Cimento* **30A**, 533 (1975); *Phys. Letters* **65B**, 148 (1976); *Lett. Nuovo Cim.* **16**, 449 (1976); P. Caldirola and E. Recami, ref. (24); M. Pavšič and E. Recami, *Nuovo Cimento* A **36**, 171 (1976); P. Caldirola, M. Pavšič and E. Recami, *Nuovo Cim.* B48, 205 (1978); *Phys. Letters* A66, 9 (1978); P. Caldirola and E. Recami, *Lett. Nuovo Cim.* **24**, 565 (1979); P. Caldirola, G.D. Maccarrone and E. Recami, *Lett. Nuovo Cim.* **27**, 156 (1980).

[55] R Mignani and E. Recami, *Lett. Nuovo Cim.* **18**, 5 (1977). Cf. also A. Garuccio, G.D. Maccarrone, E. Recami and J.P. Vigier, *Lett. Nuovo Cim.* **27**, 60 (1980).

[56] M. Pavšič, E. Recami and G. Ziino, *Lett. Nuovo Cim.* **17**, 257 (1976); M. Pavšič and E. Recami, *Nuovo Cimento* A36, 171 (1976); A46, 298 (1978); E. Recami, *Lett. N. Cim.* **21**, 208 (1978); **22**, 591 (1978); P. Castorina and E. Recami, *Lett. N. Cim.* **22** 195 (1978); G.D. Maccarrone and E. Recami, *Nuovo Cim.* A (1980), to appear.

[57] Cf. e.g. p. 879 in C.W. Misner, K.S. Thorne, J.A. Wheeler, *Gravitation* (S. Francisco, Cal., 1973); Ya.B. Zeldovich, I.D. Novikov, *Stars and Relativity* (Chicago, Ill., 1971), pags. 93–94; M. Pavšič and E. Recami, *Nuovo Cimento* A36, 171 (1976); A46, 298 (1978); J. Jaffe and I. Shapiro, *Phys. Rev.* **D6**, 405 (1974); G. Cavalleri and G. Spinetti, *Lett. N. Cim.* **22**, 113 (1978); **6**, 5 (1973); *Phys. Rev.* **D15**, 3065 (1977); F. Markeley, *Am. Journ. Phys.* **41**, 45 (1973); V. De. Sabbata, M. Pavšič and E. Recami, *Lett. N. Cim.* **19**, 441 (1977); E. Recami and K.T. Shah, *Lett. N. Cim.* **24**, 115 (1979) *and references therein*.

[58] J.D. Edmonds, Jr., *Lett. Nuovo Cim.* **18**, 497 (1977).

[59] E. Recami, *Lett. Nuovo Cim.* **18**, 501 (1977).

EVANDRO AGAZZI

TIME AND CAUSALITY

AN ACCEPTED APPROACH TO THE PROBLEM

It is quite customary, in the conceptual analysis of causality, to consider this notion as essentially or intrinsically connected with the notion of time, such a connection being the following: if we claim that an event *a* is the cause of another event *b*, we inevitably mean by this that *a* temporally precedes or at least does not temporally succeed *b*.

The most classical and thoroughly argumented codification of such a way of conceiving the causal relation may perhaps be found in Kant's *Critique of Pure Reason* and, more precisely, in that part of the 'Transcendental Analytic' in which he discusses what he calls the 'Analogies of Experience'. The second of these analogies states the 'causality law' in its very classical form: "All changes happen following the law of the link of cause and effect". Now, when he goes on to give a foundation to this claim, Kant remarks that there are cases in our experience in which we can arbitrarily change the order of our perceptions (when looking at a house, for example, we can begin by looking at it from the roof down to the ground or vice versa); while in other cases the order of our perceptions cannot be arbitrarily reversed, but appears 'necessarily' imposed on us by the nature of the object itself (when watching a ship, for example, going downstream, we cannot help seeing it first in an upstream and then in a downstream position). Kant finds in this necessity (by which the perception of something temporally *'follows* according to a rule' the perception of something else that precedes it in time) the crucial evidence for claiming the intervention of a non-empirical yet *a priori* concept, i.e., the link of causality. He can therefore conclude: "Hence the succession is absolutely the unique empirical criterion of the effect vis-à-vis the causality of the cause, which precedes it".

We may then conclude that, according to Kant, the *asymmetric* relation, thanks to which we say that *a* is the cause of *b*, is expressed by the fact that *a* necessarily precedes *b* in time. We could symbolize all that, as a first approximation, by writing (with the symbols taking their obvious meaning):

299

Maria Luisa Dalla Chiara (ed.), Italian Studies in the Philosophy of Science, 299–321.

$$aCb \Leftrightarrow t_a < t_b. \tag{1}$$

A straightforward confirmation of this position comes from the 'third analogy of experience' examined by Kant, where he claims that all substances, insofar as they can be simultaneously perceived in space, must be considered as being in a state of universal reciprocal interaction. This means that the very fact of the *simultaneous* perceivability introduces a quite new category, i.e. that of reciprocal interaction, which gives rise to a symmetric relation aRb expressible, in our notation, by:

$$aRb \Leftrightarrow t_a = t_b. \tag{2}$$

It is not difficult to see that Kant implicitly accepted, in this way of analyzing the problem, the plan of discussion that had already been adopted by Hume (and also by other authors in the philosophical tradition). The difference in treatment may be sketched as follows. Hume, together with tradition and common sense, accepted that:

$$aCb \Rightarrow t_a < t_b \tag{3}$$

and hence accepted the causal relation to be a sufficient reason for admitting temporal antecedence of a with respect to b, but he correctly argued that $t_a < t_b$ is not in itself a sufficient reason for also claiming aCb. As a matter of fact, it would be a very rough logical fallacy to infer from $t_a < t_b$ and $aCb \Rightarrow t_a < t_b$ the conclusion aCb.

On the other hand, he argues, all we can be granted from experience can be at most $t_a < t_b$ and we shall therefore never be able to conclude correctly from this to aCb.

What Kant says does not in any essential point modify this Humean analysis, but it introduces a new element in the discourse, i.e. the consideration that it is a fact of common experience that the asymmetric relation $t_a < t_b$ is sometimes apprehended as purely contingent, while at others it is thought of as being *necessary*. This kind of 'intellectual experience', as we could call it and which is phenomenological in some modern sense of this word, enables Kant to state that something new is at work and to call the transcendental conditions of the *a priori* into play. The result is that, *for such cases*, a biconditional may be properly substituted for the simple conditional, which yields (1) in place of (3). It seems therefore that, in order more faithfully to reproduce the core of Kant's conclusion, we should employ in addition the modal operator of 'necessity' and rewrite (1) in the more proper form:

$$aCb \Leftrightarrow \Box(t_a < t_b). \tag{1*}$$

After this improvement, the fallacy of *post hoc, ergo propter hoc* vanishes, as it becomes: *necessarie post hoc, ergo propter hoc* or, if we prefer to put it in a form more suitable for an application to scientific problems: *lawfully post hoc, ergo propter hoc*.

What we are going to do now is to criticize this solution of the question, but not because we find it untenable within the adopted viewpoint. On the contrary, the Kantian solution is possibly the best, if one accepts characterizing causality by means of a temporal analysis, but our claim is precisely that one should rather try to develop a purely logical analysis of this concept and eventually come to some temporal considerations if they turn out to be involved in it. Indeed, the Kantian solution can be efficient only if one shares the fundamental thesis of the 'critical philosophy', according to which, when our intellect perceives something with the character of *necessity*, then some *a priori* structure of the intellect itself must be involved, rather than the pure 'habit' of which Hume speaks. But, if causality as such does not have an intrinsic link with time, we could escape Hume's criticism simply by denying the plan he selected for discussing the matter, i.e. the plan of a temporal analysis of the causal relation.

A LOGICAL ANALYSIS OF THE CAUSALITY RELATION

The concept of cause has had a manifold characterization throughout the history of philosophy (let us recall, for example, the Aristotelian doctrine of the 'four causes'), but the most prominent meaning which has been attached to it, and maybe the only one that is preserved nowadays, is the one according to which the cause somehow 'produces' its effect. Unfortunately, it is exactly this 'production' that cannot be directly perceived by experience: we can have empirical evidence of *a* and then another empirical evidence of *b*, but we never experience the 'producing' of *b* by virtue of *a*. As a matter of fact, it was this impossibility of experiencing the transition, the productive link, that induced Hume to shift the question to the plan of a temporal analysis, where all we can state is that *b* comes *after a*, because all we can pick up from experience is the *existence* of *a* and *b* as separate entities or events.

We shall surely not overlook this crucial point and we shall also concentrate only upon the 'existence' of *a* and *b*, leaving 'production' out of consideration.

As our purpose is to perform a logical analysis, we shall first try to translate the 'ontological' link between events a and b into a logical link between two sentences describing them and, for the sake of simplicity, we shall suppose these to be first order sentences: $A(a)$ describing a and $B(b)$ describing b. Moreover, if we do not intend to speak of a contingently causing b, but rather to give evidence of a law-like relation between them, we should consider $A(a)$ and $B(b)$ rather as instantiations of two open sentences $A(x)$ and $B(y)$, and look for a suitable logical link between $\exists x A(x)$ (which is true when some event takes place realizing the conditions that express the existence of the 'cause') and $\exists y B(y)$ (which is true when some event occurs realizing the conditions that express the existence of the 'effect'). We shall put, for brevity:

$$\alpha \equiv \exists x A(x), \qquad \beta \equiv \exists y B(y),$$

and now we may ask the question, what logical connection must subsist between α and β in order to express that a is the cause of b. The most immediate answer, which seems to be implicit in what we have just said, and which is actually the one that is commonly adopted in the discussion of this topic, is the following: the logical connection required is the conditional sentence: $\alpha \rightarrow \beta$.

Despite its *prima facie* soundness and obviousness, such a logical translation of the causal link is misleading, because it conceives the cause as a *sufficient* reason for the existence of the effect, while it must be understood as a *necessary* one.

This point deserves some clarification, as common sense seems quite ready to accept that a cause is but a sufficient condition for the existence of an effect, in the sense that, for example, the fact of suspending a chair from a thin thread is surely a sufficient cause for the thread's breaking, but another cause would have acted equally well (e.g. by suspending a heavy suitcase from it or by stretching it hard with our hands etc.).

This reflection is defective for several reasons. First of all, the example does not provide us with a law-like case of causation, but simply with a contingent instantiation of some possible law-like statement, which could be something of this kind: 'The application to a thread of a force which is greater than its rupture coefficient causes its breaking.'

In this formulation, one can see that the event which acts as *the* cause is described by a generic statement (corresponding to the 'open sentence' $A(x)$ of our proposal), while every instantiation which fits in the sentence may be considered as *one* possible example of *the* cause. In other words, we are interested in the problem of answering the *general* question: 'when can *a* be claimed to be *the* cause of *b*', and not in the contingent question of finding out which particular events have caused which other particular events (a question that cannot be answered simply by logical means, but only by an additional *ad hoc* empirical inquiry).

Let us notice that we need not claim that every 'effect' must necessarily have just one single possible cause, because we may easily accept that *the* cause can split into a certain number of definite possibilities, so that our α can very well have the form:

$$\alpha \equiv \exists x(A_1(x) \lor A_2(x) \cdots \lor A_n(x))$$

Once this point is cleared up, we may proceed and remark that what we really mean when we say that a certain event *a* is *the* cause of an event *b*, is that *b* would not have happened if *a* were not there, and this is true not only in general, but also in every particular and contingent case. For instance: if I claim that having hit a piece of glass with a hammer was *the* cause of its being broken into pieces, I surely intend that if the first event had not happened, my glass would still be safe and intact before me. In other words, when an event *a* is stated as *the* cause of an event *b*, the *existence* of *a* is understood to be a *necessary* condition for the *existence* of *b*.

It follows from what we have said that, contrary to the currently accepted view, the proper way of translating the causal relation into a logical link should be the following:

$$aCb \Leftrightarrow \beta \rightarrow \alpha. \tag{4}$$

A couple of comments can be useful at this point. As we pointed out just now, it makes complete sense to speak of a causal relation even in the case of a purely contingent single event (like the one of the broken glass) and in this case we can see that the causal link may be logically translated by means of a counterfactual conditional. In fact the *meaning* of a sentence declaring that *a* is the cause of *b* seems to be expressed in the most immediate way by stating that *b* 'would not' have happened 'if' *a* had not happened. This is true, but remains only at a quite superficial level of analysis because, even in the case of the single contingent event, if we speak of *the cause* we surely *mean* that, although the circumstances involved were particular and even accidental, their global result was nevertheless *necessary*, once they were given.

The meaning of this necessity, in turn, can be correctly expressed by saying that we claim that, *whenever* the very same circumstances should come together again, the same result would occur as well, and this amounts to considering our event as an *instantiation* of a class of events which behave in the same way as our present and actual example. This consideration clarifies why we have to introduce the generic sentences α and β even when we speak of particular specific events a and b.[1]

Let us now come to the aspect of necessity. One might be inclined to believe that a suitable solution for taking this into account is that of 'strengthening' the intuitive definition

$$aCb \Leftrightarrow \alpha \rightarrow \beta$$

by writing e.g.

$$aCb \Leftrightarrow \Box(\alpha \rightarrow \beta).$$

But this view is misleading, as it would identify the particular kind of necessity involved in causality with a *logical* necessity, which is surely not the case. For the same reason, it also appears that the idea of adopting some different kind of *logical* connection between α and β (as it could be, for example, by using an 'entailment' relation) would equally miss the point. As a matter of fact, only the existence of events is being taken into account, and existent states of affairs are expressed by means of descriptive sentences; it follows that only truth-values of such sentences should be considered and, therefore, only truth-functional connectives should play a role. But then it remains that the most appropriate way of expressing the necessity of a for the occurence of b is exactly the conditional $(\beta \rightarrow \alpha)$.

As is well known, some undesirable consequences may arise from the employment of conditional sentences, owing to the fact that a conditional is trivially true when its antecedent is false, which would mean in our case that we would have to say that a is the cause of b even when b does not occur. It is interesting to remark that in our case this circumstance does no harm. In fact, in order for the question 'whether a is the cause of b' to make sense, b must be the case, as it would be meaningless to look for the cause of a non-existing event. This implies that, at least when we propose that a be the cause of b, our conditional $\beta \rightarrow \alpha$ has a true antecedent. On the other hand, once our definition

$$aCb \Leftrightarrow \beta \rightarrow \alpha \tag{4}$$

is stated, we need not care about β being always true, because all we require is that no case should happen in which β is true and α is false (corresponding to the fact that the cause has to occur whenever the effect is present); on the other hand, we should not worry if β is sometimes false and α is true: this fact would still make our conditional true and correspond to the not at all unrealistic situations in which the cause a, although present, is sometimes 'prevented' from producing the effect because of some intervening accidental disturbances. A really paradoxical situation would occur only if β were 'always false', but this will never be the case for us because, as we just pointed out, β must be true *at least* when we propose to attribute a as the cause of b. On the other hand, this does not confine our definition of causality to 'real worlds': also in the case of some 'possible world' it makes sense to propose a as the cause of b only if b is conceived as an event taking place in that possible world. Only in empty domains may no relation of causality be introduced, as no relations whatever can be properly defined under such conditions.

The mention of different 'domains', which has come into the discussion here in a rather incidental way, is instead of primary importance for clarifyng the notion of causality, especially if we have in mind the application of this concept in some exact contexts, for example, those of the different sciences.

We have already pointed out that the kind of necessity implicit in the meaning of causality is not a logical one, but rather a 'factual' necessity expressible, as such, in a purely truth-functional way. This situation seems rather complicated for, on the one hand, necessity is usually understood as meaning some kind of 'unrestricted validity', or validity under all possible circumstances, and this, on the other hand, seems to lead inevitably to identify it with logical necessity. There is, however, a way out of this difficulty, if one considers that 'unrestricted validity' does not coincide with 'validity in all possible worlds', but can simply be conceived as 'unrestricted validity within a certain domain'. This appears to be a good candidate for the meaning of 'factual necessity' (at least within the restricted limits of our discussion) and, in addition, it corresponds very well to the fact that every science always aims at establishing true (and also universally true) statements only within its specific domain of competence. Causal sentences do not constitute an exception and, as a consequence, we can propose that our definition of causality be expressed by a conditional which has to be 'always true' inside a certain domain of discourse.

Taking all this into consideration, it is not difficult to refine the definition of causality expressed in (4) by making some explicit reference to the domain of objects in which we claim the causal relation to hold. Let ω be our domain of events: it is clear that a and b must belong to ω and, moreover, that our conditional sentence must be true in this domain. These considerations therefore yield the following formulation:

$$aCb \Leftrightarrow \underset{\omega}{\models} \beta \to \alpha \text{ with } a \in \omega \text{ and } b \in \omega \tag{5}$$

Writing (5) in the expanded way that takes into account the previously given definition of α and β, we have:

$$aCb \Leftrightarrow \underset{\omega}{\models} \exists y B(y) \to \exists x A(x), \qquad \text{with } a, b \in \omega \tag{5*}$$

where $A(x)$ and $B(y)$ are open sentences describing the general situations of which a and b are particular instantiations. By standard logic we obtain from (5*) the equivalent expression:

$$aCb \Leftrightarrow \underset{\omega}{\models} \neg \, [\neg \exists x A(x) \wedge \exists y B(y)] \qquad a, b \in \omega \tag{6}$$

which expresses in the most intuitive sense that we accept a to be *the* cause of b in our domain of events if and only if it is never the case that $\exists y B(y)$ can hold true while $\exists x A(x)$ does not: we are not allowed to consider an event as *the* cause of another one if the latter can exist even when the first does not take place (the condition of non-triviality, as we noted, is given by the fact that at least $B(b)$ holds true in our domain).

The combined presence in (5) of the conditional $\beta \to \alpha$ and of the explicit mention of the 'validity in ω', expressed by $\underset{\omega}{\models}$, brings to light two distinct elements which are usually confused in the discussions on causality, and which we have not yet carefully distinguished in the present analysis either: they are necessity and lawfulness. These two concepts tend rather frequently toward a kind of confluence and become practically synonymous in several contexts: in empirical sciences, when we claim that a certain state of affairs b 'necessarily' follows from another state of affairs a, we usually mean that b follows from a 'according to a law'. It is in this sense that the concept of 'physical necessity', or more generally of 'factual necessity', is sometimes employed and we did the same in our previous discussion. This confluence of meaning is to be found also in Kant, who in general maintains that *a priori* structures of pure reason are involved whenever necessity comes into play and, on the other hand, he often qualifies this necessity as happening 'according to a law' (as he does, for example, in the very case of causality, where he speaks of something which 'follows according to a law' to something else). Although it can be admitted that in several cases such an overlapping of meanings does not create misunderstandings, it must nevertheless be recognized that the two concepts are in themselves different. In fact, the notion of necessity (as we already remarked) can receive, as a consequence of logical analysis, a linguistic formulation as a 'necessary condition for', which is expressed by means of a truth-functional conditional. But then the question arises, on what *grounds* do we claim this linguistic expression to be true, and here we can sometimes say that it is true because of some *logical* reason, or else because of some *factual* (i.e. *empirical*) *law*. Let us suppose, for instance, that we claim β to be a necessary condition for α, which we express by writing $\alpha \rightarrow \beta$. Let us now assume that $\alpha \equiv p \wedge q$ and $\beta \equiv q \wedge p$, then we have

$$(p \wedge q) \rightarrow (q \wedge p)$$

and we can say that our expression is true *by pure (sentential) logic*. If, on the contrary, we assume, for example, that $\alpha \equiv \forall x P x$ and $\beta \equiv \exists x Q x$, we have

$$\forall x P x \rightarrow \exists x Q x,$$

which is surely not true by pure logic; nevertheless it may happen that it is true (under suitable interpretations of P and Q) within a particular science. In this case, we should say that our expression is true by some *empirical law*, which is *valid in the domain* ω of that science (either because our expression is itself such a law or because it logically follows from some such laws). We express this fact by writing

$$\underset{\omega}{\models} \forall x P x \rightarrow \exists x Q x$$

It follows from the above that we can have cases of 'necessity without lawfulness' (where a conditional expression is claimed to be true without reference to any empirical law, but for example simply by pure logic); cases of 'lawfulness without necessity', i.e. without the stated law expressing a 'necessary condition' statement (which happens when the sentence involved, though being true by some empirical law, does not have the form of a sentential conditional); cases of 'necessity together with lawfulness' (like the one just exemplified above).

The case of causality belongs to the third category and this completely clarifies why we pointed out that the causality link, though including the idea of necessity in its meaning, does not express any form of 'logical necessity': the idea of necessity involved can be made explicit, as we noted, by saying that the cause is a 'necessary condition' for the effect and this yields the linguistic formulation of this link by means of a suitable sentential conditional; the fact that this conditional is claimed to be true not by virtue of pure logic is expressed by adding the requirement of lawfulness. Let us remark that this consideration suggests not using the rather ambiguous locution of 'factual necessity' in discussions of this kind: the notion of 'factual validity' or of 'validity in a domain' appears to be much more recommendable. In addition, identifying lawfulness with 'validity in a domain' has the advantage of purifying the concept of empirical law of any normative flavor, which would be quite out of place. As a matter of fact, scientific laws are descriptive and not prescriptive statements: they tell 'how things are' and not how they 'ought to be', they express states of affairs and not imperatives.

If we now consider the Kantian proposal in the light of the analyses developed thus far, we find some additional reasons for dissatisfaction with it: considering (1*), which we assumed to express in a formal way the interpretation of causality proposed by Kant, we can see that the definiens, not being in the form of a conditional statement, fails to express the idea of a 'necessary condition'. On the other hand, Kant stresses that the temporal antecedence of cause with respect to effect has to be perceived with some kind of necessity, which he also expresses with the words 'according to a law'. As he did not clarify what kind of law has to be envisaged and, as we, on the other hand, know from the context of his philosophy that he usually refers to some *a priori* (i.e. to some 'law of reason') when confronted with what he calls necessity, it does not appear too unfaithful with respect to his thought to express this by means of the modal operator of necessity $\Box$.

On the other hand, it is clear not only to us, but also in a Kantian perspective, that a sentence like $t_a < t_b$ cannot be considered true by *logical* (i.e. formal logical) *necessity*. This conclusion would suggest dropping the modal operator and considering the sentence as belonging to the class which we called 'lawfulness without necessity', or equally as being simply 'valid in a domain'. But now we shall ask what this domain ought to be, and it appears from the terms occurring in the sentence that this domain is that of *time*. Thus the Kantian proposal amounts to identifying causality with some *law of time*. This fact seems quite unsatisfactory, because (not only generally, but also according to Kant) the causality relation is meant to hold as a relation to objects and not as a relation to time: consequently, we should expect it to be defined by means of some relation between objects (these objects being possibly 'indexed' on time), whereas we have here a relation between time instants (indexed on objects). We certainly do not disregard the possibility that some sentence about time instants (indexed on objects) could follow as a logical *consequence* of a law governing objects, but this is just the step which is missing in Kant's proposal. Notice that this has nothing to do with the fact that, according to Kant, (universal) laws governing objects are 'imposed' on them by pure reason: they still remain laws governing objects and not laws governing time, as he seems to claim.

Anyhow, we are not interested in a thorough discussion of the Kantian solution of the problem of causality: we have devoted some attention to it only to see how delicate a question that of the 'necessity' involved in the causality concept is and to find a confirmation for its not being correct to locate the core of this concept in some structure of time.

This does not imply, of course, that causality has 'nothing to do' with time. Quite the contrary, although it is true that one can clarify this concept by simply referring to *existence*, leaving *production* (of the effect) out of consideration, it is no less true that the idea of production, which intuitively accompanies the concept of cause in such a natural way deserves some consideration and this might well call time into play. But before coming to this point we need a number of preparatory steps.

TAKING TIME INTO CONSIDERATION

After the logical analysis of the intrinsic nature of the causal link, we shall proceed now to investigate what possible *consequences this analysis may* have for the temporal situation involved. We shall therefore consider time, for the limited purposes of this investigation, simply as an ordered 'index set' T of instants, on which the relation $t_i \equiv t_j$ obviously means that t_i does not follow t_j, while by $t_i = t_j$ we mean that the instants t_i and t_j coincide, and by $t_i < t_j$ that t_i strictly precedes t_j.

Let us now consider (6) at a given *arbitrary* instant t_0. We shall say that a is the cause of b at time t_0 (which we shall indicate by $a\underset{t_0}{\mathbf{C}}b$) if and only if the right hand side of (6) holds true in ω at time t_0 (which we shall indicate by $\underset{\omega t_0}{\models}$). We have therefore

$$a \underset{t_0}{\mathbf{C}} b \Leftrightarrow \underset{\omega t_0}{\models} \neg\, [\neg \exists x A(x) \wedge \exists y B(y)] \tag{7}$$

The information provided by (7) is not at all trivial, despite its *prima facie* obviousness. Our attention is called to the fact that, at *every time* t_0

(remember that t_0 was designated as an *arbitrary* instant) *a* can be thought of as the cause of *b* if and only if both are present. This implies that, if we intend to claim that *a* is the cause of *b* at any time (i.e., independently of the particular instant considered so that we can drop the mention of t_0), we should write

$$aCb \Rightarrow t_a = t_b \tag{8}$$

Now, (8) seems to be in the most striking contrast with the common evidence of all those cases in which we believe we are right in maintaining that the cause of something takes place *before* its effect. This is true, but this puzzling situation only indicates that the concept of cause needs to be investigated further, in order to account for this two-fold behavior with regard to time.

As a matter of fact, the *situation* we are facing now was not at all unknown in the philosophical tradition. Scholastic philosophy, in particular, used to distinguish between the *causa essendi* and the *causa fiendi* and it explicitly remarked that only the *causa fiendi* can be prior to the existent effect, while the *causa essendi* is simultaneous with it. For example, we can say that Leonardo is the *causa fiendi* of the 'Gioconda', which he painted several centuries ago, and whose existence continues even after the death of its creator. Leonardo was the cause of the 'Gioconda's' *coming to existence*, while he is no longer the cause of its persisting in existence. If, on the contrary, we consider a body hanging by a thread, we can say that the mechanical resistance of the thread is the *causa essendi* that the body does not fall to the ground, and such a cause is acting all the time the body remains suspended, the proof being that if we eliminate the cause by cutting the thread, the body will immediately fall (if no other remedy is simultaneously provided).

The moral of all this is the following: as long as an event *a is really acting as the cause* of an event *b*, they must be simultaneous in time as expressed by (8), while the accepted use of common language also admits of *a* as being the cause of *b* also when *a* has already ceased to act as the cause of *b* (as in the case of Leonardo, who acted as the cause of the 'Gioconda' in a very strict sense only while he was painting it and is surely not acting as the cause of the 'Gioconda' now, although in a *quite different* sense we may still consider him as the cause of this picture).

This distinction of two possible meanings of the concept of cause thus plays a useful role in the clarification of our problem, but some more

light can be shed by a further philosophical consideration, which concerns
the exact place where the causal principle comes into play in both general
philosophy and philosophy of science.

CAUSALITY IS PROPERLY CONCERNED WITH CHANGE

As for general philosophy, we must remark that the causal principle does
not properly concern *existence as such*, but rather *mutation or change*.
If we believe that the principle of causality concerns pure existence, its
formulation should be: 'everything existing must have a cause'. One
sometimes actually finds such a formulation in books, but it is, philosophically speaking, untenable. In fact, if every entity *x* must have a cause,
this again is an entity *y*, which in turn must have another entity *z* as its
cause (according to the alleged principle) and so on *ad infinitum*. As is
well known, practically every ontology finds out a way of stopping such an
infinite regression, by putting somewhere a 'first cause' which is an entity
that has no cause behind itself. But this simply means that the above formulation of the causality principle was wrong, as it is not true that *everything* existing must have a cause, the 'first cause' being an exception.

If, on the contrary, we regard causality as having to do with *change*,
the principle receives its correct formulation: 'every change has a cause'
and this was actually the formulation of the principle in classical philosophy (which we also find in Kant). Let us remark that this second formulation also covers the cases in which the first could in a way be properly
applied. For example, if the 'Gioconda' appears at a certain moment in
the world, we can correctly ask for the *cause* of its *existence*: this is true,
but we must not miss the all-important point, i.e. that the appearance of
the 'Gioconda' means a *change* in the previous state of the world, and we
are therefore asking for the cause of this change. It was therefore customary in traditional philosophy to say that every *contingent* being must have
a cause, but one must not forget that a being cannot be labelled contingent
'as such': contingency is not self-evident, but rather something that must
be proved, and *change* is perhaps the most appropriate way of uncovering
it. We may say that an entity is contingent because there was a time in
which it did not exist.

If causality comes into play only to explain change, it follows that,
once a cause has been attributed to an entity for its coming into being, we
need no longer look for a cause to explain its *persisting* in existence, as

persistence is exactly the contrary of change. We should rather look for a cause if this entity should cease to exist, because this fact would again mean a change.

We are now in a position of better appreciating and also of correcting slightly the ancient distinction between *causa essendi* and *causa fiendi*. It is clear, on the one hand, that the inquiry on the *causa essendi* must be split into two parts: one which concerns the *origin* of the entity under investigation and one which concerns its *persistence*. This second part, as we remarked, falls outside the proper realm of causality, when this persistence is no longer bound to the continuous presence of the cause (as in the example of the 'Gioconda'). It can remain inside the realm of causality when persistence is bound to the continuous presence of the cause (as when the rotation of an electric engine is bound to a continuous supply of electric energy to it). The first part (i.e. the one which concerns the origin of the entity) is simply the *causa fiendi* of the entity under consideration and it fully falls under the competence of the causality principle, as it concerns a *change*.

If we put ourselves in the viewpoint of classical philosophy, we may remark that if the *causa fiendi* is not itself uncaused, it is not believed (according to that view) to be 'the true cause' of the entity, this quality being attributed only to the first cause. But we are not interested here in the analysis of this question.

Rather what deserves to be further specified is the following: when a causal connection is supposed to be necessary for the *persistence* of the effect, this happens because we think, in a way, that the effect must be 'continuously generated' by the cause, so that we could speak here of the continuous action of the *causa fiendi*. But we could also represent the situation in another way and say: as long as the cause (*fiendi*) *persists*, the effect also persists because *no change* occurs, and no further cause must be involved to explain this persistence.

From every side therefore we reach the same conclusion: the only proper logical place for the cause is *change* and in it we have to do only with the intervention of an *instantaneous causa fiendi*. Let us notice that this point clarifies the rather vague intuitive notion of the cause 'producing' the effect, including in it only those aspects which are not affected by any anthropomorphic feature. The cause 'produces' its effect simply in the sense of accounting for its 'coming to existence' as a necessary condition for that.

All this discussion confirms our (8) and states that cause is *asymmetrically* directed towards effect as a necessary condition for it, but this ontological and logical asymmetry does not involve any temporal asymmetry but, strictly speaking, implies rather a simultaneity or concurrence of both. (Let us remark, parenthetically, that we introduced some major modifications in the traditional notions of *causa fiendi* and *causa essendi*. In particular, considering the first as being related to the moment of coming into existence of an entity, we put it as simultaneous with its effect, while the tradition considered it as antecedent. This derives from the fact that we concentrate the reference of the notion of causality on to *change*, while the traditional view referred it to *existence* in general. We explained why we consider it more appropriate to restrict causality – within the general domain of existence – only to change).

A DIACHRONIC APPROACH TO THE QUESTION

The philosophical considerations of the preceding section enable us to overcome some difficulties, which could arise from the analysis of (8). In fact, the following seems to be a quite obvious objection: having assumed considering the relation between a and b at a given instant t_0, it is only too easy (and quite trivial) to conclude that, in order for a to be the cause of b exactly *at the given time* t_0, both must be present then. In other words, such a simultaneity is simply the consequence of having artificially imposed the condition of considering the matter synchronically. If instead we had approached our problem diachronically, i.e. by considering time-intervals instead of time-instants, we would have had enough 'room' to see the cause occurring prior to the effect, and in such a way the familiar antecedence of cause with respect to its effect would have been accounted for.

We are going to take this objection seriously and apply our definition of causality not to a time instant t_0, but to a time *interval* $\bar{\imath}$. Such an application yields: 'a is the cause of b in $\bar{\imath}$, if and only if it is not the case that b occurs in $\bar{\imath}$ while a does not occur'. But now it is easy to see that the desired goal of firmly establishing the antecedence of a with respect to b is still far from being reached. In fact, it may happen that, in our interval, there is an instantiation of b and that all instantiations of a come before t_b, but it might also happen that all instantiations of a come concurrently with b or even *after* t_b. In all three cases, our definition would be satisfied, because, in none of them, would it be the case that b occurred in $\bar{\imath}$ without *a* occurring in *t*, but they would correspond respectively, to the cause coming before, concurrently, and after the effect.

Let us try to discard the third possibility by a closer investigation of the case in which it might occur, i.e. of the case in which we find *b* occurring prior to *a* in our interval. How could we maintain (without disregarding our definition) that in this case *a* could not be the cause of *b*? A satisfactory answer seems the following: if we consider our interval *ī*, we find that there was a moment in it, at which *b* was present, while *a* was *not yet* present, and this already shows that *a* is not a necessary condition for *b*. This clarification might be satisfactory, but unfortunately it is so only insofar as it abandoned the consideration of the time interval *ī* and came back to an instantaneous way of thinking: not only because all the attention is concentrated on the instant t_b at which *b* occurs, but especially because the exclusion of *a* as a necessary condition for *b* is pronounced on the basis of *a* not being present *at that moment*, which is tantamount to applying (8) in this so-called diachronic approach.

An attempt to remain faithful to the diachronic viewpoint in trying to discard the possibility of retro-causation could perhaps be made by introducing the difference between *actual* and *possible* events, according to the following definition, which involves time: we call *actual*, with respect to an instant t_0, those events which actually occurred at t_0, and *possible* events those which did not occur. In this sense, we can say that all future events (with respect to t_0) are only possible, while past and concurrent events (with respect to t_0) include both actual and possible cases. Leaning on this distinction, we could say that, within a time interval *ī*, *a* is the cause of *b* if it is not the case that *b* occurs without *a* being among the actual events in this interval, with respect to the instant t_b at which *b* occurs. This solution has the nice feature of preventing the cause from coming after the effect, and it is, on the other hand, not an *ad hoc* escape. In fact, if causality (according to our previous analysis) implies that the *existence* of *a* is a necessary condition for the existence of *b*, such an existence must be an *actual* one, as possible existence by necessity implies the intervention of *logical* features which are foreign to causality as such.

If we look closer at this proposal, we see that it actually introduces a substantial modification into the notion of '*a* being the cause of *b* in the time-interval *ī*', because (by limiting the consideration to actual events with respect to t_b), it amounts to saying that our notion is rather that of '*a* being the cause of *b* in the time interval *ī*, whose upper limit is t_b'. But, even with the proposed adjustments, the consideration of time in-

tervals still does not provide us with a useful tool for mastering those
cases in which it is commonly stated that a certain cause has occurred
sometimes before its effect. As a matter of fact, it might happen that a
does not occur within a time-interval preceding t_b simply because this was
taken 'too short', while the same a might occur within a time interval
which is 'sufficiently long'. In such a condition, we either relativize the
definition of causality to different time intervals (which has no practical
or scientific counterpart and would be quite arbitrary without any criterion for assigning the 'suitable' intervals), or we keep the definition of
causality general and we simply take intervals into consideration in order
to 'apply' it to particular cases. The last alternative, however, falls exactly
in the epistemic difficulty just mentioned, namely that of finding out the
proper 'length' of the interval in which the existence of the cause must
be ascertained.

Still, the main reason for being sceptical about considering time-intervals in discussing causality is the following: in our logical analysis of this
notion, we proposed that a is the cause of b, if and only if it is *never* the
case that b occurs and a does not occur, which is tantamount to saying
that a must occur *whenever* b occurs. The idea of considering time intervals
was suggested by the desire of safeguarding the antecedence of a with
respect to b, but now let us see whether this would be compatible with
our definition. It is clear from it that it would not be enough to find the
above condition satisfied in a *single* time-interval, or even in *some* intervals
(this would not correspond to the 'whenever' requirement); on the other
hand, we already noticed that it is not possible to fulfil the condition in
all time intervals (remember what we said about 'too short' intervals).

It follows that, if we want to apply our definition to events occurring
in time, reference to time-intervals is not suitable in itself and the original
idea of referring to time-instants is the only one which seems acceptable.
We shall therefore abandon the diachronic approach and see whether the
'instantaneous' viewpoint can be submitted to closer scrutiny, in order to
provide us with an acceptable interpretation of those cases as well in
which the cause seems to occur before the effect.

THE DYNAMIC APPROACH

In order to make some progress in our analysis, we must remember that,
when we speak of a being the cause of b, we understand this as a *general*
statement, meaning that in *all* cases in which an event occurs which is

instantiated by *b*, an event which is instantiated by *a* must also be present. This fact contains two useful indications. The first is precise information about the *relevant* instants to be taken into consideration: they are the instants at which *b* occurs. The second is that such instants are meant to be indefinitely repeatable. This leads us to what we could call a 'dynamic' viewpoint, that is to say, to a viewpoint according to which events are not simply existing, but rather 'coming into existence' in time, so that it makes sense to say that they do not exist at a certain instant t_0 while they do exist at a certain different instant t_i and they might no longer exist at another instant t_j. As is quite clear, we make a transition here from the pure and simple consideration of existence, to the consideration of *change* (understood as an alternating of existence and non-existence). If we now apply our definition of causality to the instant at which *b* 'comes into existence' (i.e. to the instant at which this change occurs), we see that *a* can be said to be the cause of this coming into existence if and only if (i) *a* is present at this instant (simultaneity of cause and effect) and (ii) if this is the case *whenever* such a change occurs (generality). It is easy to recognize that here we are bringing about the old notion of *causa fiendi*, with the additional feature, however, that it is simultaneous with the effect and this now makes the correct interpretation of (8) quite precise. Let us now see how this fact does not imply any difficulty even if we have in mind some familiar examples in which it is customary to speak of a cause occurring *before* its effect. We shall devote specific attention to this problem in the next section, but we can already make a few remarks here simply by resorting to our dynamical viewpoint. Included in this viewpoint is the fact that not only may a state of affairs 'come into existence', but also that it may 'go out of existence' or 'no longer be in existence'. Let us consider a state of affairs *a*: if it is *still in existence* when *b* occurs, it is quite reasonable to admit that *a* may be the cause of *b* and this preserves simultaneity of both at time t_b. But if *a* is *no longer in existence*, it would be really hard to admit that something non-existing can be the cause of anything else. Our effort at clarification will therefore concentrate on this second point.

THE SENSE IN WHICH THE CAUSE MAY COME BEFORE THE EFFECT

The preceding considerations prevent us, strictly speaking, from admitting retro-causation as well as post-causation. But, as we noticed, there are a lot of instances in every day life and in scientific practice in which we currently admit that a cause a produces its effect b only after a certain time, when the cause is *no longer in existence*. In order to clarify this point, we must stress again that causality always concerns change: if we keep this in mind, we can see that, in all cases in which a is alleged to be the cause of b and to occur before b (i.e. when a does not immediately 'give rise' to b), there is always some *intermediate* chain of changes taking place, which ends with b. To this chain of changes, a chain of *intermediate* causes does correspond, lying between a and b, and the last change (the one which yields b) is not determined by a or, at least, not by a alone. In such a way, the cause producing the last change is the actual *causa fiendi* of b and is simultaneous with b: it may happen that a still belongs to this cause (if the cause is 'complex' and a is still in existence and in action), but it may also happen that a is already out of play at that moment.

If, in many practical cases, we do not usually take the intermediate changes (and the corresponding intermediate causes) into consideration, it is because we consider a as the *relevant* cause and the intermediate causes as purely subsidiary or instrumental to the possibility of its operating. In other words, we underline very strongly the necessity of a in order for b to happen (which is the core of the link of causality, as we saw it), while considering the occurrence of the subsidiary conditions to be more accidental.

There is no reason why we should reject such a commonly adopted way of conceiving cause, provided we are conscious of the *elliptic* sense involved in it, which we could express by stating, e.g., that a is the *relevant cause* of b when it is a necessary condition for b to occur, but we understand that some other conditions must be fulfilled as well, leaving it open whether they occur simultaneously with a or during some subsequent time.

We can therefore distinguish two cases:

(a) When the cause a is directly and immediately responsible for the event (change) b, which we should continue to express by aCb, with:

$$aCb \Rightarrow \models_{\omega} t_a = t_b. \tag{8}$$

(b) When a is simply the *relevant* cause for the occurrence of event (change) b, which we should express by $a\bar{C}b$, with

$$a\bar{C}b \Rightarrow \models_{\omega} t_a \leqslant t_b. \tag{9}$$

It must be stressed again that cause plays a role only when change occurs and this is true not only in philosophy, but in science as well. So Newtonian *force* is involved as a cause of *acceleration*, which is a *change of velocity*, but not of velocity itself, which needs no force to be produced. Aristotle, on the other hand, supposed that a force must be given as the cause of velocity, but this is not really an exception to the rule that force does not produce velocity as such. In fact, according to the Aristotelian viewpoint, *rest* was supposed to be the natural condition of every body, so that its acquiring a non-zero velocity needed to be caused, like every *change*. Under this assumption, the Newtonian viewpoint would also lead us to the same conclusion: in fact the transition from a zero velocity to a non-zero velocity (i.e. from rest to motion) involves an instantaneous *acceleration* that calls a force into play in order to causally yield it.

Let us consider, instead, the alleged retro-causal action of the electro-magnetic forces which should be responsible for a pre-acceleration of a charged particle, according to a certain interpretation of Dirac's equation for the motion of this particle in one dimension:

$$m\left(\frac{dv}{dt} - \frac{2}{3}\frac{e^2}{mc^3}\frac{d^2v}{dt^2}\right) = f(t).$$

We can note, as Grünbaum correctly pointed out, that such external forces $f(t)$ cannot be responsible for this pre-acceleration, because the change which turns out to be determined by external forces (as one can see by some mathematical developments of the given equation) is the change of the acceleration itself.[2] It follows that the acceleration represents the attribute which undergoes change and which is itself, therefore, quite independent of such forces, while they can be held responsible for the *change* of acceleration, which does not occur *before*, but concurrently with their occurrence.

To summarize the results of our inquiry, we can claim that a logical analysis of the causal relation shows that time is not primarily implicated in it, while it *entails* that, if an event *a* may be thought of as being the cause of an event *b*, it must *stricto sensu* be held to occur at the same time as *b* or, in a wider sense, it may occur at most before *b*. Which one of the two senses is intended must be derived from the context, having borne well in mind, firstly, what the domain of events is to which we are referring and, secondly what exactly the change is for which the cause *a* is proposed to be specifically responsible. This can be rather obvious in several cases, but sometimes such an analysis may unfortunately appear rather arduous.

THE QUESTION OF 'FINAL CAUSES'

From the considerations of the preceding sections we may draw some useful conclusions about the *vexata quaestio* of the so-called 'final causes'. In the philosophical tradition it was customary to understand by this some 'goals' or 'final states of affairs' introduced to explain certain kinds of changes which seem to occur 'in order to reach' those goals. The interest of the question, in the present context, lies in the fact that such causes would provide an example of retro-causation, or of cause coming *after* its effect. From a purely logical viewpoint, there seems to be some plausibility in favor of these causes, because they may be presented in the light of a necessary condition for some event to occur. For instance, in the simplest case of purposive behavior, we could say that if I take a taxi in order to catch a train at the station, the fact of catching the train is the cause of my taking a taxi and I could even express this fact by means of a counterfactual conditional by saying: 'If I did not have to catch a train, I would not take a taxi'. An immediate objection could be that it is not necessary or imperative for me to take a taxi to go to the station, as I could, for example, walk or go by bus, but this is not the important point and we shall assume, for simplicity, that taking a taxi is the only way for me to go the station. Even with this simplification, it is easily seen that the above counterfactual conditional cannot be translated in the form of a truth-functional conditional expressing the 'necessary condition' requirement. In this case, putting β for the sentence expressing the fact of taking a taxi and α for the sentence expressing the fact of going to the station, we should have: $\beta \rightarrow \alpha$. But this is not true, because it may happen sometimes that I take a taxi not to go to the station, but to go to the university, for example, and in this case β would be true and α false (or, if we prefer, it is not the case that *whenever* β is true, α is true as well). In order to overcome this difficulty, I should impose on my taking a taxi some much more restrictive conditions, for example, that of taking a taxi and following some specific itinerary leading from my place to the station: in this case not only the *final* state of affairs would be involved, but an entire *pattern*, of which this state of affairs constitutes only the final event. This is already an interesting remark, but it is still not enough to overcome the difficulty. In fact, I might sometimes take a taxi, give the driver the itinerary to follow and nevertheless be unable to reach the station and catch the train because of some accidental reasons. In this case too, β would be true and α false. A better consideration of this example shows where the weak point lies: the 'cause' of my taking the taxi is not the *existence* of the station, train, itinerary to follow, but my *intention* to go to the station and catch the train by following that itinerary, and this intention must be *present* at the moment I take the taxi, and therefore it does not constitute any *future* state of affairs with respect to it.

An analogous line of reasoning may also be repeated with respect to non-intentional behavior, e.g. in the case of an animal embryo developing to the stage of the adult organism. In this case, we would say that the 'pattern' is inscribed under the form of genetic information codified in the DNA of the embryo cells and that this information (together with some suitable environmental conditions) is the cause of its developing according to the pattern up to the stage of the adult organism, if no accidental disturbances come into play.

It follows from the above that, in the case of the so-called final causes, not the final event, but an actual pattern terminating with this event could be taken as the cause of the change and it must be *present* concurrently with the change, be it under the form of an intentional design, or of some stored information. If we indicate this pattern 'leading' to α by $P(\alpha)$, we could express this particular kind of finalized causation C_f under the form:

$$aC_f b \Leftrightarrow \underset{\omega}{\models} \beta \Rightarrow P(\alpha) \tag{10}$$

and it is clear that this does not violate the rule that the cause is, strictly speaking, simultaneous with its effect.

These reflections should justify, on the one side, the use of the language of final causality in all those domains in which intentional behaviors or at least goal-seeking processes are patently involved, like psychology, history, sociology, economics, biology, systems theory, etc. without implying, on the other side, an alteration in the essential features of causality. In particular, it is only too true that, in the case of human events the future often has a much more significant, stimulating, and productive power than the past or even the present, but this is only possible if this future takes the form of something *actual* (be it a pattern, or a project, or a design) having the possibility of affecting changes in the present.

NOTES

[1] Let us stress again that we are looking for a logical analysis of the *meaning* of causality, which is a *semantic* and not an *epistemic* aim. It is therefore not surprising if in many cases it remains difficult to assign the cause of an event even if our concept of a cause

is clear. Yet conceptual clarification, though not sufficient, seems to be a necessary prerequisite of every correct epistemic enterprise.

[2] For an analysis of this question see A. Grünbaum, 'Is preacceleration of particles in Dirac's electrodynamics a case of backward causation? The myth of retrocausation in classical electrodynamics ', in *Philosophy of Science* **43**, 165–201 (1976), which also includes the relevant mathematical discussion.

GIULIANO TORALDO DI FRANCIA

THE CONCEPT OF PROGRESS IN PHYSICS

It is said that Czar Nicholas I once issued an ordinance in which he forbade the use of the word 'progress'.[1] I do not feel I have much in common with the Czar of all the Russias and, on the other hand, I do not have the misfortune to be one of his unhappy subjects. Because of this, I have no reason to indulge in the intellectual habit[2] of shamefully avoiding the word 'progress'. I shall say more. I believe that progress does exist and that consequently it is worth pursuing. If for no other reason, I believe that it is progress that Czars can no longer promulgate such edicts. The attitude of those 'progressives' who do not believe in progress, for me remains an enigma. He who does not believe in progress can be at the most a conservative (every change is a worsening). But in reality the step from that position to obscurantism does not seem very long.

The scruple I mentioned above leads more than a few modern scholars to replace the expression 'progress of science' with 'growth of science'. It is not very profitable to quibble about terms. If there is general agreement about that expression, I shall accept it as well. But I reject the ideology behind it. I cannot admit that Galileo's science was merely 'bigger' than Aristotle's. Otherwise I could not explain how it was capable of opening the way to a philosophical and social upheaval, let alone a greater assertion of man's intellectual autonomy. Above all, I could not explain the cause of the bitter battle the reactionaries of that time joined against Galileo; a battle on a par with that joined against Darwin by the reactionaries of the last century (or should I say of our century as well?).

At this point I fear I may have convinced the reader that he is dealing with a man of other times who is about to inflict on him an ode to the magnificent and progressive destinies of humanity! Rest assured it is not so. It is one thing to believe in the validity of a given concept and another to lack critical sense in analyzing it.

At any rate, my task is made easier by the fact that I am not going to concern myself with the general progress of humanity but rather with the progress of a well-defined branch of science, namely physics. Further, I shall leave to others the task of analyzing a number of important aspects of the problem, such as the historical one, the sociological one or the

Maria Luisa Dalla Chiara (ed.), Italian Studies in the Philosophy of Science, 323–339.

problem of technological fall-out. I shall concentrate my attention solely on the *cognitive* aspect of physics.

Someone might naturally object that in this way we are shirking the most important problem, namely the relations between the progress of science (let's call it X) and the progress of society[3] (let's call it Y). But it seems to me that if we want to clarify the nature of these relations, which we shall indicate schematically with

$$X \leftrightharpoons Y \tag{1}$$

we must first of all agree on a reasonable meaning for X and Y. Otherwise we will not escape those very painful and futile discussions between those who *want* X and Y to mean one thing and those who *want* them to mean something else, the whole thing being conditioned simply to the more or less tendentious conclusions each one wants to draw. If we do not succeed in reaching at least a certain degree of intersubjectivity, we are only losing time.

Now the modest task I am setting for myself is to show that even if reaching intersubjectivity on the definition of Y seems a desperate undertaking, we can have some hope, on the other hand, in the case of X. If we were to succeed in understanding only something about what X is, we would still have made some advance in treating the problem (1). As we shall see, we are dealing more with promising indications and programmatic directives than with solutions. We are dealing with points of view which deserve to be scrutinized and discussed.

The problem of defining and measuring progress in physics has some peculiar characteristics which distinguish it quite clearly from the analogous problem posed by the other sciences.

First of all, there is a specific reason which makes it very important to succeed in giving a reasonable evaluation of progress in physics. It is a question of the extremely high cost of basic research, bringing us to the frontiers with the unknown, that is, of that very research which for the present and for the immediate future has an almost exclusively cognitive value. Of course, research is expensive in all the modern sciences. But one has the impression that it is *most* expensive in physics, often by orders of magnitude that make it possible only for superstates or international organizations.

In this situation even a purely cultural good, such as knowledge of the world, must be subject to some rules which are valid for material goods and services. To the society which pays and bears the sacrifices, we must be able to say *what* it is paying for. One must be able to assure society that when it pays more, it obtains, at least presumably, a larger quantity of the good it wants to buy.

Here then is an indispensable reason for trying to *measure* the progress of science. Until we know how to do this, even in an approximate fashion, political planners will continue to make important choices without having any objective element on which to base them. At best, one will trust the subjective evaluations of 'experts'; but more often than not one will yield purely and simply to the pressures of interest groups.

On the other hand, the problem of this measuring appears less desperate in physics than in other sciences, because of the very nature of this discipline which is always intent on achieving quantitative rigor. As we shall try to show, the problem of measuring progress in physics can become a problem very analogous to those of physics itself. In any case there is a kind of circularity by which the progress of physics helps us better to understand progress in physics.

At this point, having announced that physics has particular characteristics which distinguish it from the other sciences, it might be well to say what we mean by physics. I am not a 'fan' of definitions, but in this instance I think it appropriate to pinpoint the object of our discussion.

As is well known, many people, when faced with the difficulty of defining physics, get out of it by pointing to physicists and what they do. Physics, they say, is what physicists do. I do not deny that this position makes some sense and that, like all methodologies which start with the historical data, it has its dignity as well. But for our purposes this is not adequate. The physicist does many things and not everything that he does is physics. I am leaving out the banal observation, for example, that he is not doing physics when he eats breakfast. But, I wonder, is he doing physics when he asks for research grants, when he participates in a competitive examination for a chair, when he teaches the history of physics? We see that here, as happens in various other cases, the purely extensional definition is not possible because the *extension* of the concept is in its turn definable only by means of an *intension*.

In my opinion, physics is characterized by *method*. Physics is everything that can be treated with that method.

One should now go on to describe and characterize the method. And this could be done but, as is obvious, this would lead us to a rather lengthy discussion and take us far away from our principal theme. Suffice it to say that we are referring to the quantitative experimental method which arose with Kepler and Galileo and to the improvements it has undergone in almost four centuries of application. It is really its quantitative aspect which seems to offer some footholds for solving our problem.

What does the method of physics allow us to do? It is not a bad idea to pause a bit at this question, because from the mistaken answers frequently given to it, hopeless confusion has arisen.

The method allows us to ascertain the validity or falsity of certain *assertions* which we make concerning the external world.[4] It is a method for accepting or discarding propositions, not for *attaining* them, *discovering* them or *devising* them. Unfortunately (or fortunately) the straight and well-defined path to scientific discovery does not exist. It is certainly true that for a long time even the greatest scientists deluded themselves that they knew this path. Even Newton thought he could proceed "from effects to their causes and from particular causes to more general ones until reasoning reaches the most general cause of all." But the historical critique of science has cleared up this misunderstanding.

It is interesting to read what G. Holton writes with regard to John Dalton's formulation of the atomic theory[5]:

Initially, Dalton had been interested in meteorology, and particularly in the problem of why the gaseous constituents of the atmosphere are so thoroughly mixed despite the differences in their specific gravities. In reading Newton's *Principia*, Dalton had found the proof that *if* a gas consisted of particles repelling one another with a force proportional to the distance between them, then it would have to exhibit a reciprocal relationship between pressure and volume such as had been found by Boyle in actual experiments on existing gases. Dalton took this statement to be a *proof* of the proposition that real gases do indeed consist of particles endowed with the stated forces. Next he accepted the postulate, in natural accord with the contemporary caloric theory of heat, that each gas particle is surrounded by a sphere of caloric fluid, a fluid endowed with the quality of self-repulsion. Thirdly, he announced, partly on the basis of his own experiments, that the individual particles of one pure gas must differ in size from those of another gas. Finally, Dalton concluded that the thorough mixture of components in the earth's atmosphere was now explainable because mutually-repulsive contiguous particles of several different sizes would not be in equilibrium in strata. It is now well known that this work led Dalton to the epochal concepts of the chemical atom, atomic weight, the Law of Mulitple Proportions, and so forth, but it is worthy of note that *each and every one of his steps as just given was factually wrong or logically inconsistent.*

Holton correctly distinguishes between two types of science, which he calls S_1 and S_2, and it is very dangerous to confuse them. S_1 represents science *in fieri* while S_2 is the science we find in scientific books or articles. The method of physics is applied to S_2 and not to S_1.

No one begins a scientific article in this way: "The night of 26 August I had an idea. I got up at 7 in the morning and began to study its consequences. I did the first calculation wrong but then, once I spotted the error, I got an interesting result [...]". Someone, with moving ingenuousness, is even sorry that scientific works are not written in this way and accuses the authors of falsity who present things as if they had received them ready-made! But it is understood, and with the best of reasons, that S_2, not S_1, is found in scientific journals. Far be it from me to disdain historical facts and their research; I shall limit myself merely to saying that history is something else, it corresponds to a different interest, and above all, to another method.

More serious is the fact maintained by some that S_1 and S_2 cannot be separated, in that S_1 influences S_2. The statement can be either obvious or false. It is obvious if it is asserted that on S_1 depends the *content* of S_2, that is, what propositions are considered in it. But it is false if we want to make the *validity* of the propositions of S_2 depend on S_1. In particular I believe that the negative conclusions concerning the progress of science, which some want to draw from the fact that S_1 is anything but linear and is not lacking in set-backs, are absolutely unjustified.

Let us suppose that a climber, on returning from an expedition in the Himalayas, states 'I have climbed Mount Everest'. This assertion cannot be refuted by objecting that he followed a zig-zag path, that at a given moment he turned back for a stretch, that toward the end he thought he had reached the peak when he noticed he had another 100 meters to go[6], and so on. His statement: 'I have climbed Mount Everest' is absolutely independent of all this. If it is true, it is true! And it remains true even if by chance there were two rival groups of climbers involved in reaching the summit by different routes.[7]

Very similar considerations must be made for the *motivations* of science as well. Science can be done for curiosity, for the ambition to excel, to earn a living, to make a career, to increase surplus value and military power. These sociological, economic, and political data are very interesting and should be studied. The politician on the one hand, and the scientist on the other, can no longer pretend to ignore them as they once did, and must bear them in mind in planning their own action. But it is false to assert that the validity of the results of S_2 depends on them.

When people asked Hilary why he had climbed Everest, he answered, "Because it is there". Whatever other reason may have been given, the statement: 'I have climbed Everest' would have remained equally valid. The anodyne answer of the climber probably meant this as well.

After making these things clear, let us turn to our specific theme. What increases or changes with the progress of physical science? The answer many people have given or give to this question is that it increases our knowledge of the regularities of the universe. But this is too generic an answer. What are these regularities? Few are concerned with providing a rigorous *explicatum* for this *explicandum*.

Naturally in discussing these things, it is necessary to reckon with K.R. Popper,[8] who represents one of the most able and sure voices heard in recent decades concerning the criticism of science. And let us recognize that it is not possible to do him justice in a few lines. Having said this, I hope I will be forgiven if I give a somewhat picturesque description of his conception. It seems to me that the progress of science as presented to us by Popper is *cemeterial* progress. Science is to a great extent a graveyard where falsified hypotheses lie buried. The larger the cemetery the greater the progress of science because the number of surviving possibilities diminishes. The scientist has a strong instinct of an assassin and has the function of falsifying hypotheses, that is, of sending them to the cemetery. Popper actually allows him to attack those which seem most likely to succumb because if they pass the test, they will be more strongly *corroborated*.

Modern physicists do not generally share this point of view, and do not express themselves in this way. They speak more of *confirmations* than of falsifications. Nevertheless, under certain aspects their general conclusions are not very far from those reached by the Popperian school. I shall describe very briefly and schematically the essential lines of one rather diffuse way of seeing.[9]

First of all one can never insist enough on the fact that the enunciation of a physical theory must always be accompanied by the (explicit or implicit) indication of its *domain of validity*. The domain is established with experience and is constituted by a determined *class of phenomena*, by an allowed *range* for each of the parameters in play, by the *precisions* of the measuring instruments used. *Extrapolation* is admitted only for heuristic purposes. The assertion that a theory is valid *in the absolute* does not make much sense. If we are determined to speak of theories in the absolute, it is difficult to avoid the disconcerting conclusion, once discovered by R. Carnap, that the probability that a general theory is true is zero.

Let us suppose that at a given period we have a theory T_n and a domain D_n in which we know the theory is valid (Figure 1). We then perform experiment *A outside of* D_n (that is, referring to a broader class of phenomena, or widening the range of at least one of the parameters or im-

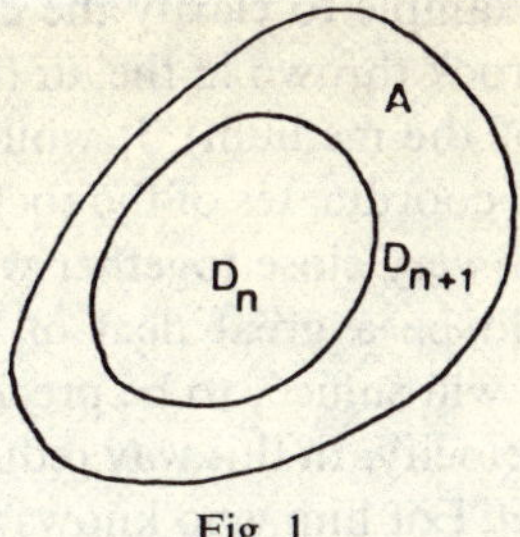

Fig. 1

proving the precision of the instruments). It may be that the result of A agrees with T_n. We shall then say that we have broadened D_n, recognizing that it includes A. On the other hand, it may be that the result does not agree with T_n. We should then expect that a theoretical physicist with imagination will think of a new theory T_{n+1} such that (1) it is valid within D_n, and (2) it agrees with A. T_{n+1} has a domain of validity D_{n+1} larger than T_n which is valid only in D_n. And the process can continue with T_{n+2}, T_{n+3} ..., valid in D_{n+2}, D_{n+3} ... Thus we have an *historical ladder* of theories.[10]

Note that the chronological succession between theory and experiment is not necessarily that described; rather it often tends to be reversed. A T_n is known to be valid in D_n and a theoretician, driven by motives of unity or symmetry[11] or others, works out a T_{n+1}, also valid in D_n, but which predicts at least one phenomenon A outside of D_n. We then do an experiment to see whether the latter actually confirms this prediction. If the answer is in the affirmative, T_{n+1} is accepted as the next rung of the ladder of theories.

The historical ladder of theories can convince us that *there is progress in physics*. But it cannot tell us *how much it is*. How can we *measure* the successive domains D_n, D_{n+1} ...?

A suggestion concerning the path to follow can be found in this way. We extract information[12] about the universe surrounding us by means of our senses, sharpened perhaps by instruments. Now we are immediately convinced that the world which can be experimented upon through our senses is extremely *redundant* with respect to the real world. In other words it can furnish us with an amount of information far in excess of that necessary to know the complex of real phenomena. Eliminating the redundancy and reducing the amount of information to that which is necessary and sufficient is the task of physics.

Let us take a simple example to clarify the concept. Suppose we want to know the motion of a rock thrown in the air (for simplicity's sake, leaving aside the resistance of the medium). It would seem necessary to take experimentally the spatial coordinates of the rock at each instant, or better in a succession of instants very close together which can be differentiated by our clock. This would be a great deal of information. But physics teaches us that much less will suffice; to be precise, it is sufficient to know the initial direction and velocity. In this way redundancy is eliminated, and the information is *codified*. For him who knows the *code* (or the law of the motion of heavy bodies) these initial data suffice to reconstruct the entire trajectory, or the complete information.

One should not think that these considerations concern only the evolution of phenomena in time. It is true that for the physicist the universe is a material structure in space-time and that the research into the regularities and redundancies refers to this complete structure. But even the universe taken at a given instant presents a degree of structure such that it makes our channels of information extremely redundant. To take a particularly instructive example, let us take the case of the visual universe as it appears to us at a given instant of time.

The human eye can distinguish two shining points when their angular distance is greater than or equal to about 1', or 3×10^{-4} radians. Squaring this we get a solid angle of about 10^{-7} steradians. Now, the total solid angle, subtended by a sphere which surrounds us is $4\pi = 12.56$ steradians. Dividing by the elementary solid angle 10^{-7} just found, we are convinced we can see in a static scene about 10^8 distinct points. Now let us suppose that at each point 100 bright levels and different colors[13] can be distinguished. For each point then we have 100 possible cases. If we imagine that these cases are all equiprobable, verifying one of them leads us to information equal to $\log_2 100 \simeq 7$ bits.[14] Multiplying by the number of distinguished points, that is, by 10^8, we conclude that from a static visual scene we can deduce about 10^9 bits of information. This would be actually the quantity of information which the scene would give, if the world were made of clear, dark, variously colored points, distributed absolutely at random. It would take an extremely long time to record and elaborate all this information, since, roughly speaking, we can do it only at the rate of 100 bits per second.

Fortunately, the visual world is anything but made at random. The information of 10^9 bits is enormously redundant and much less is sufficient for us to realize what surrounds us. Think, for example, of the thief who, caught red-handed, jumps out of the window; he gives a very quick look around and almost always escapes in the right direction without even tripping up.

That visual information was very redundant and that it could be codified in a particularly effective way was discovered by our ancestors about 20,000 years ago, when they invented drawing. In fact drawing is nothing other than a code which, starting from very few data (the contours of objects), allows us to extract a great deal of the total available information. If the visual world were made haphazardly, it would not be possible to draw.

It is clear, then, that the discovery of *regularities* and therefore the progress of physics consists in the discovery of the redundancy of the world of sense data and in its progressive elimination by means of suitable codes.[15] And since we know how to measure redundancy, we can hope in this way to arrive at measuring the progress of physics. Nonetheless the way is long and anything but simple.

It is easy enough to establish how much information is extracted from the measurement of a physical quantity. For example, let us measure the temperature of some water, about which we know only that it is liquid (and thus above 0°C) and that it is not boiling (and thus below 100°C). Let us suppose that the thermometer at our disposal is calibrated so that its precision $\varepsilon = 0.01°C$ throughout the scale. The possible results are $100/\varepsilon = 10^4$ and since they are equiprobable, each one bears information of $\log_2 10^4 \simeq 13$ bits. We know then how much information we gain when we perform the measurement.

Note parenthetically that when ε tends toward zero, information tends toward infinity. If information were *free* (and if we had the absurd wish to leave aside the dictates of statistical thermodynamics and quantum mechanics) we could think it possible to reach this absolute exactness. But L. Brillouin has demonstrated[16] that every bit of information costs at least $k \ln 2$ thermodynamic units (k being Boltzmann's constant), that is, it increases at least as much the entropy of the system in which the measurement is being carried out. So to start with, we should have a system which is very far removed from the condition of equilibrium, which is absurd. This is one of the reasons why the finite calibration ε cannot be eliminated from physics and must be taken into account in establishing the limits of the validity of theories.

More complicated is the problem when we are concerned not with a single measurement but with a *physical law* whose informational content

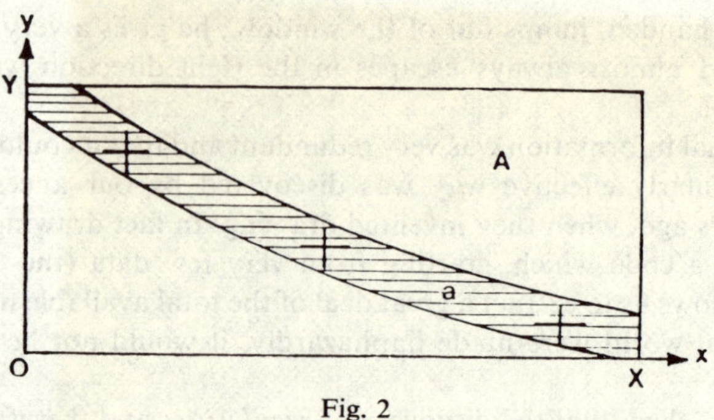

Fig. 2

we want to evaluate. L. Brillouin[17] believed he was able to resolve the question in the following way.

Let us suppose we have two physical quantities x and y of which the second is a function of the first, and we want to determine experimentally this functional dependency. Let X be the range of x, and Y, that of y (Figure 2). Let ε_1 and ε_2 represent the accuracies with which we measure x and y respectively, and let us suppose that they are constant over the entire domain XY. As we know, every experimental 'point' will be represented by a cross whose arms are ε_1 and ε_2 long respectively. Some of these crosses are represented in the figure. The domain occupied by all the experimental crosses is represented by the shaded strip.

We now observe that the number P_0 of possible *a priori* cases (that is, before doing the experiments) for the values which can be found for the pair xy is equal to the total area $A = XY$ divided by the area $\varepsilon_1\varepsilon_2$ of the so-called *resolving power*. Therefore information I_0 which *could* be carried by a measurement, or our *a priori ignorance*, is given by

$$I_0 = \log P_0 = \log (A/\varepsilon_1\varepsilon_2). \tag{2}$$

Having performed the experiments and constructed the diagram in Figure 2, the possible domain has become only that of the strip whose area we shall indicate with a. The number of possible cases has become $P_1 = a/\varepsilon_1\varepsilon_2$ and therefore our *a posteriori ignorance* is

$$I_1 = \log P_1 = \log (a/\varepsilon_1\varepsilon_2). \tag{3}$$

This means that the discovery of the law represented in Figure 2 has brought information given by

$$I = I_0 - I_1 = \log (P_0/P_1) = \log (A/a). \tag{4}$$

If ε_1 and ε_2 are not constant but vary from point to point on the diagram, it will be necessary to take this into account in the computation of P_0 and P_1, or in measuring the areas. It is also evident that the results do not depend on having used Cartesian coordinates for the representation of Figure 2 (in fact in transforming the coordinates, ε_1 and ε_2 are also transformed). Finally, we note that it is very easy to generalize these considerations to the case in which physical law concerns more than 2 magnitudes $x, y, z \ldots$ We do not have to carry out this explicitly. Expression (4) certainly represents an important step forward for our purposes. Nevertheless, the assertion that it provides us with the measurement of the *informational content of a physical law* seems to me too optimistic. In reality (4) tells us only how much information the law gives us with regard to a *single* physical system to which it is applied. To find the informational content of the law *tout court*, it would be necessary to multiply by the number of systems of that type which really exist in the universe!

To see to what paradox the neglect of this observation can lead, let us reason in this way. Let us suppose that the diagram in Figure 2 represents pressure as a function of the volume of a real gas, say, He^4 at constant temperature. Let us suppose further that our instruments are so perfected as to allow us very great precision, as much as will be necessary for our argument. The information expressed by (4) will also be very great. It could happen that the PV law found for He^4 furnishes us with more information concerning the universe than the law of universal gravitation $F = Gm_1m_2/r^2$ verified with the precision possible today! This does not seem really acceptable.

It is clear that we should be able to attribute a *weight* to each law indicating its generality or better its real incidence in the universe. Information, I, given by (4) should then be multiplied by that weight. Unfortunately, we are not in a position to do this today. This problem will deserve careful examination in the future.

Let us pass to another difficulty arising from this way of posing the problem. Our knowledge of the physical world appears to us in two fundamental forms: *nomological* and *factual*. The first, as the name says, relates to the laws of physics; it does not concern a single object or a historical situation, but classes of objects and situations. The second, on the other hand, concerns well-determined objects or situations *una tantum* and not foreseeable with any law; one can only take note of them. Our knowledge of universal gravitation is nomological while that concerning the existence of the planet Mars is factual.

It would be rather difficult to evaluate the informational content of knowledge of fact. But it can be objected that only nomological knowledge belongs to physics while that of fact belongs to other disciplines such as: history, geography, geology, biology, astronomy. And in effect we usually stick to this division of contents.

But this is not as obvious as it may seem. First of all we can observe with E.P. Wigner[18] that it is

impossible to adduce reasons against the assumption that the laws of nature would be different even in small domains if the universe had a radically different structure.

This might seem a mere scruple of a philosophical nature, which is far from being supported by reality. But this is not the case.

Everyone knows that the second law of thermodynamics represents one of the crucial pillars supporting our interpretation of what goes on around us. In the form given it by R. Clausius, it says that the entropy S of an isolated system cannot decrease and that once it has reached its maximum S_M, it remains stationary. Evolution then is *irreversible*.

We know the statistical interpretation of thermodynamics established by J.C. Maxwell and L. Boltzmann. The same *macrostate* can be realized with an enormous number of different *microstates*. If one takes a microstate at random which corresponds to a macrostate of entropy $S < S_M$, there is an enormous probability (or practical certainty) that the microstate will evolve toward conditions of increasingly greater entropy until it reaches S_M.[19] But, observed J. Loschmidt, since, by inverting the velocities of all the molecules, we get from any microstate another microstate corresponding to the *same* macrostate, irreversibility is an absurdity. Then came H. Poincaré's famous theorem that when one starts from a given microstate and lets the system evolve, if only one waits long enough, one returns as close as one wants to the starting microstate. But, observed E. Zermelo, how can a system which passes through the same states be irreversible?

After the basic contributions of P. and T. Ehrenfest, J.W. Gibbs and various other writers, the problem can be considered clarified in the following way. If we trace the curve of entropy of an isolated system as a function of time, we find a practically horizontal line (Figure 3) at height S_M. In reality we are not dealing with a straight line. There are continuous small fluctuations below S_M which are practically unobservable. Every now and again there is a larger fluctuation such as that indicated by A in the figure. Nevertheless such fluctuations are so rare that, in order for

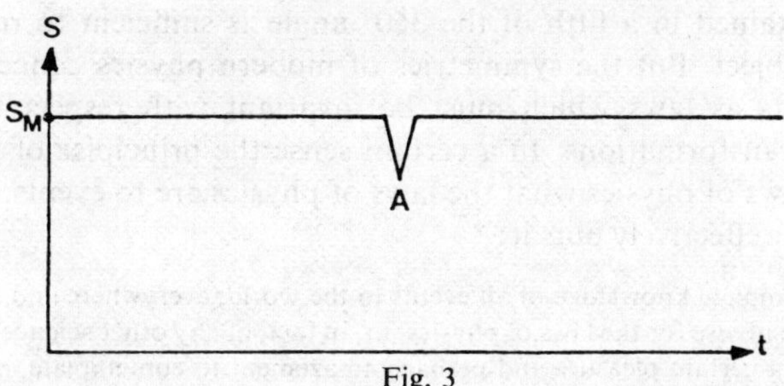

Fig. 3

one to occur, starting from S_M, one would normally have to wait an inconceivably long time, on the order of the age of the universe or greater! Therefore if we start from the system with $S < S_M$, we are virtually certain that the system will remain indefinitely in this state of equilibrium. On the other hand, the case is different in which we start from a point like A, with $S < S_M$. In this case it is extremely more probable (that is practically certain) that we go back up rather than continuing to go down. Then we are certain that entropy will increase. And yet, as can be proven, the curve does not present systematic asymmetries, in the sense that it does not favor the positive direction of time with respect to the negative one! The second law of thermodynamcis is valid because *in fact* we are in a universe in a condition of entropy[20] much smaller than the maximum. It is a question of *initial conditions* and not of *law*.[21]

This result is disconcerting. If the validity of so important and general a law as the second law of thermodynamics depends essentially on initial conditions of fact, in what case will we be sure of dealing with purely nomological questions? How can we rule out that one day a general law of our universe won't be found that depends on the fact that the planet Mars exists?

The answer that can be given is that since we live in *this* universe and are concerned with it, it matters little if our laws would no longer be laws in other 'possible universes'. I agree that there is a certain reasonableness in this answer, but it does not seem to me to clear up all the doubts.

We must now face the problem of the informational status of the *superlaws* of modern physics. I am alluding to the so-called *symmetries* or laws of *invariance* or *conservation*.

When symmetry is applied to an object, there are no serious interpretative problems. It is clear that if the object presents, let us say, pentagonal symmetry, such as the starfish (and if the symmetry is perfect), the information contained in a fifth of the 360° angle is sufficient to reconstruct the entire object. But the symmetries of modern physics concern not so much objects as laws which must be invariant with respect to certain groups of transformations. In a certain sense the principles of symmetry are to the laws of physics what the laws of physics are to events.

As Wigner effectively puts it:[22]

If we had a complete knowledge of all events in the world, everywhere and at all times, there would be no use for the laws of physics, or, in fact, of any other science. . . . They might give us a certain pleasure and perhaps amazement to contemplate, even though they would not furnish new information. [Similarly] if we knew all the laws of nature, or the ultimate law of nature, the invariance properties of these laws would not furnish us new information.

And so, as laws furnish us with information precisely because they allow us not to investigate experimentally all the events of a given series, but to rise from the knowledge of some of them to all the others, so the principles of symmetry spare us the work of investigating laws. For example, invariance through spatio-temporal translation tells me that from the law that holds true *here* and *now* I can extract analogous laws which are valid in all other places and all other times. The relativistic invariance tells me that from the laws which are valid in my (supposedly inertial) laboratory I can extract analogous laws which are valid in any other laboratory in uniform rectilinear motion with respect to mine. And so on.

We could perhaps say the following. Spatio-temporal invariance multiplies the informational content of a law by the number of points of the chronotope which can be distinguished from one another by us.[23] Analogously the relativistic invariance multiplies the informational content of a law by the number of different (vectorial) velocities which can be distinguished by us between 0 and c. Invariance under rotation multiplies by the number of different directions which can be distinguished by us in the 360° angle. And so on.

But here too, we encounter the problem of *generality* and *weight*, that is, the problem already seen for laws. For example, it would seem that the CPT invariance must multiply the informational content of the laws by 2. Is it possible that such a modest factor corresponds to such a general principle? And what can be said of the *approximate* symmetries or of those, like parity conservation, which are not valid in a certain class of phenomena? It is difficult to see how they can fit in our context.

It should not be passed over in silence that today different classes of concepts go under the name of symmetries or invariances which it might be well to keep distinct. There are geometric or kinematic invariances such as those we have mentioned, there are intrinsic symmetries which concern the properties of particles and their quantum numbers, there are dynamic invariances. And what can be said of the invariances applied to particular phenomena, such as gauge invariance which holds true in electrodynamics? Perhaps it should be conceived as invariance in the choice of the system of reference, analogous to invariance with respect to the choice of general coordinates at the basis of general relativity; and this last, as V. Fock[24] indicates, is different from all other kinds of symmetries. And what about scale invariance which today is the subject of particular discussion? How should we place it if it were to reveal that it is valid in continuous intervals separating the different levels of discrete spectra in which physics seems structured (Chinese boxes)? But here we are decidedly going into S_1 (science *in fieri*) and for it, as we have noted, we cannot construct theories.

After laws and superlaws would come full-fledged *theories* such as, for example, general relativity or quantum mechanics. Will it make sense to speak of the informational content of quantum mechanics? I maintain yes, but I add immediately that today I would not even know where to begin to tackle the problem of a quantitative evaluation.

In conclusion, we can acknowledge that the 'progressology' of physics is still at the beginning and is quite far from giving clear and complete answers to the problems we may wish to pose it; nevertheless, it does not seem at all that such problems make no sense or cannot be confronted. It will take much reflection, critical analysis and, perhaps, also a bit of imagination. But probably we will be able to reach some concrete results.

NOTES AND REFERENCES

[1] E.H. Carr, *What is History?* (New York, Knopf, 1963), p. 148.

[2] Who knows why, this is sometimes called a modern habit. In reality, in one guise or another, it is as old as the world.

[3] I am not unaware that between science and the society in which it originates there is a close relation, which gives rise to reciprocal conditioning. This is obvious, so obvious, in fact, that I am even somewhat irritated to have to assure people that I am aware of the fact. But I must do this because, as is known, there are more than a few who, starting from this obvious premise, think they can draw absurd conclusions or at any rate conclusions not universally shared, after which they attributed disagreement to the fact that others are unaware of the premise.

[4] What these assertions are like is expounded, for example, in M.L. Dalla Chiara and G. Toraldo di Francia, 'A logical analysis of physical theories', *Rivista del Nuovo Cimento* 3, 1 (1973).

[5] G. Holton, *Thematic Origins of Scientific Thought, Kepler to Einstein* (Cambridge, Mass., Harvard University Press, 1973), pp. 385–386.

[6] In the mountains it is not rare to experience that phenomenon which T.S. Kuhn, in taking a concept of Gestalt psychology and applying it to the development of science, has called 'Gestalt switch'. See T.S. Kuhn, *The Structure of Scientific Revolutions* (Chicago, University of Chicago Press, 1970), Chapter 10.

[7] It is clear that I want to allude to Lakatos' conception of *research programs*. See I. Lakatos, 'Falsification and the methodology of scientific research programs' in *Criticism and the Growth of Knowledge*. Proceedings of the International Colloquium in the Philosophy of Science, London, 1965, volume 4. Edited by Imre Lakatos and Alan Musgrave (Cambridge, Cambridge University Press, 1970), p. 91.

[8] K.R. Popper, *Logic of Scientific Discovery* (London, Hutchinson, 1959; 3rd revised edition , 1968); *Conjectures and Refutations: the Growth of Scientific Knowledge* (New York, Harper, 1963); *Objective Knowledge* (Oxford, Clarendon Press, 1972).

[9] For details see: G. Toraldo di Francia, 'Induction in physics', *Rivista del Nuovo Cimento* **4**, 144 (1974).

[10] Note that we are dealing only with a rational reconstruction or *internal history*, which does not necessarily coincide with actual history. In this regard see: I. Lakatos, 'History of science and its rational reconstruction' in *Boston Studies in the Philosophy of Science*, **8** (Dordrecht and Boston 1971), p. 91.

[11] This is the case, for example, with Einstein, who was induced to devise the theory of relativity much more by the 'asymmetries' of the electrodynamics of the time than by the Michelson-Morley experiment.

[12] We shall make use here of some notions concerning information theory. See, for example, J.R. Pierce, *Symbols, Signals and Noise: the Nature and Process of Communication* (New York Harper, 1961).

Let us remember that in addition to ordinary statistical information theory, there is a so-called *semantic information* theory proposed by Y. Bar-Hillel and R. Carnap, 'Semantic information', *British J. Phil. Science* **4**, 147 (1953), and carried further by J. Hintikka, 'On semantic information' in *Physics, Logic and History* (New York, Plenum Press, 1970), p. 147. Perhaps such a theory could have some interesting conceptual side for our purposes, but unfortunately it seems absolutely impossible to apply it in practice in order to draw some concrete results from it.

[13] The estimate errs much on the side of cautiousness. But bear in mind that the eye's capacity to distinguish different levels and colors in very small areas is much smaller than is the case in extended areas.

[14] As is usually done in information theory, we are using logarithms with base 2.

[15] Here is the link with Popper's falsificationist conception. To falsify a scientific theory means to reduce the number of possibilities for the physical world and thus to gain information.

[16] L. Brillouin. *Science and Information Theory* (New York, Academic Press, 1956), chapter 12.

[17] L. Brillouin. 'Observation, information, and imagination' in *Information and Prediction in Science* (New York, Academic Press, 1965), p. 1.

[18] E.P. Wigner. *Symmetries and Reflections* (Bloomington, Indiana University Press, 1967), p. 3, note.

[19] In a rather simplified manner, this statement expresses the content of Boltzmann's famous H theorem.

[20] Talking about the entropy of the universe and its maximum is anything but rigorous. But we haven't the time here to go into the necessary details and we must be content with suggesting an intuitive idea. For a good historical and conceptual discussion, see A. Grünbaum, *Philosophical Problems of Space and Time. Boston Studies in the Phil. Sci.*, **12** (Dordrecht and Boston, Reidel, 1973), Chapters 8 and 10.

[21] See H. Mehlberg. 'Physical laws and time's arrow' in *Current Issues in the Philosophy of Science* (New York, Holt, Rinehart and Winston, 1961). See also O. Costa de Beauregard, 'Information and irreversibility problems' in *Time in Science and Philosophy* (Amsterdam, Elsevier, 1971); by the same author, 'No paradox in the theory of time anisotropy' in *The Study of Time* (Berlin, Springer, 1972), p. 131.

[22] E.P. Wigner, *op. cit.*, pp. 16 and 17.

[23] Naturally a system to which a physical law is applied does not occupy only one point. But for each system I can refer to a characteristic point, for example, the center of gravity. Two systems of identical structure, translated with respect to one another, are different for me, only if I succeed in distinguishing the center of gravity of one from the center of gravity of the other. This depends on the precision my measuring instruments allow me.

[24] V. Fock, *The Theory of Space, Time and Gravitation* (New York, Pergamon Press, 1959). E.P. Wigner, *op. cit.*, p. 23.

M. PIATTELLI-PALMARINI

EQUILIBRIA, CRYSTALS, PROGRAMS, ENERGETIC MODELS, AND ORGANIZATIONAL MODELS*

INTRODUCTION

The processes by which an organization can be set up and maintained, can grow, reproduce itself and give rise to other organizations, are now the object of study of autonomous scientific disciplines. According to the late W. Ross Ashby, the great English cybernetician, the distinctive trait of this new science is to "presuppose without, however, specifying" the physical nature of the systems it studies, since its real vocation is to look for the general laws of organization and not the rules of production, transport and utilization of energy. It is in practically identical terms that Ludwig von Bertalanffy defines 'general systems theory'; and the emphasis placed on the non-directly material or energetic nature of the phenomena dealt with often reappears in the works of other cyberneticians such as Gordon Pask, Heinz von Foerster, Donald D. McKay and Norbert Wiener. Here is a source of misunderstandings: moreover we could see here an indication of how theoretical preoccupations go beyond the framework of these disciplines. Every theoretical model presupposes in fact the material nature of its realizations without having to specify it, and every theoretical model establishes relations between abstract invariants. The principle of virtual work, Lagrange and Hamilton–Jacobi equations, the second principle of thermodynamics, Maxwell's equations all presuppose the material nature of concrete instances to which they are applied, without, however, specifying it. It is as much a question of general laws of organization of physical systems – if, by organization, we understand the ordered unfolding of a dynamic compatible with the constraints acting on the system – as of mechanical, energetic, geometrical and other constraints. The Le Châtelier-Braun principle, to which we shall return in what follows, has been initially interpreted as a general law of self-compensation, applying to every material system and then literally, as a principle of self-organization. What constitutes the characteristic feature of the new science (information theory, general systems theory, self-organizing systems theory, cybernetics) is rather the *physical interpretation of variables*

---

* This paper was presented at the 94th Congress of the French Association for the Advancement of Science (AFAS), Brussels, 10 July 1975.

341

which appear in its equations. By limiting ourselves to examples (which are, moreover, not at all numerous) of the general laws established by these disciplines, the truly new thing that emerges from this is that the specified observables are, for example: the substitution frequencies of one symbol for another (Shannon's law of channel noise), the number of acceptable responses elicited by external disturbance (Ashby's law of minimum requisite variety), the redundancy of a coding system or the relative number of assimilable new components in terms of a given organization (Foerster's criterion of self-organization: see Atlan, 1972). It is not the degree of abstraction which has increased, for the definition of the probability of equivocation on a signal is no more abstract than the definition of entropy, the notion of information no more abstract than that of energy. Nor is the operational translation of concepts more laborious, for measuring the quantity of information in bits [binary digits] is hardly more complicated nor more conventional than measuring temperature. What has changed is the general theoretical framework, the basic concepts which regulate the chain of deductions, the nature of the essential parameters to which we propose to reduce every observed phenomenon. The focal concerns of this new science are exclusively those occasional or accessory concerns of many other present-day or ancient sciences. Biology is *also* concerned with the phenomena of regulation, compensation, repair and communication (between cells, organs, individuals), but the theory of communications and cybernetics is concerned *only* with these mechanisms of self-regulation evidenced by biological systems. Electronics and circuit theory are *also* concerned with showing up the phenomena of suppression, attenuation, distortion, inversion, retroaction, mutual perturbation in the production and propagation of currents and the differences of potential. It is only with information theory, however, that we study exclusively the relations between the *forms* of electrical signals emitted and received by distinct apparatus, connected by transmission channels with assigned characteristics. In the first case attention was focused on currents, propagation waves and electromagnetic fields. In the second, the principal interest turns on the relations between the signal sent and the signal received, the actual signal and possible signals. The very concept of message is based on these relations.

Although some definitions and declarations of principle of these new disciplines often lend themselves to misunderstandings, it is clear nonetheless that conspicuous sectors of science are today interested *exclusively* in something other than transformations of matter and energy flows. Their common domain of action and reflection is order, organization, self-regulation, *whatever the particular system where these phenomena manifest themselves*. Physics, chemistry, biology and the human sciences *per se* are also sometimes concerned with these phenomena; they have all contributed to establishing criteria thanks to which one can often recognize without fail the matter under consideration. The problem of order and organization, an interest in the relations of difference and their stability, pervade all of science from its birth, but if one wanted to find a founding hero for this current of thought, it would not, as a matter of fact, be either Galileo or Newton or Carnot or Darwin. Rather the founding hero would be Gutenberg, who provided mankind with a machine made of particular materials, which uses energy, but whose *raison d'être* is the reproduction of an order and the replication of a system of differences. What makes the reproduction of messages expedient as well as conceptually remarkable – what we might call the heritage of Gutenberg – is probably the true line of demarcation between classical science and the new science. The normative ideal of the former was equilibrium, that of the latter is transmission and reproduction: it has been necessary to change the basic observables, to shift from a constitutive interest in movement, forces and energy, to one in codes and programs; in *meaning*. A transition science, straddling the old and the new preoccupations, is biology, and molecular biology in particular (J.P. Changeux, 1975). The range of such a shift in basic concerns is, nevertheless, for the reasons we have just outlined, transdisciplinary. The examples we have chosen to develop here seemed to us particularly significant, but they are not at all exhaustive.

Modern epistemology has removed the partitions separating disciplines and has shown the common substructures constituted by normative models, cognitive strategies, heuristic rules, recurrent themes which, as Michel Serres admirably described it, flow between the different fields of knowledge and let themselves be 'translated' into particular principles and laws proper to each discipline. Within this new epistemology one finds oneself practicing a thematic analysis of the sciences (in the sense now rigorously specified by Gerald Holton) somewhat as M. Jourdain once discovered himself speaking prose. Equilibrium is precisely one of these 'themes', a 'preconception' among those which one often meets in scientific activity and which, according to Holton (1973, p. 23). "appear to be unavoidable for scientific thought, but are themselves not verifiable or falsifiable."

Just like continuity and discontinuity, symmetry and asymmetry, plenum and void, equilibrium is a condition that exists prior to observation, an *a priori* criterion which decides what to look at and how to measure. By applying itself to the study of paths, forces, changes of state, classical science of necessity had to compare observables with a class of reference invariants, reduce the variables to commensurable intervals, and measure magnitudes in terms of normalized scales. The new science, in applying itself instead to evaluating the powers of self-selection of systems with respect to their possible configurations and to the creation/destruction of constraints, no longer has any profound relation to the postulate of equilibrium. The new implicit foundations are the notions of program and dynamic memory; these are in fact the 'themes' of the new science. These considerations of the notion of equilibrium are in effect intended to prepare the field for a parallel analysis of new 'themes'. The point of articulation between classical themes (particularly the notion of equilibrium) and new themes, is constituted, as we shall see, by biological macromolecules: the nucleic acids and the proteins. The collective term encompassing them, that of 'informational' macromolecules, reveals by itself alone that a transition has occurred. This thematic transition is the object of my paper. To describe it better, we should start from the development of the notion of equilibrium, not so much in terms of a detailed historical genealogy nor in terms of a close criticism of texts, but rather in terms of representative examples that cut across time and disciplines.

FROM THE NOTION OF EQUILIBRIUM TO THE IDEAL OF EQUILIBRIUM

I shall limit myself here to noting the normative role played by equilibrium-oriented frames of thought (expressed by a multiplicity of distinct laws or principles which are, however, closely related, e.g., the principle of least action, the principle of the conservation of energy, the law of homeostasis and homeorhesis[1], the law of optimal adaptation, balanced polymorphisms, etc.) in areas from linguistics to economics, physics to psychoanalysis, population genetics to psychology. However, I shall develop only those aspects more particularly concerned with physics, chemistry and biology.

The criterion of equilibrium cuts across these 'model' sciences where it represents the possibility condition for grasping change, discontinuity, tensions, exchanges and morphogeneses. It is through the laws of equilibrium that the field of the possible is structured, that the actual and the virtual are divided, that variables are indexed, that metrical scales (constructed on equipartitions of a gap) find their operational definition.

If at present we still remain framed by an equilibrating thought, it is because the richness of this notion has shown itself to be very great, so great that the theories of disequilibrium, irreversibility and self-organization emerge with great logical and ontological difficulties. The notion of equilibrium, or rather the multiple strategies of equilibrium, aim at closing the set of possible transformations of a system, at normalizing states, at conceiving change always in relative terms, at imposing the minimization of all gaps in real time or, asymptotically, in infinite time. The balance-sheet, the metaphor of the scales, hence of the archetypal instrument of equilibrium, is also the logical operator of structuration of a catalogue, that through which 'items' become factors, stresses become systems of forces, displacements are taken as global transformations.

To arrive at a general law of equilibrium means being able to enclose a system within its own lifetime, a normalized time which has lost every trace of the period of constitution of the system, when elements could be added to it or taken away from it, before the forces had acquired a direction and a meaning, before articulations had become constraints.

A state of equilibrium, even ideal, even statistically defined, allows one to attach reference values to variables, to effect the simple transformation $X = x - x_0$ (where x_0 = value at the equilibrium), hence to exploit best our perceptual structure which detects relative lack of alignment with better precision than absolute changes by several orders of magnitude. The quantitative is only a protocol of comparison of the qualitative, the synopsis of a reference standard and an actual pattern, or a series of synopses up to the exhaustion of the residual through gearing down the standard. The alignment of reference marks is the source of quantity. The stability of the standard and its unaltered permanence in time represent, perhaps, the most worn but also the most fundamental criterion of equilibrium. The very practice of measurement presupposes a local equilibrium between the measuring system and the system measured.

The necessity of a standard-system, of a standard state, will constitute the first strategy of equilibrium, the strategy of fractional synoptic disalignment which allows one to take in diachronic and continuous changes. The knowledge of first the physical world and then the living world had to conform to this normative ideal of the stable, of the equilibrated, of the balanced, on and by which all transformation becomes measurable.

FROM STATIC EQUILIBRIUM TO DYNAMIC EQUILIBRIUM

> Once an area of knowledge has been reduced
> to a self-regulating system or 'structure', the
> feeling that one has at last come upon its inner-
> most source of movement is hardly avoidable.
>
> Jean Piaget, *Structuralism*, p. 14

Treatises on mechanics open with a chapter on statics, where the notion of vector (hence transport, indeed, effort and stress) is introduced in order to establish an algebra of mutual compensations, where one defines forces in terms of their possible effects of deformation or displacement in order to define their conditions of non-effect, in order to compose a null vector or one exactly compensated by the reactions of constraints.

The axiomatic approach in teaching physics prepares the mind to grasp movement, the disequilibrium of forces in terms of the preformative conceptual structure of equilibrium. The works of analytical mechanists complete this strategy with notions of '*vis viva*' and virtual displacement. The famous principle of virtual work represents the culmination of the equilibrating thought of the eighteenth century.

Newton's second law had set up an equation between two heteronomic universes, that of forces as defined by statics, and that of trajectories and kinematic magnitudes, thanks to the mediating concept of mass, dependent on the integration of a geometric notion with a hard-core materialist notion (density multiplied by the volume). Newtonian physics is a 'vectorial' physics (cf. Elkana, 1974a). In analytical mechanics, these two universes are unified, because both are subsumed under the notion of a variable distribution within a single, more abstract universe, where forces and masses in movement can be permuted conceptually and quantitatively. Equilibrium is found again at a more general level in terms of the assimilation of stresses or static forces to physical work extracted from bodies in movement, representing a virtual force, actualized each time this movement is stopped, hindered, channeled. Between statically measured force (dynamometer, springs, weights, pulleys, strings, etc.) and masses in movement, interconversion was possible without loss, without residue (the conceptual role of friction arose only later), without conditions other than the congruence established by the three laws of dynamics.

Rest and movement – Galileo's transformation group was there to demonstrate it – were only two points of view. The general equilibrium of dynamic systems was assured by the invariance of laws despite the eventual displacement of the observer. But what interests us in this too well-known story is finally the underlying notion of a group of possible transformations.

We are dealing with a group admitting translations, rotations, precessions and every possible combination of these spatial transformations, forming altogether a not very large class. The transformation of macroscopic work into heat by friction for the moment remains excluded from it, and every variation in the internal state of systems (fusion, evaporation, etc.) remains equally excluded from it.

The set of forces is catalogued and closed. The possible imbalance between electrical charges and forces resulting from it constitute yet another chapter apart. Mechanical equilibrium and the principle of least action impose a drastic reduction of observable phenomena, of interactions having acceptable status. The dynamic principles of conservation (momentum, angular momentum, position of the center of mass for isolated systems) depend on a structuring of the set of equivalencies, and they give a synthetic formulation to a principle of equivalence. As Piaget remarks:

the assimilation of linear and angular velocities involves at once an assimilation as far as common spatio-temporal relations are concerned and an accommodation to these distinct situations.

It is through the history of the sciences as well as through genetic psychology that we find, with dynamics and its laws of equilibrium, the beginning of a long process of perceptive assimilation and formal identification between magnitudes and phenomena which are at first glance distinct, indeed opposed (force and rest, stasis and movement, straight line and circle). It was only in the middle of the nineteenth century that classes of apparently very varied phenomena were made commensurable, permutable, and finally perceived as fixed transformations between elements of a unique, closed set, established once and for all: the set of movements, microscopic and macroscopic, actual or virtual, susceptible of being logically grasped and synoptically perceived as transference without loss, from one to the other, of a generalized and constant motor aptitude in the universe. It is the general concept of energy unifying dynamics, thermodynamics, electric currents and 'animal heat' which is being formed throughout the nineteenth century and which becomes up to our day the variable of minimization par excellence, the one to which we refer when we speak of equilibrium.

STAGES IN THE CONSTRUCTION OF THE NOTION OF EQUILIBRIUM[2]

With his *Réflexion sur la puissance motrice du feu et sur les machines pro-pres à développer cette puissance* ['Reflections on the motive power of fire, and on machines fitted to develop that power'], Sadi Carnot (1824) opened the way to a formal and very general description of energy transformations.

We are dealing with a reflection on the *intrinsic limits* of power (energy) which can be extracted from any machine. As we know, it results in a law of limitation applicable to any given machine.

In order to consider in the most general way the principle of the production of motion by heat, it must be considered independently of any mechanism or any particular agent. It is necessary to establish principles applicable not only to steam-engines, but to all imaginable heat-engines, whatever the working substance and whatever the method by which it is operated. (*Reflections*, p. 6).

The basis of this theory is that: "wherever a difference in temperature exists, production of motor power can exist". The theory hinges on the relation Q/T which opens the way to the definition of entropy as a function of state given by Clausius (1850–1854): a quantity, whose variations are subject to universal rules and which, with Kelvin (1851), define a *natural*, unique sense of *spontaneous* transformations. Carnot defines as "*inadmissable*" perpetual motion of the second kind which one would want to be

capable of starting from rest all the bodies of nature if they should be found in that condition, of overcoming their inertia; capable, finally, of finding in itself the forces necessary to move the whole universe, to prolong, to accelerate incessantly, its motion. (*Reflections*, note to page 12).

In the posthumous 'Notes' which follow the *Reflections*, he assimilates the notion of heat to that of "vibratory motion of molecules" and abandons the conception of caloric. A nomologically inevitable change followed from the conclusion:

We can establish the general proposition that motive power is, in quantity, invariable in nature; that it is, correctly speaking, never either produced or destroyed. It is true that it changes its form—that is, it produces sometimes one sort of motion, sometimes another—but it is never annihilated. ('Selections from the Posthumous Manuscripts of Carnot' in *Reflections*, p. 67)

To exclude all exhaustion of motor power implies recognizing at the same time the possibility of its transformations and of posing a universal principle of conservation. Equilibrium and conservation are indissolubly linked; together they constitute the abstract invariant which henceforth unifies microphysics and macrophysics. In an article published in 1845 in *The Philosophical Magazine*, following joint works carried out with his French colleagues, James Prescott Joule asserted that "any theory which, when carried out, demands the annihilation of force, is necessarily erroneous" because he firmly believed that "the power to destroy belongs to the Creator alone."[3]

An example of simultaneous discovery which has become 'classic' (Kuhn 1955), the principle of the conservation of energy was firmly established in 1847, thanks above all to Wilhelm von Helmholtz's famous report, 'Über die Erhaltung der Kraft' ['The conservation of force'] where, in terms of the relation to the actual formulation, the purely semantic ambiguity of the word 'Kraft' was the only obstacle. Motion, heat, thermal effects of electric currents, friction, changes in chemical states were revealed as distinct forms of the same basic entity, conserving itself quantitatively through all its avatars. Plant photosynthesis is foreshadowed and the concept of a caloric equivalent of feeding in animals is set forth in full, following the experiments of Dulong and Despretz. Helmholtz rightly remarks that

The object of this investigation was to lay before physicists as fully as possible the theoretic and practical importance of a law whose complete corroboration must be regarded as one of the principal problems of the natural philosophy of the future (in Elkana, 1974a, p. 129).

The universality of this principle soon allowed one to apply it to more and more vast classes of natural phenomena, above all in chemistry and physiology. The task was completed by Le Châtelier and Claude Bernard about 1885.

The general law of displacements of equilibria was definitively formulated by Le Châtelier in 1884[4] and in 1888 in 'Recherches sur les équilibres chimiques':[5]

Under the influence of a change in any one factor of equilibrium, every system in a state of chemical equilibrium undergoes a transformation in one direction such that if this transformation had taken place by itself, it would have produced a change opposite to the direction of the factor in question. Factors of equilibrium are temperature, pressure and electro-motor force, corresponding to the three forms of energy: heat, electricity and mechanical energy.

Extrapolations from the Le Châtelier-Braun principle to contexts more general than that of chemical equilibria are part of the 'Zeitgeist' and go

back to the years preceding its formulation. In 1885 Léon Frédéricq wrote
in the *Archives de Zoologie*:

The living being is so designed that each disturbing influence by itself provokes the
setting into activity of the compensating apparatus which must neutralize and repair
the damage.[6]

In addition, Spencer advocated the extension of a similar criterion in
First Principles:

among the involved rhythmical changes constituting organic life, any disturbing force
that works an excess of change in some direction, is gradually diminished and finally
neutralized by antagonistic forces; which thereupon work a compensating change in
the opposite direction, and so, after more or less of oscillation, restore the medium
condition. And this process it is, which constitutes what physicians call the *vis medicatrix naturae*.[7]

Spencer saw here an analogy with the '*vis medicatrix Naturae*' postulated
by doctors.

In his *Introduction à l'étude de la médecine expérimentale*, Claude Bernard in 1885 elaborated the concept of 'internal environment' and cleared
away the last obstacle to the conception of a true *self-organization* at the
heart of *natural phenomena*. In 1929 Cannon created the concept and term
'homeostasis'. The beginning of the twentieth century in fact saw the Le
Châtelier principle assume a growing importance and become one of the
great principles of experimental science having *ontological* value. An entirely different weight is also given to the concept of system. W.D. Ban-
croft wrote in 1911:

The broadest definition of [the theorem of Le Châtelier] is that a system tends to change
so as to minimize an external disturbance.[8]

The emergence of the generalized notion of system is also referable to
Chwolson (*Traité de physique*, vol. 3^2, Paris, Hermann, 1910, p. 477):

In a body or a system every external action engenders changes directed in such a way
that following this change, the resistance of the body or system with regard to the external action is increased. . . . the Le Châtelier-Braun principle can be considered as a
kind of adaptation of non-living matter.

A formulation even more consonant with contemporary ecosystemic
concepts is given by Jacob Löwy in 1911:

If the equilibrium of a natural complex (system of masses, organisms, systems of ideas) is disturbed, we witness an adaptation to the stimulus producing the disturbance such that said stimulus decreases progressively until the initial equilibrium or a new equilibrium is re-established.[9]

The heuristic power of the principle of generalized dynamic equilibria and the idea of system which it materialized were so overwhelming that experimental physiology during the first decades of the twentieth century often confined itself to an analysis of the details of such equilibration.

However, there was a weak point in this conception, where the physiological phenomena and the organization of the living being called on the reproductive order, on chemical memory, on the architectural specificity of certain structures, on the molecular individuality of certain components, particularly of genetic 'units' and proteins.

The feed-back model, the reservoirs which discharge and receive unceasingly through a complex circuit of compensation, epitomized by the metaphor of the flame, slowly began to be replaced by the model of the crystal.

THE LIMIT OF THE NOTION OF EQUILIBRIUM AND THE DEVELOPMENT OF THE NOTION OF CRYSTAL-PROGRAM

The historical development of these two models is admirably reconstructed in two texts to which we are going to refer: Langley's collection *Homeostasis, Origins of the Concept* and J. Lorch's paper, 'The charisma of crystals in biology' in Elkana 1974b.

The history of the divergence of these models and analogies is long, complex and far from being resolved today. On the contrary, the transition from models of equilibrium to those based on messages, codes and programs, from the metaphor of the flame to that of the crystal constitutes in our opinion perhaps the nodal point of the epistemology of modern biology.

Starting with two quotations, one from Walter Cannon (1925), the other from Salvador Luria (1975), the contrast should appear obvious right away. In 'Some general features of endocrine influence on metabolism':

. . . the liver is an organ of central importance not only for carbohydrate, but also for protein metabolism. It shares with other parts the power of constructing protein from amino acids, but when large amounts of protein are fed it lays by a reserve or deposit to a much greater degree than do the other parts. This reserve can be called out by sympathetic stimulation and by adrenin.[10]

This concept, nowadays devoid of meaning, of a reserve of proteins, is dictated to Cannon by a straightforward analogy with the equilibrating function of the liver in the glycolitic functions. In effect, after having revealed the fundamental roles of proteins in biological organization (structure, reparation, blood coagulation) and after having ascertained that the metabolic processes involve a disintegration and a loss of available proteins, he concluded:

Since there is an elaborate arrangement for storing excess of carbohydrate and for liberating it as required, *it would appear reasonable to look for a similar arrangement for storing indispensable protein* against a time of need. The process of deaminization and the excretion of extra nitrogen indicate that accumulation of protein must be limited. The existence of a nitrogen equilibrium, capable of being altered and adapted to the requirements of the organism, implies that there prevails an orderly regulation for riddance or retention and the maintenance of a steady state [italics mine].[11]

The reasoning is extremely clear and the heuristic power of the model of equilibrium is completely explicit. The normative 'pre-concept' guides the theoretical construction. The analysis seemed plausible at the time but quite profound conceptual changes make it unacceptable today. Proteins do indeed have a very particular biochemical individuality and a specificity in their chemical sequence which excludes every direct transformation between proteins. Each protein must be constructed end to end starting from its primary components. A passage in a college text of molecular biology addressed to MIT students – S.E. Luria's *36 Lectures in Biology* – sets forth clearly the present day state of the problem:

In order to synthesize DNA, the appropriate enzymes plus the properly activated nucleotides are not enough. There must also be DNA to act as the tape, the template to direct the synthesis of the new DNA. Likewise, in order to synthesize protein, in addition to amino acids attached to the proper carriers and various enzymes, coenzymes, and other factors, a molecule of RNA is needed to serve as a template. A mixture of all the precursors and enzymes will not make the coded substances in the absence of the appropriate templates.

The template does not provide energy, or chemical reactivity, or pieces or synthesis; it provides *information about the order* in which the pieces are to be assembled. [It may seem that the template, a piece of DNA or RNA with bases in a specific order, may provide a certain amount of *negative entropy*, that is, of order. But this is not thermodynamic order: any sequence of 30 nucleotides or 30 amino acids has as much chemical and physical order as any other. What a template provides is biological order, that is, the selection of certain useful sequences among all the possible ones. In reality, however, the enzyme machinery itself has evolved in such ways that without templates the syntheses would proceed very slowly if at all.] (p. 95; italics are Luria's)

The half-century separating these two models of the synthesis of basic biological materials did see the occurrence of an unprecedented scientific revolution in the conception of the living being. Obviously it would be impossible to reconstruct this change in paradigm here, even in broad outline. However, among the most revealing symptoms of this change in basic models we can cite the lectures given by Erwin Schroedinger in February 1943 at the Dublin Institute for Advanced Studies and subsequently collected in the volume, *What is Life?* The unique physical constitution of the living cell is seen by Schroedinger as amenable to an analysis of its basic molecular components, their arrangement and reproduction "in relation to the statistical point of view" (p. 4). From this point of view the fundamental processes of life appear based on substances that differ "so entirely from that of any piece of matter that we physicists and chemists have ever handled physically in our laboratories or mentally at our writing desks" (p. 5). The profound difference in behaviors must be based on a specificity of structures on which these laws and these regularities are based. "In physics we have dealt hitherto only with *periodic crystals*", but, anticipates Schroedinger, "the most essential part of a living cell – the chromosome fibre – may suitably be called an *aperiodic crystal* . . . which, in my opinion, is the material carrier of life." (p. 5; italics are Schroedinger's).

With our present day vocabulary we can say that Schroedinger was looking for a stock of information, a program. The aperiodicity of the chromosomal crystal (which Schroedinger, like the vast majority of his colleagues, believed was made of proteins) assured a microscopic reservoir of information which was both adequate and reproducible. The term 'information' was not yet in use (information theory came into being five years after his lectures), but the concept is very clearly formulated there:

A well-ordered association of atoms, endowed with sufficient resistivity to keep its order permanently, appears to be the only conceivable material structure that offers a variety of possible ('isomeric') arrangements, sufficiently large to embody a complicated system of 'determinations' within a small spatial boundary. Indeed, the number of atoms in such a structure need not be very large to produce an almost unlimited number of possible arrangements. (p. 65).

Ten years later the physical reality of these models was revealed in detail by Crick, Watson and Wilkins (DNA) and by Perutz and Kendrew (globular proteins).

The model of the aperiodic crystal as applied literally in the case of proteins and in a rather logical sense for nucleic acids forms the hinge between the notion of structure and that of program. The quotation from Schroedinger is nothing but an operational definition of the genetic program. Today we speak of informational macromolecules among which are also included certain synthetic polymers. The transition from the model of the flame, from 'steady state', hence from the structure-producing process to that of the aperiodic crystal, a process-governing structure, is the foundation of present day molecular biology. The hesitation with which many molecular biologists and fundamentalist geneticists today greet models based on dissipative structures bears witness to a profound theoretical divergence which is extremely interesting for epistemology. It seems that the primacy of the structure over the order of processes comes up against insoluble problems when it is a question of explaining the very origin of the structures. In other respects the logical priority of the self-organizing process over the constitution of reproducible structures (Prigogine, Eigen) for the moment comes up against the problem of the extreme specificities which should have resulted from it. At the heart of this theoretical debate there is, in our opinion, the problem of a true science of organization only in its infancy.

THE ORIGIN OF THE METAPHORS OF THE NEW SCIENCE

One of the most distinctive traits of the matter composing living organisms and particularly *informational macromolecules* (DNA, RNA, proteins) is that of forming itself by a process of *ordered* and specific *polymerization*. A limited variety of elementary components (the monomers) are assembled one after the other by means of successive addition by forming long chains where at each position chemical laws tolerate in principle any instantiation of such variety. Hence one has laws of non-specific chemical addition (any sequence whatsoever is structurally equivalent to any other) and of *invariant stereochemical conformations through a change in sequence*. This is particularly well realized in DNA and RNA where, if we leave aside interactions at great distance, however necessary they are to furnish a biologically functional architecture, the famous double helix structure (or simple helix in the case of RNA and certain forms of DNA) is compatible from the point of view of chemical affinity and the energetic stability *with any sequence whatsoever* selected at random. It is the bearer of reproductive variability. On this point the case of the proteins is particular since in an aqueous solution the requirement of possessing a stable structure is precisely met only by a very restricted number of all possible amino acid sequences. If we consider the overall number of theoretically conceivable different sequences (hence of different chemical individualities) for a 'small' protein of 150 residues, we get the dizzying figure of 20^{150}. Now, the biochemistry of proteins shows us that only an infinitesimal number of possible combinations will have a stable *structure*. If to this very selective criterion of having a stable structure we add the further constraint that a viable protein must realize a given stereoscopic interaction (support, framework, 'reconnaissance' or catalysis), the number of acceptable substitutions between component units (or 'residues') becomes extremely limited. Now, it is exactly these constraints which act on random mutations by selecting those that can become stable through genetic transmission, which can be reproduced and integrated to the genuine individuals bearing the mutation. As a first approximation (by ignoring the effects of the superstructure of DNA) we can say that *every* mutation is compatible with the DNA/RNA structure if we consider its local conformational stability. However only a very restricted number of the ensuing sequence substitutions is *acceptable for a protein*, and this at the *simple level of physico-chemical interactions which determine its stable structure*. It is precisely at this level that we pass in the living organization from mere combination to meaning, and the models of perpetual renewal (the flame) must give way to informational models. The process of 'molecular epigenesis' (Monod) is the theoretical and operational level where one requires notions of reproductive order, genetic program, template-surface, prefixed morphogenetic sequence. The molecular chreods,[12] to use Waddington's term (Waddington, 1957) are dictated by a highly specific sequence which no chemist could hope to realize even after millions of random attempts. The problematic horizon has changed totally in terms of classical physics and chemistry. What has occurred is a *change in the logical space* where we perceive phenomena. We can visualize this change by referring to the three-dimensional space of classical mechanics.

Every rigid body is displaced according to transformations of the Galilean type (group S_3) applicable in reversible sequences. The configuration at instant t_1 is congruent with the configuration at instant t_0 by a successive application, without predetermined order, of translations and rotations. If we introduce constraints, after having established a system of Lagrange coordinates, which take into account all the residual degrees of liberty in the system, the classical dynamic evolution and the point(s) of equilibrium will be defined in terms of the principle of virtual works.

However, a dynamic system can present constraints possessing a hierarchical structure, that is, equilibrium configurations which are metastable for certain of its constituent parts such that displacements, even minimal ones, can *make* new states of equilibrium *accessible*. These newly permissible states can be characterized by energies which are much lower than that of the preceding metastable state. In such a system, the state of minimal energy (or a limited set of such minimal states) can be reached only by an *ordered* system of transformations; by ordered is meant a sequence of *non permutable* dynamic transformations.

Between the form of constraints and the order of transformations a relation of *congruence* must be established. This projective relation (mapping) evolves in time and in principle can change with the very transformations of the system. The constraints can also be of a statistical nature, determined by microscopic fluctuations, geometric or energetic and often they are both at once. From a vectorial space and from a space of phases we pass to an *informational space* (Eigen 1971). In this abstract space of variable geometry, presenting strong local anisotropies, the only 'fully licensed' moving spots are representative of systems which embody in themselves the ordered list of viable transformations. We are no longer in the presence of moving bodies immersed in fields of force as in classical physico-chemistry; the nature of real space can no longer take into account the regular, reproducible and complex trajectories deployed there. With the study of systems equipped with an internal program, we have shifted to another type of space, a space of programs. For the notion of a field of force acting from the outside we must substitute that of an ordered sequence of transformations governed by an internal code. Even in embryology, in spite of the growing abstraction of the concept of 'gradient', we now prefer the richer notion of 'membrane code' (J.P. Changeux). Code, program, expression, translation are the key metaphors of the new biology. Perfectly *translatable* into more classical physico-chemical terms, they are not *reducible* to concepts of force, affinity and gradient. The space of representation has progressively and almost insensibly changed. Living beings are intelligible today because they are immersed at once in real space and informational space. From the epistemological point of view, this is the most remarkable result of the molecular revolution in biology.

CONCLUSIONS

A model, above all when we are dealing with conceptual models as abstract as the ones we have outlined here, is at once a condition of possibility and a rigid selector of observations. The models based on equilibrium have allowed a progressive enlargement of classes of equivalence established between dynamic phenomena. The spaces of basic representation have become more and more abstract, embracing classical mechanics, statistical microphysics, thermodynamics and transformations of state in chemical and biological systems. Once this work of unification has been accomplished, certain characteristic traits of living systems are revealed as not being reducible to basic concepts of classical physico-chemistry although they may be translatable in terms of particular properties of the matter composing them. The notion of the aperiodic crystal and later of the informational macromolecule has engaged biology in another order of models and has engendered other constitutive metaphors (code, program, memory). The present day hybridation between these two orders of models presents remarkable theoretical problems all while appearing acceptable in current scientific practice. The first results of a science still in its infancy, dedicated to the study of self-organizing systems, leave one hope for a future which will transcend the process/structure dichotomy. The present day discrepancy between certain aspects of molecular biology and certain aspects of the thermodynamics of irreversible processes (notably the theory of dissipative structures and theories on the origin of life) show the interest of an epistemological reflection on the validity and mutual compatibility of the respective basic models.

NOTES

[1] The neologism *chreod* has been coined by C.H. Waddington from the Greek words for 'fixed' and 'path'. In embryology proper a chreod is a stabilized morphological pathway, an orderly sequence of developmental stages that invariably tends to manifest itself even if the environmental parameters vary within certain definable limits. In Waddington's terms *homeorhesis* is the mechanism through which a developing oragnism is geared to a specific chreod, while *homeostasis* is the mechanism through which a steady state, once attained, is kept as a standard reference. (Cf. C.H. Waddington, *The Strategy of the Genes*. London, 1957.)

[2] For a detailed analysis of the historical development of this notion, we refer the reader to the works of Langley (1973), A.J. Lotka (1924; reprinted 1956; see in particular chapter 22), Elkana (1974a) and Cardwell (1971).

[3] pp. 382–383 of 'On the changes of temperature produced by the rarefaction and condensation of Air', *Philosophical Magazine* **26**, 369–383 (1845).

[4] 'Sur un énoncé des lois des équilibres chimiques', *Comptes rendus de l'Académie des sciences* **99**, 19, 786–789 (1884).

[5] *Recherches expérimentales et théoriques sur les équilibres chimiques*. (Extrait des *Annales des Mines*, March-April 1888.) Paris, Dunod, 1888. The quotation on page 347

is taken from this work, pages 210 ("Under the influence ...") and 33–34 ("Factors of equilibrium. . .")

[6] 'Influence du milieu ambiant sur la composition du sang des animaux aquatiques', *Archives de zoologie expérimentale et générale'*, 2nd series, **3**, (1885), p. XXXV.

[7] *First Principles* (London, Williams and Norgate, 1862), p. 458.

[8] p. 92 of 'A universal law', *Journal of the American Chemical Society*, **33**, 2, 91–120 (1911).

[9] *Kosmos*, 1911, p. 331, cited in Lotka (1956), p. 283.

[10] *Trans. Cong. Am. Physicians and Surgeons* **13**, 31–53 (1925). Reprinted in Langley, L.L., ed., *Homeostasis, Origins of the Concept* (Strousburg, Dowden, Hutchinson and Ross, 1973), pp. 223–245. (The quotation is taken from page 44 of the *Transactions* and *page* 236 of Langley.)

[11] *Ibid.*, p. 42 of the *Transactions*, p. 234 of Langley.

[12] See note 1.

BIBLIOGRAPHY

Atlan, H. (1972), *L'organisation biologique et la théorie de l'information*. Paris, Hermann.

Cardwell, D.S.L. (1971), *From Watt to Clausius: The Rise of Thermodynamics in the Early Industrial Age*. London, Heinemann.

Carnot, S. (1824), *Réflexions sur la puissance motrice du feu et sur les machines propres à développer cette puissance*. Paris, Bachelier. Reimpression with added notes, Paris, A. Blanchard, 1953. [*Reflections on the motive power of fire*. Edited by E. Mendoza. New York, Dover, 1960. This is a slightly corrected republication of *Reflections on the Motive Power of Heat*, translated and edited by R.H. Thurston. London, Macmillan, 1890.]

Changeux, J.P. (1975), Inaugural lecture held on 16 January 1976 at the Collège de France, chair of 'Cellular communication'.

Eigen, M. (1971), 'Self organization of matter and the evolution of biological macromolecules', *Die Naturwissenschaften*, **58**, 10 October issue, pp. 465–523.

Elkana, Y. (1974a), *The Discovery of the Conservation of Energy*. Cambridge, Mass., Harvard University Press.

Elkana, Y. (ed.) (1974b), *The Interaction between Science and Philosophy*. Atlantic Highlands, Humanities Press. Lorch's paper, 'The charisma of crystals in biology', which is cited in this essay, is on pp. 445–461.

Haraway, Donna Jeanne (1976), *Crystals, Fabrics and Fields; Metaphors of Organicism in 20th Century Developmental Biology*. New Haven. Yale University Press.

Holton, G. (1973), *Thematic Origins of Scientific Thought, Kepler to Einstein*. Cambridge, Mass., Harvard University Press.

Kuhn, T.S. (1955), 'Energy conservation as an example of simultaneous discovery', in Clagett (ed.), *Critical Problems in Science*. Madison, University of Wisconsin Press.

Langley, L.L. (ed.) (1973), *Homeostasis, Origins of the Concept*. Stroudsburg, Dowden, Hutchinson and Ross.

Lotka, A.J. (1924), *Elements of Physical Biology*. Baltimore, William and Wilkins, 1925. Reprinted as *Elements of Mathematical Biology*. New York, Dover, 1956.

Luria. S.E. (1975) *36 Lectures in Biology*. Cambridge, Mass., M.I.T. Press.

Piaget, J. (1974), *Le structuralisme*. Paris, Presses Universitaires de France. [The French edition first appeared in 1968; this has been translated into English by Chaninah Maschler as *Structuralism*. New York, Basic Books, 1970.]

Piattelli-Palmarini, Massimo (1979) 'How hard is the 'hard core' of a scientific research program?' in Piatelli-Palmarini (ed.), *Language and Learning: the Piaget-Chomsky Debate*. Cambridge, Mass., Harvard University Press, 1980.

Schroedinger, E. (1944), *What is Life?*. Cambridge, At the University Press. The references in this paper are to the 1967 edition.

Waddington, C.H. (1957), *The Strategy of the Genes*. London, Allen and Unwin.

PART III

HISTORY OF THE SCIENCES

FRANCESCO PATRIZI: HEAVENLY SPHERES
AND FLOCKS OF CRANES

1. In this essay I shall aim primarily at two goals. The first is to rediscuss a chapter in the history of the fortunes of Copernicanism. The second is to emphasize, through the close examination of a few texts, the meaning of some of the astronomical and cosmological doctrines in the work of Francesco Patrizi of Cherso (1529–1597).[1] By means of this two-fold discussion I think that I will be able to contribute to the destruction or weakening of some overly schematic descriptions which still appear in the work of many historians of astronomy.

The following list includes only some of the principal theses advanced by Patrizi concerning the astronomy and cosmology of his time:
   (1) the polemic against a purely 'hypothetical' astronomy separated
      from physics;
   (2) the denial of the physical existence of the celestial spheres;
   (3) reflections on the infinity of the universe;
   (4) the ethereal constitution of the heavens;
   (5) the claim to the sphericity of the heavens;
   (6) the claim to the circular movement of the heavens;
   (7) the continuity of the heavens;
   (8) the rejection of instruments in astronomical research;
   (9) the assertion of the earth's centrality and the explanation of gravity;
   (10) evaluation of astrology;
   (11) the denial of the incorruptibility of the heavens;
   (12) the distinction between physical space and mathematical space;
   (13) the rejection of hypotheses in astronomy.
The theses sustained by Patrizi concerning points (8) and (13) made people think that he should be ranged among the most resolute deniers of any possible astronomical science. I shall try to show that although this thesis is substantially true, the problem is not formed in such a way as to allow quick solutions. To this end it is appropriate to pay particular attention to the doctrines numbered (1) and (2).

2. In Book XII of the *Pancosmia* devoted to the treatment of the number

Maria Luisa Dalla Chiara (ed.), Italian Studies in the Philosophy of Science, 363–388.
Copyright © 1980 by D. Reidel Publishing Company.

of the heavens, Patrizi deals with the theme of the variety and plurality of astronomical theories which had been rivals in the field since antiquity. The Pythagoreans spoke of spheres, eccentrics, and epicycles; Eudossus rejected eccentrics and epicycles but assigned 26 spheres to the universe; Callippus added another seven bringing the total to 33; Aristotle added another 22 spheres; Hipparchus resurrected epicycles; Ptolemy attributed two motions to the eighth sphere; Thābit ibn Qurrah added a tenth mobile sphere above the ninth; Averroes returned to eight heavens; Alpetragius placed the driving force of the universe in a ninth sphere, outside that of the fixed stars. The variety of theories is not smaller in the modern era: "in our time three new astronomical theories have arisen, very different from those of the ancients and very different from one another." The first is that of Copernicus which revived Aristarchus' doctrine and "overturned the whole of ancient astronomy and the world order"; the second is that of the two scientists from Verona, Giovanni Battista della Torre and Gerolamo Fracastoro who brought back into use the homocentric spheres, increasing their number to 77; the third is that of Tycho Brahe who elaborated a different hypothesis from that of Copernicus, which left the earth at the center of the universe but moved the spheres so that the sphere of Mars intersected the sphere of the sun at two points.[2]

Patrizi's discussion of astronomy rests on two fundamental presuppositions: first, that a physical description of the real world is possible; and second, that all the 'monstrosities' of ancient and modern astronomy derive from having accepted the absurd idea that the heavenly bodies do not move freely in a fluid space but are fixed in solid and real spheres.[3] The variety and irreconcilability of the various descriptions of the universe created an untenable situation. Faced with that variety one was led to think that the theories were either all false or all true.

Patrizi leans toward a radical pessimism: if we turn to the calculation of celestial movements, tables and ephemerides we find that the oldest are obsolete and fail to take into account either new phenomena or new observations; on the other hand modern astronomers make calculations based on the Copernican hypothesis and assert that they have obtained true tables, but if we listen to the new astronomer, Tycho Brahe, these tables are also defective.[4] Patrizi has the impression that astronomy has given up describing the world, that astronomers have worked out increasingly complicated machines with the sole aim of 'saving appearances' and that the cosmos in that way had been led back to a kind of *novum aliud Chaos* [another new Chaos].[5]

3. Assertions of this kind arise from an attitude of uncertainty and bewilderment in the face of the variety of the solutions offered by cosmology. As we know, the choice between the Earth and the Sun as the center of the universe did not concern only problems in astronomy, nor was it ever represented as merely the choice between two different technical solutions to the problem of planetary motion. At the same time that the traditional, millenary system seemed destroyed or at any rate inadequate, the new cosmologies, which were rivals for the field, gave first of all the sensation of a radical uncertainty and of an end. We are dealing with attitudes destined to last a very long time in European culture. The 'uncertainties' when faced with the Copernican system were characteristic not only of Francis Bacon (between 1612 and 1620) but also (between 1625 and 1660) of Mersenne, Gassendi, Roberval and Pascal. And to document the bewilderment resulting from the variety of solutions it would be well to recall not only the often cited lines of John Donne's 'Anatomy of the World' (1611) but also those of Milton's *Paradise Lost* (1665) and finally those of Mauduit which bring us to the last 20 years of the seventeenth century:

Chacun en sa manière a bâti l'univers L'un par un Ciel qui meut tous les cieux qu'il enserre Fait tourner le Soleil à l'entour de la Terre L'autre fixe le Ciel et par un tour pareil Il fait rouler la Terre à l'entour du Soleil. Un autre survenant, par une adresse extrême Forge des deux premiers un mitoyen système.[6]	Every one has built the universe in his own way One, with a Heaven which moves all that it encloses Makes the sun turn 'round the earth; Another fixes the heavens and with a similar round Makes the earth turn 'round the sun. Another unexpectedly, by a great sleight of hand Makes a system midway between the first two.

But Patrizi's attitude does not spring from this soil. Astronomy appears to him as a kind of knowledge which *does not present its hypotheses as corresponding to the truth of things*, and astronomers, as builders of theories who are not concerned with whether they are true or not, but only with whether they are more or less suitable for celestial calculations and the construction of astronomical tables. Astronomy builds increasingly perfect machines which solve the difficulties which astronomers, insofar as they are mathematicians, continually set for themselves, but it has given up speaking of the real world, it has given up being physics or natural philosophy. The differences between the various world systems leave the same astronomers cold. What sense does it make to fight hypotheses such as these? which do not have and do not intend to have any connection with the real world? which are not presented as 'true' when they are advanced?

Patrizi does not accept the ancient thesis (formulated by Proclus, Simplicius, Philoponus, taken up again later by Thomas Aquinas) which had been reformulated by Andreas Osiander in the preface to Copernicus' *De revolutionibus*:

the hypotheses are not articles of faith, but bases for calculation, it does not matter whether they are false, it is enough that they reproduce exactly the phenomena of the motions . . . Since it can in no way assign true causes to celestial movements, astronomy's task is to imagine and invent hypotheses with whose aid such movements can be calculated exactly.[7]

Unlike Bruno, Kepler and Galileo, Patrizi saw in Copernicus a 'traditional' astronomer who was not concerned with the correspondence of hypotheses to reality and who moreover remained tied to the old and unacceptable thesis of the physical existence of the celestial spheres.

4. As has been seen, Patrizi was convinced that all the 'monstrosities' which filled astronomy derived from the assertion of the real character of the spheres and the consequent conception of the celestial bodies as objects attached or welded to them:

Then as now all the study and effort of astronomers has been directed (as is usually said) toward 'saving phenomena'. All phenomena are observed first through seeing celestial objects, second through considering their movements. Almost all astronomers considered it certain that the stars were fixed in the heavens. In depending on this presupposition they filled the sky with innumerable chimeras.[8]

In order to get out of the difficulties of astronomy, to overcome the contradictory character of the theories and the non-correspondence between theories and observations, to overcome all at once the dramatic situation deriving from the co-existence of incompatible doctrines, Patrizi recalls the hermetic vision of the world and offers a solution of disarming simplicity. It allows one to overcome all at once all the insoluble problems over which natural philosophers exerted themselves for millenia and is perfectly in line with that Platonic philosophy which was wickedly outlawed because of "monks in convents who adored Aristotle".[9] In order to transform a confused and chaotic situation into an intelligible one, it will suffice to eliminate the spheres – which are the chimerical presupposition of the old and new astronomy – and conceive of the celestial bodies as divine and living beings, endowed with an intelligence which is guided and supported by a mind identical to the world order:

Since [the spheres and the *fixio* (attachment) of the celestial bodies to them] were insufficient to account for the phenomena, one ended up with many frenzies and oddities, the upheaval of the entire cosmos and almost another new chaos. But with that vain and impossible presupposition is eliminated from astronomy everything will again be intelligible. One will give a free course to the stars in the heavens and will be able to account for all the phenomena where the stars are conceived – as in reality they are – as being carried by their own spirit, moved by the soul, governed by the order of the intellect . . . Since they are animals, and divine animals, the stars must have a divine soul and life and intellect.[10]

The attribution of a "free course in the heavens" to celestial bodies, their identification with "divine animals" date back to Hermes Trismegistus, Zoroaster and Plato. Aristotle also conceived of the stars as animated, attributing life and action to them even if, in nailing them to orbits, he then made them dead and immobile.[11] Having recourse to souls, to the intellect and minds means freeing oneself from the astronomers' disputes, taking away the meaning of discussions about orbits, eccentrics, and epicycles. If the individual celestial bodies are endowed with souls, if their intelligence is linked to divine intelligence and dependent on it, it is not absurd to think that each of the stars moves on its own, still maintaining "the same place, order and distance."[12] The "vector spirit" of each star is in proportion to that of the other celestial bodies, the individual intellects act in accordance with the mind which has regulated the universe from the beginning:

The order among the stars, their position, and the distance between them do not originate in the fact that the stars are nailed to orbits, but in the fact that we are dealing with animals and divine animals . . . Why have good astronomers thought that the world order would turn into Chaos if they had not attached the stars to spheres as nails to a board? if they had not conceived of them as without life instead of as rational and intellectual beings?[13]

From this point of view, the irregularities of the planetary motions as well as the perfect order of the celestial movements turn out to be easily explainable. While it counters the astronomy of the hypotheses builders with a vision of a living cosmos, Patrizi provides for the elimination of eccentrics and epicycles and insists on the accord which flourishes among all the minds of the universe.

The irregularities of the planetary movements "which the astronomers attribute to eccentrics and epicycles are attributed by us to nature." The *natura* of which Patrizi speaks is defined in the *Panarchia* as *principium corporis* (principle of the body) and in that text he also says that nature derives from the soul, the soul from the mind, the mind from life, life from being, being from unity; unity from the One.[14] It is the spirits, the souls and the minds "which govern the celestial movements to conserve the harmony of the cosmos": for this reason it is possible to speak of the cosmos as a *system* and to conceive of the universe as a *harmonic frame*. When the intellect of every one is in harmony with that of others or depends on a higher intellect, do not men and soldiers succeed in proceeding regularly in order? and do not flocks of cranes perhaps fly in order in the sky? and do not herds of deer and elephants follow their leaders in order?[15]

5. The substitution of souls for spheres brought about by Patrizi is expressed in a work published in 1591.[16] It will be well to remember that Kepler's *Mysterium cosmographicum* was published in 1596. In that book Kepler abandoned the purely 'astronomical' terrain of a *description* of celestial motions in geometrical terms and passed from a kinematic to a dynamic conception of such motions by posing the problem of their *cause*. In the search for this cause Kepler fully supported the thesis of motor intelligences or souls and identified life, motion and soul of the world. The sun, which is the center of the planets' orbits, is the seat of light, of motion, and of the soul of the world. Kepler added to this assertion on the basis of a choice between the following two assertions: either the souls which move the planets become weaker and weaker the greater their distance from the Sun *or* there is a single soul which pushes each planet with greater and greater strength the closer the planet is to the sun and it becomes weaker as the distances increase.[17]

Unlike Patrizi, however, Kepler was deeply interested in the power of the motor souls, their quantitative definition, and the relations between the various powers: these aspects of his discussion transform his hermetic-neoplatonic metaphysics into a fundamental chapter in the history of astronomy.[18] The *spiritus, mens, anima architecta* appeared to him for many years as the supernatural, immaterial element at the origin of celestial movements. When he faced the problem of the *novum sidus* [new star] in the *De stella nova in pede Serpentarii* (1606) he assigned the task of purifying the vapors emanating from the ether and forming new celestial bodies with them to the *anima mundi* [soul of the world]. In the *Astronomia nova* (1609) Kepler referred again to the motor virtues characteristic of the single planets – which are like the boatmen who pilot boats – and attributed to the planets themselves – although they lack eyes – the capacity to 'perceive' the variations in the apparent diameter of the solar disk. In the same book Kepler considered it "not very probable" that a soul could give a body the movement of translation and he posed a series of questions.[19] How can a spiritual force increase and diminish without growing weak and finally disappearing in the course of time? Besides implying a multiplication of souls, does not the animistic hypothesis imply the existence of the spheres? If solid orbits do not exist–and Tycho Brahe has shown that they do not–how can a spiritual force lead the body of the planet through the spaces of the cosmos?[20]

On this point Patrizi, like Bruno, had been of a different opinion: exactly *because* the stars are intelligent animals there is no need to admit the existence of celestial spheres.

6. "Copernicus maintained that the planets, like the other stars, were carried along by the orbits and fixed in them."[21] Patrizi was convinced (and in this he was absolutely right) that Copernicus was the defender of the idea that the celestial spheres were physical entities. He was also convinced (and in this he was decidedly wrong) that Tycho Brahe also supported this doctrine:

None of the ancient systems is as monstrous as these two [Copernicus' and Brahe's]. All the systems – these two and the ancient ones – start from a single presupposition: that the stars in the heavens are fixed in orbits like knots in a board. In fact it was in view of this adherence that the spheres were first devised.[22]

According to Patrizi this basic error has conditioned the entire history of astronomy: "toto ergo errarunt coelo et philosophi et astronomi omnes qui stellas fixas uti nodos in tabulis esse docuerunt" [therefore both the philosophers and astronomers who taught that the stars were fixed in the whole sky like knots in boards were in error].[23] Patrizi returns indefatigably to this theme: stars and planets have been conceived as being *incapable of moving themselves* through space. All astronomy, ancient and recent, has been dominated by this false presupposition and has been vitiated by this dogma. The stars said to belong to the eighth sphere always appear in the same order and always at the same distance: from this single 'appearance' it was deduced that they were nailed to a sphere.[24] Until then men were incapable of conceiving the movements of celestial bodies within a 'liquid heaven'. From this arose the variety and contradictions of the theories regarding the solar system:

Such a great conflict of opinions stemmed from the fact that the astronomers believed that the planets as well as the other stars were carried along by the spheres and were attached to them.[25]

As in the Chaldeans' teachings, the sky is instead only *lumen* [light] and neither solidity nor hardness can be attributed to it. If, as Plotinus would have it, the sky is *calor* [heat] or *ignis* [fire] it will still lack solidity and hardness. If it is *fluor* [flux] (which Moses identified with water and Hermes Trismegistus with a moist nature) it will still be neither hard nor solid. And if there is no solidity and hardness in the heavens, every *fixio* [adherence] will be impossible. And this in fact, as the senses attest, always takes place between solid and hard parts, as in the case of knots in wood or nails driven into beams or between metals and minerals and the flesh and bone parts of animals. There is no other possibility of adherence:

and therefore there has never been nor is there now any *fixio* between the heavens and the stars and the stars are not fixed to the heavens nor the heavens to the stars. And it is not true that the heavens carry the fixed stars with them and cause them to revolve. And all the philosophers and astronomers who taught that the stars were attached to the heavens like knots in a board were wrong.[26]

For Patrizi, the stars "fly within a liquid sky" (*sinemus nos sidera per liquidum volare coelum*).[27] The image of the flight of the stars brings us to that of the flight of the flock of cranes. Since they are provided with *animus* [soul] and conceived as living, autonomous beings joined to a common *mens*, the stars need no material support. The presence of a *common intelligence* prevents the flight from being transformed into a disordered wandering of bodies in a free space, guarantees the presence of a structure of the world, and is enough to avoid a new Chaos.

7. In all the textbooks in the history of science the denial of the physical existence of the celestial spheres and the assertion of the 'fluid' nature of the sky is traced back to Tycho Brahe. The observation of the comet of 1577 – as C.D. Hellman, for example, has written[28] – was enough to definitively convince astronomers like Brahe, Christopher Rothmann and Thaddaeus Hegek of the validity of that denial and that thesis. Contemporary historiography (save in some of its particularly backward areas) considers the problems of the 'chronological priority' of so-called scientific 'discoveries' with appropriate diffidence and does not believe that the question of their 'contemporaneity' makes much sense. But it is one thing to stop – in an increasingly sluggish and provincial manner – vindicating priorities

and tracing precursors; and another to accept – without discussion – the priorities established by the textbooks.

On his way home on the evening of 11 November, 1572, Tycho Brahe saw a very brilliant new star in the constellation Cassiopea, directly opposite Ursa major with respect to the sun. As bright as Venus in the moments of its greatest brightness, it became less and less brilliant until it disappeared all together at the beginning of 1574. That star – wrote Kepler – "if it was the sign of nothing else and if it generated nothing else, it was nevertheless the sign and generated a great astronomer." If it was not a comet, if the star appeared in the same position against the sphere of fixed stars, then the immutable heavens – as happened during Hipparchus' life – had undergone a change and it was possible to question the contrast between the immutability of the heavens and the mutability of the sublunary world.[29] The observation of the comets of 1577 and 1585 confirmed Brahe in his hypothesis:

In my opinion – he wrote to Kepler – the reality of all the spheres (however they may be conceived) must be excluded from the heavens. I have learned this from all the comets which have appeared in the heavens since the new star of 1572 and which are in truth celestial phenomena. They obey the laws of none of the spheres but act in opposition to them . . . The motion of the comets clearly proves that the mechanism of the heavens is not a hard, impenetrable body composed of various real spheres as has been believed by many until now but is fluid and free, open in all directions such that it offers absolutely no resistance to the free course of the planets, which is regulated, in accordance with the regulating wisdom of God, without any machinery or rolling of actual spheres . . . In this way there is no real and disjoined penetration of the spheres: they do not really exist in the heavens but are admitted only for the benefit of learning.[30]

The description of the Tychonic world system and its concomitant denial of the physical existence of the spheres are contained, as is known, in the eighth chapter of the *De mundi aetheri recentioribus phaenomenis liber secundus* published in Uraniborg in 1588:[31]

Among some of the great corollaries of the present composition I have decided to add a broader explanation of this new disposition of the celestial orbits toward the end of the work. In it I have clearly shown, with recourse to the motion of the planets, that the machine of the heavens is not a hard and impervious body constituted of a series of real spheres (as has been believed up till now by most people) but a very fluid and simple body which is open everywhere, has no obstacles, and in which the orbits of the planets are free and devoid of the work and *circumvection* of spheres.[32]

In 1593 as has been seen, Patrizi in Book XII of the *Pancosmia* referred to the Tychonic world system and described it rather approximately. In the same work the belief in the spheres was also erroneously attributed to Brahe. Patrizi had reached the denial of the spheres and the assertion of the fluidity of the heavens, independently of Tycho, because of his adherence to the Ficinian-Hermetic theses and articulate criticism of Aristotelianism and the Aristotelian world system. Independently of Tycho, Patrizi also considered the new star of 1572 a decisive fact: just like the new star, the planets come close to earth and draw away from it; in so doing they should cross those spheres which tradition holds to be solid and real.[33]

Patrizi erroneously attributed to Brahe the belief in the existence of the spheres. His error derived from a failure to read or a hurried reading of Brahe's work and there is precise documentation of this. The *De mundi aetherei recentioribus phaenomenis*, published in 1588, was put up for sale only in 1603. But various copies circulated among friends and correspondents. Gellius Sascerides (Copenhagen, 1562–1612), a student of Brahe, carried copies of the book to Rothmann and Maestlin. Other copies reached Thomas Savelle in Oxford, Caspar Peucer, and Giovanni Antonio Magini in Padua (in 1590).[34] On 22 March, 1592, Patrizi answered a letter of Sascerides. The content of this letter can help us understand the reasons for his approximate treatment of the Tychonic world system. Sascerides protested because in the twelfth book of the *Pancosmia* things are asserted with regard to Brahe's astronomy, which cannot be read at all in his books. With respect to the celestial spheres and the hardness of the heavens, Tycho maintained theses which are the exact opposite of those attributed to him by Patrizi (*imo contraria scripta ibi sunt sententiae de coelestibus orbibus et de coeli duritia*). Faced with these objections, Patrizi spoke with frankness: when he knew that a book had been published in which Brahe had shown a new astronomical system, he tried in vain to get a copy. Gian Francesco Pinelli informed him that Tycho had sent a copy of the work to Giovanni Antonio Magini. He wrote to Magini for a copy, or at least the description of the new system. Magini sent him some of Sascerides' letters and information about the hypothesis of the universe *delineata in orbibus* [outlined in circles]. From this he deduced the information about Tycho and his new astronomy which he published in the *Pancosmia*. But at that time he had neither the letters nor the hypothesis in his hands. This material together with all his writings and part of his books was on its way to Rome where he was about to go himself, summoned by the Pope. Once arrived in Rome, he would compare the book with this hypothesis and the letters. If he was mistaken, given that he was above all a lover of truth, he was ready to excuse himself and praise Tycho. He wanted Tycho to be informed of all this.[35]

As emerges from this letter, Patrizi worked with second-hand sources. Either he leafed through the book or, as is much more likely, he did not even see it and availed himself of small bits of information drawn from Sascerides and Magini.[36] Magini, too, as emerges from a letter written by Brahe to Kepler in 1599 erroneously believed that Tycho's orbits were real bodies ("putavit enim ex vulgata sententia reales esse in coelo orbes"). And Brahe, as emerges from that same letter, was not at all disposed to forgiveness – not even at a distance of seven years:

Patrizi has perverted things in such a way and arranged them so badly because of his ignorance and has fixed them up and distorted them contrary to my own postulates and assumptions that I failed to recognize many many ideas as my own in his summary; and I intend to protest publicly as well.[37]

8. Patrizi had already come to the denial of the spheres and the concomitant image of celestial bodies as "knots in a board" seven years previously (almost twenty years before the publication of the *Nova philosophia*). The statements in the *Obiezioni a Telesio* which were sent to Telesio on 26 June, 1572, and were published for the first time by Francesco Fiorentino do not differ from the conclusions reached in the *Pancosmia*. It will be useful to reread the observations on the second chapter of *De rerum natura*:[38]

It is not permissible to argue, as you and Aristotle and many other authors would, that the stars are placed and fixed above the body of the sky when, on the contrary, the air and waters of the earth are neither attached to nor fixed on it . . . Why should we conceive of the stars as fixed in the heavens like knots in a board and moved by the motion of the heavens, instead of as animals endowed with a will and appetites? The stars are remarkable bodies: why are they compared not to more remarkable objects which are accessible to the senses, but to knots in wood? We see that ants, worms and the lowest insects move of their own accord. Why should we believe that the stars, divine bodies in themselves, are in a lower condition? Must I not conclude from your statements that you attribute a soul to all those things to which you attribute sense? There is no doubt that you attribute sense to the heavens and the stars: sense is attributed in view of its own conservation; motion preserves itself and so the sentient stars move themselves.

Why is it not possible to conceive of the stars in motion, not as knots in a board, but as beings existing in themselves, which act by their own nature and have the sense of their own preservation? Telesio attributes motion to the sky and all the other beings, why does he not attribute it to the stars as well? In this way are not the most remarkable substances of the universe put in a worse condition than everything else? The stars always move following the same circles and those circles differ from one another: "from this it should not be concluded at all that the stars are affixed to a single body". What prevents us from thinking that they 'feel' that their own preservation lies in moving themselves along that orbit? and that for this reason they move just as they actually move? Why does Telesio conceive of the stars as knots in a board and not, instead, as celestial animals as Pythagoras, Plato and all the Platonic schools did?

Perhaps you are afraid to call those stars, to which you still attribute sense, animals. It does not seem that one should be afraid to admit that they – by their own nature – move following larger or smaller orbits which are stable in terms of place and order and which are never forgotten; to admit that these circles are completed not in the void, but in the heavens: in a much more humid and thin body than the air and water through which birds and fish move without any fear of the void.

9. In 1572 Patrizi came to deny the physical existence of the spheres and to assert the fluid character of the heavens, arguing against Aristotle and debating with Telesio. In 1591 he confirmed his conclusions, arguing against Copernicus and Brahe (or rather against the theses he had carelessly attributed to Brahe). As we know, in the *Pancosmia* Patrizi also denied the finite and spherical nature of the heavens.

Patrizi did not reach these radical conclusions by either studying the problems of astronomy first hand or by carrying out celestial observations. He reached them on the basis of his conception of astronomy, which must also be a physics (not to be reduced to pure calculating), and his philological and philosophical criticism of the positions of Aristotelian cosmology. We are dealing with a critique which is extremely careful to bring together the theoretical weaknesses and contradictions present in Aristotelianism. This critique was developed in the *Discussiones peripateticae* with that 'pedantic tendency' for which Bruno pitilessly reproached him. But the denial of the spheres in the *Obiezioni a Telesio* and the *Pancosmia* was the greatest fruit of that pedantry and was inextricably bound up with the careful discussion of the Aristotelian thesis on the inalterability and incorruptibility of the heavens.

Generation and corruption take place between opposites. The sky has no opposite, consequently it knows neither change nor corruption. This argument, writes Patrizi, has always been considered very valid: instead, it must be taken into one's hands, really weighed, and examined ("in manus sumamus, prehendamus, excutiamus"). What kind of generation is being talked about? Perhaps not that which takes place in the sublunary world or in the elements? What precisely is meant by 'the sky has no opposite'? Aristotle stated this proposition on the basis of the perception of the senses or reason? And if it is on the basis of reason, what is the sensible foundation of this rational knowledge given that, according to the theses of this same Aristotle, *nihil est in intellectu quod prius non fuerit in sensu*? The foundation of this assertion is certainly sight. But what does sight see in the sky if not visible objects or common visible objects? And what other qualities if not light and darkness from which reason draws the idea of the transparence of the spheres and the non-transparence of the stars? Did Aristotle see or not see with his own eyes what we see? If he saw what we see, how can he state with so much certainty that there is no *contrarietas* in the heavens, while *lumen* and darkness, transparency and opaqueness are present together in it? If these opposites were not in the sky when he watched it, then the sky has changed. If they were there and he saw them, why did he not admit that at least two opposites were in the sky? the light and the dark, the transparent and the opaque? the light in all the stars, the dark on the face of the moon? the transparent in the spheres, the opaque in the moon and other planets?[39]

Or are light-dark, transparent-opaque not opposites? The family of Aristotelians asserts that their master was speaking not of opposition in general, but only of active and passive opposition which is not present in the celestial substance because there is no oppostion in substances. And then one goes further, always posing new questions and showing instead that opposition is present in quantity, quality, relation, place . . . In the fourth chapter of the second book of *De coelo* Aristotle asserts that the sky has a spherical configuration on the basis of two considerations: the first figure lies within the first body and the sky is the first body; if the sky were angular or oval in shape there would be void and space beyond the sky. Furthermore: because the first sky is spherical, all the others must be spherical which are contained in and touched by the first. That which is touched by and contained in another sphere is in fact spherical. But why precisely is Saturn's sky (which is contiguous with the first spherical sky) spherical and why are all the other successive skies spherical? For what reason, through what cause and force does Saturn's sky conform to the spherical shape of the first sky? Aristotle states that this depends on the contiguity of Saturn's sky with the first sky and on the fact that the former is contained in the latter. But does this happen because Saturn's sky resists or tries to resist the first sky or because it yields to it and remains inert ("quia resistit aut restitit vel quia cedit et cessit")? Is the sky hard or fluid? If the shape of a sphere depends on the embrace of another, is it not the *impressio* and *compressio* of a first sphere on a second an *actio*? And how can one assert that there are no active and passive qualities in the sky?

Aristotle states that the sky is immutable. But how can this be asserted solely on the basis of the fact that there is no evidence of such change? If change had taken place and the first sky had stopped, all the lower things, according to Aristotelian theory, would be corrupted. And would not men have disappeared? And who would have been able to bear witness to the change, transmit the information and preserve its memory?[40]

10. In 1613, when Federico Cesi printed Galileo's letters on sun spots in Rome, one of the censors' interventions was explicitly dictated by the desire to safeguard the doctrine of the incorruptibility of the heavens. In August, 1618, Cesi still felt the need to demonstrate to Cardinal Bellarmine the non-existence of the crystalline spheres. W. Donahue, in a paper presented and discussed in the History and Philosophy of Science Department at Cambridge University in 1970, amply documented the remarkable vitality still enjoyed by the thesis of the solidity of the spheres in the 1620's.[41]

Only by clearly understanding a cultural context of this kind and by renouncing a history of science as a list of truths which immediately inspire respect in intellectual communities, can one realize the effective reasons of the polemic conducted by Galileo – in the 1630's – against the Aristotelian vision of the world. This polemic is insistent and continued: it is specifically directed against the solid and separate heavenly spheres which carry the planetary bodies along. The same arguments expounded by the Aristotelian Simplicius in the *Dialogo sui massimi sistemi* are the same ones which Francesco Patrizi had faced more than 50 years earlier.[42]

In order to realize how the polemic in favor of Copernicus, Kepler, Tycho Brahe and Galileo against Aristotle's and Ptolemy's 'world fabric' maintained its precise function even beyond the mid-seventeenth century, just open the *Academiarum examen* published in London in 1654 by John Webster, physician and chaplain to the Parliamentary forces during the Civil War.[43] To Webster, one of the decisive and distinguishing points between ancients and moderns is the belief or lack of it in the heavenly spheres:

Tycho Brahe, Copernicus, Kepler, Galilaeus, and others clearly demonstrate (beyond the refutation of Logick) that there are changes and mutations in the heavens, and so they are not incorruptible bodies as is falsely asserted . . . Another thing the Academic Masters grossly maintain: that the heavens or Orbs are hard as steel, and as transparent as glass.[44]

The text of the *Pancosmia* would not have displeased Webster: caught up with the thought of the English Paracelsians, he availed himself of Descartes as an anti-scholastic, he called to mind Pico, Della Porta, van Helmont, and saw the basis of a new culture and a new 'experimental' science in the union of Bacon's legacy with that of Robert Fludd. Seth Ward, one of the principal exponents of the 'new Oxford science' was in a good position to reproach him his sympathies for the Cabbala and make him notice the radical opposition between Baconian philosophy "based on experiments" and that of Fludd "based on ideal and mystical reasons".[45] But Ward's reaction to the vehement accusations of his adversary is characteristic. It is not true that at Oxford astronomy is taught according to the Ptolemaic system rigorously considered 'true'. Ptolemy "never medled with solid Orbes" and "medled not with physicall part at all". He only "salved [saved] the Phenomena" and, from this point of view, "there is no mathematick book in the world more learned or useful":

I believe there is not one man here, who is so farre astronomicall, as to be able to calculate an Eclipse, who hath not received the Copernican System (as it was left by him, or as improved by Kepler, Bullialdus, our own Professor, and others of the ellipticall way) either as an opinion, or at leastwise, as the most intelligible, and most convenient Hypothesis . . . . The method here observed in our Schooles is, first to exhibit the phenomena, and shew the way of their observation, then to give an account of the various Hypotheses, how these phenomena have been salved, or may be . . . If Mr. Webster have any thing to amend in this method, and will afford it our professor, I will undertake he will be thankfull for it.[46]

The simplistic and overly familiar antitheses – as has been noted by Allen Debus who has written a useful commentary on the texts of this controversy – give rise, also on the level of understanding historical processes, to bankrupt results. Pre-Newtonian science must always be portrayed as a terrain in which there are radical alternatives and contrasting theories, and in which 'advanced' and 'backward' positions constitute a very difficult knot to unravel.

11. As A.O. Lovejoy demonstrated in a book written in the thirties, none of the great sixteenth and seventeenth century astronomers advanced those innovative ideas or "truly revolutionary theses in cosmography" which became decisive characteristics of a changed vision of the world.[47] None of those theses can be found in Copernicus and each of them was rejected in one way or another by the three greatest astronomers of Bruno's age and the following generation: Tycho Brahe, Kepler, and Galileo. The new doctrines regarding the destruction of the outer walls of the medieval universe, the assertion of the infinity of the universe, the hypothesis of a plurality of inhabited worlds appears instead in different ways in Nicholas of Cusa, Palingenio Stellato and Bruno. They appear in a discussion in which Copernican, Neoplatonic and Hermetic legacies are mixed and matched in different ways.

Patrizi's vision of the universe from the privileged perspective of us moderns, seems an odd mixture. Earth, in his system, is still at the center of the cosmos and the sun rotates around the earth (as is the case in Tycho's system). But the earth is not immobile: it rotates on its axis, carrying with it the air and water on its surface (Patrizi accepts only one of the three terrestrial motions theorized by Copernicus). The stars move by themselves, are not fixed to real spheres "tanquam nodi in tabula" [like knots in a board] (as they are not for Tycho Brahe) but move because of a soul within them (as in Kepler). There are no separate zones in the heavens. The sky is one and continuous. Like all other things, it is made up of four elements (air, light, heat, *fluor*) and it moves neither as a whole nor in its separate parts. All the motions belong to the stars and (unlike what Telesio also believed) there is no bearer of these movements. The movement of the stars is apparent and depends on the daily movement of the earth on its axis. The stars are not all at the same distance from earth. They are scattered in the depths of the heavens. They are not even contained in a limited zone but are scattered in infinite depths.[48]

Patrizi's cosmological discussion touches on a series of important themes. In 1571, many years before the publication of Brahe's book, he theorized about the destruction of one of the principal pillars of the traditional world system and showed its uselessness and theoretical inconsistency. In order to reach this end he recalls, as did the young Kepler in those same years, the old hermetic-Platonic thesis of the animation of celestial bodies. Unlike what happens in Patrizi, as has been seen, the denial of the spheres, the thesis that there is a soul in the heavenly bodies, and a motor soul in the sun, are linked by Kepler not only to calculations and observations but also to a series of specific questions relating to the *modes of functioning* of those souls. From the enormous work done by Kepler on the basis of those initial 'animistic' hypotheses arose – through the twisted and difficult paths which are characteristic of the growth of scientific knowledge – a great and decisive chapter in the history of astronomy.

It would be well to remember that within this very important chapter some fundamental categories of Aristotelian physics still continued to function. For Kepler – and in this he is Aristotelian – *a continuous uniform velocity requires the application of a continuous motor force.* Kepler is not acquainted with the principle of inertia nor does he have any notion of centripetal force. The force emanating from the sun exerts no central attraction: it serves to stimulate the motion of the planets and to *maintain them in motion.* Like gigantic arms projecting from the sun, that force pushes the planets ahead on their orbits; and it does not seem in any way necessary for a centripetal force to prevent the flight of the planets along the tangent. Even in the text of the *Astronomia nova*, in which the motors of the planets appear as purely corporeal or magnetic faculties, to explain the rotation of the sun about itself "it seems that a force coming from a soul is necessary". Since there are no solid spheres, as Tycho Brahe showed – Kepler asserts – it follows that the body of the sun is the seat and source of the force that makes the planets turn in their orbits. The sun emits from its body an immaterial *species* [appearance] analogous to the appearance of light which is spread throughout the universe and which carries the planets with it.[49]

As we know, Kepler is a thinker closely tied to the mystical perspectives of Platonism. His 'modernity' – as has been emphasized many times – is connected with two precise orientations: the search for quantitative variations in the forces at work in space and time; the progressive abandonment of an animistic point of view in favor of a mechanical one. The celestial machine – we read in the *Astronomia nova* – must be compared not with a divine organism but with the mechanism of a clock. All the movements are performed thanks to a very simple magnetic force just as in the clock all the movements are caused by a simple weight. This physical conception – Kepler concludes – must be presented "by means of calculation and geometry".[50]

We do well to remember that we are dealing with assertions more than ten years later with respect to the initial questions posed by Kepler (in the same years in which Patrizi was writing) when he was still being influenced by the doctrine of the motor souls or intelligences present in every single planet. Only in 1625 – almost a quarter of a century after Patrizi's and Bruno's deaths – did Kepler (in the notes to the new edition of the *Mysterium cosmographicum*) clearly expound the transition from his theses on the *soul* to those on the motor force (*vix motrix*).

Once – writes Kepler in his notes – I firmly believed that the motor cause of every planet was a soul because I was saturated with the dogmas of Scaligero on motor intelligences. Now I have reached the conclusion instead that this force is something corporeal even if the term 'corporeal' is understood in the figurative sense (*corporeum aliquid, si non proprie saltem aequivoce*) in the same way that light is something corporeal.[51]

Many of the questions and hypotheses from which Kepler had begun – in terms of the way in which they had been formulated, the sources to which they referred and the 'occult philosophy' which they seemed to presuppose – were completely extraneous to Galileo's mentality. For Francesco Patrizi, who in Bacon's harsh judgment, "sublimated the fumes of the Platonists",[52] those specific questions and those hypotheses were also completely extraneous and for the opposite reasons. Patrizi never maintained that the study of the variation of the forces coming from the souls of the planets and the sun could have any significance; he never maintained that those motions and those spiritual forces could be relevant to geometry insofar as they were essentially "geometric things"; he never thought that any machine built by man could serve as a good model for understanding the universe. From the analytical, patient destruction of one of the most ponderous and entrenched dogmas of the cosmology of his time, he extracted only the possibility of a 'regression' toward the hermetic-magical theses of the coincidence of nature with life. Adopting this thesis as a general one, Patrizi radically eliminated all the problems of modern astronomy. They turned out to be void of meaning when the 'flight-of-the-flock-of-cranes' or 'fish-who-move-freely-in-water' model was assumed to explain both the regularities and the variations in celestial movements. Patrizi's hypothesis, unlike Kepler's, did not need to be compared with experience. The 'flock-of-cranes' model could explain anything that had happened or is happening in the heavens.

The disagreement with Kepler on this point is really irremediable. And it is in this area that we can really measure the distance separating the theses maintained by Patrizi from those integral and central to modern astronomy. Even when moving within a general, animistic hypothesis, Kepler (as we have seen) is primarily interested in calculations and figures, quantitative variations, and the specific modes of functioning of the souls which move the celestial bodies. From his point of view the virtues of the stars must be "subject to calculation and geometry" and in no instance, from his perspective, can just the directly observable motions be acknowledged as real. In Kepler's astronomy both the regularities and the variations are explained by means of recourse to abstract models, numbers, calculations, geometrical figures. Beyond what appears, beyond what is immediately "given to the senses", beyond apparent disorder, norms, regularities, and laws are tracked down. For Kepler it is precisely in this that 'philosophy' or 'philosophical astronomy' consists. In the *Apologia Tychonis contra Ursum* – in a passage pointed out by Cassirer in 1923 – Kepler in his polemic with Patrizi defended the meaning of a 'philosophical' astronomy and distinguished the "accidental courses of the planets" visible to the human eye, from their "true motions"; he distinguished the plane of what appears to the senses from that of the real orbits. Only on this plane, he concluded, is it possible to *also account for what appears*.[53]

Kepler returned to these same themes with greater breadth and decisiveness in other texts which, in supplementing the one used by Cassirer, is worth mentioning. In 1599, as we have seen, Kepler received Tycho Brahe's letter containing a harsh and irritated judgment of Patrizi's work. Four years later, in July 1603, writing to David Fabricius, the same Kepler briefly outlined Patrizi's position: according to Patrizi the celestial bodies did not revolve in circles but actually traverse lines in the heavens which appear to our sight. The definitive reasons for a radical divergence are clarified in a passage in the *Astronomia nova*:

Today some people – disparaging the effort, work, knowledge and science of 2,000 years – are still trying to resurrect that first aspect of astronomy which is completely incapable of explaining causes, which trusts only the slowest experience of sight, which cannot be explained with either figures or numbers, which is in perpetual disagreement with itself and incapable of comparing one motion to another in an interval of time . . . They offer themselves up to the admiration of the crowd and, in the presence of incompetents, not without some success. But the competents rightly think that either they are delirious or that – when they want to be called philosophers like Patrizi – they are prey to a form of lucid madness.[54]

14. On the subject of Patrizi, Kepler spoke of a lucid madness (*cum ratione insanire*). Giordano Bruno also spoke of 'madness' as well as 'presumptuous vanity'. And yet it should be clear that to trace overly rigid lines of demarcation on the basis of 'animism' would lead – as far as the twenty

year period (1571–1591) in which Patrizi wrote is concerned – to solutions
which are difficult to accept. In his self-critical comment of 1625 Kepler
undoubtedly tended to project his definitive withdrawal from the animistic
and mentalistic conceptions of the heavens back in time to the years when
he had written the *Astronomia nova*. The publication of William Gilbert's
De magnete had in fact exercised a decisive influence on his positions.
Because of that book, published for the first time in 1600 in London,
Kepler was encouraged to identify the lines emanating radially from the
anima motrix of the earth and from that of the sun with chains of magnetic
force. In *De magnete* the adamantine spheres of Ptolemaic astronomy were
defined as imaginary entities composed of a substance of which nothing
is known, as "a philosophic fable" which is the object of derision. "The
agent force", concludes Gilbert,

abides in bodies themselves, not in space, not in the interspaces. But who supposes that
all these bodies are idle and inactive, and that all the force of the universe pertains to
those spheres, is as foolish as the one who, entering a man's residence, thinks it is the
ceilings and the floors that govern the household, and not the thoughtful and provident
good-man of the house.

Gilbert too reproached Aristotle for having attributed a soul to the celestial bodies and not to earth:

As for us, we deem the whole world animate, and all globes, all stars, and this glorious
earth, too, we hold to be from the beginning by their own destinate souls governed and
from them also to have the impulse of self-preservation . . . Indeed in some plants and
shrubs the organs are hardly recognizable, nor are visible organs essential for life in all
cases. Neither in any of the stars, nor in the sun, nor in the planets, that are most operant
in the world, can organs be distinguished, or imagined by us; nevertheless, they live and
endow with life small bodies at the earth's elevated points . . . Pitiable is the state of
the stars, if this high dignity of soul is denied them, while it is granted to the worm, the
ant, the roach, to plants and morels; for in that case worms, roaches, moths, were more
beauteous objects in nature and more perfect, inasmuch as nothing is excellent, nor
precious, nor eminent that hath not soul.[55]

Patrizi's books, in which these expressions turn up almost word for
word, also moved within this perspective. Without doubt the thesis of the
physical existence of the heavenly spheres had been an essential component of traditional cosmology and was destined for a very long life
even for many years after Patrizi's death. In Thomas Kuhn's words, that
component "was the principal barrier to the success of Copernicanism".
If this is true, if it is true that "any break with the Aristotelian cosmologi-

cal tradition worked for the Copernicans",[56] then we must also leave some room – in the complicated history of the dissolution of a millenary vision of the world – for the denial of the spheres as formulated by Francesco Patrizi.

Like Bruno and even more than Bruno, Patrizi was no stranger to the great tradition of Aristotelianism. In his work the hermetic and Platonic theses were not simply opposed to those of Aristotelianism: they led to a subtle discussion, were used to dismantle the great Aristotelian machine piece by piece. From this point of view the *Discussiones* have an exemplary value and a characteristic course. The Aristotelian passage, quoted in Greek and translated into Latin is taken apart, then every part is finally dismantled and compared with other pieces which in their turn are the fruit of a previous disassembly. Something of this way of proceeding also comes through from Bruno's very ferocious judgment:

we cannot assert that [Patrizi] has understood Aristotle well or ill, we can only say that he has read and reread him, sewn him up and torn him to bits, compared him with a thousand other Greek authors, his allies and opponents alike.[57]

To Bruno this way of proceeding seems capable of producing only "a very great loss", only a demonstration of "folly and presumptuous vanity". But in the very dense pages of the fourth book of the *Discussiones* Aristotelian philosophy was criticized from within with singular force: to the point of bringing about a crisis in conceptual instruments, techniques of understanding reality, fundamental categories.

In this Bruno was right: to dismantle a philosophy, to take it to pieces without any intention of putting it back together is not the best way to understand it. But this kind of machine for making mincemeat of a philosophy – as represented by Patrizi's *Discussiones* – did not fail to bear historically significant fruit.

15. As specialists know – but it is well to mention it again in referring to late sixteenth century Italian philosophy – gathering those fruits was not an easy undertaking. In a span of ten years (during the Papacy of Clement VIII) Francesco Patrizi's *Nova philosophia*, Bernardino Telesio's *De rerum natura*, Giordano Bruno's and Tommaso Campanella's *opera omnia* were placed on the *Index*; investigations were conducted against Giambattista della Porta and Cesare Cremonini, Francesco Pucci was condemned to death, and Giordano Bruno burned at the stake.[58] The lean pages of the *Emendatio*, recently published by P.O. Kristeller, integrate material already published by T. Gregory and are important for clarifying the positions assumed by Patrizi after the accusations made against him in November 1592. The list of theses sustained by Patrizi, his declarations of being disposed to abandon and cancel them from his book throw a new light on the tragedy of Italian science and the difficulties encountered by some of the 'revolutionary ideas' of modern cosmology:

Assertion that there is only one heaven. With Moses and often with the Acts of the Apostles and Chrisostomos. And it is a matter of controversial things for theologians. In fact some claim two heavens, others three, others (with the astronomers) eight, still others (with other astronomers) nine, ten, eleven with the empyrean. Tell me what I must do, if you wish I shall wipe it out . . . *That the earth rotates.* I asserted this on the basis of many reasons and with philosophical authorities. Nevertheless I did not state that the earth leaves its medium and natural place. Basilio says that it is firm (*eam stare*) if it is not placed outside of its place. In this way all the places of the Holy Scripture used against me are saved. Nevertheless I will take them out if you order it.

Assertion that beyond the world there is an infinite space filled with the *lumen* of the stars and this *lumen* extends to infinity. I do not know whether there is anything to the contrary in the Holy Scriptures or whether it is prohibited by the Sacred Councils or theologians to maintain it. In fact that space above the sky is not limited by any other body. If the *lumen* penetrates all the transparent bodies offering resistance (like the sky, air, water, glass, crystal) it will be able to pass even more easily through that which, like the empty space, offers no resistance. If you so command, I shall take this out too.[59]

NOTES AND REFERENCES

I have used the following editions of 'collected works' in this paper.

Brahe, T., *Opera omnia*, edited by J.L.E. Dreyer. 15 vols. Copenhagen, Libraria Gyldendaliana, 1913–29.

Galilei, G., *Le opere*. Edizione nazionale. 20 vols. in 21. Florence, Tip. di G. Barbèra, 1890–1909.

Kepler, J., *Gesammelte Werke*, edited by Walther von Dyck and Max Caspar. (19 vols.) Munich, Beck, 1937–; and *Opera omnia*, edited by C. Frisch. 8 vols. Frankfurt am Main and Erlangen, Heyder and Zimmer, 1858–1871.

I have used the following editions of Patrizi's work:

Discussionum peripateticarum tomi quatuor. Basel, 1581 (indicated below as *Discussiones*).

Nova de universis philosophia. Venice, 1593 (indicated below as *Pancosmia*).

Emendatio in libros suos novae philosophiae, edited by P.O. Kristeller in *Rinascimento*, 2nd series, **10**, 1970, pp. 215–18 (indicated below as *Emendatio*).

Lettere ed opuscoli inediti. Critical edition by D. Aguzzi Barbagli. Florence, Istituto nazionale di studi sul Rinascimento, 1975 (indicated below as *Lettere*).

Among the recently published texts are: T. Gregory, 'L'*Apologia* e le *Declarationes* di F. Patrizi' in *Medioevo e rinascimento: studi in onore di B. Nardi*. 2 vols. Florence, Sansoni, 1955. See volume I, pp. 387–442; *La poetica*. Critical edition by D. Aguzzi Barbagli. 3 vols. Florence, Istituto nazionale di studi sul Rinascimento, 1969; *L'amorosa filosofia*. Edited by J.C. Nelson. Florence, Felice Le Monnier, 1963. A good bibliography is contained in *Onoranze a F. Patrizi da Cherso. Mostra bibliografica*. Trieste, 1957.

[1] For some studies in English, see: L. Thorndike, *History of Magic and Experimental Science*. 8 vols. New York, 1923–1958. Volume 6: New York, Columbia University Press, 1941. pp. 373–77; B. Brickman, 'An introduction to Francesco Patrizi's 'Nova de universis philosophia'. Ph. D. dissertation, Columbia University, 1941; P.O. Kristeller, *Eight Philosophers of the Italian Renaissance*. Stanford, Stanford University Press, 1964. pp. 110–26. Some important information is contained in F.A. Yates, *Giordano Bruno and the Hermetic Tradition*. London, Routledge, 1964 (but it is not true, as she asserts, that Patrizi never referred to Copernicus' work).

[2] *Pancosmia*, f. 90r–91r.

[3] *ibid.*, f. 91r.

[4] *ibid.*

[5] *ibid.*

[6] P.M. Mauduit, *Melange des diverses poésies divisez en quatre livres*. Lyon, 1681. p. 180. Cf. J. Donne, 'An anatomie of the world. The first anniversary'. London, Nonesuch Press, 1962, p. 202:

> And new Philosophy calls all in doubt,
> The Element of fire is quite put out;
> The Sun is lost, and th'earth, and no mans wit
> Can well direct him where to looke for it.
> And freely men confesse that this world's spent,
> When in the planets, the Firmament
> They seeke so many new; then see that this
> Is crumbled out againe to his Atomies.
> 'Tis all in peeces, all cohaerence gone;

and J. Milton, *Paradise Lost, VIII*, 72–84:

> From man or angel the great Architect
> Did wisely to conceal, and not divulge
> His secrets to be scanned by them who ought
> Rather admire; or if they list to try
> Conjecture, he his fabric of the heav'ns
> Hath left to their disputes, perhaps to move
> His laughter at their quaint opinions wide
> Hereafter, when they come to model heav'n
> And calculate the stars, how they will wield
> The mighty frame, how build, unbuild, contrive
> To save appearances, how gird the sphere

With centric and eccentric scribbled o'er,
Cycle and epicycle, orb in orb.

and P. Rossi, *Aspetti della rivoluzione scientifica*. Naples, Morano, 1971. pp. 153–63.

[7] Letter from Osiander to Copernicus, 20 April 1541, cited in J. Kepler, 'Apologia Tychonis', *Opera*, I, 246; A. Osiander, 'Ad lectorem', in Copernicus, *De revolutionibus*. Thorn, Copernican Society, 1873, pp. 1–2. ['To the reader . . .' in *On the revolution of the heavenly spheres*. Translated by Charles Glenn Wallis. Great Books of the Western World, volume 16, pp. 505–509. Chicago, Encyclopaedia Britannica, 1952].

[8] *Pancosmia*, f. 89v.

[9] *Lettere*, f. 178.

[10] *Pancosmia*, f. 91r.

[11] *ibid.*, f. 89v, 90r, 114v.

[12] *ibid.*, f. 90r.

[13] *ibid.*, f. 90v.

[14] *Panarchia*, f. 4v.

[15] *Pancosmia*, f. 90v.

[16] The *Nova de universis philosophia* was printed in Ferrara by Benedetto Mammarelli at the end of 1591. There are copies with a false title-page with the name of the publisher Meietti of Venice and the date 1593. On this question, cf. P. Zambelli, 'Aneddoti patriziani', *Rinascimento*, 2nd series, 7, 1967, pp. 309–18 where there is also mention of an edition giving the name of the publisher as Mammarelli and the date as Ferrara 1640 with no indication of the author and the generic title *Naturale magia: opus a variis probatissimis auctoribus in unum collectum*.

[17] Kepler, *Mysterium cosmographicum*, XX, *Werke*, I, p. 70.

[18] A. Koyré, *La révolution astronomique*. Paris, Hermann, 1961. pp. 151–52. (*The Astronomical Revolution*, translated by R.E.W. Maddison. Paris, Hermann; London, Methuen; Ithaca, Cornell University Press, 1973. pp. 151–52).

[19] Kepler, *Werke*, I, p. 268; *Astronomia nova, Werke*, III, pp. 254, 261. Cf. Koyré, *La révolution astronomique*, p. 223. (*The Astronomical Revolution*, pp. 222–23).

[20] Kepler, *Astronomia nova, Werke*, III, p. 236. Cf. Koyré, *La révolution astronomique*, p. 190. (*The Astronomical Revolution*, pp. 188–89).

[21] *Pancosmia*, f. 104v.

[22] *ibid.*, f. 91r.

[23] *ibid.*, f. 89r.

[24] *ibid.*, f. 90r.

[25] *ibid.*, f. 106r.

[26] *ibid.*, f. 89r.

[27] *ibid.*, f. 92r.

[28] C.D. Hellman, *The Comet of 1577: Its Place in the History of Astronomy*. New York, Columbia University Press, 1944.

[29] T. Kuhn, *The Copernican Revolution*. Cambridge, Mass., Harvard University Press, 1957. pp. 205–208. J.L.E. Dreyer, *Tycho Brahe*. Edinburgh, A. and C. Black, 1890; New York, Dover, 1963. pp. 39–96.

[30] Brahe to Kepler, in Kepler, *Opera*, I, 44, 159.

[31] On the lack of the publication of the first and third books see Dreyer, *Brahe*, pp. 162–63.

[32] T. Brahe, *De mundi aetherei recentioribus phaenomenis*. Frankfurt, 1610. p. 149.

[33] *Pancosmia*, f. 104r–104v, 150r–151r. Cf. Thorndike, *History of Magic and Experimental Science*, **VI**, p. 376.

[34] Dreyer, *Brahe*, pp. 181–82; J.A. Gade, *The Life and Times of Tycho Brahe*. Princeton, Princeton University Press, 1947. pp. 87, 141. For the correspondence between Magini and Sascerides and between Magini and Tycho Brahe, cf. A. Favaro, *Carteggio inedito di Ticone Brahe, Giovanni Keplero e di altri celebri astronomi e matematici dei secoli XVI e XVII con Giovanni Antonio Magini*. Bologna, N. Zanichelli, 1886. Other useful bibliographical information can be found in D. Aguzzi Barbagli's notes in *Lettere*, pp. 79–82.

[35] *Lettere*, pp. 79–82.

[36] From the text of the letter it does not seem at all possible to deduce, as Aguzzi Barbagli on the other hand does in *Lettere*, p. 80, that Patrizi drew his information *directly* from the copy of *De phaenomenis* sent to Magini.

[37] Brahe, *Epistolae astronomicae, Opera* **VIII**, pp. 206–07, or Kepler, *Gesammelte Werke* **XIV**, p. 92.

[38] Published in F. Fiorentino, *B. Telesio ossia studi storici sull'idea della natura nel Risorgimento italiano*. 2 vols. Florence, Le Monnier, 1872–1874. See volume II, pp. 376–77.

[39] *Discussiones*, pp. 429, 430.

[40] *ibid.*, pp. 431, 435.

[41] W. Donahue, 'The place of the stars in the early seventeenth century'. Mimeographed. The letter from Cesi to Bellarmino and Bellarmino's answer are in C. Scheiner, *Rosa ursina*. Bracciano, 1630, pp. 777–783. On the question and intervention of the censors, see W.R. Shea, 'La controriforma e l'esegesi biblica di G. Galilei', in A. Bausola and others, *Problemi religiosi e filosofia*. Padua, 1975, p. 47.

[42] G. Galilei, 'Lettera a Mons. Piero Dini', 23 March 1615, *Opere* **V**, p. 299; *Dialogo sopra i due massimi sistemi del mondo, Opere* **VII**, pp. 62, 63, 66–67. (*Dialogue Concerning the Two Chief World Systems*, translated by Stillman Drake. 2nd edition. Berkeley, University of California Press, 1967, pp. 37, 38–39, 41–43).

[43] John Webster, *Academiarum examen*. London, 1654, p. 103. I am using the facsimile reprint contained (together with Seth Ward, *Vindiciae academiarum*, Oxford, 1654, and Thomas Hall, *Histrio-mastix, a Whip for Webster or an Examination of one John Webster Delusive Examen of Academies*, London, 1654) in A.C. Debus, *Science and Education in the Seventeenth Century*. London, Macdonald; New York, American Elsevier, 1970.

[44] Webster, *Academiarum examen*, p. 46, 48.

[45] Ward, *Vindiciae academiarum*, p. 46.

[46] Ward, *Vindiciae academiarum*, pp. 29–30 ff.

[47] A.O. Lovejoy, *The Great Chain of Being*. Cambridge, Mass., Harvard University Press, 1957. p. 108. See P. Rossi, 'Nobility of man and plurality of worlds' in *Science, Medicine and Society in the Renaissance. Essays to Honor Walter Pagel*, edited by A.C. Debus. 2 vols. New York, Science History Publications, 1972. See volume II, pp. 131–62.

[48] *Pancosmia*, f. 85v.

[49] Kepler, *Astronomia Nova, Werke* **III**, p. 35. For Kepler, the Sun, as the central body of the universe, *must* be in rotation. On Kepler, see G. Holton, 'Kepler's universe: its physics and metaphysics', *American Journal of Physics* **24**, 340–351 (1956).

[50] *Astronomia nova, Werke* **III**, p. 241; cf. D.P. Walker 'Kepler's celestial music', *Journal of the Warburg and Courtauld Institute* **30**, 228–250 (1967).

[51] Kepler, *Mysterium cosmographicum, Opera* **I**, p. 176, note C. The reference to Scaligerus is to *Exercitationes exotericae* **VIII**, 2, p. 673.

[52] *The Works of Francis Bacon*, edited by R.L. Ellis and J. Spedding. 7 vols. London, Longmans, 1887–1901. See volume IV, p. 359 and volume I, p. 564. See also P. Rossi, *Francis Bacon from Magic to Science*, translated by Sacha Rabinovitch. Chicago, University of Chicago Press, 1968. pp. 63–65.

[53] *Apologia Tychonis contra Ursum, Opera* **I**, p. 247. This passage is discussed by E. Cassirer in *Philosophie der symbolischen Formen*. 3 vols. Berlin, B. Cassirer, 1923–29. See volume II: *Das mythische Denken* **II**, 2, 4. (*The Philosophy of Symbolic Forms*, translated by Ralph Manheim. 3 vols. New Haven, Yale University Press, 1966. Volume II: *Mythical Thought*, pp. 138–140.)

[54] Kepler, *Astronomia Nova, Werke* **III** p. 62. Kepler's letter to David Fabricius of 4 July, 1603, is in *Werke* **XIV**, p. 431. In a letter to Herwart von Hohenburg of January 1607 (*Werke,* **XV**, p. 387) Kepler recalled the criticism of the theory of the tides in Book XXVIII of the *Pancosmia*.

[55] W. Gilbert, *De magnete*, translated by P. Fleury Mottelay. New York, Dover, 1958. pp. 309–10, 318, 319, 321–23, 338.

[56] T. Kuhn, *The Copernican Revolution*. Cambridge, Mass., Harvard University Press, 1957. p. 206.

[57] G. Bruno, 'De la causa, principio et uno', in *Dialoghi italiani*. Florence, Sansoni, 1958. pp. 260–61. (*Cause, Principle and Unity*, translated by Jack Lindsay. New York, International Publishers, 1964. p. 99). Also important on Patrizi is the trial testimony in A. Mercati, *Sommario del processo di Giordano Bruno*. Città del Vaticano, Biblioteca apostolica vaticana, 1942. pp. 55–56.

[58] L. Firpo, 'Filosofia italiana e controriforma', abstract, with additions, from *Rivista di Filosofia* **41**, 150–173 (1950); **42**, 30–47 (1951); Milan, s.d., p. 6.

[59] *Emendatio*, pp. 217–18.

MASSIMO MUGNAI

LEIBNIZ ON THE STRUCTURE
OF RELATIONS*

For more than seventy years now one of the topics discussed most by students of Leibniz's logic is undoubtedly that of the role assigned by Leibniz to relations and relational propositions. Russell and Couturat mentioned the problem at the beginning of the century and today, above all with the works of Mates, Clatterbaugh, Burkhardt and Castañeda,[1] decisive steps have been taken toward the investigation of certain aspects of the *vexata quaestio*: it is not bold to think that shortly, with the help of the publication of Leibniz's unpublished works, sufficiently firm conclusions can be drawn. In this sense – that is, along the road which can lead to similar conclusions – the contributions of Burkhardt and Clatterbaugh seem to me to deserve particular attention: in them the distinction made between the ontological and logical aspects in the problem of relations can lead to the assumption of a fruitful point of view, capable of clearing the field of difficulties which are otherwise hard to overcome. But beyond these particular results, a decisive step forward was made, in my opinion, by the acknowledgment of the very necessity of a distinction.

The fact that such a necessity also arises from an analysis of the scholastic and late scholastic logical tradition which played such a great role in Leibniz's philosophical formation, is only one further proof of the accuracy of the intuitions of Clatterbaugh and Burkhardt. A comparison between Leibniz's positions and the assertions contained in the manuals written by thinkers like Fonseca, Soto and Suárez really shows that the problem of relations involves a formal distinction of levels which is much broader than that looked at until now. Using a series of indicators which can be gleaned from reading these manuals, one can in fact pinpoint at least five 'levels' within Leibniz's theory of relations. Such a theory involves the distinction of a strictly logical plane from a grammatical plane and from metaphysical, ontological, and psychological planes. The kind of separation between each of these planes cannot always be defined with exactness: at certain points there are superimpositions and comminglings of one with the others. Let us try to examine this more closely and see

*With a few variations, this paper reproduces the text, originally in German, of a paper published in *Studia Leibnitiana* 10, 1 (1978), 1–21.

389

Maria Luisa Dalla Chiara (ed.), Italian Studies in the Philosophy of Science, 389–409.
Copyright © 1980 by D. Reidel Publishing Company.

how the above-mentioned distinction is articulated and at the same time justified.

Consider first of all the Leibnizian theory of relations according to what I, with perhaps not total precision, have called the *metaphysical plane*. In this regard we learn from the *Nouveaux essais* that relations, like truths, have their special reality and their natural place in the divine Intellect:

Relations have a reality dependent upon the mind like truths; but not the mind of men, since there is a supreme intelligence which determines them all for all time.[2]

Relations and orders have something of the essence of reason, although they have their foundations in things; for we can say that their reality, like that of eternal truths and possibilities, comes from the supreme reason.[3]

The same concept is supported in an outline of a letter to Des Bosses, dating from around 1712:

God observes not only the single monads and the modifications of each monad, but also their relations and in what the reality of the relations and truths consists.[4]

But in order to know something more precise about this topic we must turn again to the *Nouveaux essais*: here the source of such a concept is indicated in the Augustinian theory of the Divine Intellect as seat of the ideas and formal reasons of created things:

This leads us finally to the ultimate ground of truths, viz: to that Supreme and Universal Mind, which cannot fail to exist, whose understanding, to speak truly, is the region of eternal truths, as St. Augustine recognized. . . Thus these necessary truths being anterior to the existence of contingent beings, must be grounded in the existence of a necessary substance. Here it is that I find the original of the ideas and truths which are graven in our souls, not in the form of propositions, but as the sources out of which application and occasion will cause actual judgments to arise.[5]

As we can see, Leibniz in this case goes back to a commonplace of the medieval metaphysical tradition: in the Mind of God are the archetypes, the *exempla* in the Platonic sense, on whose base reality has been created: as an *Ens Realissimum* God contains in himself all the *realitates*, that is, the ideal essences of all individuals and of all that exists. Like ideas and truths, relations are an integral part of the entities conceived by the divine Intellect. But in defining, so to speak, the profound reality of relations, Leibniz at the same time furnishes useful indications for determining their ontological *status*. In another place in the *Nouveaux essais* we read in fact:

Yet, although relations are from the understanding, they are not groundless or unreal. For ... the first ... understanding is the origin of things. . . .[6]

And in one of the passages quoted above he stated, as we saw, that relations "have something of the essence of reason" while having foundation in things. Two determinations, then, are inherent in relations: they are founded in the Mind of God and are the work of the intellect; in particular, as far as this last property is concerned, one infers from Leibniz's assertions that it is valid for the Supreme Intellect as well as for the human one. In proceeding to a classification of the 'objects of our thoughts', Leibniz reaches a division into *substances, qualities,* and *relations* to which a precise ontological choice corresponds. Having assumed the individual substance or monad as the basic structure of existing reality, the *qualities* must be considered as 'modifications of the substances'; and once the substances with their modifications are given, it is the *intellect* which adds *relations* to them:

This division of the objects of our thought into substances, modes, and relations is sufficiently to my taste. I believe that qualities are only modifications of substances, and that the understanding adds thereto the relations. From this follows more than you think.[7]

The 'objects of our thoughts', then, in the case of substances and qualities, refer back to entities which in a certain sense exist in and of themselves, while relations fall exclusively within our same cognitive activity. Insofar as they are *objects* of thought, substances and qualities are also apprehended and understood by the *instrument* or the *means* of the intellect, but their ontological status makes them independent of the activity of this same instrument. Not so for relations, even if, as we have seen, this does not mean that the relations are mere subjective appearances being founded in the Divine Intellect. It could be objected, however, that substances and qualities also have the same foundation of relations: from whence comes the difference then? The difference seems to consist not so much in the foundation, which is indeed common, as rather in the different mode of relating to the foundation. The ideas of relations – if we are concerned with Leibniz's statements – seem to depend more directly than those of substances and qualities on the act of thought by which they are asserted. However, there is probably another motive for such a distinction. In medieval and late scholastic treatises one distinctive property of relation mentioned is not only that of inhering in determined *subjects* (like other qualities) but also that of having a *fundamentum.* So according

to Domingo de Soto (who harks back to St. Thomas), in order to constitute a relation, a *subjectum* as well as a *fundamentum* are necessary:

The *subjectum relationis* is that which is denominated (*denominatur*) by the relation; on
the other hand, the *fundamentum* is that which by virtue of which such a relation agrees
with the subject. . . For example: the subject of similarity is the substance which is said
to be similar, while one foundation is whiteness or shape by virtue of which the substance is similar.[8]

Thus, in a certain sense we can say that the ideas of property precede,
if not *de facto*, at least *de jure*, those of relation: from an ontological point
of view the ideas of quality have greater independence and autonomy
with respect to those of relation. As far as the human intellect is concerned,
Leibniz's assertion related above leaves no room for doubts: given substances and their modifications, the intellect *adds* the relations *to them*.
What is primary then are substances and modifications: relations have as
their presupposition and condition for their rising the existence of both.
This conclusion is also implicitly suggested by some unpublished passages
quoted by R. Kauppi in his work *Über die Leibnizsche Logik*:

The *relation* is an accident which exists in several subjects and is only a result, or it
occurs without any change taking place in them, if several things are thought together
simultaneously: it is the possibility of thinking several things together (*concogitabilitas*).

Relation is thinking two things together

Relation exists when two things are thought at the same time.[9]

These brief fragments, the date of whose composition unfortunately escapes me, assert a thesis which Leibniz seems to state not only in the
Nouveaux essais but also in the early writings of the Parisian period:

It seems then that we must say this: to relations, which are beings and also true beings
when they are thought by us, such as numbers, lines, or distances, we do not give a
number; in fact, they can be reduplicated over and over again and thus they are not real
or possible beings, except insofar as they are thought.[10]

Insofar as relations are also 'real' in a certain sense, they do not have the
same reality which can be attributed to substances and their modifications:
in advancing this thesis Leibniz seems to have a typically nominalist
position. Equally typically nominalist is his instinctive diffidence toward
the use of abstract terms which – to use his very words – should be outlawed from 'characteristic language'.[11]

This nominalist attitude stands out even more if we move from the ontological plane to the *psychological* plane. However, such a move does not imply the abandonment of a reference to ontology: in setting forth the mechanisms which allow us to gain knowledge of the ideas of relation, Leibniz furnishes us with useful data concerning their special nature. Consider the by now even too well-known sections 46 and 47 of Leibniz's fifth letter to Clarke:[12] in them Leibniz, in an attempt to put in evidence the origin of our notion of space, develops his arguments by assuming the definition of 'place' (*place*) or 'same place' (*même place*) as a point of reference. Then he works out the following example. Given a certain number of coexisting objects, if any one of them changes its own situation with respect to the others without these changing the relations among themselves, and if a new object comes to have the relation which the object which moved had with respect to the remaining fixed objects, then we say that that object has come to the 'place' of the first. This change is called a 'movement', which is found in the object and which is the immediate cause of the change. Developing this example Leibniz asserts that:

. . . 'place' is that which we say is the same to A and to B, when the relation of the co-existence of B, with C, E, F, G, etc., agrees perfectly with the relation of the coexistence which A had with the same C, E, F, G, etc., supposing there has been no cause of change in C, E, F, G, etc.[13]

An accurate analysis of this passage, however, reveals that to understand it fully, it is necessary to determine at closer range what Leibniz means by the expression *convient* ['agrees']. In fact what does it mean to say that the relation of coexistence of B with C, E, F, G, etc. 'agrees' entirely with the relation that A has with the same? In order to clarify this point we must bear in mind the distinction made by Leibniz in the same context, between 'place' and 'relation of situation' which is in the body which occupies the place. In fact:

For the place of A and B is the same, whereas the relation of A to fixed bodies is not precisely and individually the same as the relation which B (that comes into its place) will have to the same fixed bodies.[14]

That is, the relations are not *identical*: they *agree* only among themselves. As Leibniz specifies, this distinction between identity and *convenance* ['agreement'] finds its own justification in the principle according to which "two different subjects, as A and B, cannot have precisely the same individual affection, it being impossible that the same individual accident should be in two subjects or pass from one subject to another."[15] Nevertheless, if *convenance* is reduced to identity, this is due to a particular attitude of the human spirit [*esprit*] which, "not contented with an agreement, looks for an identity, for something that should be truly the same, and conceives it as being extrinsic to the subject."[16] Hence when compared with ideas of connections or relations, *convenance* assumes the same function that *ressemblance* has when compared with universals and abstracts. In fact, just as it is from considering the *ressemblance* between individual properties that we come to constitute the general idea of a given property, it is in observing the *convenance* of a series of individual relations that we succeed in forming our general idea of such relations. The general idea of a relation, once formed in our mind, indicates – to use Leibniz's own words – something which is 'outside of' the objects considered. In the letter now under our consideration Leibniz nonetheless points out the necessity of furnishing his correspondent Clarke with a further illustration of the process which leads our intellect to consider such abstractions and it is for this reason that he constructs the example made famous by Russell[17] of the two lines L and M, one of which is longer than the other:

The ratio or proportion between two lines L and M may be conceived three several ways: as a ratio of the greater L to the lesser M, as a ratio of the lesser M to the greater L, and, lastly, as something abstracted from both, that is, the ratio between L and M without considering which is the antecedent or which the consequent, which the subject and which the object. And thus it is that proportions are considered in music. In the first way of considering them, L the greater, in the second, M the lesser, is the subject of that accident which philosophers call 'relation'. But which of them will be the subject in the third way of considering them? It cannot be said that both of them, L and M together, are the subject of such an accident; for, if so, we should have an accident in two subjects, with one leg in one and the other in the other, which is contrary to the notion of accidents. Therefore we must say that this relation, in this third way of considering it, is indeed out of the subjects; but being neither a substance nor an accident, it must be a mere ideal thing, the consideration of which is nevertheless useful.[18]

The preceding *psychological* setting of the problem has changed at this point into a question of ontology, as is easy to see. Once more the interpretative key to the whole passage is given by the principle according to which the same accident cannot be in two subjects; again the vision of a reality composed of only individual substances and their accidents or modifications is confirmed. It is necessary to note that among these last Leibniz will include 'the being greater than M' proper to line L and 'the being smaller than L' proper to line M; the relation or *raison* common to both *L* and *M* is to be considered instead "a mere ideal thing"; we are dealing with a thing which the intellect imagines 'outside of subjects'. The same concepts and the same argumentation can be found in a passage in a letter from Leibniz to Des Bosses which is also well known and which has many analogies with a corresponding passage in the *Logica Hamburgensis* of Jung. For convenience' sake I am quoting below the passage from Leibniz side by side with that of Jung:

I do not believe that you will admit an accident that is in two subjects at the same time. My judgment about relations is that paternity in David is one thing, sonship in Solomon another, but that the relation common to both is a merely mental thing whose basis is the modifications of the individuals. . .[19]

And on these occasions we are speaking *as if only one relation intervened between two terms*, when instead the relations must always be understood as double. So let us say there is a *friendship between Orestes* and *Pylades* when instead a thing is the friendship which Orestes has for Pylades, another thing that of Pylades toward Orestes.[20]

In order to indicate the externality of relation with respect to subjects, and probably in order to offer a reference point to his own correspondents, Leibniz has frequent recourse to the scholastic term, 'extrinsic denomination': thus relations – insofar as one does not mean by them relative individual accidents—are extrinsic denominations. Unless, just as soon as we try to examine the meaning of this definition in the light of Leibniz's texts, we find ourselves faced with an unusual reversal of perspective: in fact, on various occasions Leibniz asserts explicitly that in reality purely extrinsic denominations *do not exist*. The contrast which is delineated in this way is, nonetheless, in Leibniz's own words, only apparent: it can be overcome by the acquisition of a superior *metaphysical* perspective. This apparent contrast hides a philosophical problem which, as Duns Scotus points out, has a long tradition. In the first Scholium to Quaestio XI on book five of Aristotle's *Metaphysics* introducing the problem of the reality of relations, Duns Scotus in fact asserts:

In the book on predicaments Simplicius discusses this question concerning this conclusion: whether the relation is one thing and carries many arguments in favor of the contrary thesis, since the relation occurs and recedes without change. He also states that this was a very strong argument of the Stoics. . . In favor of the opposite thesis he argues instead: it is not sensible to destroy the harmony which really causes us to experience pleasure; hence the cause of pleasure is real.[21]

Through Simplicius, whose opinion he reports sufficiently faithfully and whom he had very probably read in a translation of Moerbeke, Duns Scotus establishes an interesting link between the reality of relations and harmony. The reasoning he attributes to Aristotle's commentator in fact implies that to deny reality to relations means rendering problematic the reality of the harmony we perceive with pleasure in the events and structure of the world. This argument, as documented by manuals and commentaries after Scotus, seems to have met with particular success.

Domingo de Soto, reporting it in the *liber praedicamentorum* of his *Commentaria* cites St. Thomas directly together with Duns Scotus:

In the fourth place the argument principally follows St. Thomas . . . and Scotus . . . There are some relations which are called real perfections, such as the order of the universe, the union of matter and form, the inherence of accident to subject . . . relations which affirm the perfections according to the nature of the thing, independently of the operation of the intellect: hence such relations are real, and if they are real, they can be distinguished in reality. On the basis of this argument, Scotus asserts in the passage cited that to deny that relation can be distinguished is to deny the entire order to the universe.[22]

If we bear in mind the role which the concept of harmony plays in Leibniz's thought from the beginning, it is difficult to think that Leibniz did not test himself with a similar argument. All the more as his specific notion of 'harmony of the world' as union and participation of all things among themselves cannot leave relations out of consideration or fail to acknowledge their reality which is superior to that of purely mental beings. By harmony one must understand a continual connection and link among all substances: a connection whose existence constitutes one of the strongest proofs in favor of the existence of God:

For all substances should be in mutual harmony and interrelation . . . This mutual correspondence of different substances . . . is also *one of the strongest proofs of the existence of God* . . .[23]

How then do we reconcile 'ideality of relations' and reality of harmony? The problem of such an agreement hides the more general question of the reconciliation between two profound motifs in Leibniz's thought: on the one hand, the 'nominalist' inspiration tending to reduce existing reality to a set of individuals or substances which are irreducibly different in their individuality and isolation; and on the other, the image of a world in which 'tout se tient' and in which differences and multiple nuances find their profound unity at every moment. Now the solution envisaged by Leibniz to reconcile these two contrasting tendencies finds a real point of strength in the distinction between extrinsic and intrinsic denominations which we shall now try to clarify.

We have mentioned that the assertion that every relation must have its own foundation is part of the Aristotelian-scholastic doctrine. Leibniz, in perfect agreement with this doctrine, maintains in a letter to Des Bosses that it is really contradictory to think that a relation without foundation can exist.[24] If relations are considered as 'extrinsic denominations' (in the sense mentioned above), the principle which has now been expressed can be stated in the following way:

. . . there is in general no external determination which does not presuppose an internal one as a foundation . . . [25]

However, on this occasion we are dealing with a more general expression of the same principle: what Leibniz intends to assert in this case is that whatever determination or property of a given object or state of things is pointed out, it can never be so external that it is not founded on some aspect or quality within the object or state of things under consideration. The doctrine of foundation or intrinsic quality states nonetheless in synthetic form a relation which must be articulated in a more complete manner. When we read in scholastic manuals that relations must have a foundation, they mean to assert that relations can exist only in reference to some determination found in correlated subjects. From a rigorous point of view – and above all in conformity with a consequent nominalist attitude – the relation is based on a series of individual determinations implicit in correlated subjects: it falls or is modified if one or more subjects loses the property which establishes it. I would not rule out that a strong stimulus to the elaboration of this thesis came to Leibniz from reading Suárez, in particular the theory of relations set forth in the *Disputationes metaphysicae*.[26] But Leibniz goes still further and not only asserts that if only one of the individual properties on which the relation stands fails, the relation completely changes or falls with it, but also maintains that this brings about a change in the other or in each of the other correlated subjects. Given two subjects A and B between which exists a relation of size, Leibniz distinguishes the *abstract relation* of size between them from the *accidents* or *individual properties* which establish the relation in each of them. According to this conception, the relation of each individual is composed of an indefinite series of properties which, in a certain sense, express the size of this individual when compared with all the other individuals in the world. Changing the size of any one of the individuals corresponds to an alteration in the abstract relation of size with a certain number of other individuals, but contemporaneously it *involves an intrinsic change in all the other correlated individuals:*

While another becomes larger than me, some change takes place in me as well, as the denomination concerning me is changed. And in this way everything is in some manner contained in everything.[27]

The variable changing of extrinsic denominations and hence of relations, their capacity to be added or taken away, involves a parallel change in the deep reality of things. This thesis is advanced by Leibniz with sufficient clarity in the following three passages, the last of which, if I am not mistaken, is still unpublished:

. . . there are no extrinsic denominations, and no one becomes a widower in India by the death of his wife in Europe unless a real change occurs in him. For every predicate is in fact contained in the nature of the subject.[28]

§5. Ph. *There may, however, be a change of relation without any change happening in the subject. Titius, whom today I consider as a father, ceases to be such tomorrow without any change being made in himself, by the sole fact of his son's death.*
   Th. That statement may very well be made in view of things which are perceived; although in metaphysical strictness it is true that there is no entirely exterior denomination (*denominatio pure extrinseca*) because of the real connection of all things.[29]

It seems then that extrinsic denominations really belong to the relation, that is, those denominations which arise and disappear without any change in the subject but only because a change takes place in another subject. Thus a father becomes a father if a son is born to him even if by chance he is in East India and thus not directly determined by this event. So my resemblance to another originates and arises even without my changing, by the sole change in another. Nonetheless it is necessary that according to a rigorous point of view, there is no extrinsic denomination in things since nothing happens in the world in any place without actually influencing all existing things. But once this universal sympathy of all things is put aside, relations can be considered extrinsic denominations.[30]

Yet it seems that whenever the connection of all things is left out (it is taken for granted that in the final analysis and according to a metaphysical point of view, there are no totally extrinsic denominations), the terms 'extrinsic' and 'intrinsic' can be used to distinguish between their various types of relation. This is what is drawn from another passage taken from an as yet unpublished text:

The relations between two things seem to be only either rational or real or essential or existential. Real relations are either of 'position' (time and situation) or of 'influence'; when by means of one thing a change takes place or is impeded in another. So when

I say 'man is mortal' or 'animal is mortal', this is a relation of reason and gives what is commonly called extrinsic relation; the same can be said of the relation of position. Consequently if I say that Peter is 100 paces away from me, this is an extrinsic denomination, and if I move and Peter stands still, certainly Peter's distance from me changes but without any change taking place in Peter unless by virtue of the universal connection of all things.[31]

After the foregoing, there remains only to consider the problem of relations on the logico-grammatical plane; however, it is necessary to specify first of all the area within which the investigation must develop at this point. Now, in my opinion, such an area is defined by all those contexts in which Leibniz is concerned with the *form* of propositions and conclusions which even for the scholastic tradition hid relations and involved problems which were referred to relations. The contexts of this kind which are most known and discussed by what can now be called the 'trend' of studies on Leibniz's logic concern: (1) the so-called inversion of relation and (2) the propositions which can be drawn from *oblique cases*. By the expression 'inversion of relation' Leibniz intends to designate conclusions of the type: 'David is the father of Solomon, therefore Solomon is the son of David'; oblique cases, on the other hand, are all the attributions in the genitive and the other cases. The most well-known examples of oblique cases studied by Leibniz are those of the 'sword of Evander' and 'Paris and Helen'.[32] Ever since Couturat's book on *La logique de Leibniz* [Paris, Alcan, 1901] this last example in particular has been examined and variously interpreted as one of the preferred places for understanding Leibniz's theory of relations: still for more than seventy years the text in which the example appears has been discussed without anyone feeling the need to examine Leibniz's *entire* argument in the manuscript. No one has taken into account the circumstance that Couturat in publishing the passage with the example of Paris and Helen in the *Opuscules* had indicated that this passage was incomplete. In having to discuss the example of Paris and Helen in my turn, however briefly, I maintain that it is necessary, first of all, to provide the missing final part along with the part published by Couturat:

The genitive is the addition of one substantive to another by means of which what is being added to is distinguished from another. The sword of Evander, that is the sword Evander owns. Part of a house, that is, the part a house has. Reading of poets, that is, the act with which one reads a poet. One can explain in an excellent way, for example, Paris is the lover of Helen; that is, 'Paris loves *and for this very reason* (*et eo ipso*) Helen is loved. These are two propositions contained in a summarized form in one.

Or: 'Paris is a lover and for this very reason Helen is loved.' The sword is the sword of Evander; that is, the sword is equipment insofar as Evander is owner. The poet is read insofar as this or that person is reading. In fact, if you do not resolve the oblique case into more propositions, you will never go on to devise new ways of reasoning, without being forced to, as Jung was.[33] Or so: 'Paris loves Helen', that is, 'Paris* maintains that Helen will be happy with him.' The sword of Evander, that is, the sword which is subject insofar as (*quatenus*) Evander is owner. Subject or subjected thing is what undergoes an action insofar as [*quatenus*] another acts, and insofar as the same is just. Thus, reduced to the nominative we have: the sword is Evander's, that is, 'if the sword undergoes an action and Evander acts, Evander is for this very reason (*eatenus*) just.' Or: '(if) the sword undergoes an action insofar as (*quatenus*) Evander acts (then in this respect) Evander is just.' Those conjunctions which connect propositions are that necessary; indeed the same 'insofar as' (*quatenus*) should be explained more broadly in this way: if the sword of Evander undergoes an action because (*quia*) Evander acts, Evander is not therefore unjust. Indeed it would be necessary in the explanation to go back to the definition of cause and effect. But as this is not always necessary, it will be useful in the characteristic to devise certain signs which represent once and for all such a resolution, in order that a troublesome resolution not be always necessary. The only requirement is that the equality of the calculation always be preserved, so that provable propositons are evident from the calculation itself.

'The sword of Evander is beautiful.' That is 'that sword is beautiful which, if it undergoes an action since Evander acts, Evander is not therefore unjust.' From this we can understand how much commitment is required by the reconstruction of the true characteristic, from which all the oblique cases and all the inflections must be far removed, as if they were forbidden by an edict.

Every (personal) verb is preceded by the nominative. Sometimes it is followed by the nominatives, as 'be sober!'; 'the quadrigae [four-horse chariots] are an instrument'.

Every infinite or impersonal verb presupposes the expressed or suppressed accusative. 'I know that I love you.' 'I want to know', that is, 'I want that I know' (*cupio me scire*). All men who strive to be superior to other animals.

Every active verb, whether finite or infinite, always governs the accusative (*perhaps I would prefer to say 'or can govern': as to live life, to wage a struggle* [einen Kampf kämpfen].[34]

In rational grammar neither the oblique cases nor other inflections are necessary . . .[35]

Leibniz's principal preoccupation as shown in this passage is that of discovering a method which will allow the reduction of oblique cases – which are essentially cases in the genitive – to straight cases (in the nominative). The presupposition hidden behind a similar aim is – as we read in the last lines of the text – that "In rational grammar neither the oblique cases nor other inflections are necessary." In rational grammar, that is, the standard form of a proposition will be given by a subject, a copula

---

*In Leibniz's manuscript 'Paul' appears instead of 'Paris', probably by mistake.

and an attribution to the nominative. In a certain sense this is the *grammatical* aspect of the question. However, the more strictly logical plane is welded to it: given the connection in the genitive of the two terms, *sword* and *Evander*; given further the proposition *Paris is the lover of Helen*, Leibniz thinks of resolving them into more equivalent assertions. For both examples the method of reduction put into operation by Leibniz presents common elements: first of all, the use of locutions such as *eo ipso, quatenus, eatenus,* etc. which scholastic and late scholastic logical tradition calls *dictiones reduplicativae* or *reduplicantes*. According to Pedro da Fonseca:

Those 'statements' are called reduplicative *which consist of some one repeating or duplicating dictio*, as it is called. As if we say for example: 'Man insofar as (*quatenus*) man is capable of instruction.' In fact the dictio *quatenus*, since it is suited to duplicate and repeat, is usually called reduplicating or reduplicative . . .[36]

In the second place Leibniz separates the two terms connected in the oblique case, making each of them the subject of a proposition in which a predicate, which is correlative to the predicate of the other term, is attributed to it. Thus in the passage cited we have the corresponding pairs, *lover-loved, owner-subject,* referring respectively to *Paris* and *Helen, Evander* and *sword*. In this case the *dictio reduplicativa* has the function of indicating the specific sense in which the meaning of a term or that of a proposition is considered. In general such a dictio, according to the teachings of the scholastic manuals, can express both a condition under which a term is considered or the specific sense in which a given term is considered the *cause* of a particular attribution. With his usual clarity Jung defines *causal reduplication* as follows:

Causal reduplication exists where the reduplication contains the cause for which the predicate inheres in the subject, as '*man insofar as (quatenus) he is rational, is capable of philosophizing*' . . .[37]

The fact that in the text cited above Leibniz, in order to clarify the meaning of *quatenus,* explicitly mentions the cause-effect relation, but above all the fact that he maintains that *quia* [because] can be substituted for *quatenus,* shows that in all probability he attributes a causal meaning to the *dictio reduplicans*. Yet I confess that the various logical transformations made by Leibniz in the text Couturat did not publish leave me perplexed: I believe that the effort to translate the function of *quatenus* into present day logical language is difficult. This dictio seems to create something similar to intentional contexts but the reason why Leibniz passes to a transcription, in hypothetical form, of propositions reduced to the nominative escapes me. The addition of the term *just* attributed to Evander was probably motivated by the desire to construct a complete inference, but the succeeding passage with the negative conclusion remains obscure: *Evander is not therefore unjust.* What is certain is that Leibniz, through his explicit assertion, makes room for such transformations with the intention of moving *within* the currently accepted logical tradition: compared with Jung, who was forced to devise new modes of reasoning, Leibniz vindicated the validity and vitality of tradition. Thus it is plausible to think that many explanations concerning Leibniz's logical operations and intentions can be obtained more clearly, precisely from the study of that tradition.

To conclude, let me say a few more words about Paris and Helen and the *inversion of relation.* Given the proposition *Paris is the lover of Helen,* Leibniz tries, as we have seen, to resolve it into *several propositions* and asserts that two propositions are united there in summarized form. Now, most of the time we have seen in this attempt of Leibniz the intention to reduce one relational proposition into two propositions in the subject-predicate form: besides, this is what Leibniz's text shows, but it is necessary to bear in mind that we are dealing with a *joining* of two propositions by means of *et eo ipso.* Furthermore, it seems problematic to respond to the question whether Leibniz intended with this to work out a system of total reduction of relational propositions. There is in fact at least one other unpublished text in which Leibniz shows that he had no difficulty in accepting relational propositions in the sense defined by Bertrand Russell:

Relations are either of comparison or connection. The *relation of comparison* between A and B arises from the fact that A is found in one proposition and B in another; the *relation of connection* is such that both A and B are in the same proposition (which cannot be resolved into a relation of comparison). Otherwise even the relation of comparison would be a relation of connection: a proposition including A and B, can be formed, that is, 'A is similar to B'. But this resolves it all the same into two propositions of which one speaks particularly of B, the other, particularly of A, that is: 'A is red and B is red, and thus A is similar in this respect to the same B'. By A and B are understood things or individuals, not terms. But what can be said of these propositions?: 'A exists today and B also exists today', or, 'A and B exist at the same time'? Will this be a relation of comparison or connection? The same holds true for coexistence in the same place.[38]

Undoubtedly this text has a problematic character and can state, as often happens in Leibniz's writings, a point of view which is not absolute, yet because it begins with a classification of relations which corresponds perfectly to that in the *Nouveaux essais*[39], it would deserve to be taken into due consideration.

Finally, the inversion of relation once more recalls Jung's presence. In the *Logica Hamburgensis,* the *inversion of relation* is an inference which is not proved and whose validity is taken as evidence of and in itself:[40] on at least two occasions Leibniz indicates instead that he wants to prove its conclusivity. In the first case he expresses a desire in this regard and in a certain sense he carries out a purely programmatic assertion:

The goal of our characteristic is that of using voices such that all the consequences can be constructed if they can be drawn from the same words or characters. For example, given the consequence, 'David is the father of Solomon, hence [Solomon] is the son of David', it cannot be proved a recourse to the words in which it is expressed, at least it cannot be resolved into other equipollent consequences; in general language, on the other hand, it must be proved by means of the decomposition of words into the letters which compose them.[41]

In the *Nouveaux essais,* on the other hand, Leibniz provides greater indications of the possibility of such a proof:

These inversions, compositions, and divisions of reasons which he makes use of are only the species of forms of argumentation peculiar and characteristic of the mathematicians and the matter they treat; and they demonstrate these forms with the aid of the universal forms of logic. Further, you must know that there are *good asyllogistic conclusions* which also cannot be rigorously demonstrated by any syllogism without changing somewhat its terms; and this change itself of terms makes the conclusion asyllogistic. There are several of these, as among others, *a recto ad obliquum*; for example, Jesus Christ is God; therefore the mother of Jesus Christ is the mother of God. Again, that which clever logicians have called *inversion of relation*, as, for example, this conclusion: if David is the father of Solomon, without doubt Solomon is the son of David. These conclusions do not cease to be demonstrable by the truths on which the common syllogisms themselves depend.[42]

From a logico-technical point of view we do not learn much positive concerning the eventual proof of asyllogistic inferences; however, we know that *no* traditional *syllogism* could prove them and that they are provable by recourse to truths on which the same common syllogisms depend. This appeal to a more general logic leaves open many possibilities which only an accurate and systematic examination of as yet unpublished texts can concretely verify. Yet it is difficult to rid oneself of the impression that in the purely logical problematic of relations, certain ontological options

do not interfere. And it is also true that a similar interference can be recorded with difficulty and with equal intensity on the whole range of
discussions and logical problems faced by Leibniz. In the case of 'Paris
is the lover of Helen', it is still particularly difficult not to recognize in the
project of reduction worked out by Leibniz, the wish to construct, on a
logical level, a structure of the 'underlying' reality: on the one hand Paris
with his individual affection, on the other, Helen with the correlative
affection. The *et eo ipso* in such a case would mean the connection and
reciprocal relation which, in the two subjects, would give rise to two
distinct modifications: in dealing with two *real* subjects and not pure
concepts or 'incomplete subjects', it may really be that the *et eo ipso*
expresses the reality of *harmony* at the logical level. Naturally this is little
more than a hypothesis and as such it must be verified, but I would not
rule out that, for clarifying many questions connected with Leibniz's
theory of relations, it is really *necessary* to try to test its validity.

NOTES

Below is an explanation of the abbreviations used in the notes, followed by lists of
Leibniz's papers contained in each volume which are cited in the preceding essay.

1. *Leibniz's Original Works*

PA.   Leibniz, G.W. *Sämtliche Schriften und Briefe*. Edited by the Preussische Aka-
       demie der Wissenschaften, after 1945 the Deutsche Akademie der Wissen-
       schaften. Darmstadt, Reichl, 1923–[1945]; Berlin, Akademie-Verlag [1945]–.
       This is to be a complete and critical edition.
       Reihe VI: Philosophische Schriften, volume 6: *Nouveaux essais* [given in the
       notes as PA., R VI, 6, p . . .]. Berlin, Akademie-Verlag, 1962.

 G.   Leibniz, G.W. *Philosophische Schriften*, edited by C.I. Gerhardt. 7 volumes.
       Berlin, Weidmannsche Buchhandlung, 1875–90.
       vol. 2: 'Briefwechsel zwischen Leibniz, Landgraf Ernst von Hessen-Rheinfels
       und Antoine Arnauld, 1686–1690', pp. 1–138.
       'Leibniz an de Volder', pp. 139–283.
       'Briefwechsel zwischen Leibniz und des Bosses, 1706–1716', pp. 285–521.
       vol. 7: 'Streitschriften zwischen Leibniz und Clarke, 1715–1716', pp. 347–440.

LH.   Leibniz-Handschriften. The numbering follows that of Eduard Bodemann's
       Die Leibniz-Handschriften der Königlichen Öffentlichen Bibliothek zu Hannover.
       Hanover, Hahn'sche Hof-Buchhandlung, 1895.
       All citations in this essay are from the Philosophical papers designated by
       Bodemann with the Roman numeral IV.

Jag.   Jagodinski, I., *Leibnitiana Elementa philosophiae arcanae de summa rerum*.
       Kasan, 1913.

Cout. OF.   Couturat, Louis, *Opscules et fragments inédits de Leibniz*. Paris, Alcan,
             1903.
             'Analyse grammaticale', pp. 243–44.
             'Grammaticae cogitationes', pp. 286–87.
             'Characteristica verbalis', pp. 432–435.
             'Introductio ad Encyclopaediam arcanam', pp. 511–15.

2. *English Translations*

Lang.   Leibniz, G.W., *New Essays on Human Understanding*, translated and edited
         by Alfred Gideon Langley. 3rd edition. LaSalle, Open Court, 1949, [transla-
         tion of *Nouveaux essais*, G., vol. 5].
Loem.   Leibniz, G.W., *Philosophical Papers and Letters*, translated and edited by
         Leroy E. Loemker. 2nd edition. Dordrecht, Reidel, 1969.
         This contains selections from:
         Leibniz-Clarke correspondence, pp. 675–721 [translated from G., **7**, 347–440];
         Leibniz-Des Bosses correspondence, pp. 596–617 [translated from G., **2**,
         285–521];
         Leibniz-Arnauld correspondence, pp. 331–350 [translated from G., **2**, 1–138].
  W.   Leibniz, G.W., *Selections*, edited by Philip Wiener. New York, Scribners, 1951.
         'Letters to De Volder, 1699–1703', pp. 156–181. [selected and translated from G.,
         2, 139–283].

3. *Other Works*

Jung   Jung, Joachim. *Logica Hamburgensis* [1638]. Edited by Rudolf W. Meyer.
         Hamburg, Augustin, 1957.

---

[1] Cf. B. Mates, 'Leibniz on possible worlds', in *Leibniz: a Collection of Critical Essays*,
H.G. Frankfurt, ed. (Garden City, New York, Anchor Books, 1972, pp. 335, ff); K.C.
Clatterbaugh, 'Leibniz's doctrine of individual accidents', *Studia Leibnitiana*, Sonder-
heft 4, 1973; H. Burkhardt, 'Anmerkungen zur Logik, Ontologie und Semantik bei
Leibniz', in *Studia Leibnitiana* **6**, 1, 49–68 (1974); H. Neri-Castañeda, 'Plato's 'Phaedo'
theory of relations', *Journal of Philosophical Logic* **1**, 467–480 (1972).
[2] PA., R VI **6**, 265 [Lang., 276–77].
[3] PA., R VI **6**, 227 [Lang., 235].
[4] "Porro Deus non tantum singulas monades et cujuscunque Monadis modificationes
spectat, sed etiam earum relationes, et in hoc consistit relationum ac veritatum realitas."
G., **2**, 438.
[5] PA., R VI **6**, 447 [Lang., 516–17].
[6] PA., R VI **6**, 145 [Lang., 148].
[7] PA., R VI **6**, 145 [Lang., 148].
[8] "Subjectum relationis est id quod denominatur a relatione: fundamentum vero est id
ratione cujus talis relatio convenit subjecto, quod propterea vocatur etiam ratio fundandi
relationem: ut subjectum similitudinis est substantia quae dicitur similis: fundamentum
vero est albedo aut figura ratione cujus est similis." Domingo de Soto. *In Porphyrii*

Isagogen, Aristotelis Categorias, librosque de demonstratione, absolutissima commentaria.
Venice, 1573. p. 206.

[9] "*Relatio* est accidens quod est in pluribus subjectis estque resultans tantum seu nulla mutatione facta ab iis supervenit, si plura simul cogitentur, est concogitabilitas." LH, IV, VII, C, f. 74.

"*Relatio* est duorum concogitabilitas." LH, IV, VII, C, f. 35.

"*Relatio* est secundum quod duae res simul cogitantur." LH, IV, VII, C, f. 47.

These passages are all to be found in R. Kauppi, *Über die Leibnizsche Logik mit besonderer Berücksichtigung des Problems der Intension und der Extension, Acta philosophica Fennica* **12**, Helsinki, 1960, p. 49.

As we shall see, 'the relation occurs without any change taking place', only if we are concerned with an external consideration of reality.

[10] "Videtur ergo dicendum: Relationum, quae sunt Entia, et tum vera, cum a nobis cogitantur, ut sunt numeri, lineae seu distantiae . . . . non esse numerum; nam perpetuis semper reflexionibus possunt multiplicari, adeoque nec sunt Entia realia, possibiliave nisi cum cogitantur". Jag., 79–80.

[11] Cout. OF., 243; 287; 435; 512.

[12] G. 7, 399–402 [Loem., 703–04].

[13] G. 7, 400 [Loem., 703].

[14] G. 7, 400–401 [Loem., 704].

[15] G. 7, 401 [Loem., 704].

Cf. H. Burkhardt, *op. cit.* (see note 1), pp. 54–55; B. Mates, *op. cit.* (see note 1), p. 349 and note 16.

[16] G. 7, 401 [Loem., 704].

[17] Cf. B. Russell, *A Critical Exposition of the Philosophy of Leibniz* (Cambridge, Cambridge University Press, 1900) §10.

[18] G. 7, 401 [Loem., 704].

[19] G. 2, 486 [Loem., 609].

[20] "Et de hisce interdum ita loquimur *quasi una relatio sit inter duos terminos*, cum tamen semper binae relationes sint intelligendae. Ut *amicitiam* dicimus esse *inter Orestem* et *Pyladen*, cum tamen alia sit amicitia qua Orestes erga Pyladen, alia qua Pylades erga Orestem afficitur." Jung, 42.

[21]"Hanc quaestionem circa hanc conclusionem, an relatio sit res? disputat Simplicius super Praedicamenta, et ponit rationes multas, quod non: quia sine mutatione accedit et recedit. Dicit quod illa fuit fortissima ratio Stoicorum. . . . Ad oppositum arguit: inconveniens est harmoniam destruere quae realiter delectat; ergo realis est causa delectationis." Joannes Duns Scotus. 'Quaestiones subtilissimae in Metaphysicam Aristotelis, cum annotationibus R.P.F. Mauritii de Portu Hiberni, Lib. V, Quaestio XI, Scholium I', in *In XII libros Metaphysicam Aristotelis expositio* (volume 4 of the *Opera Omnia*, edited by Luke Wadding). Lyon, Durand, 1639, p. 634 (reprinted by Olms; Hildesheim, 1968).

[22] "Quarto principaliter arguitur, ut arguit S.Th. . . . . . et Scotus. . . . Sunt aliquae relationes quae dicuntur perfectiones reales, ut ordo universi, unio materiae et formae, inhaerentia accidentis ad subjectum, ... quae omnes relationes dicunt perfectiones ex natura rei seclusa operatione intellectus: ergo illae sunt relationes reales: quae si

sint reales, realiter distinguuntur. Propter hoc argumentum ait *illic Scotus quod negare relationes distingui est negare ordinem universi"*. Domingo de Soto, *op. cit.* (see note 8), p. 206.

[23] G. 2, 115 [Loem., 341].

[24] G. 2, 420.

[25] G. 2, 240 [W., 177].

[26] Francisco Suárez. *Disputationes metaphysicae.* Mainz, 1600. xlvii, xxii, II, p. 548.

[27] "Dum alius fit me maior, crescendo, utique in me quoque aliqua accidit mutatio, cum mutata sit denominatio de me. Et hoc modo omnia in omnibus quodammodo continentur". Jag., 122.

[28] G. 7, 321–322 [Loem., 365].

Translator's Note: Loemker has: "there are no extrinsic denominations." He reads *nullae* where Gerhardt has *multae* [see Loem., 366, note 3].

[29] PA., R VI 6, 227 [Lang., 235–36].

[30] LH, IV, VII, C, f. 107v–108r.

"Videndum ergo ad Relationem proprie pertinere denominationes extrinsecae, quae scilicet nascuntur et pereunt nulla subjecti ipsius mutatione, sed tantum, quia fit mutatio in alio, ita pater fit pater nato filio etsi ipse qui forte in India Orientali agit, inde non afficiatur. Ita similitudo mea cum alio nascitur et oritur etiam sine mutatione mei, sola mutatione alterius. Fatendum tamen interim rigorose nullam esse denominationem extrinsecam in rebus, quia nihil contingit ullibi in mundo, quod non omnia in mundo existentia reapse afficiat [.] Seposita tamen hac rerum universali sympathia, pro extrinsecis denominationibus haberi possunt."

[31] LH, IV, VII, C, f. 17r.

"Videntur relationes inter duas res vel esse solum rationales vel reales, seu esse essentiae vel existentiae. Reales sunt vel 'positionis' (temporis nempe et situs) vel 'influxus'; cum per unum mutatio aliqua fit vel impeditur in aliquo. Ut cum dico Homo est mortalis, et animal est mortale est relatio rationis et hoc dat quod vulgo vocant relationem extrinsecam, quod et dici potest de Relatione positionis. Itaque si dicam Petrus distat a me 100 . . . est denominatio extrinseca; et si ego movear quiescente Petro, mutatur quidem distantia Petri a me, sed sine mutatione Petri, nisi quae fit in virtute connexionis rerum universalis . . ."

[32] Cout. OF., 287, 357.

[33] "Genitivus est adjectio substantivi ad substantivum quo id cui adjicitur ab alio distinguitur. Ensis Evandri, id est Ensis quem habet Evander. Pars domus, id est pars quam habet domus. Lectio poetarum, id est actus quo legitur poeta. Optime sic explicabitur, ut Paris est amator Helenae; id est: Paris amat *et eo ipso* Helena amatur. Sunt ergo duae propositiones in unam compendiose collectae. Seu Paris est amator, et eo ipso Helena est amata. Ensis est ensis Evandri, id est Ensis est supellex quatenus Evander est dominus. Poeta est lectus quatenus ille vel ille est legens. Nam nisi obliquos casus resolvas in plures propositiones, numquam exibis quin cum Jungio novos ratiocinandi modos finguere cogaris. . . ." Cout. OF., 287.

[34] LH. IV, VII, C, f. 267-v.

"vel sic: Paris amat Helenam, id est Petrus putat quod Helena est sibi futura jucunda. Ensis Evandri, id est Ensis qui est subditus quatenus Evander est dominus. Subditum

vel subditus est quod patitur quatenus alius agit, et quatenus is est justus. Itaque redactis
omnibus ad nominativum, fiet: Ensis est Evandri, id est si Ensis patitur, et si Evander
agit, Evander est eatenus justus. Seu (si) Ensis patitur (quatenus) Evander agit, (tunc
eo respectu) Evander est justus. Tantum ergo opus est illis conjunctionibus quae con-
nectunt propositiones imo ipsum quatenus adhuc amplius explicari deberet hoc modo:
si Ensis patitur quia Evander agit, non ideo Evander est injustus. Imo opus erit expli-
candore currere usque ad definitionem causae et effectus. Sed ne semper id opus sit, utile
erit in characteristica certa quaeque excogitari signa hujusmodi resolutionem semel
repraesentantia, ne semper molesta resolutione opus sit. Tantum servanda perpetuo
calculi aequalitas, ut propositiones demonstrabiles calculo ipso pateant.

Ensis Evandir est pulcher. id est is Ensis est pulcher, qui si patitur quia Evander agit,
non ideo Evander est injustus. Unde intelligi potest quantae molis futura sit verae
characteristicae restitutio a qua omnes obliqui omnesque flexiones exesse debent,
perinde ac si edicto vetita essent.

Omne verbum (personale) habet praecedentem nominativum.

aliquando habet sequentem nominativum, ut tu esto sobrius; quadrigae sunt in-
strumentum.

Omne verbum infinitum vel impersonale habet pro supposito accusativum expressum
vel suppressum. Scio te amare. Cupio scire id est cupio me scire. Omnes homines qui
se student praestare caeteris animantibus.

Omne verbum activum sive finitum sive infinitum semper regit accusativum [*ego
malim vel regere potest: ut vitam vivere einen Kampf kämpfen*]"

The square brackets are in Leibniz's text. I was able to decipher the bracketed words
thanks to the kind help of Professor Emanuele Casamassima.

[35] In grammatica rationali necessarii non sunt obliqui, nec aliae flexiones Cout. OF.,
287.

[36] " 'Enuntiationes' reduplicativae dicuntur, *quae constant dictione aliqua geminante,
seu Reduplicante*, ut vocant. Veluti si dicas, Homo quatenus est homo, est disciplinae
capax. Illa enim dictio *quatenus*, quia apta est ad aliquid geminandum et iterandum,
reduplicans et reduplicativa dici solet . . ."
Pedro da Fonseca. 'Institutionum dialecticarum libri octo, III, xxii', in: *Commen-
tarium . . . in libros Metaphysicorum Aristotelis.* Lyon, Giunta, 1597. p. 71.

[37] "*Reduplicatio causalis* est, ubi reduplicatio causam continet propter quam Praedica-
tum Subjecto inest, ut *Homo, quatenus rationalis est, Philosophiae capax est.*" Jung, 92.

[38] LH., IV, VII, C, f. 17r.
"Relationes sunt vel comparationis vel connexionis. *Relatio comparationis* ex eo nascitur
inter A et B, quod A reperitur in aliqua propositione, et B in alia propositione, *Relatio
connexionis* ex eo quod tam A et B sunt in una eademque propositone (quae in relationem comparationis resolvi non potest). Nam alioqui etiam relatio comparationis
foret relatio connexionis, nam formari potest una propositio comprehendens A et B,
nempe A est similis B. Sed ea resolvitur tandem in duas quarum una singulatim agit
de B, altera sigillatim de A, verbi gr. A est ruber et B est ruber, et ideo A est similis
(quoad hoc) ipsi B. per A autem et B intelliguntur res seu individua, non termini. Sed
quid de his dicemus: A existit hodie et B etiam existit hodie, seu A et B existunt simul?
An erit haec relatio comparationis, an connexionis idem est de coexistentia in eodem
loco."

[39] Cf. PA, R VI 6, 142, 275 [Lang. 144, 287–88]

[40] Cf. Jung, 115, 190.

[41] "Scopus nostrae Characteristicae est tales adhibere voces, ut omnes consequentiae quae institui possunt statim ex ipsis verbis vel characteribus emantur, verbi gratia David est pater Johannis, Ergo Salomon est filius Davidis. Haec consequentia ex his vocabulis latinis nisi resolvantur in alia aequipollentia demonstrari non potest; in lingua generali debet ex vocabulorum analysi in suas literas demonstrari posse." Cout. OF., 284.

Instead of 'Solomon' (in the bracketed text) Leibniz wrote 'John' by mistake (cf. Cout., 284, note 1).

[42] PA., R VI 6, 479 [Lang., 560].

ANDREA SANI

NECESSARY AND CONTINGENT TRUTHS
IN LEIBNIZ

In the paper, 'Vérités nécessaires et contigentes', included in Leibniz's *Opuscules et fragments inédits*, edited by Couturat, Leibniz defines necessary propositions as follows:

An *absolutely necessary* proposition is one which can be resolved into identical propositions, or whose opposite implies a contradiction.[1]

In an attempt to make this definition precise, we can say that for Leibniz, a proposition is necessary if and only if it is in the form '*A* is *A*' or '*AB* is *A*', or if it can be resolved into a proposition in one of these two forms by means of definitions and substitutions. Furthermore, since a formally identical proposition, or one which can be resolved into a statement in this form, is, for Leibniz, such that its opposite proposition implies contradiction (that is, it is in the form '*A* is non-*A*' or '*AB* is non-*A*', or can be resolved into a proposition in one of these two forms[2]), it follows from this that a necessary proposition can also be defined as a proposition whose opposite implies contradiction.

In the same paper Leibniz defines contingent propositions as 'non-necessary';[3] hence for Leibniz they are propositions which cannot be reduced to identical statements, that is, their opposite does not imply contradiction.

Having established these definitions of logical necessity and contingency, let us now tackle the question of whether true, non-necessary propositions are admissible in Leibniz's system.

A central thesis in Leibniz's thought is that in every true proposition the notion of the predicate is included in that of the subject:

The predicate or consequent therefore

always inheres in the subject or antecedent . . . the nature of truth in general or the connection between the terms of a proposition consists in this fact.[4]

Many commentators[5] maintain that this definition of truth excludes the possibility of true contingent propositions in Leibniz's system. In fact, they think that if the predicate is included in the subject in every true proposition, every truth can be reduced, by means of definitions and substitutions,

411

Maria Luisa Dalla Chiara (ed.), Italian Studies in the Philosophy of Science, 411–422.
Copyright © 1980 by D. Reidel Publishing Company.

to an identical proposition, at least by God. Hence, assuming that a proposition is, by definition, necessary if and only if it is a formal identity or can be resolved into a statement in this form, it seems that in Leibniz's system, all truths must be necessary. This interpretation leads to maintaining that when Leibniz speaks of contingent truths, he is not referring to an effective logical contingency.

In reality it does not seem to me that an accurate study of Leibniz's papers confirms these conclusions. As far as I am concerned, while maintaining that the notion of the predicate is included in that of the subject in every true proposition, Leibniz does succeed, to a certain extent, in proving that truths of fact are not necessary, neither for us nor for God.

The principal factor which renders truths of fact contingent should be sought, as Leibniz often stresses,[6] in the 'infinitary' character of their *a priori* proof. This aspect of factual truths has already been pointed out by some students of Leibniz's works, who have, however, undervalued its importance.[7]

The paper which sheds the greatest light on Leibniz's thought on this point is 'De libertate', included in the collection *Nouvelles lettres et opuscules inédits,* edited by Alexander Foucher de Careil.

In 'De libertate' Leibniz confesses his philosophical doubts and admits that with his doctrine he has come close "to the opinion of those who believe that everything is necessary":

When I considered that nothing occurs by chance or by accident unless we resort to certain particular substances, that fortune apart from fate is an empty word, and that nothing exists unless certain conditions are fulfilled from all of which together its existence at once follows, I found myself very close to the opinions of those who hold everything to be absolutely necessary.[8]

Leibniz himself asserts that the general definition of truth as the inclusion of the predicate in the subject had at first led him to doubt that contingent truths were admissible:

For I saw that in every true affirmative proposition, whether universal or singular, necessary or contingent, the predicate inheres in the subject or that the concept of the predicate is in some way involved in the concept of the subject. I saw too that this is the principle of infallibility for him who knows everything *a priori*. But this very fact seemed to increase the difficulty, for, if at any particular time the concept of the predicate inheres in the concept of the subject, how can the predicate ever be denied of the subject without contradiction and impossibility, or without destroying the subject concept?[9]

But Leibniz shows that he had drawn away from this 'fatalistic' tendency thanks to a sudden revelation:

A new and unexpected light arose at last, however, where I least expected it, namely, from mathematical considerations of the nature of the infinite.[10]

To show how he arrived at maintaining that truths of fact are contingent by means of reflecting on the nature of the infinite, Leibniz starts from a subdivision among various kinds of truth. First of all he subdivides truths into two categories: *original* and *derived*.[11] Original truths are those in which the connection of the predicate with the subject is evident in itself without having to be proved; derived truths, on the other hand, are those for which such a relation of 'inherence' must be proved. Let us examine original truths:

Original truths are those for which no reason can be given; such are identities or immediate truths, which affirm the same thing of itself or deny its contrary of its contrary.[12]

This division of original truths is linked to that expounded elsewhere, between 'affirmative identical truths' and 'negative identical truths'.[13] Affirmative identical truths are examples of the *principle of identity*, that is, they are in the form 'A is A' or 'AB is A'.[14] For example, Leibniz calls the following true propositions affirmative identical truths: "chaque chose est ce qu'elle est" [each thing is what it is] or "l'animal raisonnable est toujours un animal" [the rational animal is always an animal].[15] Negative identical truths referred to in 'De libertate' are examples of the *principle of contradiction*, that is, they have the form 'A is not non-A' or 'AB is not non-A'. The following is a negative identical truth: "un rectangle équilatéral ne saurait être non rectangle" [an equilateral rectangle would not be a non-rectangle].[16] I think that these truths are called 'negative identical' because they can be easily decomposed to the principle of identity: 'A is not non-A' which, for Leibniz, is in fact equivalent to 'A is non-non-A'[17] which in its turn equals 'A is A', since it equals "not non-A is A."[18] In general then, identical propositions are defined as 'original truths' because (as has already been shown) they are 'immediately true': Leibniz says that the relation of 'inherence' between subject and predicate "s'y peut montrer à l'oeil" [can be shown to the eye].[19]

In addition to original truths, there are derived truths in which the subject-predicate connection is not evident but must be proved. Derived truths, like original truths, are also of two kinds:

There are in turn two genera of derivative truths, for some can be reduced to [original] truths; the others [admit] in an infinite progression of resolution.[20]

Let us begin with the derived truths which can be resolved into original ones. These truths are not formally identical, that is, they are not in the form '*A* is *A*' or '*AB* is *A*'; nevertheless they can be reduced to statements in this form. According to Leibniz a proposition can be resolved into an identical statement by means of 'demonstration'. On more than one occasion Leibniz explains what he means by 'demonstration';[21] by considering the examples proposed in the *Opuscules*,[22] we can say that for Leibniz, to demonstrate a proposition '*S* is *T*' means to substitute the subject *S* with the term which defines it; this last, in its turn with the 'definiens', going forward with the definitions and substitutions until it reaches a proposition in the '*T* is *T*' or '*YT* is *T*' form. In other words, according to Leibniz, with a demonstration one starts from a non-formally identical proposition and arrives at an identity by means of an 'analysis of the concept of the subject', that is, by means of successive definitions directed to bring out concepts which 'enter' into the notion of the subject,[23] and by means of successive substitutions of terms which denote these concepts in the place of the subject (substitutions allowed by virtue of the rule of the 'substitution of equivalents').[24] It is clear that when, in the course of the analyses of the concept of the subject, one arrives, through repeated definitions, at the concept of the predicate, by substituting the term which denotes this concept for the subject, the starting proposition is resolved into an example of the principle of identity and is demonstrated. Thus, for example, the following true, non-formally identical proposition can be demonstrated, that is, resolved into an identical statement through a single substitution of the defined term for the defining term: 'homo est rationalis'; since the subject 'homo' can be defined in the following way: 'homo est animal rationale', if we substitute the 'definiens' 'animal rationale' for the subject 'homo', we resolve the starting proposition into an identical statement by means of a single substitution.

Therefore derived truths which can be resolved into original truths are those true propositions which, while not formally identical, can be resolved into identity by means of demonstration. Consequently, such true propositions are necessary by definition as original truths.

What is important to point out for the purposes of the question concerning the plausibility of non-necessary truths in Leibniz's system is that in 'De libertate' he admits, next to original truths and those reducible to original ones, true propositions (that is, propositions such that in them the predicate is included in the subject) which, however, *cannot be reduced to identity*. They are singular true propositions whose subject is an 'individual substance' and whose predicate is its 'accidental attribute'.

Let us see why, according to Leibniz, such truths are not reducible to identical propositions.

Here in 'De libertate' as in the 'Generales inquisitiones de analysi notionum et veritatum',[25] in a paper included in Grua,[26] and in some fragments of the *Opuscules*,[27] Leibniz formulates a parallel between the analysis of this kind of true propositions and the irrational proportions of mathematics.[28] According to Leibniz, because of the immense variety of things, because of the infinitesimal division of bodies and the infinite relations which exist between every existing thing and all the other existing things of its universe, individuality contains the infinite, and hence the complete notion of every single substance (which includes all its predicates) contains an infinity of elements and conditions:

For example, there is no portion of matter, however tiny, in which there is not a world of creatures, infinite in number. And there is no created substance, however imperfect, which does not act upon all the others and suffer action from all the others, and whose complex concept as this exists in the divine mind does not contain the whole universe, with all that ever is, has been, and will be. And there is no truth of fact or of individual things which does not depend upon an infinite series of reasons. . .[29]

For this reason Leibniz can maintain further on in the text[30] that, as in the case of the relationship between two incommensurable numbers one never reaches a common measure by means of successive divisions and the resolution proceeds to infinity, so in the case of the analysis of a truth of fact by means of successive substitutions of equivalent terms, one never reaches terms common to both subject and predicate; that is, however much the concept of the subject is analyzed, the concept of the predicate is not reached by any finite number of steps and hence the initial proposition is never resolved into an identical truth. It is clear that if the *a priori* proof of a truth of fact is 'infinite', that is, 'non-terminable', not even God can terminate it by linking the truth to be proved to an identity statement.[31]

From the moment that, for Leibniz, a proposition is necessary if and only if it is formally identical, or can be resolved into a proposition in this form, it follows from this that truths of fact, which cannot be reduced to identities by virtue of the infinite character of their analysis, are not necessary but contingent. Infinite analysis establishes an effective 'discrimen' between true propositions:

But in proportions the analysis may sometimes be completed, so that we arrive at a common measure which is contained in both terms of the proportion an integral number of times, while sometimes the analysis can be continued into infinity, as when comparing a rational number with a surd; for instance, the side of a square with the diagonal. Just so, truths are sometimes demonstrable or necessary, and sometimes free and contingent, so that they cannot be reduced to identities as if to a common measure, by any analysis. This is the essential distinction between truths as well as proportions.[32]

In this way the initial testimony of 'De libertate' (to which I have already referred) is justified, where Leibniz points out that the solution to the problem of the contingency of truths of fact came to him from considerations on the nature of the infinite. Elsewhere too Leibniz states:

And so I think that I have disentangled a secret which had me perplexed for a long time; for I did not understand how a predicate could be in a subject, and yet the proposition would not be a necessary one. But the knowledge of geometry and the analysis of the infinite lit this light in me so that I might understand that notions too can be resolved to infinity.[33]

One of the fragments in the *Opuscules* is entitled: 'Origo veritatum contingentium ex processu in infinitum ad exemplum proportionum inter quantitates incommensurabiles' [origin of contingent truths from the process to infinity on the example of the proportions among incommensurable quantities].[34]

This analogy with mathematics and the infinitesimal calculus is pushed by Leibniz to the assertion that in the analyses of contingent truths, through a continued decomposition to infinity, it is possible to make the difference between subject and predicate as small as one wants (*even if it is impossible to annul it*). In considering the truth of fact, 'Peter denies', Leibniz says in the 'Generales inquisitiones':

But the concept of Peter is complete, and what is more it involves infinity; therefore we can never arrive at a perfect [demonstration (*ad perfectam demonstrationem*)]. However, we always more and more nearly approximate it so that the difference is less than any given.[35]

For Leibniz, to show that in a truth of fact the difference between subject and predicate can be made smaller than any given difference, that is, that one comes closer and closer to resolving the proposition into identity, means '*proving a priori*' the truth itself. Even if these propositions *are not* '*demonstrable*' (that is, they cannot be resolved into identical statements), they can be '*proved a priori*':

Each true proposition can be proved, since the predicate is in the subject, as Aristotle says, or the concept of the predicate is involved in the concept of the subject completely understood; in any case the truth must be [provable (*opportet veritatem posse ostendi*)] by an analysis of the terms into their values or into those terms which they contain.

A necessary true proposition can be proved by reduction to identities, or of its opposite to contradictories; whence the opposite is called impossible.

A contingent true proposition cannot be reduced to identities; notwithstanding it is proved, by its being shown by a continued gradual analysis that it approaches identities continuously, but that it never reaches them.[36]

According to Leibniz a truth which is proved *a priori* by means of a continuous decomposition to infinity *is 'certain' but not 'necessary'*, since it cannot be resolved into an identical proposition: in fact in referring to the proof of a proposition with negative predicate in the form '*A* is non-*B*', Leibniz states that:

That *A* contains non-*B* is proved either by demonstration or by a perfect analysis, or only by an analysis continuable into infinity or always imperfect. Thus, it is certain, but not in fact necessary, because it can never be reduced to an identity or its opposite to a contradiction.[37]

The difference between the '*certainty*' of truths of fact and the '*necessity*' of truths of reason to which Leibniz often refers in his more mature writings[38] is founded logically in this way: necessary truths can be resolved into the principle of identity while those which are only 'certain' can only be brought closer and closer to identical propositions without ever reaching statements in this form. In calling 'certainty' also 'determination' or 'inclination', Leibniz states:

. . . there is the same proportion between necessity and inclination as there is in the Analysis of Mathematicians between the exact equation and the limits which give an approximation.[39]

In the 'Generales inquisitiones' Leibniz also says that the *a priori* proof of contingent truths which shows their 'certainty' but not their 'necessity', is accessible only to God, the only Being whose Mind is capable of following the analysis of these propositions to infinity:

Whence it belongs to GOD alone, who embraces the whole infinite with his mind, to know the certainty of all contingent truths.[40]

Note that God can only 'make reason', in the sense set forth above, of a contingent truth; that is, by succeeding in 'seeing' every step of an infinite analysis, He can show that a contingent truth can be brought closer and closer to an identical proposition (thus grasping its 'certainty'), but *He cannot demonstrate it*.[41] Not even He is allowed to resolve a true factual proposition into the principle of identity. Leibniz says this clearly in 'De libertate', where he states:

In contingent truths, however, though the predicate inheres in the subject, we can never demonstrate this, nor can the proposition ever be reduced to an equation or an identity, but the analysis proceeds to infinity, only God being able to see, not the end of the analysis indeed, since there is no end, but the nexus of terms or the inclusion of the predicate in the subject, since he sees everything which is in the series.[42]

And toward the end of the paper we find:

All the more are contingent or infinite truths subject to the knowledge of God and known by him, not by demonstration—for this would involve contradiction—but by an infallible vision.[43]

God does not demonstrate contingent truths (which would really be contradictory) but he knows them *a priori* by means of an 'infallible vision' which in 'De libertate' is not clearly specified. In what such divine *a priori* knowledge consists is explained by Leibniz, as has been seen, in the 'Generales inquisitiones', from which it emerges that God, unlike man, is capable of proving truths of fact by showing that by means of a resolution continued to infinity, they can be brought continually nearer the principle of identity.

These ideas which Leibniz developed starting from 1686 seem then to offer a solution to the problem of truth and contingency. However, some obscure points persist. For example, B. Mates[44] notes that while Leibniz rigorizes the concept of 'demonstration', that is, *a priori* finite proof, he does not explain clearly how the decomposition of terms in the infinite analysis of a contingent truth takes place.

On the other hand Poser[45] points out that nowhere does Leibniz indicate the metrics with whose aid the quantitative or even only the qualitative difference between the notion of the subject and that of the predicate can be determined: the central point of the analogy set up by Leibniz between the analysis of contingent truths and the infinitesimal calculus is not adequately illustrated.

While these obscurities in Leibniz's arguments remain fixed, it seems to me verified nonetheless that what establishes the contingency of singular true propositions for the philosopher is their infinite analysis insofar as it *impedes a decomposition to identical propositions*.

NOTES

Below is an explanation of the abbreviations used in the Notes, followed by lists of Leibniz's papers in each volume which are cited in the preceding essay.

1. *Leibniz's Original Works*

G.   Leibniz, G.W., *Philosophische Schriften*, edited by C.I. Gerhardt. 7 volumes. Berlin, Weidmannsche Buchhandlung, 1875–90.
    vol. 1: 'Leibniz an Conring', pp. 173–75.
    vol. 3: 'Leibniz an Coste', pp. 400–04.
    vol. 5: *Leibniz und Locke* (*Nouveaux essais sur l'entendement*) Book IV, chapter II: 'Des degrés de notre connaissance', pp. 342–56.
    vol. 6: *Essais de Théodicée*, pp. 1–471. Including the appendix: 'Remarques sur le livre de l'origine du mal, publié depuis peu en Angleterre', pp. 400–436.
    vol. 7: 'Praecognita ad Encyclopaediam sive scientiam universalem', pp. 43–48.
F. de C.   Foucher de Careil, Alexandre, *Nouvelles lettres et opuscules inédits de Leibniz*. Paris, Durand, 1857.
    'De libertate', pp. 178–85.
Cout. OF.   Couturat, Louis, *Opuscules et fragments inédits de Leibniz*. Paris, Alcan, 1903.
    'Origo veritatum contingentium ex processu in infinitum ad exemplum proportionum inter quantitates incommensurabiles', p. 1.
    'Conséquences métaphysiques du principe de raison', pp. 11–16.
    'Vérités nécessaires et contingentes', pp. 16–24.
    'Sur les qualités sensibles', pp. 186–87.
    'Principia calculi rationalis', pp. 229–31.
    'Specimen calculi universalis', pp. 239–43.
    'Notes de logique', p. 258.
    'Generales inquisitiones de analysi notionum et veritatum', pp. 356–99.
    'Primae veritates', pp. 518–23.
    'Sur les propositions contingentes', p. 405.
Grua   Grua, G., *Leibniz: Textes inédits d'après les manuscrits de la Bibliothèque provinciale de Hanovre*. 2 vols. Paris, Presses Universitaires de France, 1948.
    'De contingentia', **1**, pp. 302–06.
    'Conversation sur la liberté et le destin', **2** pp. 478–86.

2. *English Translations*

Loem.   Leibniz, G.W., *Philosophical Papers and Letters*, translated and edited by Leroy E. Loemker. 2nd edition. Dordrecht, Reidel, 1969.
    'On freedom' [translation of 'De libertate', F. de C., 178–85], pp. 263–66.
    'First truths' [translation of 'Primae veritates', Cout. OF., 518–23], pp. 267–71.
O'B.   *Gottfried Wilhelm Leibniz's 'General Investigations Concerning the Analysis of Concepts and Truths', a Translation and an Evaluation* by Walter H. O'Briant.

Athens, Georgia, University of Georgia Press, 1968 [translation of 'Generales inquisitiones de analysi notionum et veritatum', Cout. OF., 356–99].

Far.   'Observations on the book concerning 'The origin of evil' published recently in London', pp. 405–41, an appendix to Leibniz's *Theodicy: Essays on the Goodness of God, the Freedom of Man and the Origin of Evil*, edited by Austin Farrer, translated by E.M. Huggard. New Haven, Yale University Press, 1952 [translation of 'Remarques sur le livre de l'origine du mal . . .', G., **6**, 400–36, an appendix to *Essais de Théodicée*].

Park.   Leibniz, G.W., *Philosophical Writings*, edited by G.H.R. Parkinson, translated by Mary Morris and G.H.R. Parkinson. London, Dent, 1973.
'Necessary and contingent truths' [translation of 'Vérités nécessaires et contingentes', Cout. OF., 16–24], pp. 96–105.
'Metaphysical consequences of the principle of reason' [translation of 'Conséquences métaphysiques du principe de raison', Cout. OF., 11–16], pp. 172–178.

W.   Leibniz, G.W., *Selections*, edited by Philip Wiener. New York, Scribners, 1951.
'Principles of a logical calculus' [translation of 'Principia calculi rationalis', Cout. OF., 229–31], pp. 26–28.

Lang.   Leibniz, G.W., *New Essays Concerning Human Understanding*, translated and edited by Alfred Gideon Langley. 3rd edition. La Salle, Open Court, 1949. [Contains 'Of the degrees of our knowledge', translation of 'Des degrés de notre connaissance', G. **5**, 342–56].

[1] Cout. OF., 17 [Park., 96].

[2] Cout. OF., 186.

[3] "That which lacks such necessity I call contingent. . .", Cout. OF., 17 [Park., 97].

[4] Cout. OF., 518–19 [Loem., 267].

[5] Cf. for example, A.O. Lovejoy, *The Great Chain of Being*, Cambridge, Mass., Harvard University Press, 1936; W. and M. Kneale, *The Development of Logic*, Oxford, Clarendon Press, 1962; R. Kauppi, *Über die Leibnizsche Logik mit besondere Berücksichtigung des Problems der Intension und Extension, Acta Philosophica Fennica* **XII**, Helsinki, 1960, pp. 94–101.

[6] Cf. Cout. OF., 1; 18 [Park, 97–98]; 388 [O'B., 64–65]. Grua **1**, 303–04; F. de C., 179–80 [Loem., 263–64].

[7] For example, Couturat, after having shown that Leibniz tries to differentiate between contingent truths and necessary truths by virtue of their infinite analyses, asserts: "Is the difficulty thus really resolved? We are far from asserting this. But here (as in our book) we are writing as an historian not as a critic" (L. Couturat, 'Sur la métaphysique de Leibniz', *Revue de métaphysique et de morale* **10**, p. 11). More recently E.M. Curley has maintained that he sees no plausible reason for the connection set up by Leibniz between the contingent character of truths of fact and their infinite analysis:

> Leibniz clearly attached great importance to the fact that contingent propositions involve an infinite analysis, and he seems to have thought that their contingency in some way depended on the fact that their *a priori* proof was impossible for a finite intellect. At present I can see no reason

for this at all. It appears to be a mistake. (E.M. Curley, 'The root of contingency' in *Leibniz: a Collection of Critical Essays*, edited by H.G. Frankfurt. New York, Anchor Books, 1972. p. 92.)

[8] F. de C., 178 [Loem., 263].

[9] F. de C., 179 [Loem., 263–64].

[10] F. de C., 179–80 [Loem., 264].

[11] F. de C., 181 [Loem., 264–65].

[12] F. de C., 181 [Loem., 264].

[13] Cf. Cout. OF., 518 [Loem., 267]; G. 5, 342–43 [Lang., 404–05].

[14] G. 5, 343 [Lang., 405].

[15] G. 5, 343 [Lang., 405].

[16] G. 5, 343 [Lang., 405].

[17] Cout. OF., 378.

[18] Cout. OF., 230 [W., 26].

[19] Cout. OF., 186.

[20] F. de C., 181 [Loem., 264].

[21] Cf. F. de C., 181 [Loem., 264–65]; Grua 1, 304; G. 7, 44 where Leibniz says: "To [demonstrate] a proposition means to make clear the inclusion of the predicate or consequent in the subject or antecedent by means of a resolution of the terms into other terms equivalent to them." Bear in mind that for Leibniz "resolution is the substitution of the definition for the defined." (Cout. OF., 258).

[22] Cf. Cout. OF., 11 [Park., 172]; p. 17 [Park., 96–97]; 240.

[23] Leibniz says that "demonstration is nothing if not a chain of definitions" ["Est enim Demonstratio nihil nisi catena definitionum."] G. 1, 174.

[24] "By substituting equal things [for equal things], equality remains" [Mettant des choses égales à la place, l'égalité demeure.], G. 5, 394.

[25] Cout. OF., 388 [O'B., 64–65].

[26] Grua 1, 303–04.

[27] Cout. OF., 1; 17–18 [Park., 96–98].

[28] This analogy set up by Leibniz between the incommensurable relations of mathematics and truths of fact has been brought to light by Couturat in the *Logique de Leibniz d'après des documents inédits*. Paris, Alcan, 1901. pp. 211–213.

[29] F. de C., 180–81 [Loem., 264].

[30] F. de C., 183–184 [Loem., 265–66]; cf. also Cout. OF., 17–18 [Park., 96–98].

[31] As we shall see, God can only prove, by means of a resolution to infinity, that a truth of fact can be brought closer and closer to an identical proposition but can never be reduced to it.

[32] F. de C., 183–84 [Loem., 265–66].

[33] Cout. OF., 18 [Park., 97].

[34] Cout. OF., 1.

[35] Cout. OF., 376–77 [O'B., 52].

[36] Cout. OF., 388 [O'B., 65].

[37] Cout. OF., 387 [O'B., 64].

[38] Cf. for example, in the *Essais de Théodicée*, G. 6, 127 (Far., 98–99) and in the letter from Leibniz to Coste, G. 3, 401.

[39] Grua, **2**, 479. At this point the distinction Leibniz makes between '*reasons which incline*' and '*reasons which necessitate*' (cf., for example, G. **6**, 414 (Far., 418–20) and Cout. OF., 405) seems clear to me. Leibniz considers every true, non-formally identical proposition (that is, not true of and in itself) as linked, by means of the rule of the substitution of equivalents, to a series of other propositions logically antecedent to it which constitute the 'reasons' of its truth. Now if a statement whose reason we want to render is a 'vérité de raison', Leibniz says that the propositions (that is, 'reasons') on which it depends, 'necessitate' that such a proposition be true, while if the statement in question is a 'vérité de fait', Leibniz says that the 'reasons' on which it depends 'incline without necessitating'. In fact, the series of propositions connected to a truth of reason terminates in a finite number of steps in the principle of identity (foundation of every necessary truth) and hence 'necessitates' the truth of reason insofar as it 'transmits' to it the necessity of the identical axiom; instead, the series of propositions connected to a truth of fact is 'infinite', that is, it never terminates in the principle of identity and hence, as it is not in possession of any necessary element, it cannot 'transmit' any necessity of the truth of fact. Yet such a series of 'reasons' 'inclines' the proposition which depends on it to be true, since it 'converges' on the principle of identity understood as the (never reached) limit of the same series. (In the following passage of the 'Remarques sur le livre de l'origine du mal' ['Remarks on the book on the origin of evil'], placed as an appendix to the *Essais de Théodicée* [*Theodicy*], Leibniz explicitly connects the infinitary character of the series of propositions to which a contingent truth is connected, with the fact that this series 'inclines without necessitating':

> . . . when in making the analysis of the truth submitted one sees it depending upon truths whose contrary implies contradiction, one may say that it is absolutely necessary. But when, while pressing the analysis to the furthest extent, one can never attain to such elements of the given truth, one must say that it is contingent, and that it originates from a prevailing reason which inclines without necessitating. (G. **6**, 414 [Far., 419])

[40] Cout. OF., 388 [O'B., 65].

[41] On this point cf. also W.E. Abraham, 'Complete concepts and Leibniz's distinction between necessary and contingent propositions' in *Studia Leibnitiana* **1**, Heft 4, 1969, pp. 263–79.

[42] F. de C., 182 [Loem., 265].

[43] F. de C., 184 [Loem., 266].

[44] B. Mates, 'Individuals and modality in the philosophy of Leibniz', in *Studia Leibnitiana* **4**, Heft 2, 1972, p. 99.

[45] H. Poser, 'Zur Theorie der Modalbegriffe bei G.W. Leibniz', *Studia Leibnitiana Supplementa* **6**, 1969, pp. 134–35.

MIRELLA CAPOZZI

KANT ON MATHEMATICAL DEFINITION*

I

I.1. DEFINITION AND THE THEORY OF SCHEMATISM

What does Kant mean by 'definition'? Strictly speaking,

> (∗) to define is "to present the complete, original concept of
> a thing within the limits of its concept" (A727 = B755).

That is to say, the concept, must be presented with all the characteristics
(*Merkmale*) which are sufficient for its distinction and which can be seen
to belong to it directly, i.e. without any proof. In other words, the *original*
concept must be presented with *completeness* and *precision* (A727 = B755
footnote; *Logik* §99). According to Kant only a few concepts are definable
in the sense of the quoted statement (∗) and these few are mathematical
concepts.

However, throughout the *Critique*, definition is linked to the notion of
meaning. For instance, in speaking of categories, Kant maintains that
they are ways of giving an object "its *meaning* [...] on the basis of some
functions of the understanding, i.e. of *defining* it" (A245. Italics mine).
This raises the problem of how the sense (∗) of definition fits into Kant's
theory of meaning, for surely he did not believe that concepts which are
not definable in the sense (∗) are meaningless. To solve the problem, we
shall start by explaining what giving a meaning to a concept represents
for Kant, and then we shall compare the sense (∗) of definition with such
an explanation.

Kant says that a concept has a meaning when "it refers to an object"
(B303), i.e. when it is "made real" through a reference – either *a priori* or
a posteriori – to an intruition. The way such a reference is established is
described in the theory of schematism. (Incidentally, a connection between
definition and meaning seems to be admitted by Kant even independently
of any reference of concepts to objects, as in the case of the pure "logical
meaning" (A147 = B186) of unschematized categories. This question will
be dealt with later on.)

423

Maria Luisa Dalla Chiara (ed.), Italian Studies in the Philosophy of Science, 423–452.
Copyright © 1980 *by D. Reidel Publishing Company.*

The theory of schematism, although apparently devised to explain how categories refer to intuitions, is an all-inclusive theory of the relations which concepts in general have with intuitions (and vice versa). As is well known, schematism is a procedure of the faculty of imagination (*Einbildungskraft*) meant to provide concepts with schemata that can act as intermediaries between them and "images", i.e. intuitions. Schemata and images are carefully distinguished by Kant. A schema is an *a priori* determination of time-intuition that depends on the "productive" *Einbildungskraft* and relates concepts to images; an image, on the other hand, depends on the "reproductive" *Einbildungskraft* and is the end-product of the whole procedure of schematization.

Not only does this general theory not ignore the differences between the various types of concepts, but it even enhances them by making it easier for us to understand what differentiates definable concepts from all the others. In fact, while every schematizable concept has a meaning, schemata are not all of the same kind and only some of them can guarantee a strict definition of the concepts they refer to.

Thus "the schema of a pure concept of understanding [the transcendental schema of a category] can never be brought into any image whatsoever" (A142 = B181). This means that categorial concepts, although capable of referring to objects, and hence meaningful, have no completely and precisely determinate meaning. Therefore, in the strict sense (∗) of definition, categories can have an 'exposition' but never a 'definition' (A729 = B757; A244 – 5).

As regards empirical concepts, Kant is so certain of their meaningfulness that he maintains that "many empirical concepts are employed without question from anyone. Since experience is always available for the proof of their objective reality, we believe ourselves, even without a deduction, to be justified in appropriating to them a meaning, an ascribed [*eingebildete*] significance" (A84 = B117). But empirical concepts pose a problem opposite to that posed by categories: their Schemata can be reduced to innumerable images, but such schemata bear too close a resemblance to common empirical examples and therefore cannot achieve precision and completeness greater than that obtained by exemplification.[1] So it is easy to understand why empirical concepts fail to satisfy sense (∗)'s requirements for definition and are only capable of a "designation" that makes them explicit but never properly defined (A728 = B756; *R* 2917).

Finally the *Schematismuskapitel* deals with "pure sensible concepts" which, judging from Kant's examples, are concepts of figures in space, i.e. geometrical concepts. Their schemata are "a product and, as it were, a monogram, of pure *a priori* imagination, through which, and in accordance with which, images themselves first become possible" (A141 = B181). Now, it is the monogramatic nature of these schemata that makes it possible to attribute a complete and precise meaning to pure sensible concepts. In fact, such schemata, unlike those of pure concepts, can be reduced to images; and such images, unlike those of empirical concepts, are completely 'prefigured' by the *typus* of their monogramatic schemata.[2] Therefore pure sensible concepts are the only ones to be strictly definable among meaningful concepts.

This shows that Kant's theory of definition is consistent with his theory of meaning, which in turn is modelled on the division of concepts into (*a priori* or *a posteriori*) *given* concepts and *made* concepts (*Logik* §4). Within this framework, sense (∗) of definition fixes a set of conditions satisfiable only by concepts that are not given "either by the nature of understanding or by experience" (A729 = B757), but are arbitrarily (*a priori*) made, i.e. mathematical concepts. Then one might conclude that it is their specific *origin* that distinguishes concepts definable in sense (∗) from all the others.

This conclusion, however, does not really explain how mathematical concepts differ from other concepts with respect to what they have *in common*, i.e. searching for and finding a meaning. Furthermore it remains to be seen why Kant speaks also of 'nominal definitions' and, in referring to categories, as we have already mentioned, of a 'logical' meaning. Therefore it seems necessary to go deeper into his theory of definition so as to disentangle these questions which are not merely questions of terminology.

I.2. THE ANALYTIC-SYNTHETIC AND THE NOMINAL-REAL DISTINCTIONS OF DEFINITIONS. EXTENSION AND INTENSION. CONNECTION BETWEEN THE TWO DISTINCTIONS OF DEFINITIONS

To start with, Kant is perfectly aware of the difficulties met by any theory of definition that has to make use of the German language, which has a single term (*Erklärung*) to indicate 'definition', as well as 'exposition' and 'explication' (A730 = B758). Nevertheless, following a deeply rooted philosophical tradition, he does give a classification of definitions in his lectures on logic, namely in Jäsche's *Logik* and in other *Vorlesungen* on the same subject.

Specifically, definitions are said to be (1) either analytic or synthetic, and (2) either nominal or real.

Analytic Definitions are those of *a priori* or *a posteriori* given concepts (*Logik* §§100, 101).

Synthetic Definitions are those of *made* concepts (*concepti factitii*) (*Logik* §§100, 102). But, since concepts made by an empirical synthesis of given phenomena are not properly definable (*Logik* §103), synthetic definitions are possible only for concepts made *a priori* by means of a construction, i.e. mathematical concepts.

Obviously the above distinctions between definitions is a faithful replica of the classification of concepts in terms of their *origin* and as such, does not say more than what we already knew from our previous analysis.

The distinction between nominal and real definitions is a more complex one, due to the many senses and functions the terms 'nominal' and 'real' have in Kant's statements on definitions. I have singled out three of them for each of the terms in question:

Nominal Definitions: (I) contain the meaning arbitrarily given to a certain name (*Logik* §106), i.e. give a "conventional meaning" to a name (*R* 3006);

(II) are those whose concept sufficiently indicates all that one *subjectively* knows of a certain thing;[3]

(III) indicate only the logical essence (*logische Wesen*) of the concepts they define (*Logik* §106; *R* 3966).

Real Definitions: (I) like all definitions perform the primary function of 'giving a name' to concepts in order to distinguish them one from the other (*Logik Philippi*, p. 460);

(II) are capable of expressing *all* that *objectively* belongs to the defined concepts;

(III) contain the real essence (*Realwesen*) of the defined concepts and are sufficient for the knowledge of the object (*Object*) according to its internal determinations (*Bestimmungen*) insofar as they exhibit the possibility of the object (*Gegenstand*) on the basis of its internal characteristics (*Merkmale*) alone (*Logik* §106; *R* 3006).

Since nominal and real definitions in sense (I) amount essentially to the same thing, let us start directly with sense (II). Here Kant is clearly considering definitions from the viewpoint of the relation between the subject and the object of knowledge and therefore from the viewpoint of the *Materie* of the definienda. This means that a concept that is *given* to oneself can have a nominal definition, for it is sufficient that its definiens contain what one *subjectively* knows of the concept, i.e. its *form*. But since one cannot make explicit *all* that *objectively* belongs to a concept, unless one has actually *made* it, a real definition is possible only for a concept whose definiens was *originally* thought as belonging to it. In a case like this, however, a nominal definition will coincide with a real one.

It may be easily seen that here Kant distinguishes once again between definitions by appealing to the *origin* of the definienda: *a priori* (*RR* 2914, 2918, 2926, 2995) or *a posteriori* (*RR* 2934, 2936, 2945; *Logik Philippi*, p. 460) given concepts can be only nominally defined, while *a priori* made concepts can have real definitions.

In order to appreciate the distinction between nominal and real definitions in sense (III) above, we must recall some aspects of Kant's theory of concepts. This theory, in accordance with traditional logic, arranges concepts in a hierarchical order in which subordinated concepts are the extension (*Umfang*) of higher concepts and higher concepts are the intension (*Inhalt*) of subordinated concepts.[4] Consequently any higher concept is contained *in* the subordinated concepts (as in the case of 'metal' which is contained in 'gold', 'copper', 'silver' etc.) while it contains them *under* itself (*Logik* §7).

Now the logical essence, which is dealt with by nominal definitions in sense (III), corresponds to the *Inhalt* of a concept inasmuch as the latter contains all the necessary *Merkmale* (i.e. the characteristics that must always be found in the represented thing; *Logik*, Ak. IX, 60–61) of that concept. Therefore a nominal definition characterizes the definiendum in terms of the *essentialia* that are its intension. To do this, no intervention of (either *a priori* or *a posteriori*) intuition is necessary, for all that is in question in a nominal definition is the logical aspect of the concept. Thus, when Kant speaks of a logical meaning independent of intuition of *a priori* given concepts (categories) he has in mind their logical essence, which can be the object of a nominal definition. In this sense we can explain how definition and meaning are linked also from a purely logical point of view, thus clarifying the expression, 'logical meaning', mentioned in I.1.

A real definition in sense (III) aims instead at the *Realwesen* of a concept and has to explain its *possibility* on the basis of the internal *Merkmale* alone.

Now, to know the *Realwesen* of something means to know the predicates on which (as principles of determination) depends all that belongs to the existence of that thing.[5] Therefore a real definition is such as to enable one to decide, when confronted with any object, whether that object falls under the concept of the definiendum or not. Thus a real definition allows one to use the defined concept as an *Erkenntnisgrund* with respect to all concepts constituting its extension. Obviously empirical concepts cannot be so determined as to make them applicable (as principles of determination) to the *totality* of their extension, since one cannot determine completely and precisely the bounds of this extension. In short, one cannot state a rule that allows one to decide, for example, whether, for any possible *x*, '*x* is a metal' is true or false.

Would it then be right to conclude that only non-empirical concepts have real definitions? The answer is only partially yes. Kant in fact requires of real definitions that they show the *possibility* of the *definienda* with respect not only to their form but also to their *Materie*. This restricts the range of real definitions to *a priori* made concepts, which are the only ones whose possibility *der Materie nach* can be shown simply on the basis of their internal *Merkmale*, without any appeal to actual experience. This means that really defined concepts also have a kind of *existence* that can be established *a priori*. For, in every real definition the definiens contains a clear mark by which the "objective reality" of the definiendum is proved beyond doubt (A242, footnote). Thus a real definition (1) enables us to *recognize* all that belongs to the definiendum as the range of its application, i.e. as its *Umfang*; (2) actually *produces* the concept it defines. Clearly this production is not to be understood as the production of a particular instance of the concept, but rather, by involving only *pure* intuition, as the setting of a rule for producing (or recognizing) any of its instances.

All this provides conclusive evidence for the insufficiency of the origin as a criterion for judging the objective reference of the defined concepts. Luckily, though, this does not create any opposition between the analytic-synthetic and the nominal-real distinctions between definitions. On the contrary such distinctions are closely related, for nominal definitions indicate the logical essence of concepts only on the ground of an analysis of their *Inhalt*.[6] On the other hand, analytic definitions of given concepts are nominal in sense (II) and (III): in sense (II), because the analysis of either the coordinated or the subordinated *Merkmale*[7] of given concepts results in what one *subjectively* knows about them; in sense (III), because no analysis of *a priori* given concepts can lead to their *Realwesen*, while any analysis of *a posteriori* given concepts is aimed only at their logical *Inhalt* (if one analyses the *Merkmale* which are subordinated *a parte post*).

Similarly real definitions are possible only for *a priori* made concepts, i.e. for synthetically definable concepts. Notice that in this case the two distinctions between definitions actually reduce to nothing. An *a priori* made concept has not only a real but also a (complete and precise) nominal definition: for, in "arbitrary concepts" what is subjective and what is objective coincide (*R* 4016), and both synthesis and analysis of coordinated or subordinated (*a parte ante* as well as *a parte post*) *Merkmale* are complete and precise.[8] Conversely, a synthetic definition is possible only for those concepts whose subordinated and coordinated *Merkmale* can be exhaustively analyzed. In short the peculiar capability of *a priori* made concepts to have real and synthetic definitions enables them to satisfy our sense (∗)'s requirements for definitions even when they are analytically and nominally defined. For this reason synthetic and real definitions have a special status and, in certain contexts, seem to be the only ones to 'deserve' the name of definitions.

All this leaves no doubt that Kant's attention to the *Inhalt* and *Umfang* of concepts goes beyond a dutiful respect for traditional logic, since it stems from his interest in *meaning* considered as the objective reference of concepts (as is especially clear from real definitions in sense (III). It is in this perspective that real definitions become 'the ideal', as it were, which any definition tries to approach but only definitions of *a priori* made concepts can reach.[9]

I.3. BACK TO SCHEMATISM. THE SPECIAL STATUS OF
MATHEMATICAL DEFINITIONS

We have seen what Kant means by 'nominal definition' and by 'logical meaning' of concepts. Thus we are left with only one of the questions mentioned at the end of Section I.1, namely how concepts definable in sense (∗) differ from the other meaningful concepts with respect to the way they establish their own meaning.

To find an answer to this question, we shall briefly reconsider schematism from the viewpoint of real definitions. It is obvious that real definitions, involving both concepts and intuitions, are the result of a schematization. It is not so obvious, however, that they indicate the meaning of the definienda not only through, but also *in*, their schemata: a real definition, as we have seen above, corresponds to the setting of a rule (a schema) to which *all* instances of the definiendum must conform, rather than to the production of any of its instances.

As regards empirical and *a priori* concepts, we have already shown that their undefinability in sense (∗) depends on the example-like nature of the schemata of empirical concepts and on the inability of transcendental schemata to be reduced to particular images. Now we can say with more precision that what makes it impossible to define such concepts is the impossibility of defining them *in* their schemata. Conversely, completeness and precision of real definitions are a consequence of the schematic nature of their objective reference or, as Kant says, of the "schematic construction" (*Entdeckung*, Ak. VIII, 192 n.) made by the subject.

However, the priority of meaning over origin as a criterion for classifying definitions is no reason to underestimate the latter's importance, for both criteria are grounded in the same theory of concepts which provides their vocabulary as well as the evidence for their connection. We simply want to stress that, given this common background, the priority of the meaning-criterion is the true Kantian contribution to the theory of definition.

This is largely a result of Kant's ever present reflection on the nature of mathematics and, in particular, on mathematical definition. If references to mathematical definitions have been kept to a minimum in the preceding pages, this is not only because enough emphasis has been put on Kant's identification of sense (∗) with the mathematical sense of definition, but also in order to avoid some necessary distinctions at this stage within the sphere of mathematical definition (see Part III). Moreover all that really needs to be stressed at present is that mathematical definitions can be established as the model on which Kant's theory of synthetic and real definitions has been built up, so much so that the growing importance of the meaning-criterion can be fully appreciated through an analysis of the development of Kantian thought on mathematical definitions from some of his earlier writings to his major work. In Part II we shall briefly do just this.

II

II.1. EARLY CLASSIFICATION OF CONCEPTS AND DEFINITIONS. THE ORIGIN-CRITERION

The best evidence that the meaning-criterion is the *novum* of the *Critique* is the fact that in earlier writings, origin was Kant's sole criterion for classifying concepts and definitions. The so-called *Prize Essay*, published in 1764, begins in fact by stating that mathematics always forms and defines its concepts by means of an "arbitrary connection" of concepts (Ak. II, 276).

This statement has often been interpreted in two contradictory ways: either as a forerunner of Kant's critical doctrine of mathematical definition in A713 = B741 or as an *unicum* Kantian thought in showing a conventionalist-combinatory approach to definitions in mathematics.

Both interpretations miss the point. The *Prize Essay* is a forerunner of the *Critique*, but only insofar as it maintains that mathematical concepts are *made* concepts and that their definitions are, accordingly, synthetic. However, it does not contain anything similar to the critical idea that mathematical concepts must be *constructed*, in the sense that one should exhibit *a priori* the intuition corresponding to them (A713 = B741). The *Prize Essay* lacks a theory about the objective reference of concepts, i.e., it lacks precisely the meaning-criterion.

On the other hand the *Prize Essay* is not so different from the *Critique* as to justify its being considered as a unique concession to conventionalism in Kant's philosophical career. In the first place Kant's anti-conventionalism in mathematics is one of the very few unchanging features of his thought, whatever the stage of his philosophical development.[10] In the second place the term '*willkürlich*', as L.W. Beck rightly points out, does not mean 'random',[11] but rather 'independent from experience'. Finally, Kant speaks of an arbitrary connection of *concepts*, which means that some well-individuated concepts are *given* as the only ones an 'arbitrary' connection may start with. Kant goes so far as to mention these concepts: they are the concept "of quantity in general, of unity, of plurality, of space . . ." that must be assumed for *allgemeine Arithmetik, Zahlwissenschaft, Geometrie* (Ak. II, 279). Thus the arbitrariness of connection is restricted to a content-sphere, precisely determined by a set of given concepts and by a number of fundamental propositions (*Grundsätze*) expressing their general properties and regulating their use.[12]

However, it has been claimed that the conventionalist-combinatory interpretation is supported also by the *Prize Essay*'s mention of the capability, specific to mathematics, of representing concepts *under*, and not only through, signs (Ak. II, 278) and of operating with them. This is certainly so, but one should not overlook Kant's care in explaining that (1) between geometrical concepts and their representations there is a mutual relation of similarity (*Ähnlichkeit*; Ak. II, 292); (2) arithmetical (*Buchstabenrechnung's*) signs are sufficient for certainty (ibid.); (3) algebraic symbols must undergo an interpretation (Ak. II, 278). Moreover all these signs are "visible" (*sichtbare*, Ak. II, 279), i.e. sensible, signs carrying with them the immediate evidence of sensibility, especially when they are similar to the concepts they represent.

Now, apart from contradicting the conventionalist interpretation, does all this (the restriction of content and the connection with sensibility) amount to a meaning-criterion in disguise? I do not think so. The question of meaning is surely covertly present in the *Prize Essay*, but the latter does not contain (nor does it need) a proper meaning-criterion. On the one hand, the restrictions on the content of mathematical definitions play the role of formation-rules of mathematical concepts and, as such, constitute a criterion concerning their *origin*. On the other hand, the metaphysical background of the *Essay*, as we shall see, is sufficient to guarantee the meaningfulness of arbitrarily defined concepts. Therefore there is no need to specify rules for assigning a meaning to concepts in general and to mathematical concepts in particular. Thus, there being no *rules*, there is no *criterion* to speak of.

To better understand this situation it may be useful to look at another essay of the same period where, in a much wider context, Kant investigates the modal status of mathematical concepts; the *Beweisgrund*. There he claims that mathematical concepts belong to the sphere of possibility since they are neither 'necessary' in themselves, nor actually 'existing' (Ak. II, 75). It is easy to see that this agrees perfectly with the *Prize Essay*'s thesis that mathematical concepts are neither 'given' nor 'found', but are made by their definitions.

Kant's next step is to analyze some of these concepts, coming to the following conclusion: in all possible concepts one can distinguish *das Formale* (or *das Logische*) and *das Materiale* (or the *data*) of possibility (Ak. II, 77–78). On this ground he argues for the dependence of possibility on existence, maintaining that, should *all* that exists be taken away, nothing would be left, not even the *data* for possible concepts like the mathematical ones (Ak. II, 78).

Obviously Kant does not mean that, in order to make (or to define) a possible concept such as that of a right triangle, its *Materialia*, namely 'right angle' and 'triangle', should actually exist. He simply says that something must be supposed to exist and that, in the case of geometry, this is space (Ak. II, 80–81). Thus, while from the viewpoint of the *Prize Essay*, it is sufficient to consider space as a primitive, from the viewpoint of the *Beweisgrund* it is necessary to attribute to space a kind of existence.

Therefore the connection between mathematical concepts and existence can be traced back to the primitive terms of mathematics and to universal propositions about them. Thus we know that and where the question of meaning is at work at this stage of Kant's philosophy, but we cannot say that he has a full-fledged theory about it.

II.2. FROM THE ORIGIN-CRITERION TO THE MEANING-CRITERION

The priority of existence over possibility has, from a gnoseological point of view, quite a few consequences, the most interesting being that the way one comes to know what exists, i.e. experience, must have a privileged status among all *modi cognoscendi*.

Once this autonomous result of Kant's research on the theme of knowledge came into contact with the influence of Hume in the late sixties, a period of radical empiricism was almost bound to ensue.

I shall not try to give a reconstruction – detailed or otherwise – of this period. Nevertheless I believe that the next few remarks will at least hint at the reasons why this period marked the beginning of the emergence of meaning as a criterion.

The sudden elimination of metaphysics from Kant's philosophy, caused by his empiricist 'crisis', made it impossible to take for granted any longer the connection between mathematics and the existing world. And this applied to primitive and derivative terms alike. Kant then, in order to safeguard the objective reference, the meaningfulness, of mathematics did not hesitate to link it directly with existence by means of experience: mathematics renounced its apriority (*RR* 3750, 3738, 3923).[13]

What is most relevant here, apart from the extravagance of the solution with respect to the critical doctrines, is that the objective reference of mathematical concepts had gained such an importance as to make Kant (temporarily) overlook his own assertions about the non-empirical character of mathematics.

As is well known these – for Kant – extremist views were to be abandoned mostly because the sure possession of a meaning *via* induction was counterbalanced by the loss of universality of mathematical concepts as well as of mathematical-physical laws. Many intellectual "capsizings" (*Umkippungen*), recorded in the *Nachlass*, were still to be brought about by tentative solutions of this problem before the revolution that led to the *Critique*.[14] The importance of meaning as a test for all concepts, however, was to remain unquestionably established.

Surely the origin-criterion regained its value, but only after it had been deprived (at least with respect to mathematical concepts) of any 'combinatory' connotation it might have had in earlier times. This was possible because Kant had finally stated the *rules*, i.e. the criterion, for the meaning of concepts in the – unfortunately too short – *Schematismuskapitel*.

III

III.1. MATHEMATICAL DEFINITION FROM A SYSTEMATIC VIEWPOINT: SOME OPEN QUESTIONS

In Part I and Part II we have dealt with Kant's theory of mathematical definition in general. Now we shall discuss a suspicion of one-sidedness that can be raised about the theory itself: it is a fact that in most critical and pre-critical writings, Kant refers to definitions of geometrical non-primitive terms and acknowledges a special status for monogramatic schemata of geometrical concepts.

This suspicion of a geometrical bias is all the more perplexing if confronted with Kant's clear awareness of the systematic character of mathematics. For he conceives mathematics as a system with respect to (1) its subject-matter, and (2) its internal structure. From both these viewpoints no privileged position is assigned to geometry, nor *a fortiori* to definitions of its derivative terms.

We can substantiate Kant's systematic view of mathematics as regards its subject-matter with evidence provided by the notes of his lectures taken by Herder in the years 1762–1763. The two series of notes that constitute the mathematical part of Herder's *Nachschriften* agree in indicating quantity in general as the subject-matter of mathematics, and in distributing the matter as indicated in Table I.[15]

The slight differences between the two Series are easily overcome since the subdivision of the Mathesis universalis into Series I corresponds to that of Series II if one drops common calculus and all parts of analysis except algebra. Likewise the subdivision of the Mathesis specialis into Series I corresponds, in its 'pure' part, to the subdivision relevant to space in Series II, geometry being precisely the special part of mathematics dealing with pure space. All this shows that the different parts of mathematics are organically related to each other so that all Kantian assertions about mathematics, including those about definitions, should be meant to refer, unless otherwise stated, to the whole of mathematics.

As regards the systematic aspect of mathematics from the viewpoint of

TABLE I

	Series I	Series II

Mathesis Universalis	(1) Common Calculus
	(2) Higher Calculus
	  (A) Buchstabenrechnung
	      [algebra of integers]
	  (B) Analysis:
	      Algebra of equations
	      Calculus of infinity
		(A) Arithmetic ("durch
		      Zahlen") ['through
		      numbers']
		(B) Algebra ("durch andre
		      Zeichen") ['through other
		      signs']

Mathesis Specialis	(1) "Pura"
	  (A) pure geometry
	  (B) trigonometry
	(2) "Impura sive applicata"
	    ['impure or applied']
		(1) of space: geometry
		               trigonometry
		(2) of time
		(3) of force

its internal structure, Kant unquestionably assigned a *deductive* character to all mathematical disciplines, independently of their capability to have 'axioms'. Therefore he could not be unaware of some of the problems concerning deductive systems, such as: Are their primitive terms introduced by definition or otherwise? What is the relation between the primitive terms and the principles of such systems?

To find an answer to these questions, and in general to dispel the suspicions of one-sidedness about Kant's theory of mathematical definitions, we shall deal separately with each part of pure mathematics according to the above subject-matter division.

III.2. MATHESIS UNIVERSALIS

Mathesis universalis (*allgemeine Grössenlehre*) is the science of quantity (*Grösse, quantitas*) and of the operations that can be performed on quantity. Its primary aim is to provide a set of universal fundamental propositions valid for all quantitative sciences. It is noteworthy that Kant does not speak here of axioms or 'principles' (*Prinzipien; Logik* §36 Anmerkung 2), but only of *analytische Grundsätze* that, according to the examples he gives, correspond to Euclid's κοιναὶ ἔννοιαι [common notions].[16] Their analyticity makes them flexible with respect to interpretation: an analytic *Grundsatz*, such as 'the whole is greater than the part', is valid for Kant (i.e. originates a true proposition) whatever its *quantitative* interpretation because it does not require any preliminary 'translation', e.g. into arithmetical terms. Obviously, once interpreted, all such propositions cease to be

analytic and fundamental and become synthetic *a priori* propositions of
the specific branch of mathematics in whose terms they have been interpreted.

However, *allgemeine Grössenlehre* has another very important function:
it is also an *allgemeine Arithmetik* (Ak. II, 282) insofar as, according to
H. Nach, it contains algebra as one of its parts. In this sense it constitutes
the instrument of any arithmetical generalization (see section III.4) and,
by substituting *Zahlen* for its *Zeichen*,[17] becomes a synthetic *a priori*
science. This special relation with arithmetic can be best explained by
reference to definitions.

The first thing to be noticed in this respect is that the subject matter and
the only primitive term of algebra is (as in the case of *allgemeine Grössenlehre*) indeterminate quantity (Ak. II, 287). This is a mathematically
undefinable concept since, according to the *Prize Essay*, it is an unanalyzable concept to be assumed in mathematics without definition (Ak. II,
279); according to the *Critique* it is a category.

However, the concept of quantity is schematizable and hence meaningful, but, given its categorial nature, its schematization is a process which
is far from being straightforward. In fact quantity "seeks its support and
sense in number; and this in turn in the fingers, in the beads of the abacus
or in strokes and points which are placed before the eyes" (A240 = B299).
Thus two steps are involved here: from categorial quantity to concepts
of number (i.e. to concepts of quantity determinate with respect to unity);
from concepts of number to sensible intuition. Both steps entail the number-schema, "a representation which comprises the successive addition of
homogeneous units" (A142 = B182). But while the whole process, necessary to indicate the objective reference of quantity, does not define it, the
second step is sufficient to produce the determinateness of strict definition.
Of course, since this determinateness depends on the fact that the number-
schema does not apply directly (as a transcendental schema) to the category, but rather to concepts of number, the latter, and not *unbestimmte*
quantity, are capable of strict definition (see Section III.3).

As for algebraic operations, Kant does not define them but simply
mentions some "easy and certain rules" (Ak. II, 278) that regulate their
use. He does not state these rules clearly but refers to them as if they were
able to define algebraic operation-signs implicitly. Obviously such implicit definitions do not have sufficient precision and completeness and
therefore are not strict definitions. Furthermore the rules that constitute
these implicit definitions are analytic and can be assigned an objective

reference only through an interpretation. Indeed they require an arithmetical interpretation: they have to be given numerical values if they are to be found true and synthetic *a priori*.

This is why, insofar as it is an algebra, *allgemeine Grössenlehre* has a special relation with arithmetic, so much so that also a rule like *"Gleiches zu Gleichem addiert gibt gleiche Summe"*, which is proved by purely algebraic means (*H. Nach*, p. 23) and is independent of the definition of addition of specific numbers (cf. section III.4), cannot be synthetic unless it is assigned an arithmetical meaning by transforming references to indeterminate quantity into references to determinate quantities.[18]

Now, since insufficient determinateness is responsible for the undefinability in sense (∗) of quantity, we must ask whether arithmetic proper (*Zahlwissenschaft*), which is the science of quantity determinate with respect to unity, admits synthetic and real definitions and, if so, what are and how are the primitive terms introduced which enter into the definitions of numbers.

III.3. ARITHMETICAL DEFINITIONS. NUMBERS AND PRIMITIVE TERMS

The schematization of quantity considered above shows that number is a concept that belongs to the category of totality (B111) and synthesizes (relates) unity and plurality. Therefore numbers, being the result of a relation, i.e. of an intellectual act, are intellectual concepts: as Kant says, number is a concept "se quidem intellectualis" (*Inaugural Dissertation*; Ak. II, 397). But have we not seen that concepts definable in sense (∗) have monogramatic schemata and therefore are pure sensible concepts?

Let us recall Kant's requirements for definability as summarized in A729 = B757: "there remain, therefore, no concepts which allow of definition except only those which contain an [1] arbitrary [2] synthesis [3] that admits of *a priori* construction". Now concepts of number satisfy:

Requirement [1]: "no other purely arbitrary concepts of pure reason can rise in us but those [obtained] through repetition, consequently concepts of number and quantity" (*R* 3973);[19]

Requirement [2]: "our counting, as is easily seen in the case of larger numbers, is a synthesis according to concepts, because it takes place according to a common ground of unity as, for instance, the decade" (A78 = B104);

Requirement [3]: in any rational knowledge "only relations are to be considered, and these are either given [...] or made up [*gedichtet*]. But we can make up no other relation whose possibility we can be sure of, but those made up according to quantity through repetition in *Zahlwissenschaft*" (*R* 3940). Moreover "our reason contains nothing but *relationes*. Now, if these are not given through the relations according to space and time in experience, or through the repetition and the synthesis of many into one in pure mathematics, then these are not *relationes* that refer to objects but only *relationes* of our concept according to our reason" (*R* 3969).[20]

All this amounts to saying that the concepts of number are *a priori* made (*gedichtet, willkürlich*) and therefore satisfy the origin-criterion for synthetic definitions. At the same time the emphasis on 'repetition' and on 'synthesis of many into one', as the only relations (apart from those given in experience) which refer to objects, shows that the *Zahlwissenschaft's* concepts also satisfy the meaning-criterion for real definitions. In short, concepts of number may be constructed because their *Inhalt* as well as their *Umfang* can be completely and precisely determined by our schema of quantity (the number-schema), just as happens for pure sensible concepts, thanks to their monogramatic schemata.

This does not mean that there are no differences between concepts of number and pure sensible concepts. When it comes to definition, definitions of numbers, unlike geometrical definitions, must always be technically *genetic*, i.e. they must show the rule that generates them also in their form. This depends on the fact that the number-schema, unlike monogramatic schemata, is not only the same for all number-concepts, but also provides them with a meaning in one way only (A165 = B205−6), namely by considering them under the form of *one-dimensional* time.

It is worth noticing that Kant's early works already contain, in addition to a definition of number that echoes the traditional Euclidean one,[21] a definition founded on the repetition of unity. But, while in the pre-critical period the notion of repetition seems to depend entirely on the notion of measure,[22] in the critical period the notion of repetition becomes the element that – within schematism – determines the reality of the definitions of numbers using, rather than depending on, the number-measure relation. This supports our thesis that the critical *novum* consists in the emphasis put on the reality of mathematical definitions which partly overshadows the attention paid to the origin of the definienda.

As for the question of primitive terms, it arises as soon as one realizes that, although the definitions of numbers are definitions of *a priori* made concepts, nowhere does Kant say that such concepts are made *ex nihilo*.

On the contrary, concepts of number originate from a synthesis of unity and plurality and so their definitions presuppose unity (not to be confused with the number one) and plurality.

Again, if we look at Kant's pre-critical theories we find that the *Prize Essay* actually indicates unity and plurality (*Einheit* and *Menge*) as the unanalyzable (*unanflösliche*) concepts of arithmetic (Ak. II, 279–80), while *H. Nach* (p. 21), referring to unity, says that in mathematics it must be considered as a "bekandter Begrif".

In the *Critique*, too, these concepts are simply 'assumed', but now for the reason that they are categorial, i.e. *a priori* given, concepts. As such they either have a nominal (and for mathematics useless) definition, or are founded in *a priori* intuition through schematism. In the latter case, since their schema is transcendental, they cannot have real and synthetic definitions, but are nevertheless quantitatively meaningful concepts and therefore can be used in mathematics.

III.4. ARITHMETICAL OPERATIONS. THE QUESTION OF AXIOMS

Zahlwissenschaft is not only the science of numbers, but also the science of the operations that can be performed on numbers. Therefore we must also investigate whether Kant believes them to be strictly definable. Certainly he does not identify such operations with those of the *allgemeine Grössenlehre* since they have no implicit definitions. In fact in *H. Nach* (p. 22) "addiren" is explicitly qualified as the operation that "finds one number (the sum), which is equal to different numbers (of the same kind but *not* of the same magnitude) taken together".[23] Thus arithmetical operations do refer to something precise that, in the case of addition, is the taking together, the synthesis, of different numbers into one. By the way, a synthesis is involved not only in addition but also in subtraction (Ak. XI, 555; *Zweig*, 129) and presumably in all four fundamental operations of arithmetic.

But does this objective reference of arithmetical operations make them strictly definable? Let us consider for instance the case of addition. We may notice that (1) it is an operation schematized by the number-schema since it refers to the counting of units; (2) like the concepts of number, it can determine its *Umfang a priori*, completely and precisely through construction. Therefore the definition of addition as the taking together of units does satisfy the requirements for strict definability. This can be gathered also from the fact that this operation is subject to the restrictions imposed on numbers by their origin and meaning. For any addition of numbers can take place *in one way only* and no addition of an actually infinite number of addenda (as well as no addition of addenda of infinite magnitude) is possible because it would have no objective reference.

We may generalize this conclusion by saying that the four arithmetical operations are strictly definable. But, for this very reason, they also have a narrow range of application and no universality apart from the universality of use. We must remember in fact that an arithmetical operation for Kant is only an operation performed on *specific* numbers. As he so forcefully maintains, 7 + 5 = 12 is a synthetic *a priori* proposition, but is also "singular" and a "number formula" (A164 = B204).

This brings us directly to the question why, for Kant, "arithmetic has no axioms" (Ak. XI, 555; *Zweig*, p. 129; A164 = B204). In order to understand the true meaning of this statement we have to examine it in the light of both the relation between *allgemeine Arithmetik* (algebra) and arithmetic, and the influence of intuition on the foundations of arithmetic.

Above all the algebra-arithmetic relation is, as we have mentioned before, a relation of interpretation of the first in terms of the second. It is by interpretation that algebra acquires an arithmetical meaning. In particular arithmetical operations provide a meaning for algebraic operation-signs. But the relation between algebra and arithmetic is a *mutual* relation: on the one hand arithmetical operations give a meaning to the algebraic ones, on the other hand the rules that constitute the implicit definitions of algebraic operations can also be applied to the arithmetical operations. Indeed arithmetic, insofar as it belongs to the general system of mathematics, presupposes an *allgemeine Arithmetik*. In *H. Nach* (p. 27) Kant gives a proof of 8 + 4 = 12 which makes essential use of the algebraic rule "*Gleiches zu Gleichem addiert gibt gleiche Summe*". Algebra therefore is important for the enlargment of arithmetical knowledge: as Kant says, "algebra is an ampliative science" (Ak. XI, 554; *Zweig*, p. 129).

That the above-mentioned proof of 8 + 4 = 12 is to be found in a pre-critical text does not detract from the role played by algebra in arithmetic, also in the later critical theory. This is something to be stressed because some of the *Critique's* remarks on 7 + 5 = 12 have misled many of Kant's readers into believing not only that all such operations require is some counting on one's fingers, but also that either Kant did not have a theory about operations on larger numbers or he considered small numbers (and operations on them) to be very different from larger numbers (and operations on them).

Now Kant has never expressed himself so ambiguously as to justify such misinterpretations. For him there are no differences between small and larger numbers, so much so that, in order to explain the true nature of the number-schema, he refers to the case of larger numbers (A140 = B179). Furthermore, in contrast to his critic, Eberhard, who maintained that the concept of a chiliagon is an idea of reason because it could hardly be immediately represented in intuition, Kant argued that the *number* of sides of a geometrical figure is irrelevant as far as its mathematical nature is concerned "for [...] the construction of the object can be completely prescribed" (Ak. XI, 47; *Zweig*, p. 149).

But if Kant does not distinguish between numbers according to their magnitude, neither does he distinguish between operations to be performed on numbers, whatever their magnitude. This explains how algebra can be the common background for all arithmetical operations. Let us consider for instance an equality like $1000 + 2000 + 1 = 1000 + 2001$. This would appear to be established as a correct equality only by calculating each of its members and then appealing to the following analytic *Grundsatz* of *allgemeine Grössenlehre*: things equal to a third thing are equal to each other. However, the algebraic rule "*Gleiches zu Gleichem addiert gibt gleiche Summe*" allows us to establish the above equality, independently of the computation of its members, by considering it as one of its instances. The possibility of this indirect proof, naturally enough, shows all its importance where larger numbers are involved because it eliminates the difficulty of practical computation.[24]

This does not mean that *Zahlwissenschaft* does have axioms (synthetic *a priori* universal propositions) after all. For as soon as an arithmetical meaning is given to propositions such as "*Gleiches zu Gleichem addiert gibt gleiche Summe*", they cease to be both analytic and universal just like analytic *Grundsätze*. Nevertheless their very existence provides sufficient evidence for a connection between algebra and arithmetic. Such a connection prevents one from attributing to Kant a conception of arithmetic as consisting only of infinitely many, mutually independent, singular number-formulae.[25] On the contrary, to paraphrase a famous Kantian dictum: algebra without arithmetic is empty, arithmetic without algebra is blind.

Finally, the fact that arithmetic has no axioms makes it clear that Kant never considered it to be the science of time in the same sense in which geometry is the science of space: time intuition has no direct influence on the properties of numbers (Ak. XI, 556; *Zweig*, p. 130) and "the objects of arithmetic and algebra, according to their possibility, are not subject to temporal conditions" (*R* 13). This is perfectly consistent with the Transcendental Aesthetic where axioms of space correspond to geometrical axioms, but axioms of time[26] have nothing to do with arithmetic. Thus the lack of a special object of arithmetic and of axioms about such an object entails, from a structural point of view, what we have seen to be typical of arithmetical definitions from a gnoseological point of view, i.e. the necessity for them to be genetic with respect to their form as well as their content. In geometry on the other hand, as we shall presently see, even non-genetic definitions are acceptable because axioms are sufficient to guarantee the reality of geometrical concepts.

III.5. DEFINITIONS IN GEOMETRY

Geometry is a more specific science than arithmetic because it is the part of pure mathematics that has a special object, namely space and spatial relations (Ak. II, 397, 403; A163 = B204). Now space cannot be mathematically defined: in the *Prize Essay* Kant calls it an *unanflöslich Begriff* to be assumed in mathematics simply according to its common representation (Ak. II, 278); in the *Critique* the concept of *geometrical* space is said to require no philosophical legitimization (A87 = B120). Furthermore, consistent with Kant's general thesis, that *a priori* given concepts can have an 'exposition' rather than a definition, the Transcendental Aesthetic gives an exposition but does not define space. Naturally there are no doubts about space's objective validity and meaningfulness (*objective Gültigkeit* and *Sinn und Bedeutung*) because one can establish its necessary application to the objects of experience (A156 = B195). Therefore, being a mathematically undefinable but meaningful concept, space is the true geometrical *primum*.[27]

However, traditionally, geometrical primitives are identified with notions like 'point', 'line' etc. How does Kant deal with them? The answer to this question emphasizes the difference between quantity and space, and consequently between (universal or particular) arithmetic and geometry. In fact space, inasmuch as it is the object of geometry, "contains more" than the simple form of intuition because it contains the *formal* intuition (*formale Anschauung*) which provides the unity of representation (B161, Note). In other words the object of geometry is not so much the form of our sensibility as the organization of spatial patterns according to such form.

This implies that, while one may properly speak of an indeterminate quantity, one could not speak of an *empty* space. For concepts (and categories) depend on the subject's activity in order to acquire their form, while no intuitive (*a priori* or *a posteriori*) representation is originally given) to the subject independently of its form.[28] Thus, on the one hand, indeterminate space cannot be represented at all, on the other hand, space is necessarily presupposed by all its determinations. In Kant's own words this means that "space should be properly called not compositum but totum, since its parts are possible only in the whole, not the whole through the parts. [. . .] If I remove all compositeness from it nothing remains, not even the point. For a point is possible only as the limit of a space" (A438 = B466–68).[29]

Thus Kant avoids the usual negative definitions of traditional geometrical primitive terms (a negative definition "is not a real definition", *Logik*, §106 *n*, *R* 2916). At the same time he obeys the Aristotelian imperative that one should prove the existence of terms introduced in geometry. For Kant proves the existence of non-primitive terms by means of real synthetic, i.e. constructive, definitions and assumes primitive terms together with their existence. He does so (1) by arbitrarily choosing on which simple parts to build up geometry, since simple spatial parts are not given *per se*; (2) by indicating ostensibly (and so showing the existence of) the chosen parts.[30]

That geometrical derivative terms for Kant have synthetic and real definitions is a well-established fact that needs no further explanation. Rather, it is interesting to consider the question of axioms with respect to definitions.

To start with we may notice that for Kant geometrical axioms "express the conditions of sensible *a priori* intuition" (A163 = B204) and "can be represented in intuition" (*Logik* §35). Furthermore axioms are immediately certain, synthetic *a priori* universal principles (A732 = B760; A299 = B356). Now even a superficial examination of Kant's examples of axioms makes it clear that they correspond to *some* of Euclid's *postulates*. For instance the Kantian axiom "two straight lines do not enclose a space" (A163 = B204) has been proved since Proclus' times to be reducible to Euclid's first postulate.[31] Clearly then Kant considers as axioms (i.e. as *Principia* and not merely as *Grundsätze*) those Euclidean postulates that express general properties of geometrical space and retain the classical function of extra-logical restrictions on the arbitrariness of geometrical definitions. The fact that some Euclidean postulates are not among

Kant's examples of axioms can be explained on the grounds that their originary function as existential assumptions on non-primitive terms is made redundant in the *Critique* by the schematization involved in the definition of concepts. This is evident in the case of Euclid's third postulate.[32]

Now, granted that the impossibility of giving a real definition of a figure enclosed between two straight lines (A220 = B268) can be established simply by acknowledging that it is impossible to construct it (and therefore not necessarily by appealing to the above axiom), the following question could be raised: Could geometrical axioms be dispensed with? The answer is plainly No.

First of all, Kant is perfectly aware that science is not so much concerned with the way its objects are obtained as with the properties of these objects (Ak. XI, 40–48; *Zweig*, pp. 145–6). But the properties of geometrical objects are established by proving theorems about them, and in turn proofs appeal to the general properties of space expressed by the axioms. Secondly, the above question seems analogous to that about the dispensability – from an allegedly 'Kantian' point of view – of propositions like "*Gleiches zu Gleichem addiert gibt gleiche Summe*", granted that numerical propositions, e.g. $1000 + 2000 + 1 = 1000 + 2001$, can and should be calculated independently of it. But, similar to what has already been pointed out for arithmetic, universal propositions (and *a fortiori* geometrical axioms) guarantee the constructibility and provability of properties of mathematical concepts that are highly complex and difficult to construct practically. This means that whoever poses the above question is concerned more with the explicit formulation than with the assumption of a set of axioms, for it is evident that he tacitly assumes them in all the 'easy' cases of construction.

What is more, the question reflects a viewpoint which is so little 'Kantian' as to overlook the fact that Kant, although convinced of the Euclidean character of man's intuition (which, because of its uniqueness, would justify the lack of an explicit formulation of the axioms), could certainly accomodate a different geometry within his philosophical framework. Such a geometry (or geometries) would have the nature of a 'game' (A239 = B298; A175 = B196) and would be incapable of application to reality, but still would not be absurd. This cliam is indirectly supported by the fact that arithmetic, which has no axioms, does not allow anything of that kind: there is just *one* arithmetic.[33]

This structural difference between geometry and arithmetic is responsible for the fact that arithmetical propositions are universal only with respect to their use. In fact axioms in general not only express extralogical restrictions on the *Willkürlichkeit* of definitions, but also guarantee the universality of the definienda. Therefore, having no axioms, arithmetic, on the one hand, has to show in each case the possibility of its objects through a technically genetic definition; on the other hand, it cannot state *a priori* the conditions of their intrinsic universality.

Geometry is a different case. As mentioned before, for Kant there is no need to define geometrical concepts in a strictly genetic form: "If a circle is defined as a curve all of whose points are equidistant from a midpoint, is not this concept given in intuition? And this even though the practical proposition that follows, viz., *to describe a circle* (as a straight line is rotated uniformly about a point), is not at all considered".[34] Obviously in the case of complex geometrical concepts a strictly genetic form of definition is used, but this means only that in defining such concepts one follows more closely their schemata. Indeed genetic definitions of complex figures are a matter of practical utility rather than a necessity. What is more, all geometrical definitions, unlike arithmetical definitions, provide their definienda with an intrinsic universality; for instance, once one triangle is given by means of a definition, so are all triangles.[35]

In spite of these differences between geometry and arithmetic, geometry is a science only insofar as it is a part of mathematics, i.e. insofar as spatial figures undergo a quantitative determination. Therefore not only the analytic *Grundsätze* of *allgemeine Grössenlehre*, but also the rules and operations of particular arithmetic, apply to geometry. This is what Kant means when he says that "*Zahlwissenschaft* [. . .] *ist* [. . .] *das Instrument der ganzen Mathematik*" (*H. Nach*, p. 17). Geometry's peculiarity consists in its being a quantitative science whose objects are considered "together with their quality" (A720 = B748). The schemata of geometrical figures are in fact monograms, i.e. 'figurative' schemata.

According to Kant, though, the relation between geometry and arithmetic is neither a mere matter of mathematical architectonic, i.e. of an 'external' connection represented by their sharing the language of numbers, nor does it go in one direction only (i.e. from arithmetic to geometry). In a letter to A.W. Rehberg (Ak. XI, 207–10; *Zweig*, pp. 166–169) Kant emphasizes once again the "necessary synthesis [*Zusammenhang*] of inner sense with outer sense" whenever *quantification* is involved: "time [. . .] has to be imagined as a line, if it is to be quantified, just as [. . .]

a line can be quantified only by being constructed in time". But what is really noteworthy in this letter is that Kant assumes a connection between arithmetic and geometry stricter than the general *Zusammenhang* of spatial and temporal intuitions. For a quantity like $\sqrt{2}$, although not representable as a "complete numerical concept (the relationship of $\sqrt{2}$ to unity)", can have a geometrical representation in the diagonal of a square of side 1.

Thus the 'image' of $\sqrt{2}$ is not simply an (*a priori* or *a posteriori*) *spatial* image, as is the case for 'complete numerical concepts', but is a certain *geometrical* construction. Hence, on the one hand, we may associate the diagonal of the square to $\sqrt{2}$ only because the diagonal is considered *sub specie quantitatis* as a mathematical object (see *R* 13); on the other hand it is the association of the diagonal to $\sqrt{2}$ that makes the latter strictly definable. Without this association $\sqrt{2}$ would still be mathematically meaningful, but not strictly definable since its objective reference would consist in a non-terminating (hence never *completed*) calculation.

IV

CONCLUDING REMARKS

Three general remarks can be added as a conclusion.

(1) Kant's theory of definition agrees with his general view that truth is neither the result of an immediate mind-world *adaequatio* nor the product of a convention. First, this explains why so much care is taken by Kant in stating the criteria for the meaningfulness of concepts: all significant concepts must have an object reference, but this reference has to be mediated by schemata. Now, while all concepts that do not have such a reference are not significant (which does not mean that they are absurd), only mathematical concepts whose objective reference is exhibited *a priori in* their schemata have a strictly defined meaning. Secondly, this also explains Kant's refusal to reduce mathematics to logic: all mathematical concepts must show their objective reference, albeit without any intervention of experience.

The construction of mathematical concepts satisfies these requirements. In this sense intuition in mathematics is not a synonym for 'visualization': the completeness and precision of meaning, which are obtained by construction, are attained by all concepts whose *Umfang* is given in its *totality* by their schemata. This means that constructibility is a property not only of concepts that are visualizable by the 'mind's eye', but also of those mathematical concepts whose complexity and abstraction could hardly allow any visualization.

(2) The second remark concerns the attention that has been paid in this paper to Kant's theory of concepts as it is expounded in his logical *Vorlesungen*. I believe that this theory can serve not only as a terminological aid, but also as a powerful conceptual tool for deeper insight into Kant's thought in general.

For example, by bringing forward this paper's approach to the theory of mathematical definitions through Kant's theory of concepts, one could reconsider the old theme of synthetic *a priori* judgments from a new perspective. For, granted that every definition is a judgment, every strict mathematical definition is a synthetic *a priori* judgment. Twenty years ago L.W. Beck already moved along these lines, but it seems to me that more attention should be paid to the *Inhalt* and *Umfang* as being responsible for the analyticity and syntheticity of concepts and judgments. In this way an investigation into the theory of concepts would not be a substitute for but a complement to the study of schematism since, in considering the extension and intension of concepts, one not only goes beyond their hierarchical ordering but also looks at the theory (schematism) concerning the way their intension and extension could be known. And this is of great importance for a philosophy that hinges essentially on an epistemology.

(3) The third remark is more pertinent to mathematics. We have seen how the theory of mathematical definitions applies to the different parts and levels of mathematics. This has brought to light how important Kant's assessment of mathematics as a system is for his conception of mathematical knowledge. In fact it constitutes both the starting-point for his reflections on the entire subject and the testing-ground for the results of such reflections. Furthermore all this should put the elementarity of Kant's mathematical examples into the right perspective, since this has given way to speculations about the limits of his actual knowledge of mathematics. As we have seen, the consideration of mathematics as a system emphasizes the quantitative, i.e. conceptual, aspect of all its parts. Therefore, on the one hand, the elementarity of the examples is not the consequence of an alleged necessity to visualize mathematical concepts; on the other hand, as we have seen, non-elementary parts of mathematics (e.g. what at the time was considered to be algebra) and classes of numbers other than natural numbers can be accomodated within Kant's general conception of mathematics.

NOTES

List of abbreviations of references used in this essay.

Ak. (followed by the number of the volume in Roman numerals and by the indication of the page) stands for *Kants gesammelte Schriften*, under the editorship of the Königliche Preussische Akademie der Wissenschaften, Berlin; from vol. 23, under the editorship of the Deutsche Akademie der Wissenschaften, Berlin; from vol. 24, under the editorship of the Akademie der Wissenschaften, Göttingen, (G. Reimer), W. de Gruyter, Berlin, 1902–.

A, B, as usual, indicate the two major editions of the *Kritik der reinen Vernunft*. Quotations are from N. Kemp Smith's English translation (*Critique of Pure Reason*. London, Macmillan, 1968) with minor changes.

Beweisgrund: Der einzig mögliche Beweisgrund zu einer Demonstration des Daseins Gottes, (1963), Ak. II, 63–164.

Entdeckung: Über eine Entdeckung nach der alle neue Kritik der reinen Vernunft durch eine ältere entbehrlich gemacht werden soll, (1790), Ak. VIII, pp. 185–251.

Gedanken: Gedanken von der wahren Schätzung der lebendigen Kräfte und Beurtheilung der Beweise, deren sich Herr von Leibniz und andere Mechaniker in dieser Streitsache bedient haben, nebst einigen vorhergehenden Betrachtungen, welche die Kraft der Körper überhaupt betreffen, (1747), Ak. I, 1–182.

H. Nach: I. Kant, *Aus den Vorlesungen der Jahre 1762 bis 1764. Auf Grund der Nachschriften J.G. Herders*, edited by H.D. Irmscher, *Kant-Studien*, Ergänzungshefte **88**, Cologne, Kölner Universitäts-Verlag, 1964.

Inaugural Dissertation: De mundi sensibilis atque intelligibilis forma et principiis, (1770), Ak. II, 385–420.

Logik: Logik. Ein Handbuch zu Vorlesungen, (1800), edited by G.B. Jäsche, Ak. IX, 1–150.

Logik Philippi: Kants Vorlesungen: Vorlesungen über Logik, Ak. XXIV, pp. 303–496.

Naturgeschichte: Allgemeine Naturgeschichte und Theorie des Himmels oder Versuch von der Verfassung und dem mechanischen Ursprunge des ganzen Weltgebäudes, nach Newtonischen Grundsätzen abgehandelt, (1755), Ak. I, 215–368.

Prize Essay: Untersuchung über die Deutlichkeit der Grundsätze der natürlichen Theologie und der Moral, (1764), Ak. II, 273–302.

Prol.: Prolegomena zu einer jeden künftigen Metaphysik, die als Wissenschaft wird auftreten können, (1783), Ak. IV, 253–383.

R (RR): Reflexion(en) followed by a number relative to their progressive ordering in Kant's *Handschriftlicher Nachlass*, Ak. XIV-XVIII.

Zweig: A. Zweig's Engl. transl. of Kant's *Briefwechsel* (Ak. X-XII), *Kant. Philosophical Correspondence 1759–99*, Chicago, University of Chicago Press, 1967.

[1] See *Kritik der Urtheilskraft*, Ak. V, §59: the intuitions needed to prove the reality of concepts "in the case of empirical concepts [...] are called examples, in that of pure concepts of understanding [...] are called schemata". Cf. also A138 = B177. On this question cf. W.H. Walsh, 'Schematism', *Kant-Studien* **49** (1957), reprinted in R.P. Wolff (ed.), *Kant. A Collection of Critical Essays*. London, Macmillan, 1968, pp. 75–6.

[2] Cf. H.J. Paton, *Kant's Metaphysics of Experience*. 2 vol. London, Macmillan, 1951, I, p. 36.

[3] Cf. *Logik Philippi*, p. 456: definitions are nominal "wenn der Begriff zureichend ausdrückt alles was ich von der Sache erkenne".

[4] The notions of *Umfang* and *Inhalt* are obviously connected with the notions of *étendue* and *compréhension* of Port Royal Logic.

[5] *Logik*, Ak. IX, 61: "Zum Real-Wesen des Dinges (esse rei) die Erkenntniss derjenigen Prädicate erfordert wird, von denen alles, was zu seinem Dasein gehört, als Bestimmungsgründe, abhängt". See also *R* 2321.

[6] Cf. *RR* 2994, 2995.

[7] Cf. *Logik*, Ak. IX, p. 59.

[8] For a very thorough exposition of this point see J. Proust, 'Analyse et définition chez Kant', *Kant-Studien* **66** (1975), especially pp. 17–27.

[9] This privilege of real definitions is clear from the fact that in the case of empirical concepts even an incomplete and imprecise real definition is much more useful (and used) than an equally incomplete and imprecise nominal definition. In the case of categories, although one could try to give them nominal and analytic definitions (A244–5), what is important is the procedure of schematization which shows the way categories refer to objects, and hence shows their reality. One might say that the schematization of categories is a fruitful but only tentative 'real definition' of them (A731 = B759 Footnote).

[10] Already in his earliest writings Kant stressed the 'hypothetical' (i.e. axiomatic) nature of any mathematical system (cf. for example *Gedanken*, Ak. I, 139–40). But he did not let this interfere with his effort to give mathematics an objective foundation, although at first he supported this effort by metaphysics.

[11] L.W. Beck, 'Kant's Theory of Definition', *Philosophical Review* **65** (1956), reprinted in R.P. Wolff (ed.), *Kant. A Collection of Critical Essays, op. cit.*, p. 31.

[12] These fundamental propositions, which are "*augenscheinliche Sätze*", correspond, judging from Kant's examples, to both critical *analytische Grundsätze* and axioms (Ak. II, 281: "Das Ganze ist alle Theilen zusammen genommen gleich; zwischen zewi Punkten kann nur eine gerade Linie sein). Probably this depends on the fact that Kant is referring here to mathematics in general.

[13] *R* 3970: "Alle *principia primitiva* sind entweder elementarsätze und analytisch oder *axiomata* und sind synthetisch. Die rationale sind analytisch, die empirische synthetisch, imgleichen mathematische"; *R* 3738: "Alle analytische Urtheile sind rational und umgekehrt. Alle synthetische Urtheile sind empirisch und umgekehrt . . ."; *R* 3923: "Einige Grundsätze sind analytisch und betreffen das formale der Deutlichkeit in unserer Erkenntnis. Einige sind synthetische und betreffen das materiale, als da find die arithmetischen, geometrischen und chronologischen imgleichen die empirischen".

[14] The turning-point in this transition from extreme empiricism to criticism can be seen in *R* 3932: "Die ideen und regeln der Vernunft werden auch in dem Verhältnisse empirischer Begriffe gebrauchet, und dieses ist ihr natürlicher und richtiger Gebrauch; sie sind alsden aber auf *iudicia empirica primitiva* gegründet, die nur durch induction allegemein sind. Aber eben diese Urtheile der Vernunft, so fern sie rein sind, sollen an sich allgemein seyn". Then Kant asked the fundamental question: (*R* 3926) "Wie werden empirische und synthetische [Urtheile] allgemein?".

[15] See *H. Nach*, pp. 17–18; 29–30. This division of mathematics was common to the major mathematical textbooks of the time, such as C. Wolff, *Auszug aus den Anfangsgründen aller mathematischen Wissenschaften*, Halle 1717 which Kant used for his lectures.

[16] Kant's examples of analytic *Grundsätze* are:

(1) "2 Grössen einer 3. gleich sind gleich sich selbst" (*H. Nach*, p. 23; Ak. XI, 556; *Zweig*, p. 130);

(2) "Gleiches, zu Gleichem hinzugethan oder von diesem abgezogen, in Gleiches gebe" (A164 = B204);

(3) "$a = a$, das Ganze ist sich selber gleich" (B17; *R* 4634; *H. Nach*, p. 23; *Prol.* §2);

(4) "$(a + b) > a$, d.i. das Ganze ist grösser als sein Theil" (B17; *Prol.* §2; Ak. VIII, 196);

(5) "Das Ganze ist allen Theilen zusammen genommen gleich" (Ak. II, 281). Here *Grundsätz* (3) is a special case of Euclid's common notion (4); *Grundsätz* (5) corresponds to an axiom introduced by Clavius. All the other *Grundsätze* correspond exactly to the Euclidean common notions (cf. T. L. Heath's Commentary to his translation of Euclid, *The Thirteen Books of the Elements*, New York, Dover Publications, 1956², vol. I, pp. 221–232).

[17] See Ak. XI, 40–8; *Zweig*, p. 145: "The mathematician cannot make the smallest assertion about any object whatsoever without exhibiting it (or, if we are considering only quantities without qualities, as in algebra, exhibiting the quantitative relationships for which the symbols stand) in intuition". Cf. also *R* 13.

[18] "*Gleiches zu Gleichem addiert gibt gleiche Summe*" is a proposition belonging to algebra not to be confused with the analytic *Grundsätz* of the *allgemeine Grössenlehre* listed as (2) in note 16 above. In algebra "addiren" refers to an operation to be performed on numbers, although not on determinate numbers. The "hinzutun" of the analytic *Grundsätz* (2) refers instead to an operation that could be performed, for example, also on spatial (or geometrical) entities and therefore not necessarily on numbers. S. Stenlund has recently maintained that, if the operation of addition is for Kant a non-defined primitive and its properties are exposed as axioms (that define it implicitly), then formulae like 7 + 5 = 12 are synthetic, where 'synthetic' means 'non-decidable from definitions alone' (see S. Stenlund, 'Analytic and Synthetic Arithmetical Statements', in S. Stenlund (ed.), *Logical Theory and Semantic Analysis*, Dordrecht, Reidel, 1974, pp. 199–211). Now, although Stenlund is right in saying that for Kant addition has an implicit definition, he does not distinguish between algebraic and arithmetical addition. Indeed he seems to have in mind only addition of specific numbers, as in 7 + 5 = 12, which for Kant has an explicit definition (cf. III.4).

[19] The German text is: (*R* 3973) "Keine andere blos willkürliche Begriffe der reinen Vernunft können in uns entstehen, als die durch die Wiederholung, folglich der Zahlen und Grösse".

[20] Original text: (*R* 3940) "In allem Erkenntnisse der Vernunft sind nur Verhältnisse zu betrachten, und diese sind entweder gegebene [. . .] oder gedichtet. Dichten aber können wir keine Verhältnisse, (von) deren Möghlichkeit wir überzeugt seyn können, als den Grösse nach durch Wiederholung in der Zahlwissenschaft"; (*R* 3969) "Unsere Vernunft enthält nichts als *relationes*. Wenn nun diese nicht gegeben sind durch die Verhältnisse nach Raum und Zeit in der Erfahrung, auch nicht durch die Wiederholung

und die Zusammensetzung des Eins aus vielen bey der reinen Mathematik, so sind sie keine *relationes*, welche auf objecte gehen, sondern nur Verhältnisse unserer Begriff nach Gesetzen unserer Vernunft".

[21] Se *H. Nach*, p. 21: "Ein Ganzes, aus vielen Dingen von einerlei Art, deutlich ausgedrückt, ist Zahl".

[22] For the pre-critical connection between number, repetition and measure see *H. Nach*, p. 17: "das Maas der Grösse (Vielheit) ist die Einheit, oder Eins"; "wenn ich deutlich ausdrücke, wieviel in einer Sache das Eins enthalten, so ist das eine Zahl".

23 Here the term 'addiren' has a definite meaning and, in this sense, constitutes the arithmetical interpretation of the term 'hinzutun' used in algebra.

[24] My argument is related to that of D. Prawitz in *Meaning and Proofs: on the Discussion of the Conflict between Classical and Intuitionistic Viewpoints*, Institute of Philosophy, University of Oslo, 1976, pp. 24 ff. Prawitz gives very significant examples involving, for instance, the commutativity law of addition, while in Kant we find only less relevant uses of algebraic laws, such as that quoted in the text. Nevertheless it is interesting to note that a former student of Kant, J. Schultz, while remaining faithful to Kantian philosophy, provided an axiomatization of algebra in *Prüfung der Kantischen Kritik der reinen Vernunft*, Königsberg 1789–1792, and in other writings on mathematics. His axioms include the commutativity law of addition. G. Martin, who first become interested in Schultz's works, in *Arithmetik und Kombinatorik bei Kant* (Itzehoe 1938), Berlin, De Gruyter, 1972, goes so far as to maintain that Schultz was directly influenced by Kant on these questions. But, whether Martin is right or wrong, his thesis is unnecessary for the soundness of my argument.

[25] See G. Knauss, 'Extensional and Intensional Interpretation of synthetic propositions *a priori*', in L.W. Beck (ed.), *Proceedings of the Third International Kant Congress*, Dordrecht, Reidel, 1972, p. 358: "the number-formulae form an infinite set of individuals which cannot be deduced from each other".

[26] Cf. A31 = B47: time "has only one dimension; different times are not simultaneous but successive [. . .]". In the *Inaugural Dissertation* (Ak. II, 397) space is the subject-matter of geometry and time is the subject-matter of pure mechanics, not of arithmetic. Cf. also *Prol.* §10.

[27] In the *Prize Essay* (Ak. II, 277) Kant mentions, in addition to space, another mathematically undefinable concept, namely similarity (*Ähnlichkeit*). Indeed he insists that one should not follow C. Wolff's attempt to provide a mathematical definition for it. Now, while E.W. Beth (*The Foundations of Mathematics*, North-Holland, Amsterdam, 1959, p. 45) believes that Kant misses a good chance here to do something 'salutary' for mathematics, G. Martin in *I. Kant. Ontologie und Wissenschaftstheorie* (Berlin, De Gruyter, 1969, p. 309) seems convinced that Kant considers similarity to be implicitly defined by its properties. However, since Kant does not expound such properties, it seems more reasonable to say that he had in mind simply a 'common notion' of similarity and did not consider it to be either explicitly or implicitly definable.

[28] On this question see K. D. Wilson, 'Kant on Intuition', *The Philosophical Quarterly* **25**, 252ff. (1975).

[29] Cf. A25 = B39: the parts of space "cannot precede the one all-embracing space as being, as it were, constituents out of which it can be composed; on the contrary they can be thought only as *in* it". See also *Inaugural Dissertation*, Ak. II, 403 *Note*, and

A169 = B211. The pre-critical *R* 3926 considers the concepts of 'part' and 'whole' –
along with the concepts of unity, plurality, space and time - as two of the basic notions
of ontology. Therefore the part-whole relation may be included among the "few fundamental concepts of space" that, according to the *Prize Essay* (Ak. II, 282), mediate
the application of universal *Grössenlehre* to geometry. But also in pre-critical writings
space was prior to traditional primitive notions.

30 "Kant emphasized that space was an individual, the notion of which, understood in a
way analogous to ostension, and the same ostensive understanding, would be necessary
for the [...] primitives of geometry" (C. Parsons, 'Kant's Philosophy of Arithmetic', in
S. Morgenbesser, P. Suppes, M. White (eds.), *Philosophy, Science and Methodology*,
New York, St. Martin's Press, 1969, p. 574). Kant linked geometrical primitives to
knowledge of an immediate kind even before developing the Transcendental Aesthetic,
i.e. when he was still speaking of a *recta ratio* which enables us to single out such primitives (see e.g. Ak. I, 394, 403, 409; Ak. II, 11, 16). On this last point cf. G. Tonelli,
Elementi metodologici e metafisici in Kant dal 1745 al 1768, Turin, Edizioni di Filosofia,
1959, p. 148.

31 See T.L. Heath's Commentary to Euclid, *op. cit.*, vol. I, p. 232. Kant does not give a
detailed list of geometrical axioms, but only mentions some of them. For instance, in
addition to the axiom quoted in the text: (A) "Between two points there can be only one
straight line" (Ak. II, 281; *Inaugural Dissertation*, §15 C; A163 = B204, A300 = B356;
this axiom corresponds to Euclid's first postulate and presupposes the definition of
'straight line'); (B) "In a triangle the sum of two sides is greater than the third" (A25 =
B39; in Euclid this is Proposition 20 of Book I); (C) "Space has three dimensions" (Ak.
II, 281; *Inaugural Dissertation* §15 C); D) "Three points lie always in a plane" (A732 =
B761).

32 Euclid's third Postulate states: "To describe a circle with any center and distance".
See Note 34 below.

33 This depends also on the strict connection between arithmetic and logic. The latter
was identified by Kant with the science of expressing the true laws of human thinking and
was therefore *unique*.

34 Letter to K.L. Reinhold (Ak. XI, 40–48; *Zweig*, p. 146). In a letter to Marcus Herz
(Ak. XI, 48–55; *Zweig*, p. 155) Kant says: "I may always draw a circle free hand on the
board and put a point in it, and I can demonstrate all properties of the circle just as well
in it, presupposing the (so-called) nominal definition, which in fact is a real definition,
even if this circle is not at all like one drawn by rotating a straight line attached to a
point". Furthermore Kant considers the proposition "to inscribe a circle" as a "practical
corollary of the definition (or so-called postulate)", and indeed this proposition corresponds to Euclid's third postulate (cf. Note 32). In Kant's own terminology, then,
postulates are accessories, as it were, of definitions losing part of their ancient prestige.
Further evidence for the irrelevance of the exterior form of mathematical definitions is
provided by the *Prize Essay* where, on the same page (Ak. II, 278), a strictly genetic
definition and a non-genetic one are given by Kant as paradigmatic instances of mathematical definitions.

35 In arithmetic, not only definitions, but also postulates are universal only with respect
to their use. For instance propositions like 7 + 5 = 12, which are called postulates by
Kant (Ak. XI, 555–6; *Zweig*, pp. 129–30), have this kind of universality (A164–B205).

VITO MICHELE ABRUSCI

'PROOF', 'THEORY', AND 'FOUNDATIONS' IN HILBERT'S MATHEMATICAL WORK FROM 1885 TO 1900

PREMISE

Within the *general* aim of establishing *whether* and *how* the most typical aspects of David Hilbert's 'philosophy' of mathematics are present in his mathematical works (referred to here by title, followed by a bracketed number corresponding to its number in the bibliography, pp. 486–87), we shall limit ourselves in this paper to considering *in particular*:

–the first period of Hilbert's mathematical work (1885–1900) which ended with his first foundational works (for example, *Grundlagen der Geometrie* ['Foundations of Geometry'] [12]) and with his first works on philosophical and methodological problems and themes (for example, in 'Mathematische Probleme' [15]);

–the aspects of Hilbert's philosophy of mathematics concerning the notions of 'proof', 'theory', 'foundation' (Hilbert's position regarding these notions was more stable in later years, if compared, for example, with his position on the notion of 'finitary methods').

During the period under consideration Hilbert devoted himself first to the theory of *algebraic invariants* (1885–1893) and then to (algebraic) *number* theory (1893–1898); between 1898 and 1900 he devoted himself to *geometry* and problems of the *calculus of variations*. (See *Weyl*, 1944; *Hasse*, 1932; *van der Waerden*, 1933; *Hellinger*, 1935; *Bernays*, 1935.)

1. ON THE CRITERION OF SIGNIFICANCE IN MATHEMATICAL PROOFS

In the nineteenth century in particular, important developments in mathematical theories were determined by what Hilbert[1] called *Beweisgründen* (that is, principles, concepts, methods) used to solve or even only to attempt to solve certain mathematical problems, particularly when *new Beweisgründen* were introduced which proved *useful* for organizing one or more theories or for further investigations.

Hilbert, who often remembered this characteristic of the history of mathematics,[2] during this period solved problems, common to different

453

Maria Luisa Dalla Chiara (ed.), Italian Studies in the Philosophy of Science, 453–491.

theories, by means of new and very useful *Beweisgründen*. We can recall:

(a) His two proofs for the theorem of the existence of a complete system of invariants for every algebraic form (see Technical Note I, p. 475. Their importance, both historical and conceptual, lies not so much in the *solution* to that central problem of the theory of algebraic invariants, as in their *Beweisgründen*: i.e., the basis theorem established by Hilbert (in the first proof), the treatment of the system of algebraic invariants of an algebraic form as a field of algebraic functions (in the second proof).

(b) His new proof of the theorem of decomposition of the ideals of a field into prime ideals, with which he aims expressly at finding new bases for his further investigations into the theory of algebraic number fields.[3]

(c) His proof of the Dirichlet principle in the calculus of variations (see Technical Note III, pp. 476–77) with which he gives a "broadly generalizable" procedure "for the solution of problems surrounding it".[4]

Hilbert seems to aim at *significant proofs*, having this criterion for significance: a proof in a theory is significant if it contains *new* and *useful Beweisgründen* for organizing the theory and for further research.

On the Criterion of Simplicity in Proofs

An analogous criterion is that of simplicity, which Hilbert *propounds* in 'Die Theorie der algebraischen Zahlkörper' [8]: a proof in a theory is simple if it contains *generalizable* and *useful Beweisgründen* for further research.[5] Hence the simplicity of a proof depends in the first place not on the intuitiveness or on the degree of self-evidence of the content of its *Beweisgründen*, but on their capacity for explanation and unification in theories. With this criterion Hilbert deems 'not simple' both Kronecker's (finitary) proof and Dedekind's (abstract) one for the theorem of univocal decomposition of a field's ideals into prime ideals;[6] over them he prefers a new proof which "offers multiple bases for the further development of field theory".[7] We suspect that in general Hilbert's concept of 'simplicity' in proofs is not well understood. In fact, since, *from the finitistic point of view*, the simplicity of a proof is proportional to the simplicity of the self-evidence of its premises and inferences, one has been led to believe that this is really Hilbert's concept of simplicity in mathematical proofs: but in general the self-evidence of the most general, the most unifying, and logically the strongest *Beweisgründen* is not simple! We believe instead that even when Hilbert presented the problem of a "criterion of simplicity in mathematical proofs"[8] in 1917 in 'Axiomatisches Denken' [19], he was aiming at *rigorously* establishing a criterion according to which the simplicity of a proof in a theory is proportional to the generalizability and utility of its premises and its inferences. A confirmation: Hilbert (still in 'Axiomatisches Denken' [19]) says that his first proof for the existence of a complete system of invariants "satisfies our demands for simplicity and clarity",[9] even if it is a proof of a purely existential assertion and is carried out using transfinite methods. Analogously Hilbert had emphasized in 'Die Theorie der algebraischen Zahlkörper' [8] how, before the introduction of general and abstract concepts and methods in number theory, and notwithstanding the greater simplicity of number theoretical concepts and principles, number theoretical proofs were anything but simple, requiring "disproportionately great efforts to prove elementary facts".[10] Therefore the simplicity of the intuitive content of concepts or principles must be distinguished from the 'structural' simplicity of proofs; and it must be borne in mind that the use of more intuitive principles (as in number theory without general and abstract principles, hence in finitistic number theory) in general does not produce simple proofs and often not even significant proofs. Hence recent developments in proof theory toward a systematic examination of the 'structural' complexity of proofs[11] seem capable of contributing to an adequate response to the "problem of a criterion of simplicity of proofs" proposed by Hilbert for proof theory from 1917 on.

What is a Proof: To the Root of Proof Theory

In order to understand fully the role which Hilbert acknowledges in mathematical proof activity, it would be well to specify what he understands (in this period) by the term 'proof': the (informal) logical decomposition of a problem to certain assumptions[12], (the proof of the acceptability of the assumptions[13]), the (informal) logical deduction of assumptions.[14] Logical decomposition and logical deduction are somewhat informal in this sense: 'α decomposes to $\alpha_1, \ldots, \alpha_n$ (or α is deduced from $\alpha_1, \ldots, \alpha_n$')' is considered as a proof transition if α is a logical consequence of $\alpha_1, \ldots, \alpha_n$. So, proof activity includes three essential moments in the development of mathematical theories. Note that in *Hilbertian proof theory* the use of *new Beweisgründen* to prove one theorem of a theory is expressed with the *introduction* of new axioms in the theory and with the *formal proof* in the extension so obtained. And in fact in 'Neuebegründung der Mathematik' [20] Hilbert assigns to *proof theory* the task of rigorizing logical deduction from established assumptions as well as the (more important) one of rigorizing the introduction of new axioms:

the development of mathematical science takes place through the interchange of these two moments: obtaining new 'provable' theorems from axioms by means of formal proofs . . . and introducing new axioms together with the proof of their consistency.[15]

Hilbert's strong interest beginning from this period in examining and evaluating mathematical proof activity is unquestionable. However, we are not dealing with a strictly 'logical' interest, that is, one directed toward the examination and evaluation of logical deduction from premises, or that of the logical decomposition to premises;[16] it is instead an interest directed predominantly at the mathematical *Beweisgründen* of proofs, which we can call 'mathematical interest in proofs'.

2. ON RIGOR IN PROOFS

With the aim of expounding the theory of algebraic number fields "according to a logical development and from unitary points of view", Hilbert presents the *results* of the theory, taking the greatest care to establish their *Beweisgründen* exactly.[17] Afterwards, with the aim of a 'full understanding' of the most important results of nineteenth century geometrical research (according to Hilbert, these concern the provability or non-provability of geometric assumptions *under* certain given assumptions), and in general of all mathematical problems concerning the possibility or impossibility of a solution,[18] Hilbert maintained that it was necessary for every such proof (and every such presentation of problems) to satisfy this requirement: that all its *Beweisgründen* (assumptions) be *expressed* in some assertions so that of their intuitive content only what has been explicitly formulated in those assumptions is used in the proof. This is the demand for 'rigor' as Hilbert describes it in the *Grundlagen der Geometrie* [12] and in 'Mathematische Probleme' [15].[19]

Rigor, Axiomatization, Formalization

If a theory is worked out following the formal axiomatic method (as, for example, in the *Grundlagen der Geometrie* [12]), then 'rigor' is assured for *all* the proofs and for *all* the presentations of problems of the theory.[20] But the significance of this concept of 'rigor' emerges from the fact that for Hilbert an essential condition for the axiomatization of a theory is that each of its proofs satisfy the demand for rigor. Further he demands rigor for *each* proof (and in each presentation of problems) in *each* mathematical or physical theory, even if, and especially if it is not axiomatized.[21]

From a logical point of view, a rigorous proof of a mathematical theorem turns out to be "an [informal] logical deduction from certain explicitly established and formulated mathematical premises", as Hilbert himself says in 'Mathematische Probleme' [15].[22] The analogy with the concept of 'formalized proof' (which is obtained by simply 'formalizing' the logical inferences) and hence with formalization is not fortuitous. With reference to this concept of 'rigor' we already find methodological attitudes in this period which became typical of his position toward formalization in the twenties. For example, Hilbert maintains that: 'nothing is lost' with rigor (that is, using the content of mathematical concepts and methods only when that is explicitly formulated); each concept, method, and principle can always be adequately rigorized;[23] the 'cause' of the advantages and significance of a mathematical problem lies in the relations which link it to mathematical theories precisely by means of its *Beweisgründen* which must be rigorously expressed.[24]

The Influence of Weierstrass

The attention paid by Hilbert in 'Mathematische Probleme' [15] to the *usefulness* of rigor in mathematics and physics and the examples adopted there, bear witness to a profound link with certain essential aspects of a concept of rigor which finds its typical representative in Weierstrass, namely the discovery of correct mathematical formulations of principles, the discussion of their mathematical formulation more than their intuitive content, the conviction that rigor not only does not block but develops theories. Remember that Weierstrass's mind greatly influenced the (German) mathematical world at the end of the nineteenth century.[25] (See also Section 9 below on Hilbert and Weierstrass.)

3. PROOFS AND THE DEMANDS FOR CONSTRUCTIVITY

Is a mathematical proof *only* a logical deduction of an assertion from 'established' mathematical premises? – or does it also demand: (a) the (effective) construction of the entities whose existence is being asserted in the theorem; (6) the (effective) determination of the properties whose validity is being asserted in the theorem?

From his works on the theory of invariants on, we can draw this response from Hilbert: in order for an assertion to be a mathematical theorem, it is sufficient that it be correctly deduced from established mathematical premises.

Hilbert's Attitude

In fact his proofs for the existence of a complete system of invariants for every algebraic form (see Technical Note I, p. 475) offer a typical instance of the debate between the constructive approach and the non-constructive one. The first proof is non-constructive because Hilbert shows only that the *existence* of such a system is a *logical consequence* of certain (new) premises which he establishes; on the other hand, the second proof furnishes a (finitary) method by means of which, for every given algebraic form, its complete finite system of invariants can be *determined*. Paul Gordan (1837–1912), perhaps the principal figure in invariant theory at that time, harshly criticized Hilbert's first proof and was not disposed to accept it as a mathematical proof in the full sense:[26] although he recognized the "correctness" of that proof, he noted 'a gap' in that Hilbert had limited himself to proving the existence of that system of invariants "without taking care to establish their propriety" and "without giving an upper boundary for their number and weight".[27] Even if he put forward the 'gaps' of his first proof[28] in similar terms, Hilbert never considered it 'superseded' or 'corrected' by the second proof; for him both proofs are mathematical proofs in the full sense and with autonomous value.[29]

Some Reasons for Hilbert's Attitude

We think these reasons, among others, are possible: there are two distinct problems (the *existence* and the *determination* of the complete system of invariants) and the first proof is the simplest and most suitable solution to the first problem as the second proof is for the second problem;[30] *or*, both proofs contain important new *Beweisgründen*, etc.

In any case it would be well to bear in mind that the (old) conviction according to which mathematical knowledge is not necessarily a mathematical construction had strong support in the nineteenth century from the development of mathematical theories. As Hilbert himself mentioned in 'Mathematische Probleme' [15] with significant interest, notable results, including constructive ones, were obtained precisely as a consequence of (attempts at) proving (abstract) or purely logical non-constructive assertions (for example, as a consequence of abstract methods, results of impossibility, investigations on the relations between different theories, etc. were obtained etc.)[31]

Some Methodological Considerations

In 1917 and 1922[32] Hilbert devoted himself to commenting on his proofs in the theory of invariants in the context of his new foundational research.

He specified in 'Die logische Grundlagen der Mathematik' [21][33] that his first proof was non-constructive *because* it contained transfinite inferences; we can further specify that in his first proof he used the *tertium non datur* for infinite domains together with a (weak) principle of choice.[34] In addition, in 'Die logische Grundlagen der Mathematik' [21] and in 'Axiomatisches Denken' [19] he developed these considerations more or less widely and clearly, some of them revealing less 'known' aspects of his methodological-foundational position:

(a) The entire episode of his two proofs reinforces the *conjecture* that "a finitary proposition can always be proven even without using transfinite inferences".[35]

(b) This conjecture (note carefully: *not* a 'conviction') is similar to the conjecture that "theories can decide every one of their problems";[36] and hence must be rigorously established (in an affirmative or negative sense) with a mathematical proof.

(c) In any case this conjecture does not admit of an affirmative answer for every theory (fixed system of principles, concepts, methods); this is born out by the fact that "completely new considerations have been necessary as well as new principles to establish that the determination of the complete system of invariants demands only fixed operations whose number is finite and below a boundary determined before the calculation".[37] Thus Hilbert seems to maintain that it is possible and natural that, given a theory T, 'true' finitary propositions formulated in the language of T are provable finitarily only in certain extensions of T; and that in particular it is possible there are non-provable finitary truths in 'ordinary' formal systems of arithmetic.[38]

(d) It is necessary to distinguish two different meanings of 'constructive (or finitary) proof'; according to the first meaning, a 'constructive' proof must not contain non-constructive (non-finitary) principles; according to the second meaning, the entities and properties established in the theorem must be determined in a constructive (finitary) manner. For example, as Hilbert himself pointed out,[39] Gordan did try to obtain a 'constructive' proof of the theorem on the complete system of invariants, but 'constructive' in the second meaning of 'constructive'; and it is for this reason that he confined himself to *adding* a certain effective determination of the system of invariants to Hilbert's proof *without removing* from the proof those transfinite principles which guaranteed the existence of a complete system of invariants for each algebraic form;

(e) as already noted in paragraph 1, a transfinite or non-constructive proof (like Hilbert's first proof on invariants) can be 'simpler' than a finite or constructive proof (like Hilbert's second proof[40]).

4. MATHEMATICAL INVESTIGATIONS OF MATHEMATICAL PROOFS

At the end of the *Grundlagen der Geometrie* [12] and in his letter to Frege, Hilbert set forth the aims proposed in the *Grundlagen*: to establish exactly "the axioms, hypotheses, or means necessary to the demonstration of a truth of elementary geometry";[41] to investigate and determine the possible *Beweisgründen* of each known geometric truth; to investigate the possibility or impossibility of doing certain geometric 'proofs'; to give, in substance, an objective equivalent to the (subjective) choice between geometric principles. (See also paragraph 9.) We are dealing with investigations which have as their object (geometric) *proofs* insofar as they are rigorized and studied under a 'mathematical' interest (cf. paragraphs 1 and 2). A certain awareness of moving to a 'metatheoretical level' of treatment was already present in Hilbert in 1898; as Blumenthal testifies[42], Hilbert obtained notable results in the *Grundlagen der Geometrie* [12] precisely because he was aware of "carrying out purely logical operations which had nothing to do with the intuition" of the theory. The systematic study of (rigorized) proofs as an essential means for investigating mathematical theories (cf. paragraph 9) already seems to be a characteristic of Hilbertian 'metamathematics'.

Investigation of the 'Resolvability of Every Mathematical Problem'

In the following years[43] Hilbert forcefully expressed his conviction that methodological and philosophical questions on mathematics can and must be defined by means of mathematical investigations (cf. paragraph 9) which have mathematical proofs as their principal object. An apparently analogous conviction emerges from this passage of Hilbert dating from 1900:

historical experience together with philosophical motives supports the conviction held by every mathematician *but not yet supported by a proof* that every well-defined mathematical problem must perforce be susceptible of a precise resolution, either in the form of an answer to the question raised, or with the proof of the impossibility of its solution and hence the inevitable failure of all attempts in connection with it.[44]

Perhaps even an eventual proof of this conviction (to which Hilbert looked forward) should be based on an accurate metatheoretical investigation which follows closely the analysis of the 'resolution of a mathematical problem' made by Hilbert before the passage quoted above[45] and which can be summarized as follows:

(a) it is necessary to 'rigorously' formulate a problem's presuppositions in order to have a 'well-defined' and hence resolvable problem;[46]

(b) given a 'well-defined' problem, in order that an (affirmative or negative) answer to it be a 'rigorous resolution' of the problem, it is necessary and sufficient (by Section 3) that it be logically deduced from the problem's presuppositions;

(c) if the resolution turns out to be difficult, it is advisable to either *generalize* or *particularize* the problem, that is, to consider it either as "a special case of a more general problem" or as dependent on the solution of more particular problems;

(d) if (*'de facto'*) a positive or negative answer to the problem turns out to be impossible from its presuppositions, such an impossibility (*'de iure'*) must then be rigorously proved; this too is a 'rigorous resolution' to a 'well-defined problem'.

When Hilbert spoke in 1900 or in later years of the "resolvability of very mathematical problem", we believe he was clearly referring to 'well-defined problems' and intended to incorporate (d) as well among the 'rigorous resolutions'; in general, it does not seem that Hilbert referred to the 'resolvability of every problem of a theory' within the theory itself, or to the deductive completeness of mathematical theories.

5. ON INTUITION IN MATHEMATICAL THEORIES

In this period Hilbert's ideas on the role of intuition in mathematical theories can be condensed into these three concepts.

First: intuition (of its object) is the *starting point* for the development of any mathematical theory. Consider, for example: 'Über die Theorie der algebraischen Invarianten' [6][47] where with the term 'naive' Hilbert indicates the period of development of a theory based essentially on intuition; the *Grundlagen der Geometrie* [12] where a significant motto of Kant is quoted at the beginning of the volume;[48] and (widely) 'Mathematische Probleme' [15].[49] Indeed Hilbert seems to 'simply' acknowledge a historical and epistemological reality: in fact, whatever its origin and nature, 'there is' some (type of) intuition. Elsewhere Hilbert, as a mathematician, was not interested in establishing precisely the epistemological status or the sources of mathematical intuition in general, not even in that intuition present in particular mathematical disciplines;[50] it is worth noting that he never worked on theories of the intuitive stage (with the sole exception of his proof theory).

Second: intuition is like a *'given'* which the theory must express and analyze. So the entire theory can be conceived as an (attempt to give a) rational interpretation of intuition, and the principles, concepts and methods of the theory, as the rational instruments with which the truths of intuition are understood and interpreted. For example, Hilbert says that "fundamental homogeneous facts of our intuition"[51] are expressed with the groups of geometric axioms in the *Grundlagen* and that this entire work "constitutes an analysis of our spatial intuition".[52] Furthermore, the choice and the formulation of the axioms in the *Grundlagen der Geometrie* [12] and in 'Über den Zahlbegriff' [13] bear witness to this close tie between intuition and the theory's conceptual apparatus: as Schmidt observes, in each case he "prefers perspicacity and conceptual intuition to logical economy" and "introduces as fundamental concepts those concepts which intuitively and conceptually appear immediately as autonomous".[53] (See paragraphs 7 and 8.)

Third: intuition is an *essential instrument for the resolution of mathematical problems* at *every stage* of the development of mathematical theories. For example, Hilbert says in 'Mathematische Probleme' [15] that the 'symbols' of the theories represent "mnemonic signs of intuition" and are such insofar as they are indispensable for at least one first approach to the solution of any new problem.[54] Significantly Hilbert illustrates this role of intuition "recalled by symbols" by referring precisely to *geometry* and *arithmetic*, that is, to the two disciplines for which he had just presented his 'new' axiomatic method, ordinarily understood as being 'opposed' to intuition. But as early as 'Über den Zahlbegriff' [13] Hilbert had recognized the "eminent pedagogical and heuristic value of the genetic method"[55] of establishing and constructing theories, that is, the method of building 'less intuitive' concepts from 'more intuitive' ones by means of 'contentual' [*inhaltlich*] logical operations.

On Hilbert's Attitude Toward Intuition in the Twenties

To conceive this attitude as characterized by the 'negation' of (non-finitary) intuition in mathematics would mean admitting an untested 'revolution' into Hilbert's approach. We believe instead that the principal characteristics of Hilbert's 'new' approach to mathematical intuition in the twenties can be grasped in more profound mathematical and philosophical considerations. In fact Hilbert always admits the *'de facto' presence* and *usefulness* of all mathematical intuitions; but the question is raised: are these an *irreducible* source of (mathematical) knowledge? Above all, he maintains that mathematical practice suggests a negative answer to the question, and he tries to establish it definitively by means of an appropriate mathematical proof. Afterwards, the development of investigations in this direction showed him the *necessity* of accepting at least one kind of *'a priori'* (necessary, irreducible) intuition as a source of mathematical knowledge beyond experience and reason: finitary intuition.[56]

6. RELATIONS BETWEEN EXPERIENCE AND MATHEMATICAL THEORIES

A Historical Observation and a Foundational-Epistemological Attitude

In 'Mathematische Probleme' [15][57] Hilbert records two historical facts: mathematical theories have shown themselves to be essential in the physical and natural sciences, and several developments in mathematical theories (sometimes their very origin) have been produced by the natural sciences in which mathematics was present. Hilbert's comment is that this "ever-recurring exchange"[58] between experience (natural sciences) and (mathematical) thought is also *essential* for the development and constitution of mathematics, in a strong sense: in fact, he asserts that the eventual "interruption of the flow of new material from the external world" by means of the natural sciences would lead "in the final analysis to the rejection of concepts such as that of continuum or irrational number".[59] Hence for Hilbert concepts like that of continuum 'exist' insofar as they are produced and used in this exchange between experience and thought, and are only 'explained' (clarified) by their logical definitions (for example, in terms of whole or rational numbers). From this it follows that for him an epistemological critique of the logical definition of such a concept cannot lead to its denial and this is in fact the position assumed by Hilbert in this period toward Kronecker[60] and in the twenties toward the predicativists and intuitionists.[61] More than from Platonism, this position can be derived from a 'realistic' approach to applied mathematics and to those concepts which are necessary and useful in scientific explanation.[62]

464   VITO MICHELE ABRUSCI

The Usefulness of the Axiomatic Method

The observations just made illustrate why the axiomatic method seemed
to Hilbert to "correspond better in fact to what is given by experience and
intuition",[63] precisely because it introduces and treats concepts without
essential reference to their logical or intuitive origin.

With reference to the experience-mathematics relation, the axiomatic
method offers other advantages for Hilbert. Since it allows one to make a
precise distinction between purely physical hypotheses and mathematical
ones, in the natural sciences it serves to characterize rigorously the 'new'
mathematical concepts involved in them, and hence constitutes a useful
instrument for the influx of new concepts from the natural sciences to
mathematics.[64] In 'Mathematische Probleme' [15] Hilbert forcefully
proposed the problem of "axiomatizing the physical sciences in which
mathematics already plays an important role".[65] Furthermore, the physical theories themselves, once axiomatized, become a more suitable object
of study for the mathematician, who is concerned with "investigating all
the logically possible theories" and "the totality of the consequences of a
system of axioms."[66]

*On Two Old Problems: The 'Source of Knowledge' and the Relations
between Pure and Applied Mathematics*

Through the considerations set forth above, we can deduce this: Hilbert
thinks that, even if experience is not obviously the source of the contents
of mathematics, it is the perpetual source of problems for it. In fact Hilbert
sees in experience the first source of the first mathematical problems,
while other subsequent problems and *all* the answers to them are the fruit
of the 'human mind' ("the really true interrogator").[67] It is precisely on
the basis of the consideration of the concrete development of mathematics
that Hilbert constantly refuses to reduce mathematics somehow to a mere
'auxiliary' in the natural sciences (and hence refuses to exclude 'essentially'
pure mathematical problems).[68]

It would be misleading, however, to look for a 'precise and exhaustive'
answer from Hilbert to these old problems and it would also be misleading
to undervalue those answers which Hilbert nonetheless gives. In these
answers one must consider the sharp attention Hilbert pays to *not losing*
but to *improving* the instruments and procedures with which mathematics
has grown and has proved itself useful in general scientific development.
To sum up, we are dealing with one of his most profound lessons: the

achievement of a methodological and philosophical approach which explains mathematics without mutilating it but rather by strengthening it.

7. REASON IN MATHEMATICAL THEORIES

Hilbert seems to denote by the term 'reason'[69] the source of all of a theory's conceptual constructions whose content cannot be strictly reduced to the intuition (of the objects) of the theory. In any case he attributes the following to the activity of reason in a theory:

(a) the use of concepts and methods 'extracted' from another theory: for example, the use of arithmetic methods in the theory of algebraic functions (by Kronecker, Dedekind, and Weber) and the use of analytic methods in number theory;

(b) the introduction and use of 'abstract' concepts (i.e., defined by 'abstraction', or general or infinitary); for example, concepts such as Dedekind's concept of ideal, against which Kronecker harshly polemicized;[70]

(c) the study in itself of abstract or general concepts: for example, chapters on abstract algebra and Cantor set theory.

Necessity, Motives and Limits of Rational Activity

As early as this period, Hilbert maintained not only the admissibility but also the *necessity* of a certain (elevated) degree of 'abstraction' in every mathematical theory if it is so rigorously and completely developed in its foundations, at least as was done for number theory in the nineteenth century.[71] Furthermore in this period Hilbert's mathematical work itself determined important 'developments in the direction of the abstract' (in algebra); for example:

(a) the use of concepts of the theory of algebraic function fields as a basis for the theory of algebraic invariants, in 'Über die vollen Invariantsysteme' [5];[72]

(b) the treatments in itself of the general concept (of the structure) of algebraic number field and of the relative Abelian field on it, with the aim of resolving the general problem of the laws of reciprocity.[73]

Thus the development of the theories for Hilbert seems to require an increasingly heavy use of abstraction. Note that to admit this implies that if $\mathscr{D}$ is a cognitive domain, language $\mathscr{L}$ for $\mathscr{D}$ will be a predicative language such that it allows one to always raise the *order* of its concepts.

We believe that it is very interesting to show that Hilbert expressly maintains that the *reasons* for the necessity of abstraction are also the *limits* within which it must stand. The *reasons* indicated by Hilbert are: the search for (general) principles which lie at the base of already accepted (more intuitive) principles;[74] the solution of old and new problems;[75] the aim of giving the theory (and its proofs) a "more secure and continuous development" in the place of a "tortuous procedure characteristic of the first stages of a science".[76] Thus progress toward abstraction *is* closely tied to the need to give *simplicity, clearness* and *development* to theories (and their proofs). Vice versa, use of the abstract which do not correspond to these criteria are *criticized* by Hilbert as being 'superfluous' and a source of complications.[77]

In this context Hilbert emphasizes in 'Mathematische Probleme' [15] the theme of the "full liberty of logical construction in mathematics", within the sole limit of its usefulness for the theory and starting from already established concepts.[78] Hence if the fullness of logical liberty contrasts it with positions like that of Kronecker, the scope and limits he sets differentiate it from infinitary and abstract approaches like those of Dedekind and Cantor.[79]

The abstraction and full liberty of logical construction are the source of those "idealized conceptual constructions"[80] or of those "connections which are believed like dogmas"[81] on which Hilbert concentrated in his proof theory. But we want to recall that it was still in the twenties that Hilbert said expressly that the necessary condition for the justification of a mathematical method is its consistency *together* with the proof of its rational usefulness for the ends for which it has been introduced.[82]

We can draw some consequences. The prevailing presence of reason in the theories (in their advanced stages) determines (or reflects?) the conception of them not as a simple *description* of truths of cognitive domains but as a *rational interpretation* of them. Furthermore, the motives, which are also limits, of the abstraction support the conjecture (or reflect the conception?) that the abstract (the infinitary) element can be eliminated in principle (however, by obtaining less 'simple' theories and proofs).

Change of the Object of a Theory – The Existential Assumption

There is a peculiarity of mathematical theories which came out particularly in the nineteenth century and which Hilbert attributed first to the prevailing presence of reason. The theory of a cognitive domain, when it is sufficiently developed, can be used in the study of a different domain: *either* to the extent that it gives the theory of that domain new concepts, methods and results (or only new problems), *or* to the extent that (appropriately interpreted) it becomes *the* theory of that domain. We believe that it is in this respect that Hilbert in 'Mathematische Probleme' [15] speaks of the "numerous and surprising analogies and the apparently prearranged harmony that the mathematician so often perceives in the questions, methods, concepts of the various branches of his science".[83]

One well-known example of this peculiarity is the arithmetization of analysis. But we maintain that another example is still more important for the evolution of Hilbert's thought and is generally worthy of greater attention: the arithmetization of the theory of Riemann functions made by Dedekind and Weber[84] (remember that Hilbert was Weber's student at Königsberg until 1883). This arithmetization is not only presented as a 'more rigorous' treatment of the theory of Riemann functions with arithmetic methods and concepts, but it also immediately determines *new and unexpected results* by means of concepts and methods which do not even have anything to do with (that which it was previously understood to be) the object of function theory;[85] furthermore, the *arithmetic* theory of Riemann functions offers important new results for the same number theory.[86] Note that Hilbert's research in the theory of algebraic number-fields really tended (as Klein and later Weyl observed) to find analogies and reciprocal influences between number theory, algebra and function theory, following the line of investigation of Dedekind and Weber.[87]

Now, this peculiarity allows one to consider a mathematical theory as 'indifferent' to the change in the kind of objects of which it is speaking with another 'suitable' kind of objects; thus it allows one to drop the necessity that for each theory the existence of *one determined kind* of individuals must be assumed; instead it authorizes one to require only that for each theory one must assume that it is 'objective' insofar as (that is) it speaks of objects (whatever they are) "devoid of any connection with the thinking subject".[88] It is well known that this 'Platonistic' hypothesis is really the *existential assumption* underlying modern Hilbertian axiomatics: one must assume that there are *things* which are only "subjects for the predicates of the theory".[89]

8. ON AXIOMATIZED THEORIES

If the proofs of a theory are rigorized (as Hilbert requires; cf. paragraph

2), set A_T of 'known propositions' of T, and set P_T of the 'known proofs' of T satisfy these properties:

(a) A_T can be considered as a *finite* set since the language of T (cf. Section 7) allows the reduction of an infinite set of propositions to a finite set of propositions of a 'higher order';

(b) if $D(\beta, \alpha)$ is the relation "there exists a $d \in P_T$ such that d is a proof of α and in d α depends on β", then:

–for every $\alpha \in A_T$: there exists a β such that $D(\beta, \alpha)$, or there exists a β' such that $D(\alpha, \beta')$;

–for every α and β: if $D(\alpha, \beta)$ and $\beta \in A_T$, then $\alpha \in A_T$.

We believe that this scheme can be useful in understanding what Hilbert, in the period under consideration, asserts or demands with regard to the axiomatization of theories.

Completeness of a System of Axioms

At least in this period Hilbert required that the system of axioms of a theory T be a *finite, consistent* (cf. Section 10) and *complete* system of propositions.[90] The finiteness is possible for (a) and we believe we can say with exactness that for Hilbert in this period to axiomatize a theory T is to find the *minimum* number n of consistent and complete propositions $\alpha_1, \ldots \alpha_n$.[91] Now we want to specify the possible meaning of the requirement of completeness.

(a) It is not possible that it concerns a concept of 'completeness' which refers essentially and precisely to the *language* or to the *logical calculus* used in theory T, as these were not considered by Hilbert in this period (cf. paragraph 2).

(b) It could concern a completeness concept of the type: 'the axioms for T must be sufficient to settle any question (prove or refute any assertion) of the cognitive sphere of which T is a theory'. But this would imply that a system of axioms, complete in this sense, is possible only for sufficiently advanced theories while Hilbert calls for complete systems of axioms also for theories which are decidedly 'in process'.[92] Furthermore, even for sufficiently advanced theories, or ones commonly considered capable of settling every question of their cognitive sphere, Hilbert in this period practically and expressly demonstrated (his) 'mistrust' toward such a 'completeness': take the really illuminating example of the theory of algebraic invariants[93] and remember Hilbert's insistence on the use of *new Beweisgründen* to solve problems (cf. Section 1). Still further: Hilbert, in considering (cf. Section 4) even the proof of the impossibility of a solution under its presuppositions as a natural and rigorous resolution of a problem, took it for granted that, even in a proven theory, there were problems which could not be resolved by the means available in the theory. On the other hand, it was specified by Hilbert and Bernays in the *Grundlagen der Mathematik* [25] that the conviction that developed theories (like analysis) are capable of "fully representing a closed and well-defined reality"[94] should be considered an 'idealization'.

(c) It seems reasonable to consider for 'completeness' either the 'categoricity' of the axioms or the requirement that 'the axioms $\alpha_1, ..., \alpha_n$ for a theory T be such that for each proposition, $\alpha \in A_T, \alpha$ is logically derived from $\alpha_1, ... \alpha_n$'. It is in this last sense that Hilbert assures the completness of the axioms in 'Über den Zahlbegriff' [13],[95] it is this last sense of completeness for which it is an essential and sensible requirement for *every* system of axioms for *any* theory,[96] it is this last sense of completeness which, in the *Grundlagen der Mathematik* [25] is considered 'necessary' for every axiomatization.[97] On the other hand, for what seems to be Hilbert's idea of proof in this period (cf. paragraph 1), it seemed obvious to him that if α is the logical consequence of $\alpha_1, ... \alpha_n$, then α is derivable from $\alpha_1, ... \alpha_n$.

Two Meanings of Axiomatization

The axiomatization of a theory T in the sense referred to above, allows, however, of two profoundly different meanings on which Hilbert concentrated somewhat in 'Mathematische Probleme' [15] and then in 'Axiomatisches Denken' [19]:

(a) To isolate in A_T those propositions $\alpha_1, ... \alpha_n$ which are maximal in the order induced in A_T from the relation D ($\alpha \leq \alpha'$ iff $D(\alpha', \alpha)$) and to discuss their 'independence' from it in order to reduce their number.[98] This act presupposes only the 'rigorization' of the theory.

(b) To look for 'the most profound principles for theory T', that is, propositions $\alpha_1, ..., \alpha_n$ not necessarily all in A_T from which the maximal propositions (see (a) above) are derived.[99] This act is a foundational investigation, by no means trivial, and the fruit of original mathematical discoveries (cf. also paragraph 9).

Axiomatization of a Theory as Its Logical and Unitary Exposition

Axiomatizing a theory, says Hilbert, is "the best way to give it a conclusive presentation and full logical security".[100] It is in substance the crowning of his interest, present from his very first works onward, to "set forth" the content of a theory "in a logical and unitary way" in order to give it "precision and simplicity" and to use it for other theories.[101] Then it was the more general and logically stronger *Beweisgründen* present in a theory's proofs or appropriately characterized which allowed such an exposition. But even when Hilbert expounds his axioms in the *Grundlagen der Geometrie* [12] or in 'Über den Zahlbegriff' [13] for each fundamental concept of geometry and arithmetic, they are explicitly understood as the propositions from which all the known facts relative to the concept under consideration are derived.[102] Hence axioms, as understood by Hilbert, by their nature contain the strongest idealizations and abstractions present in a theory; far from being "extracts of intuition" (like Pasch's 'Kernsätze)[103] or from being the propositions most intuitively surrounding the theory's fundamental concepts" (as Gauss required in order to rigorize Euclid's axiomatics),[104] they are expressly conceived as the *strongest* propositions in the proofs surrounding a theory's fundamental concepts.

On Implicit Definition

The transition to the conception of axioms as the 'implicit definition' of a mathematical structure (though Hilbert does not use the term 'implicit definition'),[105] is completed in Hilbert within the context of the algebraic research (in the theory of algebraic number fields with which Hilbert helps to lead algebra toward the abstract. In this way the known and profound historical and conceptual link between modern axiomatics and abstract algebra is confirmed.[106]

We are referring to Hilbert's research to obtain general laws of reciprocity (see Technical Note II, pp. 475–76) and to his study[107] of the general structure of the field of algebraic numbers and the relative Abelian field on it. The general structure of an algebraic number field is introduced by Hilbert by means of logically stronger propositions established in the theory of algebraic number fields; and each particular algebraic number field is introduced by means of new and further *Annahmen* [assumptions]. In the course of these investigations, as documented in 'Über die Theorie der relativ-Abelschen Zahlkörper' [11],[108] Hilbert is fully aware that true propositions concerning structures so characterized are all—and only—those logically derived from those assumptions, and that each system of assumptions describes a 'class' of particular fields. And 'Über die Theorie der relativ-Abelschen Zahlkörper' [11] was written in 1898, the year before the *Grundlagen der Geometrie* [12].

Infinitary Thought and Finitary Thought

As early as 'Über den Zahlbegriff' [13] and 'Mathematische Probleme' [15] Hilbert concentrated on some of the epistemological consequences of his 'new' axiomatic method, consequences which will be at the basis of his future epistemological and methodological position.

What is 'mathematically' known of a mathematical structure (or of a mathematical concept) is the theory T of that structure (in short, T for Hilbert is A_T together with its logical consequences); axioms $\alpha_1, \ldots \alpha_n$ completely 'summarize' T and implicitly define the structure of which T is the theory. We believe that *from these premises* Hilbert draws the explicit conclusion that every mathematical structure *can be thought* ("insofar as it concerns mathematical ends") by means of *only one finite number* of axioms, *even if* in the (theory of the) structure there are many abstractions and idealizations.[109] And so even if the constitution of new mathematical structures from others occurrs with contentual infinitary and abstract methods, it can be thought 'completely' by means of a formal finite act of modifying the axioms of other structures.

Supported by these considerations, as early as 'Über den Zahlbegriff' [13], Hilbert drew attention to the fact that infinitary thought is *not essential* for mathematical theories; but from then on this 'non-essentiality' is not separate from the fact that every 'reliable' system of axioms requires the previous existence of a theory and infinitary thought can be essential for the development of this theory[110].

9. ON FOUNDATIONAL RESEARCH

In addition to his research into the consistency of axioms (more on this in the following paragraph) in this period, Hilbert considered as belonging to the 'sphere of foundations' of mathematical theories research in the following three categories[111]:

First: *the search for new principles*, added or modified, for mathematical theories. This was the meaning of the common expression 'new foundations' of a theory. For example, Hilbert placed in this class:

– His discovery of new principles on which to develop the whole theory of invariants.[112]

– The discovery that the theory of algebraic fields is the 'common root' of both (algebraic) number theory and Galois's theory of equations.[113]

– His proof of the Dirichlet principle, with which he established new and more profound principles of the calculus of variations.[114]

Thus this class of foundational research includes research which 'reduces' a theory's old principles (or a theory's unresolved problems) to new principles by means of 'proofs' (in the sense of logical 'decomposition' to assumptions; cf. section 1). The outcome of this research is the introduction of new principles, "the deepening of the foundations", the "logical resolution (*logisches Ausbau*)" of a theory.[115]

Second: *discussion and critique of the principles* of a theory. In 'Zum Gedächtnis an Karl Weierstrass' [9] Hilbert says that the 'critique' (*Kritik*) was the instrument with which Weierstrass attained his 'new foundation', and his 'systematic construction (*Aufbau*)' of the general theory of analytic functions, at the same time providing the theory with "certainty for its foundations . . . and clarity for its concepts."[116] With the term 'critique' Hilbert referred especially to (1) Weierstrass's investigations on the mathematical 'correctness' of the principles used in theories (for example, his critique of the Dirichlet principle, which will be discussed further in Technical Note III, pp. 476–77); (2) the rigorous definitions given by Weierstrass to the principal concepts of various theories (for example, to the concepts of 'analytic function' or 'irrational number'); (3) the analysis of the apparatus and results of a theory carried out by Weierstrass with the aim of giving it a 'complete' logical construction (*Aufbau*) (for example, the arrangement of function theory "through rigorous methods and by means of a natural continuation of the development of thoughts").[117] In his 'Zahlbericht', that is, in 'Die Theorie der algebraischen Zahlkörper' [8] published the same year that he set forth his considerations on Weierstrass, Hilbert carried out *analogous* investigations on algebraic number theory. And in the *Grundlagen der Geometrie* [12] Hilbert carried out further *analogous* investigations on geometry (or rather, on 'various' geometric theories), developing Weierstrass's 'rigor' up to his new axiomatics; hence, a typical problem examined in the *Grundlagen* is the precise role of a certain geometric principle or a certain group of geometric principles, *in the proof* of a geometric truth or a group of geometric truths.[118]

Third: *to establish relations among mathematical theories*. We have recalled several times Hilbert's particular interest in looking for and establishing relations between different mathematical theories. Hilbert saw the methodological importance of his research in 'Über die vollen Invariantsysteme' [5] in that it "places the theory of algebraic invariants under the general theory of algebraic function fields."[119] (We would be more inclined, today, to see the importance in the contribution the functions give to abstract algebra; and in Hilbert's time it would be possible to see it in the resolution of the important problem of the theory of invariants.) Hilbert's investigations into the field theory of algebraic numbers deliberately aimed at establishing relations between number theory, algebra and function theory. One of the principal characteristics of the *Grundlagen der Geometrie* [12] is precisely the systematic investigation of the relations between geometric theories, and between geometry, number theory and/or algebra,[120] using the sophisticated instruments of interpretability and relative consistency. Besides being rich in mathematical results, this class of research is considered by Hilbert as belonging to the 'foundations' of theories, as he expressly asserts in 'Die Theorie der algebraischen Zahlkörper' [8] (on the relations between algebra, number theory, function theory established or discovered by means of developments of the three theories) and in the *Grundlagen der Geometrie* [12] (on geometry).

The Non-Contentual Nature of Foundational Research

In his critique of the principles of theories, in his analysis of their meaning, in establishing a theory's position within all of mathematical science, Hilbert's approach from this period on excludes considerations which are essentially dependent on the (intuitive) content of principles or on the type of intuition present in a theory. Rather, the critique of principles is based on the examination of their mathematical formulation and their role in the theories, following the lines laid down by Weierstrass. Hilbert seems to assert in the *Grundlagen der Geometrie* [12] that the meaning of a principle is provided or at least illuminated by the system of propositions which depend on it and on its relations with other principles.[121] And the positioning of a theory within all of mathematical science is given or at least specified by the system of its relations which have been mathematically confirmed with other mathematical theories, just as the setting of number theory in 'Die Theorie der algebraischen Zahlkörper' [8] is evaluated solely on the basis of the relations pinpointed between it and the other theories and only on the basis of these relations does Hilbert see its role of 'queen of mathematics' justifiably reconfirmed.[122]. Thus Hilbert's approach is already radically differentiated from every other approach, finitary or Platonistic, based on the (intuitive) content of concepts.

Mathematical Character of Foundational Research

For Hilbert all foundational investigations are *always mathematical* investigations; furthermore, they presuppose a broad development of theories, are its consequence and make it easier. This is a conception which Hilbert expresses forcefully in 'Mathematische Probleme' [15] (after quoting Weierstrass at length):

in order to deal successfully with the foundations of a science it is necessary to have a penetrating understanding of its particular theories; only the builder is in a position to establish with certainty the foundations of a building, as he knows in detail how the building was built from top to bottom.[123]

Furthermore, Hilbert seems to maintain that the development of a mathematical theory over a mathematical 'object' provides information on the 'essence' of that object which cannot be obtained with non-mathematical epistemological and philosophical considerations; so, for example, in 'Die Theorie der algebraischen Zahlkörper' [8], he says that following the discoveries of the nineteenth century, it can be asserted that "certain periodic functions and certain functions with linear transformations are profoundly linked to the essence of number."[124] Further, in the *Grundlagen der Geometrie* [12] he asserts that his systematic investigation of all geometric theories constitutes "an analysis of our intuition of space".

These notes outline Hilbert's future, explicit conviction that mathematics had by then become capable of not only offering material for research into the foundations of his theories, but also of dealing with an increasingly extended sphere of epistemological and philosophical problems concerning mathematics and scientific theories in general – all by using its own methods.[125]

On 'Limitations in Methods'

A further element to characterize Hilbert's position in this period on the relationship between philosophy and mathematics, is provided by his attitude toward 'limitations' in the methods used in mathematical theories. In particular Hilbert has in mind: Kronecker's restrictive conceptions in number theory, the exclusive choice of the geometric path (on the part of the Italians) or of that arithmetic (on the part of the Germans) in Riemann's function theory, the choice between different 'geometries' motivated by intuitive-philosophical considerations. Hilbert appeared *hostile* to the fact that epistemological and philosophical conceptions effect 'mutilations' in mathematical methods or rigid distinctions of principle between mathematical schools; in this respect he follows Klein's attitude.[126] The choice of one method or the preference for one principle over others must remain a completely subjective fact; the task of mathematics is to investigate what exactly are the consequences and significance of a method or principle, in order to give an objective equivalent to the subjective choice.[127] In fact 'limitations' in methods and principles can have and do have an important mathematical value for Hilbert *to the extent* that they determine new mathematical results (as in the arithmetical theory of algebraic quantities, in Kronecker) or clarify the meaning of principles and mathematical methods (as in his geometry research).[128] Hence, if Hilbert seems an 'old-fashioned' mathematician in terms of his opposition to the influence of philosophical choices in mathematical practice, he appears to be ahead of his time in terms of his systematic use of 'limitations' in methods (apart from their 'philosophical' motivations), new mathematical results, and his investigation of the philosophical significance of mathematical principles.

10. THE CONSISTENCY PROBLEM AND JUSTIFICATION OF THEORIES

A Double Generalization

In presenting the problem of consistency for axiomatic systems, ('free from contradictions': *Widerspruchsfreiheit*), Hilbert made a double generalization with respect to previous discussions of the problem:

– a generalization from the problem of the consistency of 'some particular axioms with respect to other axioms', to the problem of the consistency of 'a system of axioms' (or, in other words, from the problem of whether a certain hypothesis is *consistent* with other hypotheses, to the problem of the mutual consistency of a system of hypotheses);

– a generalization from the problem of the consistency of 'some particular systems of axioms' to the problem of the consistency of 'every system of axioms' (or, in other words, from the problem of justifying a *dubious* system of hypotheses to that of justifying *every* system of hypotheses).

Hilbert's novelty seems to lie in the synthesis of both these generalizations.[129]

More than once in this period (and not only in this period) Hilbert himself mentioned how (particular) problems of consistency, although often masked, occurred in the common practice of pure and applied mathematical theories.[130] Among these we believe that the case of the Dirichlet principle is of definite interest: Hilbert did research on it precisely in the period between 1898 and 1900 and it may have contributed to the maturation of Hilbert's position concerning the problem of consistency. The Dirichlet principle (see Technical Note III, pp. 476–77) is "true on the basis of physical and geometrical intuition",[131] but when added to the other "intuitively true" principles of the calculus of variations, it leads to contradiction as Weierstrass demonstrated. In particular, Weierstrass showed that an assertion contradicting the formulation of the Dirichlet principle is derived from the system constituted of the principles of the calculus of variations (including function theory) and from the Dirichlet principle itself. A first apparent consequence of Hilbert's approach to the Dirichlet principle and his approach to the consistency problem, seems to be the same formulation given by Hilbert to the consistency of axioms in the *Grundlagen der Geometrie* [12] ("it must not be possible to derive a fact from them with logical inferences, *which contradicts one of the axioms posited*").[132] A second, more profound consequence lies in Hilbert's response to the problem raised by Weierstrass's critique of the Dirichlet principle, a response which Hilbert pointed to as the 'reason' for the importance of the consistency proof for *every* system of hypotheses even if 'intuitively true':[133] from the union of 'intuitively true' axioms (hypotheses), and hence each consistency a contradiction can arise, precisely because of the characteristic of mathematical principles, namely that they are a 'rational, possibly idealized, explanation of intuition'.

Epistemological Meaning

The problem of consistency has often been presented as a problem of the 'justification' of theories. However, one must grasp some of the particular characteristics of the sense in which the consistency proof constitutes for Hilbert a justification of a theory or hypothesis.

For Hilbert the objects discussed in a theory are 'neutral' with regard to the theory itself (cf. Section 7); thus the justification of a theory cannot be the justification of the existence of objects discussed by the theory, but rather the justification of the system of assertions which the theory is capable of making.[134] Thus, in formulating the problem of consistency, Hilbert in 1899–1900 was not presenting a 'recovery' or 'salvation' program for mathematics (we are *before* Russell's famous antinomy!), but was calling on people to turn their mathematical and philosophical attention *definitively* away from the problem of the object of mathematical theories and turn it toward a critical examination of the methods and assertions of theories (along the lines already followed by Weierstrass).

For Hilbert, the theory of a cognitive domain is the system of rational instruments by means of which the 'intuitive' truths of that domain are understood (cf. paragraphs 5 and 7), and for this rational comprehension the free use of constantly new *Beweisgründen* is necessary (cf. paragraph 1). So the justification of a theory must be a 'rational' justification (not based on intuition) and must constitute a rigorous test of acceptability of new hypotheses which guarantee their liberal use to the greatest extent. Consistency is just such a justification.

TECHNICAL NOTE I

A (finite) system of invariants $(i_1, ..., i_m)$ of an algebraic form f in n variables $x_1, ..., x_n$ is said to be *complete* if for every invariant i of f, i can be expressed as a linear combination of $i_1, ..., i_m$ in which the coefficients are whole functions of the same n variables.

Gordan (1885) and Mertens (1897) proved that for every *binary* algebraic form f (that is, for $n = 2$) there is a complete finite system of invariants of f. The proof of the theorem for any n was long sought after using the techniques of the theory of invariants perfected by Clebsch and Gordan. Hilbert solved the problem with two proofs:

– the first in 1889 in 'Über die Theorie der algebraischen Formen' [2] shows how the existence of a complete finite system of invariants for every algebraic form in the theory of invariants depends on a general algebraic theorem of forms, the so-called 'Hilbert's basis theorem' (cf. note 34 to paragraph 3);

– the second dates from 1892, in 'Über die vollen Invariantsysteme' [5]; the treatment of the system of the whole and rational invariants of an algebraic form as a particular algebraic function field furnishes Hilbert with the possibility of determining a complete system of invariants for every algebraic form, by means of the results of (Kronecker's) arithmetic theory of algebraic functions (Kronecker 1882). For interesting events concerning these two proofs, see Blumenthal, 1935, pp. 393–395, and Reid 1970, chapter 5.

TECHNICAL NOTE II

n is the *quadratic remainder* of a prime number p if n is not a multiple of p and $x^2 \equiv n \,(\mathrm{mod}\, p)$ is resolvable. In short, the laws of reciprocity of quadratic remainders for the field R of rational numbers establish:

(a) if p and q are prime numbers, $p \neq q$, and $\left(\frac{p-1}{2}\right)\left(\frac{q-1}{2}\right)$ is *even*, then q is quadratic remainder of p iff p is quadratic remainder of q;

(b) if p and q are prime numbers, $p \neq q$, and $\left(\frac{p-1}{2}\right)\left(\frac{q-1}{2}\right)$ is *odd*, then p is quadratic remainder of q iff q is not quadratic remainder of p.

Extensions of the laws of reciprocity to other algebraic number fields which are extensions of the field of rational numbers, were studied or established by Gauss, Dirichlet, Kummer. Hilbert was fully aware of the fact that in all the known laws of reciprocity, if k was an algebraic number field, the laws of reciprocity for k were rooted in a relative quadratic field $k(\sqrt{\mu})$. So in 'Über die Theorie der relativquadratische Zahlkörper' [10] and in 'Über die Theorie der relativ-Abelschen Zahlkörper' [11] he studied in depth the general theory of a relative quadratic field on any of algebraic number field k (reaching the theory of class fields in 'Über die Theorie relativ-Abelschen Zahlkörper' [11]). Hilbert's research offered his students the conceptual basis and technique to establish the 'general' laws of reciprocity, that is, the laws of reciprocity for *any* field of algebraic numbers (cf. Hasse, 1932, pp. 531–535).

TECHNICAL NOTE III

The Dirichlet principle, already used by Gauss and Thomson, is linked to the so-called 'first problem of boundary values' in the calculus of variations ("let G be a portion of (bidimensional) space, with given boundary values. To find one and only one harmonic function f on G, which agrees with the boundary values of G, and hence

$$\Delta f = \frac{\delta^2 f}{\delta x_2} + \frac{\delta^2 f}{\delta y^2} = 0").$$

The Dirichlet principle asserts that, among all the possible continuous differentiable functions u on G which agree with the boundary values of G and by which the integral

$$\iint \left\{ \left(\frac{\delta u}{\delta x}\right)^2 + \left(\frac{\delta u}{\delta y}\right)^2 \right\} \delta x \delta y$$

is defined (positive, non-infinite), there is a function f for which the integral takes the minimum value and for which, therefore $\Delta f = 0$. The understood 'empirical' meaning of the Dirichlet principle is 'true' (cf. the examples in Klein 1926, p. 259). Weierstrass 1869 showed that this principle, although intuitively 'true' and although a very simple and unifying principle in the calculus of variations, "is not sound" (*nicht stichhaltig ist*); in fact even if it is true that all the continuous differentiable functions u on G have a lower boundary, this lower boundary can be a non-continuous differentiable function, thus in the calculus of variations with the Dirichlet principle, a formulation which contradicts the Dirichlet principle, (at least) as it is formulated, is derived. (Cf. Klein 1926, pp. 264–67 on the interesting reactions of the mathematics and physics worlds after Weierstrass's 'critique' of the Dirichlet principle.) Hilbert reevaluated the Dirichlet principle in two publications [(14] and [16]) on the subject, proving *under what conditions* the Dirichlet principle could still be used to solve boundary problems.

NOTES

The page references to Hilbert's works given below are in most cases to those which are reprinted in his *Gesammelte Abhandlungen*. See the bibliography for complete information.

[1] 'Die Theorie der algebraischen Zahlkörper' [8], p. 66.

[2] Some examples cited by Hilbert himself (cf. 'Die Theorie der algebraischen Zahlkörper' [8], p. 64 and 'Mathematische Probleme' [15], p. 298): Kummer's attempt to solve Fermat's last problem and the introduction of the concept of 'ideal number' (from which came Dedekind's concept of 'ideal'); Galois's research on algebraic equations and the bases for the development of modern algebra; Gauss's research on the laws of reciprocity for biquadratic remainders and the consideration of the concept of 'field of algebraic numbers'.

[3] Cf. Hasse 1932, pp. 531–535.

[4] 'Über das Dirichletsche Prinzip' [16], p. 15.

[5] "Die Frage, welcher von mehreren Beweisen der einfachste und naturgemässeste ist, lässt sich meist nicht an sich entscheiden, sondern erst die Erwägung, ob die dabei zugrunde gelegten Prinzipien der Verallgemeinerung fähig und zur Weiterforschung brauchbar sind, gibt uns eine sichere Antwort" ['The question, which is the simplest and most natural of several proofs, cannot be decided for the most part, but the consideration as to whether the principles based on them are capable of generalization and useful for further research, or not, gives us a surer answer'] ('Die Theorie der algebraischen Zahlkörper' [8], p. 66.

[6] Blumenthal 1935, p. 387.

[7] 'Zwei neue Beweise für die Zerlegbarkeit der Zahlen eines Körpers in Primideale' [7], p. 5.

[8] 'Axiomatisches Denken' [19], p. 154.

[9] *Ibid.*

[10] 'Die Theorie der algebraischen Zahlkörper' [8], pp. 64–65.

[11] Cf., for example, Statman 1974 and Kreisel 1976.

[12] On this point of proof activity, cf. 'Axiomatisches Denken' [19], p. 148 where Hilbert speaks of "proofs . . . which are not really proofs in themselves but which make possible decomposition to certain more profound propositions".

[13] Cf. section 10.

[14] Cf. Section 2.

[15] 'Neuebegründung der Mathematik' [20], pp. 174–75: Die Entwicklung der mathematischen Wissenschaft geschieht hiernach beständig wechselnd auf zweierlei Art: durch Gewinnung neuer „beweisbarer" Formeln aus den Axiomen mittels formalen Schließens und durch Hinzufügung neuer Axiome nebst dem Nachweis ihrer Widerspruchsfreiheit mittels inhaltlichen Schließens. (cf. 'Die logische Grundlagen der Mathematik' [21], p. 180).

[16] As Blumenthal 1935, p. 422 affirms, Hilbert's 'logical' interest in proofs began soon after 1904 and mostly as a need for consistency proofs in number theory.

[17] 'Die Theorie der algebraischen Zahlkörper' [8], p. 66. In this work Hilbert, in a completely original fashion, performed the task given him by the Deutsche Mathematiker Vereinigung, of preparing a systematic exposition of (algebraic) number theory.

[18] Cf. Hilbert's letter to Frege, and *Grundlagen der Geometrie* [12], pp. 89–90 (where Hilbert mentions among the most characteristic results of nineteenth century mathematics, not only geometric results of impossibilities, but also Abel's proof of the impossibility of a general solution for fifth degree equations and Lindeman and Hermite's proofs of the impossibility of an algebraic construction of the numbers e and π).

[19] 'Mathematische Probleme' [15], p. 293 (cf. *Grundlagen der Geometrie* [12], pp. 89–90.

[20] This is a consequence of the same definition of the formal axiomatic method which is given, for example, in *Grundlagen der Mathematik* [25] I, p. 1.

[21] 'Mathematische Probleme' [15], pp. 293–294 (cf. paragraph 8).

[22] *Ibid.*, p. 293.

[23] *Ibid.*, pp. 293–94 (cf. paragraph 4).

[24] "Ein neues Problem, zumal wenn es aus der äusseren Erscheinungswelt stammt, ist wie ein junges Reis, welches nur gedeiht und Früchte trägt, wenn es auf den alten Stamm, den sicheren Besitzstand unseres mathematischen Wissens, sorgfältig und nach den strengen Kunstregeln des Gärtners aufgepfropft wird" ['A new problem, particularly when it derives from the external physical world, is like a young shoot which only grows and bears fruit when, following the strict rules of the gardener's art, it is carefully grafted onto the old stem, the more certain property of our mathematical knowledge'], *ibid.*, pp. 293–94.

[25] Cf. Pierpoint 1928, pp. 35–40.

[26] Gordan says à propos: "this is not mathematics, it is theology" (cf. Klein 1926), p. 300; Reid 1970, chapter 5).

[27] Gordan 1893, p. 131.

[28] 'Über den vollen Invariantensysteme' [5], p. 319.

[29] Cf., for example, the exposition of the entire theory of invariants in 'Über die Theorie der algebraischen Invarianten' [6], and in particular the contributions made by Hilbert himself.

[30] In 'Über die Theorie der algebraischen Invarianten' [6] Hilbert makes a clear distinc-

tion between the answer to the problem of the existence of a complete finite system (p. 382) and the answer to the problem of its effective determination (p. 377).

[31] 'Mathematische Probleme' [15], pp. 291–292, where Hilbert remarks that there are and there have been 'good' (i.e. favorable) and important mathematical problems which are not formed as problems of determining mathematical entities.

[32] In 'Axiomatisches Denken' [19] and 'Die logische Grundlagen der Mathematik' [21] respectively.

[33] 'Die logische Grundlagen der Mathematik' [21], p. 187.

[34] Hilbert proves the basis theorem (see Technical Note I, above), by means of finitistic methods in the following formulation: "for every unlimited sequence of algebraic forms $F_1, \ldots F_s, \ldots$ in n variables, there is a number m of forms $F_1, \ldots, F_m$ such that every other form F of the sequence can be expressed as their linear combination having whole rational functions in the same n variables as coefficients". But in order to use it in invariant theory, Hilbert had to extend it from the case of "unlimited sequences of forms" to that of "any system of forms, where it is undecided whether they can be ordered in a sequence or whether they are a non-numberable set" ('Über die Theorie der algebraischen Formen' [2], p. 203). Hilbert extended it with this argument: if a system of forms in n variables does not have a finite 'basis', then (using a procedure of choice between infinite sets of forms) an unlimited sequence of forms in the same n variables and devoid (by construction) of a finite basis can be constructed; but this contradicts the theorem first established; hence every system of forms in n variables has a finite 'basis' (cf. 'Über die Theorie der algebraischen Formen' [2], pp. 203–204).

[35] 'Die logische Grundlagen der Mathematik' [21], p. 187–88.

[36] *Ibid.*

[37] 'Axiomatisches Denken' [19], p. 154.

[38] We are thinking of the non-provability of finitary truths in formal elementary number theory (such as the consistency of elementary number theory).

[39] 'Die logische Grundlagen der Mathematik' [21], p. 188; cf. *Gordan* 1893, p. 132.

[40] 'Axiomatisches Denken' [19], p. 154.

[41] *Grundlagen der Geometrie* [12], p. 90.

[42] Blumenthal 1935, pp. 403–404.

[43] 'Axiomatisches Denken' [19], p. 137; 'Die logische Grundlagen der Mathematik' [21], pp. 153–155.

[44] 'Mathematische Probleme' [15], p. 297: Diese merkwürdige Tatsache neben anderen philosophischen Gründen ist es wohl, welche in uns eine Überzeugung entstehen läßt, die jeder Mathematiker gewiß teilt, die aber bis jetzt wenigstens niemand durch Beweise gestützt hat—ich meine die Überzeugung, daß ein jedes bestimmte mathematische Problem einer strengen Erledigung notwendig fähig sein müsse, sei es, daß es gelingt, die Beantwortung der gestellten Frage zu geben, sei es, daß die Unmöglichkeit seiner Lösung und damit die Notwendigkeit des Mißlingens aller Versuche dargetan wird.

[45] *Ibid.*, p. 293; pp. 296–297.

[46] We believe that in this concept of 'well-defined [*bestimmte*] problem' there is the influence of the tradition of presenting mathematical problems in mathematical journals for the appropriate 'prizes'.

[47] "In der Geschichte einer mathematischen Theorie lassen sich meist 3 Entwicklungsperioden leicht und deutlich unterscheiden: Die naive, die formale und die kritische"

['In the history of mathematical theory three stages of development at the most can be easily and clearly distinguished: the naive, the formal and the critical'] ('Über die Theorie der algebraischen Invarianten' [6], p. 383). Cf. paragraphs 7 and 8.

[48] "So fängt alle menschliche Erkenntniss mit den Anschauungen an, geht von da zu Begriffe und endigt mit Ideen" ['All human knowledge thus begins with intuitions, proceeds thence to concepts and ends with ideas'] (*Grundlagen der Geometrie* [12], p. 3; p. 2 of English edition).

[49] For example, pp. 295–296 and 301.

[50] But in 'Die Theorie der algebraischen Zhalkörper' [8], significantly, Hilbert says that the intuition of the whole number is the simplest of mathematical intuitions, so that from its beginning, number theory is characterized by the "simplicity of foundations, the precision of concepts, the pureness of truths" (pp. 64–65).

[51] *Grundlagen der Geometrie* [12], p. 4.

[52] *Ibid.*, p. 5.

[53] Schmidt 1933, p. 406.

[54] "So sind die geometrischen Figuren Zeichen für die Erinnerungsbilder der räumlichen Anschauung und finden als solche bei allen Mathematikern Verwendung. . . . Dass wir bei arithmetischen Forschungen ebensowenig wie bei geometrischen Betrachtungen in jedem Augenblicke die Kette der Denkoperationen bis auf die Axiome hin verfolgen . . . " ['Thus geometric figures are signs for the mnemonic pictures of spatial intuition and as such they are applied by all mathematicians. . . . So that in arithmetical research, as in geometry, we do not follow the chain of the thought process at each moment up to the axioms . . . '] ('Mathematische Probleme' [15], pp. 295–296).

[55] 'Über den Zahlbegriff' [13], pp. 180–181.

[56] Finitary intuition, according to Hilbert, turns out to be *indispensable* for the same description of the theoretical apparatus of mathematics, as he affirms expressly in 'Grundlegung der elementaren Zahlenlehre [24], pp. 486–487 ("die ich für die Mathematik wie überhaupt zu allem wissenschaftlichen Denken, Verstehen, und Mitteilen für erforderlich halte, und ohne die eine geistige Betätigung gar nicht möglich ist" ['which I deem necessary for mathematics as, on the whole, for all scientific thought, understanding and communication, and without which intellectual activity is not at all possible.]). See also Abrusci 1975, pp. 334–335.

[57] 'Mathematische Probleme' [15], p. 293.

[58] *Ibid.*

[59] *Ibid.*, p. 294.

[60] Cf., for example, 'Über den Zahlbegriff' [13], p. 184 ("Die Bedenken, welche gegen die Existenz des Inbegriffs aller reellen Zahlen und unendlicher Mengen überhaupt geltend gemacht worden sind, verlieren bei der oben gekennzeichneten Auffassung jede Berechtigung . . . " ['The objections made against the existence of the essence of all real numbers and infinite quantities which are generally valid, lose every justification in the conception characterized above']).

[61] Cf., for example, 'Neubegründung der Mathematik' [20], p. 158.

[62] Cf. Zassenhaus 1975, pp. 453–454 on Hilbert and Minkowski's attitude toward applied mathematics.

[63] 'Mathematische Probleme' [15], p. 301. It will be taken up again in 'Über die Grundlagen der Logik und Arithmetik' [17], pp. 137–138 and in 'Neuebegründung der Mathe-

matik' [20], p. 159 ("Der Begriff der extensiven Grösse, wie wir ihn aus der Anschauung entnehmen, ist ein selbständiger gegenüber dem Begriff der Anzahl, und es ist daher durchaus der Anschauung entsprechend, wenn wir Anzahl und Masszahl oder Grösse grundsätzlich unterscheiden" ['The concept of extensive magnitude, as we infer it from intuition, is a self-evident one compared with the concept of number, and it therefore corresponds to intuition throughout if we make the fundamental distinction between number and numerical value or magnitude'].)

[64] Cf., for example, 'Mathematische Probleme' [15] and Blumenthal 1935, p. 417.

[65] 'Mathematische Probleme' [15], p 306.

[66] *Ibid.*, p. 307.

[67] *Ibid.*, pp. 292–293; cf. Section 7.

[68] Cf. 'Mathematische Probleme' [15] and later 'Naturerkennen und Logik' [23], pp. 386–387 (one of Hilbert's last lectures).

[69] Cf. 'Mathematische Probleme' [15], p. 293 et. seq.; 'Grundlegung der elementaren Zahlentheorie' [24], pp. 485–486. For the role Hilbert assigns in this period to reason in mathematical theories, cf. the introduction of 'Die Theorie der algebraischen Zahlkörper' [8].

[70] Cf. Klein 1926, p. 323; 'Neubegründung der Mathematik' [20], p. 159.

[71] "Die arithmetische Begriffe une Beweismethoden erfordern zu ihrer Auffassung und völligen Beherrschung einen hohen Grad von Abstraktionsfähigkeit des Verstandes, und dieser Umstand wird bisweilen als ein Vorwurf gegen die Arithmetik geltend gemacht. Ich bin der Meinung, dass alle die anderen Wissensgebiete der Mathematik wenigstens einen gleich hohen Grad von Abstraktionsfähigkeit des Verstandes verlangen-vorausgesetzt, dass man auch in diesen Gebieten die Grundlagen überall mit derjenigen Strenge und Vollständigkeit zur Untersuchung zieht, welche tatsächlich notwendig ist" ['Arithmetic concepts and methods of proof require for their conception and full mastery the capacity for a high degree of abstraction in reason and this circumstance is sometimes used as a valid objection against arithmetic. I am of the opinion that all the other fields of knowledge of mathematics require/assume the capacity for at least an equally high degree of abstraction in reason, that in these fields as well the foundations are built everywhere with the rigor and completeness for analysis which is in fact necessary'] ('Die Theorie der algebraischen Zahlkörper' [8], p. 64).

[72] See Technical Note I, above. We are dealing with a "turning point in the development of algebra" consisting in the transition from the "study of the determination of all the algebraic invariants of given algebraic forms" to the "study of general arithmetic and algebraic properties of systems of rational and algebraic functions" (van der Waerden 1933, p. 400).

[73] See Technical Note II, above. This is another historical innovation: Hilbert shifts algebraic number theory from the study in itself of the single field of rational numbers and its particular extensions, to the study in itself of *extensions* on *any* field of algebraic numbers (cf. Hasse, p. 530).

[74] It is the reason Hilbert gives in 'Die Theorie der algebraischen Zahlkörper [8], p. 64 for the introduction and use of abstract concepts in number theory.

[75] This is the motive which pushes Hilbert to 'progress toward the abstract' – see notes 73 and 74.

[76] 'Die Theorie der algebraischen Zahlkörper' [8], p. 65.

[77] Remember, for example, Hilbert and Hurwitz's hostility toward and criticism of both the proofs (one finitary, the other abstract) of Kronecker and Dedekind of the decomposition theorem of the ideal numbers of a field into ideal prime numbers (Hilbert says "we found both awful" ["beiden fanden wir scheusslich"]; cf. Blumenthal 1935, p. 397.)

[78] ["Der menschliche Geist] schafft aus sich selbst heraus oft ohne erkennbare äussere Anregung allein durch logisches Kombinieren, durch Verallgemeinern, Spezialisieren, durch Trennen und Sammeln der Begriffe in glücklichster Weise neue und fructbare Probleme und tritt dann selbst als der eigentliche Frager in der Vordergrund" ['Using only logical combination, generalization, specialization, separation and collection of concepts in the most favorable way, often without perceptible external stimulation, the human spirit creates from within itself new and fruitful problems and then moves into the foreground as the actual questioner itself']. 'Mathematische Probleme' [15], p. 293).

[79] Klein 1926, p. 328 says that in his algebraic works Hilbert has "united Kronecker's approach with Dedekind's way of thinking". In addition see note 9, and remember that a polemical attitude toward the excessive influence of both Dedekind and Kronecker emerges from the Minkowski-Hilbert correspondence (cf. Zassenhaus 1975). As far as Cantor is concerned, it turns out that Minkowski, a close friend of Hilbert, was "among the first mathematicians of our generation to recognize the deep meaning of Cantorian theory", especially as mathematical investigation into the concept of the actual infinite ('Hermann Minkowski' [18], p. 360), notwithstanding his training under Kronecker. In the last years of the century Hilbert's interest in Cantor's work grew: from 'Über die stetige Abbildung einer Linie auf ein Flächenstück' [3], to a few references to Cantor in 'Über den Zahlbegriff' [13] (p. 184), to placing the problem of the continuum first on the list of mathematical problems in 'Mathematische Probleme' [15].

[80] *Grundlagen der Mathemattk* [25] I, p. 16.

[81] 'Neubegründung der Mathematik' [20], p. 161.

[82] Cf. 'Über das Unendliche' [22], p. 370.

[83] 'Mathematische Probleme' [15], p. 293.

[84] Dedekind and Weber 1882.

[85] Cf. Klein 1926, pp. 326–327.

[86] *Ibid.*, p. 333.

[87] Cf. Weyl 1944, p. 263; Klein 1926, p. 333; and Hilbert's 12th problem in 'Mathematische Probleme' [15].

[88] Bernays 1935, p. 53.

[89] *Grundlagen der Mathematik* [25] I, pp. 1–2.

[90] For example: 'Über den Zahlbegriff' [13], pp. 181, 184; 'Mathematische Probleme' [15], pp. 295 and 299.

[91] In 1898 Hilbert assigned this task to axiomatics: "Die Axiome selbst genau zu untersuchen, ihre gegenseitigen Beziehungen zu erforschen ihre Anzahl möglichst zu vermindern" ['to analyse the axioms themselves accurately, to investigate their reciprocal relations, to diminish their number as much as possible'] (from Blumenthal 1935, p. 403).

[92] For example: "wo immer von erkenntnistheoretischer Seite oder in der Geometrie oder aus den Theorien der Naturwissenschaft mathematische Begriffe auftauchen,

erwächst der Mathematik die Aufgabe, die diesen Begriffen zugrunde liegenden Prinzipien zu erforschen und dieselben durch ein *einfaches* und *vollständiges* System von Axiomen . . . festzulegen" ['. . . wherever mathematical concepts emerge from epistemological considerations or from Geometry or from theories of science, mathematics acquires the task of investigating the principles lying at the basis of these concepts and defining . . . these through a *simple* and *complete* system of axioms'] ('Mathematische Probleme' [15], p. 295). [Italics mine.]

[93] The 'disturbance' provoked by Hilbert's first proof of the existence of a complete finite system of invariants for every algebraic form (cf. Technical Note I, above) was caused by the fact that a central problem of the theory of algebraic invariants had been solved by 'going beyond' Clebsch and Gordan's 'symbolics', which were considered sufficiently rich to solve all the problems of the theory. Hilbert (the 'formalist'!) commented on this fact with these words in a letter to Minkowski: ". . . auch in unserer Wissenschaft stets nur der überlegende Geist, nicht der angewandte Zwang der Formel den glücklichen Erfolg bedingt" ['in our science too, only the reflective spirit, not the applied force of formulas, always conditions the fortunate outcome'] (from Blumenthal 1935, p. 394).

[94] *Grundlagen der Mathematik* [25] II, p. 289–290: "Wir haben in unserer Darstellung . . . der Zielsetzung der Beweistheorie von vornherein vermieden, den Gedanken eines Totalsystems der Mathematik in einer philosophischen prinzipiellen Bedeutung einzuführen . . ." ['In our representation . . . of fixing the aim of proof theory, we avoided from the first, the introduction of the thought of a total system of mathematics into a philosophical, fundamental meaning . . .'].

[95] Cf. 'Über den Zahlbegriff' [13], p. 184.

[96] Cf. notes 90 and 92.

[97] *Grundlagen der Mathematik* [25] II, pp. 289–290: "die tatsächlich vorhandene Systematik der Analysis . . . als eine solche zu charakterisieren, die einen geeigneten Rahmen für die Einordnung der geometrischen und physikalischen Disziplinen bildet" ['to characterize the actual systematics of analysis available . . . as such, which forms a suitable space for the arrangement of geometric and physical disciplines'].

[98] 'Mathematische Probleme' [15], p. 299; 'Axiomatisches Denken' [19], pp. 146–147.

[99] 'Mathematische Probleme' [15], p. 295, 307; 'Axiomatisches Denken' [19], pp. 147–148.

[100] "Über den Zahlbegriff" [13], p. 181.

[101] 'Die Theorie der algebraischen Zahlkörper' [8], p. 66; and Klein 1926, p. 231.

[102] Cf. *Grundlagen der Geometrie* [12], p. 5 ff and 'Über den Zahlbegriff' [13], p. 181 ff. In this characteristic of Hilbert's axioms of 1899 lies the full naturalness of the (logically complicated) axiom of completeness: it is a true, simple statement which can be used as a *Beweisgrund* for the concept of continuity.

[103] Pasch 1882.

[104] Cf. Enriques 1907, pp. 10–11, 13.

[105] In Hilbert's terminology, the "things" about which one speaks "have certain mutual relations" . . . whose "precise and mathematically complete description . . . follows from the axioms of geometry". (*Grundlagen der Geometrie* [12], p. 4; page 3 of the English translation.

[106] Cf. Casari 1973, pp. 7–10; Mangione 1971, p. 823 ff.

[107] ". . . man hatte sich doch durchweg auf die Betrachtung der gerade erforderlichen algebraischen Zahlkörper beschränkt. . . . Hilberts Schritt zu allgemeinen algebraischen Grundkörpern k bedeutet als neue Zielsetzung das Studium der Theorie der allgemeinen algebraischen Zahlkörper, der algebraischen und arithmetischen Gesetzlichkeiten in und über ihnen, *um ihrer selbst willen*, während die klassische Zahlentheorie nur den rationalen Zahlkörper um seiner selbst willen studiert hatte" ['. . . [until then] consideration had been limited throughout to just the necessary algebraic numbers. Hilbert's step toward general algebraic k, means, as a new goal, the study of the general theory of algebraic number fields of the algebraic and arithmetic legitimacies in and over them, *for their own sake*, while classical number theory had studied only the rational number for its own sake'] (*Hasse* 1932, p. 530).

[108] For example: "Wir machen zunächst über den zugrunde gelegten Körper k solche zwei Annahmen, unter denen die Theorie des relativquadratischen Körpers bereits . . . ausführlich entwickelt worden ist . . ." ['First we make two such hypotheses concerning the k lying at the base, under which [hypotheses] the theory of the relative quadratic is already developed in detail . . .'] ('Über die Theorie der relativ-Abelschen Zahlkörper' [11], p. 485). For another example see also p. 491 of the same work.

[109] For example: "Unter der Menge der reellen Zahlen haben wir uns hiernach nicht etwa die Gesamtheit aller möglichen Gesetze zu denken . . . sondern vielmehr . . . ein System von Dingen, deren gegenseitige Beziehungen durch das obige *endliche und abgeschlossene System* von Axiomen gegeben sind" ['Under the set of real numbers, we do not have to think of the totality of all possible laws . . . but rather . . . of a system of things whose reciprocal relations are given through the above-mentioned *finite and closed system of axioms*'] ('Über den Zahlbegriff' [13], p. 184).

[110] Cf. Section 7. Hilbert insists on the link between a system of axioms and theory in his letter to Frege [26] (against an absolute arbitrariness of axioms).

[111] The three classes, together with the research on consistency, include problems which are listed in 'Mathematische Probleme' [15] as among those belonging to the "Gebiete der Grundlagen" ['area of foundations'].

[112] 'Über die vollen Invariantsysteme' [5], p. 287.

[113] 'Die Theorie der algebraischen Zahlkörper' [8], pp. 64–65.

[114] 'Über das Dirichletsche Prinzip' [14], p. 11; 'Über das Dirichletsche Prinzip' [16], p. 15.

[115] Cf. 'Axiomatisches Denken' [19], pp. 147–148.

[116] 'Zum Gedächtnis an Karl Weierstrass' [9], pp. 332–333.

[117] *Ibid.*, p. 333.

[118] "Die vorliegende Untersuchung ist ein neuer Versuch, für die Geometrie ein vollständiges und möglichst einfaches System von Axiomen aufzustellen und aus denselben die wichtigsten geometrischen Sätze in der Weise abzuleiten, dass dabei die Bedeutung der verschiedenen Axiomgruppen und die Tragweite der aus den einzelnen Axiomen zu ziehenden Folgerungen möglichst klar zutage tritt". ['This present investigation is a new attempt to establish for geometry a *complete*, and *as simple as possible*, set of axioms and to deduce from them the most important geometric theorems in such a way that the meaning of the various groups of axioms, as well as the significance of the conclusions that can be drawn from the individual axioms, come to light.'] (*Grundlagen der*

Geometrie [12], p. 3; p. 2 of English edition).

[119] 'Über die vollen Invariantsysteme' [5], p. 286.

[120] Cf. Schmidt 1933, pp. 408–413.

[121] Cf. note 118.

[122] 'Die Theorie der algebraischen Zahlkörper' [8], pp. 68, 65–67.

[123] 'Mathematische Probleme' [15], p. 308: "In der Tat bedarf es zur erfolgreichen Behandlung der Grundlagen einer Wissenschaft des eindringenden Verständnisses ihrer speziellen Theorien; nur der Baumeister ist imstande, die Fundamente für ein Gebäude sicher anzulegen, der die Bestimmung des Gebäudes selbst im einzelnen gründlich kennt." Previously, quoting Weierstrass: "Das Endziel welches man stets im Auge behalten muss, besteht darin, dass man über die Fundamente der Wissenschaft ein sicheres Urteil zu erlangen suche" ['The final goal which must always be kept before us is that we must try to reach a surer judgement concerning the foundations of science'], pp. 307–308.

[124] 'Die Theorie der algebraischen Zahlkörper' [8], p. 68.

[125] Cf., for example, 'Axiomatisches Denken' [19], pp. 153, 155, 156; "Alles, was Gegenstand des wissenschaftlichen Denkens überhaupt sein kann, verfällt, sobald es zur Bildung einer Theorie reif ist, der axiomatischen Methode und damit mittelbar der Mathematik" ['Everything that can generally be the object of scientific thought decays as soon as it is ripe for the formation of a theory of the axiomatic method and thereby indirectly of mathematics'].

[126] Klein 1926, p. 327 opposed sharpening the distinction between the two schools (Italian and German) as he rebelled against the idea of a mathematics divided not only into 'chapters' but also into 'schools'. Remember Hilbert's attitude toward Kronecker's orientation in number theory and algebra mentioned in the preceding paragraphs. Cf. also 'Grundlegung der elementaren Zahlentheorie' [24] on the first years of Hilbert's training and on his attitude toward the influence of Dedekind and Kronecker.

[127] In the *Grundlagen der Geometrie* [12] Hilbert asserts that the aim of his research has been to establish "welches Axiome, Voraussetzungen oder Hilfsmittel zum Beweise einer elementar-geometrischen Wahrheit nötig sind" and so "es bleibt . . . dem jedesmaligen Ermessen anheimgestellt, welche Beweismethode von dem gerade eingenommenen Standpunkte aus zu bevorzugen ist" ['which axioms, hypotheses or aids are necessary for the proof of a fact in elementary geometry, and in every decision the question as to which method of proof is to be preferred, from the adopted point of view, remains open'] (p. 120; p. 107 of the English edition).

[128] Cf. notes 118 and 126.

[129] Before Hilbert, the question was posed and faced as to whether the *negation* of a 'true' principle of a theory was consistent with other principles of the theory (for example, in geometry), or whether a particular, *non-common* system of principles was consistent (for example, non-Archimedean geometry). The problem of the consistency of a system of *common* and (intuitively) *true* principles does not seem to have been formulated.

[130] Cf. Hilbert's letter to Frege [26] and 'Axiomatisches Denken' [19], pp. 151–153.

[131] 'Über das Dirichletsche Prinzip' [14], p. 11; 'Über das Dirichletsche Prinzip' [16], p. 15.

[132] *Grundlagen der Geometrie* [12], p. 19.

[133] "The very process of establishing an axiom, of naming it after its truth and concluding that it is consistent with defined concepts is one of the principal sources of errors and misunderstandings in modern physics research" (Letter to Frege [26]).

[134] "In der Tat, wenn der Nachweis für die Widerspruchslosigkeit der Axiome völlig gelungen sein wird, so verlieren die Bedenken, welche bisweilen gegen die Existenz des Inbegriff der reellen Zahlen gemacht worden sind, jede Berechtigung . . . " ['In fact when the consistency proof for axioms is completely successful, the objections which had sometimes been made against the existence of the essence of real numbers, lose every justification'] ('Mathematische Probleme' [15], p. 301; cf. 'Über den Zahlbegriff' [13], p. 184.) We believe that the assertion ('Mathematische Probleme' [15], p. 300) that the mathematical existence of a concept is its consistency must also be understood in this light.

BIBLIOGRAPHY

A. Hilbert's Works

We have listed below only those works of Hilbert which are cited in this paper; where relevant, page references are also given to the *Gesammelte Abhandlungen* (*G.A.*). 3 vols. Berlin, Springer: Vol. I, 1932; Vol. II, 1933; Vol. III, 1935. Reprinted, New York, Chelsca, 1965.

[1] 'Zur Theorie der algebraischen Gebilde I', *Göttinger Nachrichten* (1888), 450–457; *G.A. II*, pp. 176–183.

[2] 'Über die Theorie der algebraischen Formen', *Mathem. Annalen* **36**, 473–534 (1890); *G.A. II*, 199–257.

[3] 'Über die stetige Abbildung einer Linie auf ein Flächenstück', *Mathem. Annalen* **38**, 459–460 (1891); *G.A. III*, pp. 1–2.

[4] 'Über die Theorie der algebraischen Invarianten. III Note', *Nachrichten der Gesellschaft der Wissenschaften zu Göttingen* (1892), 439–449.

[5] 'Über die vollen Invariantensysteme', *Mathem. Annalen* **42**, 313–373 (1893); *G.A. II*, pp. 287–344.

[6] 'Über die Theorie der algebraischen Invarianten', Math. papers read at the International Math. Congress, Chicago 1893. New York, Macmillan, 1896, pp. 116–124; *G.A. II*, pp. 376–383.

[7] 'Zwei neue Beweise für die Zerlegbarkeit der Zahlen eines Körpers in Primideale', *Jahresbericht der Deutschen Mathematikervereinigung* **3**, 59 (1894); *G.A. I*, 5.

[8] 'Die Theorie der algebraischen Zahlkörper', *Jahresbericht der Deutschen Mathematikervereinigung* **4**, 175–546 (1897); *G.A. I*, pp. 63–363.

[9] 'Zum Gedächtnis an Karl Weierstrass', *Göttinger Nachrichten* (1897), Geschäftliche Mitteilungen, 60–69; *G.A. III*, pp. 330–338.

[10] 'Über die Theorie des relativquadratischen Zahlkörpers', *Mathem. Annalen* **51**, 1–127 (1899); *G.A. I*, p. 370–482.

[11] 'Über die Theorie der relativ-Abelschen Zahlkörper', *Acta Mathematica* **26** (1902), 99–132; *Nachrichten der K. Ges. der Wiss. zu Göttingen* (1898), 370–399; *G.A. I*, pp. 483–509.

[12] *Grundlagen der Geometrie*. Festschrift zur Einweihung des Göttinger Gauss-Weber-

Denkmals. Leipzig, Teubner, 1899. pp. 1–90. (*Foundations of Geometry*. 2nd edition. Translated by Leo Unger from the 10th revised and enlarged edition of Paul Bernays. La Salle, Open Court, 1971.)

[13] 'Über den Zahlbegriff', *Jahresbericht der Deutschen Mathematikervereinigung* **8**, 180–184 (1900).

[14] 'Über das Dirichletsche Prinzip', *J. reine angew. Math.* **129** (1905), 63–67; *Jahresbericht der Deutschen Mathematikervereinigung* **8**, 184–188 (1900); *G.A. III*, pp. 10–14.

[15] 'Mathematische Probleme', lecture given at the International Mathematician Congress, Paris, 1900. Published in the *Göttinger Nachrichten* (1900), pp. 253–297; reprinted, with some additions, in *Archiv f. Math. u. Phys.*, 3rd series **1**, 44–63 (1901); 213–237; *G.A. III*, pp. 290–329.

[16] 'Über das Dirichletsche Prinzip', *Mathem. Annalen* **59**, 161–186 (1904); *G.A. III*, pp. 15–37.

[17] 'Über die Grundlagen der Logik und Arithmetik,' *Verhandlungen des Dritten Internationalen Mathematiker-Kongresses*, Heidelberg 1904 (Leipzig, Teubner, 1905), pp. 174–185. ('On the foundations of logic and arithmetic', *The Monist* **15**, 338–352, translated by George Bruce Halsted; this paper also appears, in an English translation by Beverly Woodward, in *From Frege to Gödel: A Source Book in Mathematical Logic, 1879–1931*, edited by Jean van Heijenoort. Cambridge, Mass., Harvard University Press, 1967. pp. 129–138.)

[18] 'Hermann Minkowski', *Göttinger Nachrichten*, Geschäftliche Mitteilungen (1909), 72–101 and *Mathem. Annalen* **68**, 445–471 (1910); *G.A. III*, 339–364.

[19] 'Axiomatisches Denken', *Mathem. Annalen* **78**, 405–415 (1918); *G.A. III*, pp. 146–156.

[20] 'Neubegründung der Mathematik. Erste Mitteilung', *Abhandl. aus dem Math. Seminar d. Hamb. Univ.* **1**, 157–177 (1922); *G.A. III*, pp. 157–177.

[21] 'Die logische Grundlagen der Mathematik', *Mathem. Annalen* **88**, 151–165 (1923); *G.A. III*, pp. 178–191.

[22] 'Über das Unendliche', *Mathem. Annalen* **95** 161–190, (1926). ('On the infinite', translated by Stefan Bauer-Mengelberg, in *From Frege to Gödel: A Source Book in Mathematical Logic, 1879–1931*, edited by Jean van Heijenoort. Cambridge, Mass., Harvard University Press, 1967. pp. 367–392.

[23] 'Naturerkennen und Logik', *Naturwissenschaften* (1930), 959–963; *G.A. III*, pp. 378–387.

[24] 'Die Grundlegung der elementaren Zahlenlehre', *Mathem. Annalen* **104**, 485–494 (1931).

[25] (with Paul Bernays) *Grundlagen der Mathematik*. 2 vols. Berlin, Springer: vol. I, 1934; vol. II, 1938; 2nd ed. vol. I, 1968; vol. II, 1970.

[26] Letter to Frege of 29 December 1899, in *Nachgelassene Schriften und Wissenschaftlicher Briefwechsel*, 2 vols., Hamburg, Meiner, 1969–.Volume 2: *Wissenschaftliche Briefwechsel*, edited by Gottfried Gabriel, *et al.*, 1976,pp. 65–68.

B. Other Works

Abrusci, V.M. (1975), 'Per una caratterizzazione del programma hilbertiano', *Rivista di filosofia* **66**, no. 3, 315–338.

Bernays, P. (1922), 'Über Hilberts Gedanken zur Grundlegung der Arithmetik', *Jahresbericht der Deutschen Mathematikervereinigung* **31**, 10–19.

Bernays, P. (1935a) 'Hilberts Untersuchungen über die Grundlagen der Arithmetik', *G.A. III*, pp. 196–216.

Bernays, P. (1935b) 'Sur le platonisme dans les mathématiques', *L'enseignement mathématique* **34**, 52–69.

Blumenthal, O. (1935), 'Lebensgeschichte', *G.A. III*, pp. 388–429.

Casari, E. (1973) *La filosofia della matematica del '900*. Florence, Sansoni.

Dedekind, J. (1871), 'Über die Theorie der ganzen algebraischen Zahlen', eleventh supplement to Dirichlet's *Vorlesungen über Zahlentheorie* (Braunschweig, F. Viewig, 1894), reprinted Braunschweig, Vieweg, 1964.

Dedekind, J., and Weber, H. (1882), 'Theorie der algebraischen Funktionen einer Veränderlichen', *Journal für Math.* **92**.

Enriques, F. (1907), 'Prinzipien der Geometrie' in *Encyklopädie der Math. Wissenschaften*, III–I. Edited by W. F. Mayer and H. Mohrmann. Leipzig, Teubner, pp. 1–129.

Gordan, P. (1893), 'Über einen Satz von Hilbert', *Mathem. Annalen* **42**, 131–35;

Gordan, P. (1895) *Vorlesungen über Invariantentheorie I*. Leipzig, Teubner.

Hasse, H. (1932), 'Zu Hilberts algebraisch-zahlentheoretischen Arbeiten', *G.A. I*, pp. 528–535.

Hellinger, E. (1935), 'Hilberts Arbeiten über Integralgleichungen und unendliche Gleichungssysteme', *G.A. III*, pp. 94–145.

Klein, F. (1926) *Vorlesungen über die Entwicklung der Mathematik im 19. Jahrhundert I*. Berlin, Springer; reprinted, New York, Chelsea, 1967.

Kreisel, G. 'Notes for four lectures on proof theory'. Clermont-Ferrand, 15–16 July 1973.

Kreisel, G. (1976) 'What have we learned from Hilbert's second problem?' in F. Brouwer, ed., *Proceedings of the Symposia in Pure Mathematics* **28**, 93–130.

Kronecker, L. (1882), 'Grundzüge einer arithmetischen Theorie der algebraischen Grössen', *Jour. für Math.* **92**, 1–122 (1881/82).

Kronecker, L. (1901) *Vorlesungen über Mathematik*. Leipzig, Teubner.

Mangione, C. (1971), 'Logica e fondamenti della matematica nella seconda metà dell'ottocento' in *Storia del pensiero filosofico e scientifico*, edited by L. Geymonat. Milan, Garzanti, 1971. vol. 5, pp. 755–830.

Mertens, F. (1887), 'Beweis, dass alle Invarianten . . . ', *Jour. für Math.* **100**, 223–230.

Minkowski, H. (1973), *Briefe an David Hilbert*. Edited by L. Rüdenberg and H. Zassenhaus. Berlin, Springer.

Pasch, M. (1882), *Vorlesungen über neuere Geometrie*. Leipzig, Teubner.

Pierpoint, J. (1928), 'Mathematical rigor: past and present', *Bull. Amer. Math. Soc.* **34**, 23–53.

Reid, C. (1970), *Hilbert*. Berlin, Springer.

Schmidt, A. (1933), 'Zu Hilberts Grundlegung der Geometrie', *G.A. II*, pp. 404–414.

Statman (1974), 'Structural complexity of proofs'. Ph.D dissertation, Stanford University, 1974.

van der Waerden, B.L. (1933), 'Nachwort zu Hilberts algebraischen Arbeiten', *G.A. II*, pp. 401–403.

Weierstrass, K. (1869), 'Über das sogenannte Dirichlet'schen Prinzip', *Mathematische Werke II*. Berlin, Springer, 1895, pp. 49–54.

Weyl, H. (1944), 'David Hilbert and his mathematical work' *Bull. Amer. Math. Soc.* **50**, 612–654; reproduced in a shortened version in Reid (1970), pp. 245–283.

Zassenhaus, H. (1975), 'On the Minkowski-Hilbert dialogue on mathematization', *Canad. Math. Bull.* **18**, 443–461.

ENRICO BELLONE

THE HISTORY OF SCIENCE AS THE HISTORY OF DICTIONARIES*

SCIENTIFIC LANGUAGE AND TRANSLATION

We owe to Mark Twain's irony the following description of one way of reconstructing history:

In real life the right thing never happens at the right place at the right time:–it is the business of the historian to remedy this mistake.[1]

Twain's words will certainly console those who consider history a rubbish heap. If real events were a senseless game of caprice and tricks, then it would be the historian's duty – and the philosopher's – to intervene to restore order to chaos, remedy errors, and point out the pitfalls which have snared men in the past and destroyed theories.

Faced with a reality which has been disconnected by unforeseeable tricks, the historian in general (and the historian of science in particular) should first decide which reconstruction is most agreeable and then eliminate inconvenient facts – errors.

The result of such a two-stage operation would be a narrative which would be edifying above all for those philosophers who, having already decided how history must be, ask historians for some comforting examples.

Indeed the force of examples is undeniable. Suppose, for example, that an historian of physics wants to investigate the state of knowledge concerning the motion of fluids during the second half of the eighteenth century. He would find that in 1777 some French researchers [d'Alembert, Condorcet and the Abbé Bossut] published an important monograph on the subject [*Nouvelles expériences sur la résistance des fluides*. Paris, Jombert, 1777]. In that monograph particular experimental results are described and conjectures are made which undoubtedly concern fluid mechanics. In the same monograph, however, there are assertions which do not concern fluid mechanics. Here we read that "there is no science without reasoning or without theory": we are dealing with a sentence which can be considered valid or invalid, but which in any case has no im-

* Grateful acknowledgement is made to the Mondadori Publishing Company for allowing me to present the following sections, slightly modified, from *Il mondo di carta, ricerche sulla seconda rivoluzione scientifica* (Miland, 1976). [Eng. tr.: *A World on Paper: Studies in the Second Scientific Revolution*. Cambridge, Mass., MIT Press, 1980.]

Maria Luisa Dalla Chiara (ed.), Italian Studies in the Philosophy of Science, 493–507.
Copyright © 1980 by D. Reidel Publishing Company.

mediate connections with the laws governing the movement of water in channels. Furthermore, there are some verified historical facts which concern neither fluid mechanics nor the importance of theories, and which are, however, somehow linked to the monograph of 1777; the latter turns out to be the fruit of a work commissioned by state institutions and therefore it involves the existence of demands of a social and economic character.

Briefly, the monograph under examination is a response to demands of a practical nature, uses arguments of a philosophical nature, and is correlated to a certain level of scientific knowledge – well, which of these three aspects is the most important?

There are different answers to this same question. A first answer is the following: the historian of science must concentrate his attention on factors internal to the problem of fluid mechanics and must consequently reconstruct the state of the theory, the determination and choice of empirical data, and the interpretation of the results. Having done this, his job is over.

A second answer consists instead in asserting that to reconstruct the problem of fluid mechanics correctly, the historian's interest must be concentrated on external factors such as the needs of institutions and the philosophical and cultural stimulants which characterize the milieu in which the monograph of 1777 was worked out.

Finally, a third answer is provided by those who think of internal history and external history as complementary histories and therefore suggest fusing them together into one universal history.

It must be said that the three answers mentioned here have one problem in common: with what criteria can one reasonably trace a boundary between internal history and external history? To what extent is the pronouncement of a theorem part of internal history? To what extent are philosophical convictions about what is scientific and what is not, part of external history?

This problem is not irrelevant. Rather, on it and on its possible solutions hinges a considerable part of what is understood by the rational reconstruction of history. And at this point it may be useful to recall that the followers of the three forms of history mentioned above also pose one general question in common: what relations exist between theory and experiment? The history of science, in fact, is the history of relations between theories and experiments and cannot be reduced to a chronology of theoretical events and empirical events divided into two separate groups: and this is the source of even more serious problems.

Nature essentially does not respond to those who observe it from a distance or who touch pieces of it without being guided by certain expectations. Instead, in order to force nature to respond to us we must operate on it by posing questions – that is, it is necessary to plan experiments, manipulate techniques and machinery, to put forward conjectures and theories, to interpret answering signs which are not always clear and unambiguous. The art of interrogating nature has been enriched during the most recent centuries and today, so-called natural phenomena have very little to say to those who meet it bereft of concepts.

It is not an accident – or malice on the part of philosophy – that the reconstruction which history makes of the art of interrogating nature always had to deal with the following question: what relations exist between theory and experiment?

On the other hand one must admit that this form of questioning can give rise to discouragements of varying cultural origin. For example, who could maintain that the work of a theoretical physicist consists in a 'translation', into symbols, of knowledge already gathered and arranged by experimental physics? Is it that perhaps Kepler's calculations coincide with translating emprical laws on the planetary orbits into formulas? Does not the idea of translation hide an overly schematic representation of theoretical and experimental research?

One does not escape this schema by asserting that the theoretical physicist's work is purely inventive and that the fruits of such work are free products of the mind. Only a romantic conception can portray a Newton or a Dirac in the guise of solitary geniuses who await the moment of truth in revealing lightning flashes: even in its moments of greatest abstraction, theoretical physics, in fact, maintains a complicated network of connections and intermediaries with which it is possible to experiment and from which one can make deductions.

We can better immerse ourselves in our subject – which measures the theory/experiment relation in the realm of the history of that relation – by trying to redefine the idea that to create theories is in some way to translate. And let us consider one of the most important 'translations' which have been made since Newton's time: the construction of the electromagnetic theory of light.

Usually we read that the development of that theory took place during the last century in two phases. During a first phase some scientists – and Faraday in particular – patiently collected a large amount of data on electric, magnetic, electro-chemical and optical phenomena. During the second phase other scientists – and Maxwell in particular – translated the mass of empirical knowledge, which by then was available and arranged in tables, into mathematical formulas. And the successful outcome of the translation lay in the fact that the new theory was able to gather that mass of knowledge into compact and rigorous symbolic forms.

It is, a revolution which was on the whole rather peaceful and which seems to consist in fidelity to canons which an honest professional, endowed with a strong intelligence, must obey: in short, in our case, Maxwell made correct use of rules suited to stabilizing a genuine correspondence between the laws discovered by Faraday in the laboratory and the equations of a mathematical language already available in the textbooks.

But what happens if we really maintain – to the end – the conjecture that the theoretical physicist, Maxwell, letting himself be guided by a mechanist conception of the world and of science, translated into equations a set of statements on facts that the experimental physicist, Faraday, not being a mathematician, had already written in a language which was neither formalized nor completely clear?

If we maintain this scheme we run into two unpleasant consequences. In order to translate from one language to another one needs a dictionary. Let us suppose, for the moment, that we know what Maxwell's dictionary was and let us compare the conception of the electromagnetic field proposed by Faraday (without mathematics) and that elaborated by Maxwell (with mathematics and dictionary).

The first conception contains the assertion that an explanation of the field involves the interaction between electromagnetic fields and gravitational fields.[2] The second conception, on the other hand, contains the assertion that the interaction between electromagnetic fields and gravitational fields raises paradoxes regarding energy which have no solution.[3] According to Faraday, therefore, a field theory must take those interactions into account. According to Maxwell, on the other hand, a field theory must eliminate them.

The first unpleasant consequence of the translation argument consists in this: that the translator, appealing to the dictionary, resolutely denies that an important part of the propositions to be translated can be translated effectively without falling into paradoxes. But this then is a strange and worrisome translation. It is strange in that it raises doubts about the validity of the documents to be translated and it is worrisome in that those doubts are resolved by eliminating part of the documents. This means that Maxwell, in translating, has taken away some of the fundamental elements of Faraday's physics.

We now come to the second unpleasant consequence. It happens in fact that after the translation, some apparently elementary optical phenomena cease being elementary and appear as extremely problematic. And some not so minor areas of the translation will have to be reworked in order to resolve the new problems. And in doing this, not everything is put in place: instead other and still more serious problems in electrodynamics arise. Which means that Maxwell, in translating, has added new difficulties to Faraday's physics.

To sum up: if the construction of a theory is a translation into symbols of what is already known, the only preliminary guarantee we can ask of the translator is that he be faithful to the text and thus to experience; but the result is instead an unfaithful translation. Which is discouraging. From this one could draw the conclusion that theories are really and truly systematic betrayals and that theoretical physicists are people who favor intellectual dishonesty by pretending to work out elegant mathematical correlations between phenomena.

To avoid the two unpleasant consequences (and the universal condemnation of theories) one can follow at least two different paths.

The first path leads to the downplaying of the problem. Those accustomed to taking this path suggest that we not give excessive weight to the differences which emerge during the translation and concentrate our attention – historical and philosophical – on Maxwell's mechanist program. Legitimate questions are of the following kind: to what extent is Maxwell's physics a mechanist physics? Or: to what extent does Maxwell obey the canons of Newtonianism? In this instance playing down the problem amounts to making it banal. There are hundreds of answers to the question about the amount of Newtonianism or mechanism in Maxwell. Nevertheless none of these answers is able to get away from the metaphors of a philosophy in decline in the sense that none of them tells us anything about the actual construction process of Maxwell's theory. In this methodological banalization of the problem one simply resorts to the trick of narrating the history of Maxwell's method 'in the text' and relegating Maxwell's equations to the footnotes: one resorts to a variant of external history as history of the relations between method and society.

The second path takes into consideration the fact that one started from the conjecture that the construction of the theory (Maxwell's) must be a faithful translation of empirical knowledge (Faraday's) and it also takes into serious consideration that fact that one has reached the conclusion that faithful translation does not exist. One agrees to examine the possibility that the problems encountered derive from the dictionary, the consultation of which produced the 'transformation' of Faraday's physics into Maxwell's.

STRONG AND WEAK INTERACTIONS IN DICTIONARIES

Let us now try to characterize the dictionary whose existence we have taken for granted. We see from an analysis of Maxwell's strictly physico-mathematical records that we are dealing with a dictionary of very vast proportions. It includes an amalgam of theories: hydrodynamics, Lagrangean mechanics, and Hamilton mechanics together with theories on electricity and magnetism of the French and German schools, many areas of mathematics and some chapters on planetary science. But this list is inadequate. Maxwell also studied logic and was interested in the history of physics. He sprinkled his various works with philosophical annotations concerning the function of models and analogies in physics. He supported some ideas on molecular dynamics by appealing to theology texts and carried out experimental research.

Maxwell's dictionary is, therefore, not only vast and complicated; it is also interconnected. Maxwell's research on eighteenth century electricity and on Henry Cavendish is not unrelated to his concepts concerning the relationship between experiment and theory in physics, nor is it without reference to the laborious critical activity with which he investigated Faraday's work, contrasting it with the sophisticated theories of Poisson and German physics concerning action at a distance. Maxwell's theory of statistical mechanics, compared with that maintained at the same time by Boltzmann, concern both the cognitive power of probability calculation and the wrongness of maintaining a basic hypothesis of molecular chaos as opposed to the divine creation of molecules.

Thus, an intricate network of intermediaries crosses the various levels at which the dictionary is articulated. And these intermediaries are not of the same weight: the intermediaries which link Lagrangean mechanics and the problems of models are relatively strong while, thanks to a series of inferences which can be historically reconstructed through successive approximations, those which join the generalizations on statistical mechanics of 1879[4] and the *Discourse on Molecules* of 1873[5] (where some arguments of natural theology explicitly become a part of the problem of knowledge) are relatively weak.

It is the task of the historian of science to carry out accurate analyses of the dictionary's various levels and to complete strict evaluations of the intermediaries which form its inner texture—and it is by no means necessary, but on the contrary, dangerous to claim either to establish once and for all a line of internal/external demarcation which splits the dictionary in two before it has been studied, or to favor, philosophically, certain intermediaries to the disadvantage of others.

Until now dictionaries have been spoken of in terms of structures. But a dictionary is not only something vast, complicated and interconnected with intermediaries of varying weight. It is also something unstable.

Its parts are in fact subject to displacements in time, and the speeds of such displacements differ. In short, we are faced with dynamic situations whose structure is historically given in the sense that there are identifiable areas of the dictionary which change over the years and intermediaries which vary in intensity.

In Maxwell's case, for example, an examination of published writings, manuscripts, and private correspondence shows that large areas of the dictionary underwent drastic revisions in the course of the years. It is enough to think of the changes observable in Maxwell's conceptions of the electromagnetic field between 1854 and 1873 to realize this: all of Maxwell's theoretical physics – field theory and the kinetic theory of gases – consists in a complex of rearrangements, of a progression of deductive chains around different theories which are developing rapidly thanks to the research of other physicists and a profusion of new concepts.

And this indicates that the instability, in its dynamics, is also very rich in displacements. Certain areas of the dictionary move more rapidly than others, which means that the intermediaries between these areas are also variable in form and weight.

As we have just said, the historian tries to trace the levels and intermediaries. But this is not enough. He must also follow the dynamics of the process in which levels and intermediaries change. In short, the historian must do many things when he intends to grasp regularities. But above all there is one thing he must not do: deceive himself and others with the norm according to which he must extract from a scientist's dictionary only the method of that scientist. The dictionary is vast, complex, internally connected, unstable and out of phase: if it is made delimited, simplified, stabilized and homogenized by distilling its methodological essence with the alembics of some ready-made philosophy, then one is exercising the profession of caricaturist.

Maxwell's physics cannot be extracted from Maxwell's method since Maxwell's physics is not a corollary of the method and is not a logical consequence of some decision assumed by Maxwell from the philosophical point of view with the aim of defending Newtonian laws willy-nilly.

It is necessary to insist on this point. The historian who might try to discover the – even deep! – structure of Maxwell's science by making use of the well-known and abused category of mechanism, would certainly end up with a myth. Between the rules of mechanist philosophy and Maxwell's equations there is no direct chain of deductions, but there is all the rest of Maxwell's dictionary and the entire history of what that dictionary has become.

Whoever makes metaphorical journeys to the interior of so-called nineteenth century mechanism encounters not theories or scientists but philosophical allegories. On returning from such a journey, within the confines of a study, he would be in a unique position to draw fanciful maps without having ever faced a single exercise in the history of nineteenth century physics. Guided by illusory indications, he would have traveled over a mythological itinerary and would recount dreams. And the rules of some refined and subtle conventionalism would not save him from the *fata Morganas* of false history. The scientists whose dictionaries he should have explored were not in fact engrossed in the play of free choice between concepts and postulates. Those scientists constructed notions and theories, and one cannot choose before having constructed – even if no one constructs objects and theories well, without the knowledge of what exists behind him.

Different things, then, happen in the dictionary. The interactions between the levels are functions of time; a modification which takes place in a given area can induce slight variations in another, or radical restructurings in apparently far-off levels; presumably the phenomena observable in the dictionary have no linear course, cannot be explained in terms of trial and error strategies, and cannot be imputed to rules restricted by causality.

Furthermore, a dictionary is an open process: it interacts with other dictionaries, sometimes incorporates their relevant sections, and is restructured locally in an attempt to reach relatively stable situations. In Maxwell's case, the dictionary is not only the dynamic matrix within which a theory rationally describing the world is constructed, but is also an element in the historical articulation of the struggle for knowledge in the second half of the nineteenth century.

If all this bypasses Popperian logic or ingenious applications of classical causality, it does not mean that all this is reducible to irrationality. It means instead that the field for the historical reconstruction of non-linear processes and for the study of increasingly complex logics is indeed open and fertile.

It is therefore permissible to maintain that there are important differences between Einstein and an amoeba and that the totality of these differences is called history.

THE SECOND SCIENTIFIC REVOLUTION

It cannot be said that Galileo's productive battles have bequeathed a secure conquest to us. As is known, inheritances are sources of disputes, above all when the heirs are numerous and some of them ridden with vice: in the mutual mauling among heirs it can easily happen that the legacy is squandered.

From this point of view the case of Galileo is more than exemplary. The endless philosophical trial against Galilean science – tenaciously reopened by modern conventionalism[6] and then cleared away in the destruction of experiment[7], thanks to a qualitative dialectic which dreams of a world devoid of science[8] – coincides with the repeated and constant attempts, planned and realized by some philosophical factions, to invade the domain of scientific reason.

The fantasizing of those who think that method, and not theories, must be taught in the schools, is not eccentric, as if the method were perennial (whether one wants to condemn or exalt it) and the theories, instead, the fate of frailty. It is fantasizing which germinates the seeds of a classical metaphor of anti-Galilean philosophy, and the metaphor is the following: in the first place it is suggested that the scientific method is a totality of critical rules thanks to which it becomes possible to construct theories; in the second place it is observed that theories, as opposed to method, have a short life in the history of scientific thought; third, it is assumed that the relative longevity of the (Galilean) method is due to the philosophical root of the latter; one concludes with the thesis that the favored and important areas of scientific thought – the areas that last – are resolved in philosophy, abandoning single theories in the graveyards of Popperian history or representing them in the form of pure, provisional techniques.

We are thus faced with the criterion of general demarcation which divides the minds of men and their culture into two domains: on one side, philosophy, and on the other, science. Once it has been established that some criterion of demarcation exists, one begins to talk about the catechism of those who reduce science to philosophy or those who reduce philosophy to science: both look for the roots of one domain in the soil of the other. And the antithesis between internal and external history derives precisely from this mistake and from the consequent stimulus to look for the roots of scientific thought in philosophical thought, or vice versa.

On the other hand, the unity of knowledge is implied in the work of the historian who analyzes dictionaries, studies their structure and historically reconstructs their dynamics. From what has been suggested above, it is sufficiently plausible that a dictionary is not a paradigm adhered to because of faith or propagandistic motives just as it is not a nucleus of a research program which one or more scientists form by means of some methodological fiat and obstinately defend against every attack. The historical analysis of the dictionary does not favor the motivations of the scientist in crisis who abandons a paradigm with arguments not unlike those which trouble a betrayed lover or a thief caught redhanded; nor does it favor the whim of the methodological faction.

The recognition of the dictionary certainly leads to questions of crises. But they are local crises since they beset only determined levels of the dictionary particularly sensible to the restructurings which are occurring perforce in other levels. It is not strange that Poincaré's theorems (1892) on hydrodynamics forced the great Kelvin to abandon a theory of the structure of matter and that Kelvin saw this necessity from the perspective of a real failure (1896). It would be strange and unjustifiable to maintain that Poincaré's theorems produced a crisis and a failure in physics and consequently aroused psychological motivations suited to the emergence of a new paradigm: he who would maintain this would pay the not inconsequential price of confusing revolutions with crises.

But the recognition of the dictionary also excludes the philosophical illusions of those who envisage the following:

A brilliant school of scholars (backed by a rich society to finance a few well-planned tests) might succeed in pushing any fantastic program ahead or, alternatively, if so inclined, in overthrowing arbitrarily chosen pillars of 'established knowledge'.[9]

Facts of this kind have certainly occurred insofar as there also exists a science caged in by dogmas and, perhaps, even well-paid: but the existence of these facts must not be justified in the shadow of rationality unless one really wants an obscurantist and conservative policy as the pivot of programming research.

If, in the light of the rational reconstructions which put real history in the footnotes and internal history in the text, one arrives at the praise of the 'creative imagination'[10] foraged by a rich society, then it is true that the history written by an intransigent internalist coincides with that written by an intransigent externalist: and in this coincidence between apparently opposed positions, the dissolution of empiricism is completed and it seems that at the same time the eclipse of objective knowledge is triumphant.

In overcoming the pseudo-oppositions between internal and external history, it is recognized that the theory of knowledge of a certain physicist (or mathematician or chemist, etc.) is not defined by the sum of the rules of his method, but is provided instead by the totality of the correlations between those rules and theories and experiments which the scholar has built in time: a theory of knowledge is not a philosophy, it is an historical process.

And all this is so true that, leaving aside the fact that Galileo never wrote a 'Treatise on method', it does not follow that the law governing the fall of heavy bodies can be deduced from Galileo's method. The correlations between methodological rules and physical laws are not syllogisms but interactions which change in time and which join variable rules and laws. In fact the rules are reflections of the laws and the laws take into account the fact that nature answers our questions in not always foreseeable forms. Theories grow because nature is not an onion which science penetrates layer by layer: on the contrary, science lives with the necessity of reinterpreting knowledge already given, just as soon as one reaches a deeper level of things. Not accumulation, then, but restructuring in theories and methods.

It has been the unforeseeable answers of nature that have demonstrated the excessive claims which stand at the apex of Galileo's thought, where Galileo maintains that our knowledge of the external world is, yet, limited to some areas, but is, within those areas, a knowledge which equals that of God.[11]

In other words the strangeness of nature has taught us that there are no absolute and immutable forms of scientific knowledge. But that strangeness has also taught us that knowledge grows thanks to 'theories which violate sense' and which make use of 'sensible experiments' and 'certain proofs'; for us today, Galileo continues to repeat, against Simplicius, that reasoning's guarantee (in those areas where the dictionary operates) is mathematics: and the dialectic of sensible experiments and certain proofs is, for its part, an historical process which is not resolved in a clash to be settled in the tribunal of classical logic, nor still less in the court of crucial experiments.

For these reasons Galileo's thesis can still be maintained and can be summarized as follows: (a) "so the mathematical scientist (*filosofo geometra*), when he wants to recognize in the concrete the effects which he has proved in the abstract, must deduce the material hindrances";[12] (b) "our discourses must relate to the sensible world and not to one on paper."[13]

But it is also true that Galileo's thesis should be emended since the history of the post-Galilean physical sciences shows that the theories reflect on themselves and on those which have preceded them: objective problems arise in the paper world. And the emendation seems to praise the growing abstractness of the theories.

It was the great mathematician, Riemann, who, in the middle of the nineteenth century, defended the rationality of a body of knowledge which progressively detached itself from the raw data of the senses and everyday intuition: "Gradually our conception of nature becomes increasingly complete and exact, but at the same time it moves away more and more from the surface of appearances."[14] And it was the mathematical physicist Boltzmann who denounced those who treat theories like 'milch cows' or as simple technical instruments, and maintain that "the more abstract the theoretical investigation, the more powerful it becomes,"[15] and conquers the world.

The liberalization of the Galilean arguments – as it springs from the conceptions of a Riemann or a Boltzmann – and the growing power of theories in the world have found moments of subtle tension in contemporary physics. In the admittedly limited context of these considerations it is very appropriate in this regard to recall something written by Dirac in 1931:

The steady progress of physics requires for its theoretical formulation a mathematics that gets continually more advanced. This is only natural and to be expected. What, however, was not expected by the scientific workers of the last century was the particular form that the line of advancement of the mathematics would take, namely, it was expected that the mathematics would get more and more complicated, but would rest on a permanent basis of axioms and definitions, while actually the modern physical developments have required a mathematics that continually shifts its foundations and gets more abstract. Non-Euclidian geometry and non-commutative algebra, which were at one time considered to be purely fictions of the mind and pastimes for logical thinkers, have now been found to be very necessary for the description of general facts of the physical world. It seems likely that this process of increasing abstraction will continue in the future and that advance in physics is to be associated with a continual modification and generalisation of the axioms at the base of the mathematics rather than with a logical development of any one mathematical scheme on a fixed foundation.[16]

Following the stages of the liberalization of the themes of Galileo's thought – stages made necessary by the new worlds which the physical sciences discovered and investigated – we notice that the different areas of mathematics and physics move over the years, realizing themselves on foundations endowed with mobility and not on eternal pillars: in the recognition of such a fundamental characteristic of objective knowledge lies one of the prime factors of what must be defined as the second scientific revolution. The second scientific revolution does not consist solely in a long and complicated process of restructuring the laws of nature but bears as well a conscious insight into the profundity of such a restructuring and its historical dynamics – and it is just on the basis of such awareness that the philosophical dimension of the second scientific revolution is judged.

The conclusion which can be drawn from the above is the following. Recognition of dictionaries allows one to evaluate the terms of the inaccuracy inherent in the distinctions between the internal and external history of science in the sense that such inaccuracy seems to stem from wrongful considerations of the problem of objective knowledge. It can be suggested that objectivity does not lie in a definite method, but in the historical process through which the different forms of explanation according to rules are forced to vary as a function of the answers that nature offers to sensible questions.

According to this point of view scientific enterprise never grows in ivory towers since it is never neutral with respect to the weak intermediaries which, within the dictionaries, establish relations between theories and philosophical conceptions of the world. At the same time the reconstruction of single theories is relatively autonomous with respect to the philosophical conceptions of the world since it is entrusted to areas of locally stable rules between the dictionaries and the often surprising answers nature furnishes to sensible experiments.

It is certainly possible to adhere dogmatically to certain levels of the dictionary or, equally dogmatically, to choose to defend others. In a few words this means that there really are 'spontaneous philosophies' of scientists and they are important insofar as they stimulate certain scholars to prefer some areas of rules rather than others, or to look for certain justifications rather than others: think of the qualitative dialectic present in Bohr's thought or of the ingenious realism at work in Einstein's favorite notion of the physical world. But the possibility of reconstructing some aspects of the spontaneous philosophy of a physicist or a mathematician does not at all imply the necessity of resolving his physical or mathematical theories in that philosophy: we are dealing with problems regarding the interactions between levels of dictionaries, not problems regarding the reduction of one level to another or the elimination of certain levels which are particularly inconvenient for rational reconstruction.

NOTES AND BIBLIOGRAPHY

[1] Mark Twain's argument naturally lends itself to different readings. See, for example, G. Temple, 'From the relative to the absolute', in *Turning Points in Physics*. Amsterdam, North-Holland, 1959; G. Toraldo di Francia, *L'indagine del mondo fisico*. Turin, Einaudi, 1976.

[2] M. Faraday, 'Thoughts on ray-vibrations', *Philosophical Magazine* **24**, 136 (1846); 'On the possible relation of gravity to electricity', *Philosophical Transactions* **1**, (1851); 'On the conservation of force', *Proceedings of the Royal Society*, **2**, 352 (1857). Also worthy of mention are Faraday's considerations (1857) on the metaphysics inherent in time measurements of propagation of gravitational force.

[3] J.C. Maxwell, 'A dynamical theory of the electromagnetic field', in particular in 'Note on the attraction of gravity', *Philosophical Transactions of the Royal Society* **155**, 459 (1865).

[4] J.C. Maxwell, 'On Boltzmann's theorem on the average distribution of energy in a system of material points', *Cambridge Philosophical Transactions*, part 3, 547 (1879).

[5] J.C. Maxwell, 'A discourse on molecules', *Philosophical Magazine* **46**, 453 (1873).

[6] P. Duhem, $\Sigma\omega\zeta\epsilon\iota\nu$ $\tau\grave{\alpha}$ $\phi\alpha\iota\nu\acute{o}\mu\epsilon\nu\alpha$, *essai sur la notion de théorie physique de Platon à Galilée*. Paris, A. Hermann, 1908; translated into English as *To Save the Phenomena, An Essay on the Idea of Physical Theory from Plato to Galileo*. Translated by Edmund Doland and Chaninah Maschler. Chicago, University of Chicago Press, 1969.

[7] P. Feyerabend, 'Problems of empiricism, Part I' in *Beyond the Edge of Certainty: Essays in Contemporary Science and Philosophy*, ed. by Robert G. Colodny. Englewood Cliffs, Prentice-Hall, 1965. pp. 145–260. Part II in *The Nature and Function of Scientific Theories* ed. by Robert G. Colodny. Pittsburgh, University of Pittsburgh Press, 1970, pp. 275–353.

[8] P. Feyerabend, 'Consolations for the specialist' in *Criticism and the Growth of Knowledge*, edited by Imre Lakatos and Alan Musgrave. Cambridge, Cambridge University Press, 1970, pp. 197–230.

[9] I. Lakatos, 'Falsification and the methodology of scientific research programmes' in *Criticism and the Growth of Knowledge*, edited by Imre Lakatos and Alan Musgrave. Cambridge, Cambridge University Press, 1970, p. 188.

[10] *Ibid.*, p. 187.

[11] L. Geymonat, *Galileo Galilei*. Turin, Einaudi, 1957. Translated by Stillman Drake as *Galileo Galilei*, New York, McGraw-Hill, 1965.

[12] G. Galilei, *Dialogue concerning the Two Chief World Systems*, translated by Stillman Drake. 2nd. edition. Berkeley, University of California Press, 1967, p. 207.

[13] *Ibid.*, p. 113.

[14] B. Riemann, *Gesammelte mathematische Werke und wissenschaftliche Nachlass*, edited by H. Weber. Leipzig, Teubner, 1892.

[15] L. Boltzmann, 'On the significance of theories' in *Theoretical Physics and Philosophical Problems*, edited by Brian McGuiness with translations from the German by Paul Foulkes. Vienna Circle Collection, volume 5. Dordrecht, Reidel, 1974. p. 34. Translated from 'Über die Bedeutung von Theorien' (1890) in *Populäre Schriften*. Leipzig, J.A. Barth, 1905.

[16] P.A.M. Dirac, 'Quantized singularities in the electromagnetic field', *Proceedings of the Royal Society*, A, 1931, p. 133.

BIOGRAPHICAL NOTES

VITO MICHELE ABRUSCI was born in Bari in 1949. He graduated in philosophy from the University of Florence where he is now a teaching assistant in Philosophy of Science. He has done research in logic and the foundations of mathematics under the direction of Prof. E. Casari. At present, he is preparing a study of Hilbert's works on the foundations of mathematics and proof theory as it relates to the execution of Hilbert's program.

Publications:

'Per una caratterizzazione del programma hilbertiano', *Rivista di Filosofia* **66**, 317–338 (1975).

EVANDRO AGAZZI (b. 1934) obtained his Masters degree in philosophy at the Catholic University of Milan (1957). He then studied physics at the State University of Milan and did postgraduate research in philosophy of science and in mathematical logic at Oxford and Münster. After obtaining a *'libera docenza'* in philosophy of science (1963) and in mathematical logic (1966) he held several teaching appointments at the University of Genoa, at the Catholic University of Milan, at the Scuola Normale Superiore of Pisa (for philosophy of science, mathematical logic, higher geometry, complementary mathematics, etc.). Since 1970 he has been professor of philosophy of science at the University of Genoa, where he is chairman of the Philosophy Department. He has been visiting professor at several universities in the United States and in Europe. Besides being a member of many international scientific institutions, he was president for six years of the Italian Society of Logic and Philosophy of Science and is the present president of the Italian Philosophical Society and of the International Academy of Philosophy of Science.

. His bibliography includes over 150 titles. Among his books, of special importance are: *Introduzione ai problemi dell' assiomatica* (1961), *La logica simbolica* (1964), *Temi e problemi di filosfia della fisica* (1969), *Le geometrie non euclidee e i fondamenti della geometria* (1978, with D. Palladino). He is the editor-in-chief of *Epistemologia*, an Italian journal for the philosophy of science, and is consulting editor of *Erkenntnis*, of *Zeitschrift für allgemeine Wissenschaftstheorie*, of the *Revue Internationale de Philosophie*.

ENRICO BELLONE was born in 1938 in Tortona. He graduated in physics from the University of Genoa. Later he undertook research in the history of physics at the Domus Galilaeana in Pisa. Since 1970 he has taught the 'History of Physics' course at the Physical Sciences Institute of the University of Genoa.

Books:

Opere di Kelvin [editor, with an introductory essay] (Turin, UTET, 1971).

509

Maria Luisa Dalla Chiara (ed.), Italian Studies in the Philosophy of Science, 509–516.

I modelli e la concezione del mondo nella fisica moderna da Laplace a Bohr (Milan, Feltrinelli, 1973).
Il mondo di carta: ricerche sulla seconda rivoluzione scientifica; A World on Paper: Studies in the Second Scientific Revolution. Cambridge, Mass., MIT Press, 1980.

E.G. BELTRAMETTI received the '*laurea*' in physics from the University of Genoa in 1956, and the '*libera docenza*' in theoretical physics in 1962. His earlier work centered on problems of nuclear and particle physics; then he turned his attention to problems related to the possibility of a spin spectrum for unstable systems and to development of a generalized notion of group representations, clarifying, in particular, the group theoretical aspects of complex angular momenta. After some studies on the use of finite geometries as models of space-time, his research interests lie in the area of mathematical and logical foundations of quantum mechanics. He is currently professor of nuclear physics at the University of Genoa.

ERMANNO BENCIVENGA was born in 1950 in Reggio Calabria (Italy). He got his B.A. in Philosophy at the University of Milan in 1972, and his Ph.D. in Philosophy at the University of Toronto in 1977. He taught at the University of Milan from 1972 to 1976, and after that he was part of the faculty of the University of Pittsburgh and of Rice University in Houston. At present, he is assistant professor of formal logic and metaphysics at the University of California in Irvine. He is the author of various papers published in Italian and American journals and of a forthcoming book in Italian on the logic of singular terms. He is also the editor of the anthology *Le logiche libere* [On free logics].

SERGIO BERNINI was born in Florence in 1947. He graduated in 1971 from the University of Florence, where he studied logic and philosophy of science. At present, he teaches logic at the University of Calabria (Cosenza).

Publications:

'A very strong intuitionistic theory', *Studia Logica* **35**, 377–385 (1976).
'Interpretazione intuizionista di teorie a logica classica', *Annali dell 'Università di Ferrara*, Sez. 7, *Sc.Mat.* **22**, 21–41 (1976).
'A note on "A very strong intuitionistic theory" ', to be published in *Studia Logica*.
'Sistemi di ontologia logica', to be published in *Rivista di Filosofia*.
'Completeness proof for intuitionistic infinitary logic', to be published in *Bollettino del Dipartimento di Filosofia dell 'Universita della Calabria*.
'Hauptsatz for first order arithmetic', to be published.

ALDO BRESSAN was born 1926 in Treviso, Italy. He graduated in physics from the University of Padua. His research work concerns mainly continuous media in classical and relativistic physics, and foundations of physics which led him to work in modal logic. He first became assistant and then in 1963 professor of rational mechanics, at Padua. He was also associate professor of mathematical logic at Padua for several years.

Books:

Relativistic Theories of Materials (Berlin and New York, Springer, 1978).
A General Interpreted Modal Calculus (New Haven, Yale University Press, 1972).

PIERO CALDIROLA holds a '*laurea*' in physics from the University of Pavia, 1937. He has produced over 200 publications in a wide variety of physics specialities, from quantum theory to statistical mechanics, electrodynamics, relativity, nuclear physics and plasma physics. He always took a keen interest in the theoretical foundations of modern physics. Among the most important scientific results obtained in his activity are a formulation of Schrödinger's equation for dissipative systems, the deduction of the state equation of gases at very high pressures, various contributions to the fundamental problems of statistical mechanics (ergodic theorems, irreversibility and tendency toward equilibrium) and the study of the epistemological aspects of quantum mechanics, with particular emphasis on the theory of measurement. His latest results include a relativistic theory of the electron (in classical and quantum versions) based on the introduction of a fundamental interval of time (the so called 'chronon'), and a formulation (together with E. Recami) of a unified theory of gravitational forces in the strong interactions. From 1939 to 1946 he was assistant professor of physics, and from 1947–1949 full professor of theoretical physics at the University of Pavia. In 1950 he joined, still as professor of theoretical physics, the University of Milan, where he is also director of the Scuola di Perfezionamento in physics.

Books:

Dalla microfisica alla macrofisica (*From Microphysics to Macrophysics*), (Milan, Mondadori, 1974).

MIRELLA CAPOZZI graduated in philosophy from the University of Rome in 1972. At present she is at the Institute of Philosophy of the University of Siena with a scholarship from the Ministry of Education.

She has published papers on the origins of Kant's views on mathematics in his early writings ('Thoughts on the True Estimation of Living Forces' and 'General Natural History and Theory of the Heaven'), and a critical analysis of Hintikka's interpretation of the Kantian philosophy of logic and mathematics. She is now translating into Italian and editing Kant's *Logik* (1800).

ETTORE CASARI was born near Trento in 1933 and studied at the Universities of Pavia and Münster i.Westfalen. From 1961 to 1966, he was associate professor at the universities of Pavia and Milan; in 1966, visiting professor at the University of Münster, and soon after professor at the University of Cagliari; and from 1968, professor of philosophy of science at the University of Florence.

Publications:

Casari has written three books on logic and foundations: *Lineamenti di logica matematica* (Milan, Feltrinelli, 1959); *Computabilità e ricorsività,* (Milan, Quaderni ENI, 1959); *Questioni di filosofia della matematica,* (Milan, Feltrinelli, 1964) and edited, with introductions, two anthologies *La filosofia della matematica del' 1900,* (Florence, Sansoni, 1973); *Dalla logica alla metalogica,* (Florence, Sansoni, 1979). Selected papers: 'Sulla disgiunzione nella logica megarico-stoica', *Actes du VIIIe Con.Int.Hist.Sc.* (1956); 'Un'osservazione sulla teoria assiomatica degli insiemi', *Boll.Un.Mat.It.* (1969); 'Universali e insiemi (I, II) *Riv.di Filos.* (1969); 'Regular Models for Second Order Set Theory', *Symposia Math.V* (1970); 'Alcuni temi fondamentali della filosofia della matematica', *Atti XXIV Congr.Soc.Fil.It.* (1973); 'Axiomatical and Set-theoretical Thinking' *Synthese* (1974); 'Osservazioni logico-filosofiche sul problema dell'inferenza statistica' *Atti del Convegno sui Fondamenti dell'inferenza statistica* (Florence 1977) 'Logic and the Foundations of Mathematics' (forthcoming, Proceedings of the Conference on Modern Logic, Rome, September, 1977).

G. CASSINELLI received the '*laurea*' in physics from the University of Genoa in 1970. His research interests are mainly in the area of mathematical and logical foundations of quantum mechanics, in particular non-commutative probability theory. He is currently assistant professor of theoretical physics at the University of Genoa.

CARLO CELLUCCI was educated at the Universities of Rome and Milan, where he received a degree in philosophy, and was a research student at the University of Oxford. He has taught at the Universities of Sussex, Rome and Calabria, and is presently professor of philosophy of science at the University of Siena.

The author of a textbook on proof theory, *Teoria della dimostrazione* (Turin, Boringhieri, 1978), he has also edited two volume of readings in the philosophy of mathematics: *La filosofia della matematica* (Bari, Laterza, 1967) and *Il paradiso di Cantor* (Naples, Bibliopolis, 1979).

DOMENICO COSTANTINI was born in 1937 in Luino, Italy. He graduated in statistics from the University of Rome in 1963 where the late Prof. G. Pompilj was his thesis advisor. He pursued studies in logic and philosophy of science with Prof. L. Geymonat (1968), and from 1969 to 1975 he taught probability theory at the University of Perugia; since 1976 he has taught probability theory at the University of Bologna.

Publications:

Fondamenti del calcolo delle probabilità (Milan, Feltrinelli, 1970).
Introduzione alla probabilità (Turin, Boringhieri, 1977).
'L'esperimento e l'indipendenza nell'assiomatizzazione del calcolo delle probabilità', in *Studi di probabilità, statistica e ricerca operativa in onore di G. Pompilj* (Rome, Istituto di Calcolo delle probabilità, 1971).
'Scambiabilità e irrilevanza all'infinito', *Statistica* **37**, 347–350 (1977).
'Sul differente uso della distribuzione normale: probabilità iniziali e verosimiglianze' in *I fondamenti dell'inferenza statistica* (Florence, Department of Statistics, 1978).
'Considerazioni sulle ipotesi profonde della distribuzione Beta', *Statistica* **38**, 313–322 (1978).
'The Relevance Quotient', *Erkenntnis* (forthcoming).

MARIA LUISA DALLA CHIARA graduated in philosophy from the University of Padua in 1961. Later she studied logic and philosophy of science in Milan, under E.Casari and L.Geymonat. She was appointed professor of Logic at the University of Florence in 1970.

Publications:

Syntactic and Semantical Models of Elementary Theories (Milan, Feltrinelli, 1968).
Many papers on tense-logic, semantics of empirical theories and quantum logic, for

example: 'A general approach to non-distributive logics', *Studia Logica* **35**, 139–162, (1976). 'Quantum logic and physical modalities', *Journal of Philosophical Logic* **6**, 391–404 (1977); 'Logical self-reference, set-theoretical paradoxes and the measurement problem in quantum mechanics', *Journal of Philosophical Logic* **6**, 331–347 (1977); (with G. Toraldo di Francia) 'A formal analysis of physical theories' in *Enrico Fermi Summer School on Problems in the Foundations of Physics*, Varenna 1977. Amsterdam, North-Holland, 1979.

GIULIANO TORALDO DI FRANCIA has been President of the Italian Physical Society and of the International Commission of Optics. He is currently President of the Italian Society for Logic and Philosophy of Science and a member of the Higher Council for Education of Italy. He is a recipient of the Thomas Young medal from the Institute of Physics and the Physical Society of England, and of the C.E.K. Mees medal from the Optical Society of America. He is professor of physics at the University of Florence and Director of the Institute of Research on Electromagnetic Waves of the National Research Council of Italy.

He is the author of more than 100 papers on optics, electromagnetism, quantum electronics, foundations of physics, and philosophy of science.

Books:

Onde elettromagnetiche, (Bologna, Zanichelli, 1953); Eng. tr. as *Electromagnetic Waves*, (New York, Interscience Publishers, 1955).
La diffrazione della luce (*The Diffraction of Light*) (Turin, Einaudi, 1958).
L'indagine del mondo fisico (Turin, Einaudi, 1976); Eng. tr. as *The Investigation of the Physical World*, (Cambridge University Press, forthcoming).

GIULIO GIORELLO was born in 1945 in Milan. He received his '*laurea*' in philosophy at the University of Milan in 1968, and in mathematics at the University of Pavia in 1971. His research has been in the areas of logic, functional analysis, and probability. He has been lecturer of geometry, higher mathematics and philosophy of science at the Universities of Pavia, Catania and Milan. At present he holds the Chair of Philosophy of Science at the University of Milan.

Publications:

Essays in Epistemology – 'Il falsificazionismo di Popper', and 'Filosofia della scienza e storia della scienza nella cultura di lingua inglese' in *Storia del pensiero filosofico e scientifico*, Vol 7, L. Geymonat (ed.), (Milan, Garzanti, 1976).
'Calcolo', (with Guiseppe Geymonat) in *Enciclopedia*, Vol 2, (Turin, Einaudi, 1977).
'Mathematische Abstraktion, Dialektik der wissenschaftlichen Forschungsprogramme und Naturerkenntnis', in *Dialektik als Programm der Naturwissenschaft*, P. Plath and H.J. Sandkühler (eds.), (Cologne, Pahl Rugenstein Verlag, 1978), pp. 270–297.

SALVATORE GUCCIONE was born in 1938 in Rome. After studying law and philosophy he concluded his studies in physics, receiving his degree from the University of Rome in 1967. After joining the Institute of Theoretical Physics of the University of Naples in 1972, he began to collaborate with the Group of Cybernetics of the Institute, working especially in the fields of computational complexity and the foundations of probability as well as the foundations of physics. His research interests are presently focused also on non-classical logics with particular attention to problems related to the notions of 'vagueness' and 'precision'. He is also interested in the analysis of the science-ideology relationship seen in particular from a Marxist point of view. He is associate professor at the University of Naples.

CORRADO MANGIONE was born in 1930 and took his degree in mathematics. He has worked in algebraic logic and proof-theory. He is presently professor of logic at the University of Milan, Department of Philosophy.

Books:

Editor, *Logica e aritmetica* (Frege), tr. from German (Turin, Boringhieri, 1964).
Elementi di logica matematica (Turin, Boringhieri, 1965).
He is editor of Boringhieri's series in mathematical logic.

MASSIMO MUGNAI was born in Florence in 1947. He studied philosophy at the University of Florence. At present he is interested in the history of logic, and works at the Consiglio Nazionale delle Ricerche.

Publications:

Astrazione e realtà, saggio su Leibniz (Milan, Feltrinelli, 1976) a book on logic and language in Leibniz' philosophy.
He has also published articles in *Studia Leibnitiana* and in Italian philosophical journals.

MAURIZIO NEGRI was born in Milan in 1950 and graduated in philosophy from the University of Milan where he is now assistant professor. His area of research is proof-theory.

MASSIMO PIATTELLI-PALMARINI was born in Rome in 1942. After studying at the Universities of Florence and Bologna, he took his degree in physics at the University of Rome where he is assistant professor of physical chemistry. He has been director of the Centre Royaumont pour une Science de l'homme in Paris where he collaborated with J. Monod, and teaches at the École des Hautes Etudes. He is director of the Centro Fiorentino di Storia e Filosofia della Scienza (Florence). He has co-edited (with E. Morin) *L'unité de l'homme* (Paris, Editions du Seuil, 1974), and is editor of *Language and Learning: The Piaget-Chomsky Debate* (Harvard University Press, 1980).

CLAUDIO PIZZI was born in 1944 in Milan where he received his degree in 1969. At present he is professor of philosophy of science at the University of Calabria in Cosenza, Italy. His papers are mainly concerned with the intensional features of scientific language.

Books:

Editor of the Italian translation of Hughes and Cresswell's *An Introduction to Modal Logic* (1973).
Editor of *La logica del tempo* (1974).
Editor of *Leggi di natura, modalità, ipotesi* (1978).

ERASMO RECAMI was born in 1939 in Milan. He received his degree in physics at the

University of Milan (1964), and later received his '*libera docenza*' theoretical physics (Rome, 1971). In 1976 he directed an International Meeting at Erice on tachyons, monopoles, and related topics, whose proceedings have been recently published by North-Holland (Amsterdam, 1978). He has been a visiting scientist in the United States (Austin, Texas), USSR (Kiev), UK (Oxford), Denmark (Copenhagen), and at the International Center for Theoretical Physics (Trieste). He is Associate Professor of higher physics at the Faculty of Sciences of the University of Catania; and Collaborator at the R-3 level of the Italian National Institute for Nuclear Physics.

Referee of many international journals of physics, he is the author of about 80 scientific papers, which have appeared in various international journals of physics, dealing with relativity, quantum mechanics, elementary particle physics, and nuclear physics. In addition he has been contributing to books and various encyclopedias, and writing many historical, didactical, semi-popular articles.

PAOLO ROSSI was born in 1923 in Urbino. He studied in Florence with Eugenio Garin and later in Milan with Antonio Banfi. He has taught at the University of Milan, Cagliari, and Bologna. Since 1966 he has been professor of history of philosophy at the University of Florence.

Publications:

'The Legacy of Ramon Lull in Sixteenth Century Thought', in *Mediaeval and Renaissance Studies* V, 182–231 (1961).
Francis Bacon, From Magic to Science (Chicago, Chicago University Press, 1968).
Philosophy, Technology and the Arts in the Early Modern Era, (New, York, Harper, 1970).
'Nobility of Man and Plurality of Worlds', in *Science, Medicine, and Society in the Renaissance*, A. C. Debus (ed.) (New York, Neale Watson Academic Publications, 1972), pp. 131–162.
'Baconianism', in *Dictionary of the History of Ideas*, P. P. Wiener (ed.) (New York, Scribner's, 1974).
'Hermeticism, Rationality, and the Scientific Revolution', in *Reason, Experiment, and Mysticism in the Scientific Revolution*, M.L. Righini-Bonelli and W. Shea (eds.) (New York, Science History Publications, 1975), pp. 247–74.
Clavis universalis: arti mnemoniche e logica combinatoria da Lullo a Leibniz (Milan-Naples, Ricciardi, 1960).
Le sterminate antichità: studi vichiani (Pisa, Nistri-Lischei, 1969).
Aspetti della rivoluzione scientifica (Napoli, Morano, 1971).
Storia e filosofia: saggi sulla storiografia filosofica (Turin, Einaudi, 1974).
Immagini della scienza, (Rome, Editori Riuniti, 1977).
I segni del tempo: storia della terra e storia delle nazioni da Hooke a Vico, (Feltrinelli, forthcoming).

ANDREA SANI was born in 1953. He studied at the University of Florence where he took his degree in philosophy in 1977. He wrote his thesis on logic of modal theory in Leibniz.

GISÈLE FISCHER SERVI graduated from University of Florence in 1972 where she presented a '*laurea*' thesis on modal logic, later published in *Rivista di Filosofia* (1975). For the last two years she has taught logic in the Philosophy Department, University of Parma (Cremona).

Most of her published results are concerned with algebraic and modal logic and have appeared in *Studia Logica, Notre Dame Journal of Formal Logic* and *Le Mathematiche*.

PIERO TOSI has studied in Milan, Utrecht and Genoa under the supervision of Professors E. Agazzi and D. van Dalen. His main logical interest has been in proof theory for arithmetic, studied through a semiformal system, developed especially in his thesis 'Constructive Normal Semiformalisms' (Genoa, 1974) and extended in 'Complexity of Transfinite Induction' in *Epistemologia* 1 (1978). He is presently a scientific collaborator at the Catholic University in Milan.

INDEX OF NAMES

Des Bosses, B. 390, 395, 397, 405
Descartes, R. 97, 377
Des Coudres, T. 251
Deshpande, V. K. 295
Despretz, C. M. 349
Dieudonné, J. 132, 133
Di Jorio, M. 296
Dirac, P. A. M. 295, 318, 495, 504, 507
Dirichlet, P. G. L. 119, 478, 490
Dirichlet
   integral 120
   principle 118, 119–121, 124, 127, 454, 472, 475–476, 478, 479
Donahue, W. 376, 387
Donne, J. 365, 385
Dreyer, J. L. E. 384, 386, 387
Dugac, P. 132, 133
Duhem, P. 114, 506
Dulong, P. L. 349
Dummett, M. A. E. 72
Duns Scotus, J. 395, 396, 406
Dyck, W. von 384

Eberhard, J. A. 441
Edmonds, J. D., Jr. 290, 291, 292, 297, 298
Ehlers, J. 297
Ehrenfest, P. 334
Ehrenfest, T. 334
Eigen, M. 354, 356, 358
Einstein, A. 99, 258, 260, 279, 295, 297, 338, 501, 506
Elkana, Y. 346, 349, 351, 357, 358
Ellis, L. 178, 180
Enriques, F. ix, 129, 485, 490
Euclid 113, 125, 435, 443–444, 450, 452, 470
Eudossus 364
Euler, L. 116
Everett, A. E. 296

Fabri, E. 297
Fabricius, D. 381, 388
Fano, G. 133
Fantappié, L. 294
Faraday, M. 496–498, 506

Farrer, A. (*also* Far.) 420, 421, 422
Favaro, A. 387
Feferman, S. 14, 21, 29
Feinberg, G. 251, 295
Fenstad, J. E. 29, 152
Fermat, P. de 479
Fermi, E. ix
Feyerabend, P. K. 126, 131, 133, 506, 507
Feynman, R. P. 99, 251, 276, 283, 293, 297
Finetti, B. de 179, 180, 183
Finkelstein, D. 245, 247
Fiorentino, F. 373, 387
Firpo, L. 388
Fischer Servi, G. 72, 515
Fisher, R. A. 177, 179, 183
Fitch, F. B. 47
Fitting, M. C. 58
Fludd, R. 377
Fock, V. 198, 201, 337, 339
Foerster, H. von 341, 342
Fonseca, P. da 389, 401, 408
Fonte, G. 295
Foucher de Careil, A. (*also* F. de C.C.) 412, 419, 420, 421, 422
Foulis, D. J. 225, 227, 228, 233, 234
Fox, R. 297
Fraassen, B. C. van 44, 45, 46, 47, 48, 193, 196, 230, 234, 246, 247
Fracastoro, G. 364
Fränkel, A. A. 212
Frank, P. 296
Frankfurt, H. G. 405, 421
Frédéricq, L. 350
Frege, G. x, 90, 94, 95, 101, 102, 113, 460, 480, 486, 488, 489, 514
Friedman, H. 138, 152

Gabbay, D. M. 57
Gabriel, G. 489
Gade, J. A. 387
Galileo 323, 326, 343, 346, 366, 376, 377, 378, 380, 384, 387, 501, 503–505, 507
Galois, E. 471, 479
Galton, F. 177

Painlevé, P. 197, 198, 205, 210
Palladino, D. 509
Pantaleo, M. 295, 296
Pap, A. 73-74, 86
Papi, F. 131
Parker, L. 266, 296
Parkinson, G. H. R.   (*also* Park.) 420, 421
Parmentola, J. A. 270, 297
Parsons, C. 452
Pascal, B. 365
Pasch, M. 470, 485, 490
Pask, G. 341
Pasquinelli, A. x, xi
Paton, H. J. 449
Patrizi, F. 363-374, 376, 378, 379-385, 387, 388
Pavšič, M. 293, 296, 298
Peano, G. ix, x, 214
Peano arithmetic 138
Pearce, G. 134
Pearson, K. 177
Peirce, C. S. 18
Perelman, C. 97-98
Persico, E. x, xi
Perutz, M. 353
Peucer, C. 372
Philoponus 366
Piaget, J. 346, 347, 359
Piatelli-Palmarini, M. 359, 514
Pico *see* Mirandola, P. della
Pierce, J. R. 338
Pierpoint, J. 480, 490
Pinelli, G. F. 372
Pirani, F. A. E. 270, 297
Piron, C. 215, 234
Pizzi, C. 86, 514
Plath, P. 513
Plato 129, 367, 375
Plotinus 370
Poincaré, H. 113, 114, 116, 118, 124, 126, 128, 129, 130, 131, 132, 133, 135, 334, 502
Poisson, S.-D. 498
Polya, G. 116
Pollock, J. L. 78, 81, 86
Pompilj, G. 512

Pool, J. C. T. 220, 223, 235
Popper, K. R. 114, 115, 116, 118, 129, 133, 135, 328, 338
Porta, G. della 377, 383
Poser, H. 418, 422
Prawitz, D. 13, 16, 17, 19, 20, 29, 137, 138, 139, 142, 152, 451
Preti, G. x, xi
Prigogine, I. 354
Prior, A. N. 80, 86
Proclus 366, 443
Proust, J. 449
Prout, W. 118
Przełęcki, M. 196
Ptolemy, C. 364, 376, 377
Pucci, F. 383
Putnam, H. 187, 238-243, 245-246, 247
Pythagoras 374

Quine, W. V. O. 241

Raggio, A. 138, 152
Ramsey, F. P. 75-76, 86, 170, 180
Recami, E. 251, 280, 295, 296, 297, 298, 511, 514
Rehberg, A. W. 445
Reichenbach, H. 86, 195, 246, 247
Reid, C. 477, 480, 490, 491
Reinhold, K. L. 452
Renyi, A. 183
Rescher, N. 87, 90
Reyes, G. 107, 111
Riemann, B. 114, 118-124, 126-128, 129, 130, 131, 132, 133, 474, 504, 507
Riemann functions 467
Righini-Bonelli, M. L. 515
Rindler, W. 296
Roberval, G. P. de 365
Robinson, A. 107
Rolnick, W. B. 297
Root, R. G. 272, 297
Rosen, N. 297
Rossi, P. 386, 387, 388, 515
Rothe, H. 296
Rothmann, C. 370, 372

BOSTON STUDIES IN THE PHILOSOPHY OF SCIENCE

Editors:
ROBERT S. COHEN and MARX W. WARTOFSKY
(Boston University)

1. Marx W. Wartofsky (ed.), *Proceedings of the Boston Colloquium for the Philosophy of Science 1961-1962.* 1963.
2. Robert S. Cohen and Marx W. Wartofsky (eds.), *In Honor of Philipp Frank.* 1965.
3. Robert S. Cohen and Marx W. Wartofsky (eds.), *Proceedings of the Boston Colloquium for the Philosophy of Science 1964-1966. In Memory of Norwood Russell Hanson.* 1967.
4. Robert S. Cohen and Marx W. Wartofsky (eds.), *Proceedings of the Boston Colloquium for the Philosophy of Science 1966-1968.* 1969.
5. Robert S. Cohen and Marx W. Wartofsky (eds.), *Proceedings of the Boston Colloquium for the Philosophy of Science 1966-1968.* 1969.
6. Robert S. Cohen and Raymond J. Seeger (eds.), *Ernst Mach: Physicist and Philosopher.* 1970.
7. Milic Capek, *Bergson and Modern Physics.* 1971.
8. Roger C. Buck and Robert S. Cohen (eds.), *PSA 1970. In Memory of Rudolf Carnap.* 1971.
9. A. A. Zinov'ev, *Foundations of the Logical Theory of Scientific Knowledge (Complex Logic).* (Revised and enlarged English edition with an appendix by G. A. Smirnov, E. A. Sidorenka, A. M. Fedina, and L. A. Bobrova.) 1973.
10. Ladislav Tondl, *Scientific Procedures.* 1973.
11. R. J. Seeger and Robert S. Cohen (eds.), *Philosophical Foundations of Science.* 1974.
12. Adolf Grünbaum, *Philosophical Problems of Space and Time.* (Second, enlarged edition.) 1973.
13. Robert S. Cohen and Marx W. Wartofsky (eds.), *Logical and Epistemological Studies in Contemporary Physics.* 1973.
14. Robert S. Cohen and Marx W. Wartofsky (eds.), *Methodological and Historical Essays in the Natural and Social Sciences. Proceedings of the Boston Colloquium for the Philosophy of Science 1969-1972.* 1974.
15. Robert S. Cohen, J. J. Stachel and Marx W. Wartofsky (eds.), *For Dirk Struik. Scientific, Historical and Political Essays in Honor of Dirk Struik.* 1974.
16. Norman Geschwind, *Selected Papers on Language and the Brain.* 1974.
18. Peter Mittelstaedt, *Philosophical Problems of Modern Physics.* 1976.
19. Henry Mehlberg, *Time, Causality, and the Quantum Theory* (2 vols.). 1980.
20. Kenneth F. Schaffner and Robert S. Cohen (eds.), *Proceedings of the 1972 Biennial Meeting, Philosophy of Science Association.* 1974.
21. R. S. Cohen and J. J. Stachel (eds.), *Selected Papers of Léon Rosenfeld.* 1978.
22. Milic Capek (ed.), *The Concepts of Space and Time. Their Structure and Their Development.* 1976.
23. Marjorie Grene, *The Understanding of Nature. Essays in the Philosophy of Biology.* 1974.

24. Don Ihde, *Technics and Praxis. A Philosophy of Technology.* 1978.
25. Jaakko Hintikka and Unto Remes, *The Method of Analysis. Its Geometrical Origin and Its General Significance.* 1974.
26. John Emery Murdoch and Edith Dudley Sylla, *The Cultural Context of Medieval Learning.* 1975.
27. Marjorie Grene and Everett Mendelsohn (eds.), *Topics in the Philosophy of Biology.* 1976.
28. Joseph Agassi, *Science in Flux.* 1975.
29. Jerzy J. Wiatr (ed.), *Polish Essays in the Methodology of the Social Sciences.* 1979.
32. R. S. Cohen, C. A. Hooker, A. C. Michalos, and J. W. van Evra (eds.), *PSA 1974: Proceedings of the 1974 Biennial Meeting of the Philosophy of Science Association.* 1976.
33. Gerald Holton and William Blanpied (eds.), *Science and Its Public: The Changing Relationship.* 1976.
34. Mirko D. Grmek (ed.), *On Scientific Discovery.* 1980.
35. Stefan Amsterdamski, *Between Experience and Metaphysics. Philosophical Problems of the Evolution of Science.* 1975.
36. Mihailo Marković and Gajo Petrović (eds.), *Praxis. Yugoslav Essays in the Philosophy and Methodology of the Social Sciences.* 1979.
37. Hermann von Helmholtz: *Epistemological Writings. The Paul Hertz/Moritz Schlick Centenary Edition of 1921 with Notes and Commentary by the Editors.* (Newly translated by Malcolm F. Lowe. Edited, with an Introduction and Bibliography, by Robert S. Cohen and Yehuda Elkana.) 1977.
38. R. M. Martin, *Pragmatics, Truth, and Language.* 1979.
39. R. S. Cohen, P. K. Feyerabend, and M. W. Wartofsky (eds.), *Essays in Memory of Imre Lakatos.* 1976.
42. Humberto R. Maturana and Francisco J. Varela, *Autopoiesis and Cognition. The Realization of the Living.* 1980.
43. A. Kasher (ed.), *Language in Focus: Foundations, Methods and Systems. Essays Dedicated to Yehoshua Bar-Hillel.* 1976.
46. Peter L. Kapitza, *Experiment, Theory, Practice.* 1980.
47. Maria L. Dalla Chiara (ed.), *Italian Studies in the Philosophy of Science.* 1980.
48. Marx W. Wartofsky, *Models: Representation and the Scientific Understanding.* 1979.
50. Yehuda Fried and Joseph Agassi, *Paranoia: A Study in Diagnosis.* 1976.
51. Kurt H. Wolff, *Surrender and Catch: Experience and Inquiry Today.* 1976.
52. Karel Kosík, *Dialectics of the Concrete.* 1976.
53. Nelson Goodman, *The Structure of Appearance.* (Third edition.) 1977.
54. Herbert A. Simon, *Models of Discovery and Other Topics in the Methods of Science.* 1977.
55. Morris Lazerowitz, *The Language of Philosophy. Freud and Wittgenstein.* 1977.
56. Thomas Nickles (ed.), *Scientific Discovery, Logic, and Rationality.* 1980.
57. Joseph Margolis, *Persons and Minds. The Prospects of Nonreductive Materialism.* 1977.
58. Gerard Radnitzky and Gunnar Andersson (eds.), *Progress and Rationality in Science.* 1978.

59. Gerard Radnitzky and Gunnar Andersson (eds.), *The Structure and Development of Science.* 1979.
60. Thomas Nickles (ed.), *Scientific Discovery: Case Studies.* 1980.
61. Maurice A. Finocchiaro, *Galileo and the Art of Reasoning.* 1980.